Statistics

GUDMUND R. IVERSEN
Swarthmore College

MARY GERGEN
Pennsylvania State University, Delaware County Campus

Statistics

THE CONCEPTUAL APPROACH

Springer

Springer Undergraduate Textbooks in Statistics

George W. Cobb, Mount Holyoke College
Consulting Editor

Library of Congress Cataloging-in-Publication Data
Iversen, Gudmund R.
    Statistics : the conceptual approach / Gudmund R. Iversen, Mary Gergen.
        p. cm. —
    Includes bibliographical references and index.
    ISBN 0-387-94610-1 (hard cover : alk. paper)
  1. Statistics. I. Gergen, Mary. II. Title. III. Series.
QA276.12.I84   1997
519.5—dc   96-23148

Printed on acid-free paper.

© 1997 Springer-Verlag New York, Inc.
All rights reserved. This work may not be translated or copied in whole or in part without the written permission of the publisher (Springer-Verlag New York, Inc., 175 Fifth Avenue, New York, NY 10010, USA), except for brief excerpts in connection with reviews or scholarly analysis. Use in connection with any form of information storage and retrieval, electronic adaptation, computer software, or by similar or dissimilar methodology now known or hereafter developed is forbidden.

Production coordinated by Penny Hull managed by Bill Imbornoni; manufacturing supervised by Joe Quatela.
Photocomposed by Progressive Information Technologies, Inc., Emigsville, PA.
Printed and bound by Edwards Brothers, Inc., Ann Arbor, MI.
Printed in the United States of America.

9 8 7 6 5 4 3 2 1

ISBN 0-387-94610-1 Springer-Verlag New York Berlin Heidelberg SPIN 10491017

PREFACE

This statistics textbook is unique in its design and execution. It was created to fill a growing but previously unmet need to provide today's students with a sophisticated grasp of the nature of statistical information. It is a response to teachers who want their students to become statistically literate citizens, not (often hopelessly) amateur statisticians.

Over the years, statistical information has been liberated from the dusty archives of government agencies and academic computing centers. Statistical information now plays a part in the discussions of a broad range of topics—from national policies on health reform and defense to treatments of life expectancies, marriage, abortion, education, and sports. Statistics are regularly featured in newspapers, magazines, radio shows, and television programs; they can even be spotted on MTV and in cartoons. Statistics saturate our educational curricula, as well. In elementary school classrooms and Ph.D. seminars, statistical information has become a regular feature of instruction.

Despite this exposure, there is very little assurance that the audience for these materials is not only receptive to but knowledgeable about what is offered. When people read about the results of a research study, how can they assess whether the conclusions are valid? Do they ask: How were the variables defined in this study? What statistical methods were used? What are "statistically significant" results? What are the shortcomings of the reported results? These are some of the issues discussed in this book. It is clear that with an understanding of the main ideas of statistics, engaged citizens can grasp what the professional number crunchers have produced and evaluate the results.

This book grew out of a course designed by Gudmund R. Iversen to meet the challenges created by this greater reliance on statistical information. It was one of a series of courses designed at Swarthmore College to fulfill the mission of a liberal arts college to educate its students for the challenges of the twenty-first century. The idea was that students should not become so involved with the intricacies of a single discipline that they lose sight of the big picture. These courses were intended to educate students to understand how the major ideas of a field relate to the world. In many respects statistics seemed an ideal subject for one such course. While statistics could be a mystifying, self-aggrandized, and esoteric discipline, it could also be a key to understanding many other disciplines. The course, Stat 1: Statistical Thinking, was created to produce this understanding. The course proved to be very popular, and each year it grew in size. Over time the lecture notes for the course became more refined and extensive, and eventually the course material served as the basis for this book.

Formulas

As most statistics instructors are keenly aware, the teaching of statistics has changed dramatically. The integration of the computer into educational settings and especially the easy availability of user-friendly statistical software have made the old ways of study—in particular, memorization and manipulation of statistical formulas—no longer necessary for the vast majority of students. To be true to our objectives for this book, we have used no statistical formulas within the discourse of each chapter. Although this may seem radical, we decided with deliberation and care to deemphasize formulas by handling them in special sections at the ends of chapters.

Our experience is that statistical formulas are like an alien language. If one understands the language, the formulas add immensely to one's understanding of statistics; if not, they are indecipherable. We have seen too many students for whom the formulas became a barrier to understanding and interest in statistics, and we strongly believe it is possible to gain a deep understanding of statistical ideas without them.

Exercises

It is difficult to learn statistics by just listening to lectures and reading a textbook. Statistics is better learned by doing, so we provide a large selection of exercises. Almost all the examples and exercises use real data we have selected from books, journals and newspapers. These are data used in actual research or published reports, and together they illustrate how statistics is applied across a wide range of human activities.

The exercises are of three kinds: *Review* questions, which probe understanding of the chapter's central concepts; *Interpretation* questions, which require students to make sense of statistical information; and *Analysis* questions, which require students to analyze data and create their own solutions to problems. The Review questions serve as a check for comprehension and provide a background for class discussions. The Interpretation questions, which are verbal rather than quantitative, encourage comprehension and suggest applications to real-world issues. The Analysis questions require students to become familiar with the use of a statistical software package, either in work groups of a few students or individually. Each of the exercises provides potential topics for statistical reports.

Solutions to odd-numbered exercises are found at the back of the book, along with statistical tables useful for working on various exercises.

Acknowledgments

During the long course of writing this book, some of our students read all or parts of earlier versions of the manuscript, and we are grateful for the useful comments of Megan Falvey, Reginald Tilley IV, and especially Maya Rao. We thank Maura MacDermott for allowing us to quote from a paper she wrote in Stat 1. We also benefitted from the questions raised by students enrolled in the course over the years. The Sloan Foundation grant to Swarthmore College under their New Liberal Arts Program supported the original development of course material for Stat 1.

In addition, we wish to express our appreciation for the many insightful comments made by George W. Cobb, Mount Holyoke College; Robert W. Hayden, Plymouth State College; Thomas L. Moore, Grinnell College; Michael Orkin, California State University, Hayward;

Richard L. Scheaffer, University of Florida; and other anonymous readers whose statistical expertise greatly enhanced the text. We also wish to acknowledge our copy editor, Penny Hull, whose diligent and provocative comments spurred us on, and to Bill Imbornoni and Theresa Shields and their associates at Springer-Verlag New York for their contributions to the production of the book. Most of all, we are grateful to our publisher, Jerry Lyons, who had the faith to envision a book in the original manuscript. We could not have asked for a more creative, supportive, and enthusiastic publisher.

Finally, we are grateful for the loving support and encouragement we received from our spouses, Roberta Rehner Iversen and Kenneth J. Gergen, from the moment we began our partnership through the final proofs. We also appreciate the goodwill and support of our children—Gudmund's Eric, Gretchen, John, and Kirsten and Mary's Laura, Lisa, Michael, and Stan—and their families. Last, we want to congratulate each other on the harmonious manner in which we have managed to bring this book to life.

Gudmund R. Iversen
Mary Gergen
Swarthmore, Pennsylvania
January 1997

CONTENTS

Preface		v
1	**Statistics: Randomness and Regularity**	**1**
1.1	Statistics: What's in a word?	3
1.2	Knowing how statistics is used: Goals for the reader	4
	Understanding what can go wrong 7	
	Understanding statistical terms 9	
1.3	Central ideas in statistics	9
	Randomness and regularity: Twins in tension 9	
	Randomness in regularity 10	
	Two examples in the study of randomness and regularity 11	
	Probability: What are the chances? 13	
	Variables: The names we give things 14	
	Variables, values, and elements 14	
	Theoretical variables and empirical variables 14	
	Constants 17	
1.4	Users of statistics	17
1.5	Relationship of statistics to mathematics, pencils, and computers	20
1.6	Summary	21
Additional Readings		23
Exercises		23
2	**Collection of Data**	**29**
2.1	Defining the variables	31
2.2	Observational data: Problems and possibilities	32

		Population versus sample 32	
		Selection of the sample: Making sure the pot is stirred 33	
		Random sample: What is it? 35	
		Convenience sample: How to produce a "poor" sample 36	
		Selecting proper samples 36	
		Selection of variables on which to collect observational data 37	
	2.3	Errors and "errors" in collecting observational data	39
		Sampling error: The "error" that is not a mistake 39	
		Nonresponse error: Result of rude, rushed, and reticent respondents 41	
		Response errors 42	
	2.4	Experimental data: Looking for the causes of outcomes	45
		Experimental group and control group 46	
		Selecting the experimental and control groups 47	
		Problems with experimenting on people 47	
		Role of statistics in experimentation 50	
		Putting it all together: Does class size affect school performance? 53	
	2.5	Data matrix/Data file	55
	2.6	Summary	57
Additional Readings			59
Exercises			60

3 Description of Data: Graphs and Tables   71

	3.1	Graphs: Picturing data	73
		Creating statistical graphs 73	
		Types of graphs 75	
	3.2	Categorical variables: Pie charts and bar graphs	75
		Graphing one categorical variable 75	
		Graphing two categorical variables 77	
	3.3	Metric variables: Plots and histograms	79
		Graphing one metric variable 80	
		Graphing two metric variables 89	
		Time series plot 91	
	3.4	Creating maps from data	94
	3.5	Graphing: Standards for excellence	96
		"The least ink": Is the simplest graph best? 96	
		"Chartjunk": A new name for garbage 98	
		Data density 98	
		"Revelation of the complex" 99	
	3.6	Tables: Turning can be timely	101
	3.7	Summary	104
Additional Readings			106
Exercises			106

4 Description of Data: Computing Summary Statistics — 129

- 4.1 Averages: Let us count the ways — 131
  - Mode: The hostess with the mostes' — 131
  - Median: Counting to the middle — 134
  - Mean: Balancing the seesaw — 138
  - Mode, median, or mean? — 140
- 4.2 Variety: Measuring the spice of life — 141
  - Range: Lassoing the two extreme values — 142
  - Standard deviation: The crucial deviant — 143
- 4.3 Standard error of the means — 149
- 4.4 Standard scores: Comparing apples and oranges — 150
- 4.5 Gain in simplicity, loss of information — 153
  - Replacing the data with a graph — 153
  - Replacing the data with a summary value — 154
- 4.6 Real estate data: Out-of-sight prices — 155
- 4.7 Summary — 156

Additional Readings — 158
Formulas — 159
Exercises — 162

5 Probability — 177

- 5.1 How to find probabilities — 180
  - Equally likely events — 180
  - Relative frequency — 180
  - Using subjective probabilities — 183
- 5.2 Computations with probabilities — 184
  - Adding probabilities — 184
  - Multiplying probabilities — 185
- 5.3 Odds: The opposite of probabilities — 185
- 5.4 Probability distributions for discrete variables — 187
  - Binomial distribution — 187
  - Poisson distribution — 190
  - Hypergeometric distribution — 192
  - Displaying probabilities in graphs and tables — 192
  - Computations with probabilities — 193
- 5.5 Probability distributions for continuous variables — 194
  - Standard normal distribution: The bell curve — 194
  - The t-distribution — 196
  - Chi-square distribution — 199
  - F-distribution — 201
  - Need for normally distributed data — 202
- 5.6 Using probabilities to check on assumptions — 202
  - Is it a fair coin? — 202

Is it a fair workplace?   203
Is it an evenly split electorate?   205
5.7 Decision analyis: Using probabilities to make decisions   207
5.8 Summary   210
Additional Readings   212
Formulas   212
Exercises   216

6   Drawing Conclusions: Estimation   231

6.1 Sample statistic and population parameter   233
6.2 Point estimation   235
What is a "good" point estimate?   235
A strategic use of the point estimate: How many tanks did the Germans have?   237
6.3 Interval estimation: More room to be correct   235
Length of confidence interval   241
Confidence intervals for differences   245
6.4 Summary   247
Additional Readings   248
Formulas   249
Exercises   251

7   Drawing conclusions: Hypothesis testing   265

7.1 The hypothesis as a question   267
Null hypothesis   267
Alternative hypothesis   268
Errors in answering the question   269
7.2 How to answer the question posed by the null hypothesis   270
Probability: The p-value   271
Mechanics of hypothesis testing   273
To reject or not to reject the null hypothesis   274
Causal effect: A skip too far   275
A little statistical theory and a game on the computer   276
7.3 Significance level   279
7.4 Testing a population proportion   281
7.5 Difference between two population proportions   284
Testing the null hypothesis   285
Estimating the difference   285
7.6 Testing hypotheses versus constructing confidence intervals   286
7.7 Statistical versus substantive significance   287

7.8	Applications: When to reject the null hypothesis	289
	Psychology experiment on cooperation and competition 289	
	Community study of blue-collar workers 290	
7.9	Summary	291
Additional Readings		294
Formulas		294
Exercises		299

8   Relationships Between Variables — 313

8.1	Four questions about two variables and their relationship	316
	Question 1. Relationship between the variables in the data? 319	
	Question 2. Strength of the relationship? 320	
	Question 3. Relationship in the population? 321	
	Question 4. Causal relationship? 321	
8.2	Prediction	322
8.3	Independent and dependent variables	323
8.4	Different types of variables: Categorical, rank, metric	324
8.5	Return to the question of causality	326
	Role of other variables 327	
	Role of time 328	
	Multiple causal factors 329	
8.6	Summary	330
Additional Readings		332
Exercises		332

9   Chi-square Analysis for Two Categorical Variables — 345

9.1	Analysis of the data: Are there trustworthy differences in attitude?	347
	Bar graphs 348	
	Summary computations with categorical variables 351	
9.2	Question 1. Relationship between the variables?	351
9.3	Question 2. Strength of the relationship?	353
	Phi in the sample 353	
	Phi in the population 355	
9.4	Question 3. Relationship in the populations?	355
	Setting up the null hypothesis 355	
	Testing the null hypothesis 356	
	From chi-square to p-value 357	
	Degrees of freedom for chi-square analysis 359	
9.5	Question 4. Causal relationship?	360

9.6	Larger tables: A banquet of possibilities	360
	Question 1. Relationship between the variables? 362	
	Question 2. Strength of the relationship? 362	
	Question 3. Relationship in the populations? 363	
	Question 4. Causal relationship? 363	
9.7	Summary	364
Additional Reading		365
Formulas		365
Exercises		370

10 Regression and Correlation for Two Metric Variables   397

10.1	Question 1. Relationship between the variables?	401
	Graphing the data in a scatterplot 401	
	Learning from the scatterplot 404	
	Linear relationships 404	
10.2	Question 2a. Strength of the relationship?	406
	Is r positive or negative? Large or small? 406	
	Four scatterplots: From strong to weak relationships 407	
	Interpretation of r: An issue of inexactness 410	
10.3	Question 2b. Form of the relationship?	411
	A line through the middle of the points 412	
	How to find the regression line: The least squares principle 414	
	Predicting with regression analysis: From fat to calories 416	
	Magnitudes of effects: Interpretation of r-square 417	
	Correlation or regression? The more the merrier 422	
	Regression analysis for data on change 425	
10.4	Question 3. Relationship in the population?	426
	Confidence interval approach 426	
	Hypothesis testing using t 427	
	Hypothesis testing using F 428	
10.5	Warning: What you measure is what you get	429
10.6	How to be smart using dummy variables	431
	Categorical independent variable with two values and metric dependent variable 431	
	Categorical dependent variable with two values and metric independent variable 434	
10.7	Question 4. Causal relationship?	435
10.8	Summary	435
Additional Readings		438
Formulas		439
Exercises		442

11 Analysis of Variance for a Categorical and a Metric Variable — 473

- 11.1 Analysis of variance: Comparing the mean-ings of things — 475
- 11.2 Question 1. Relationship between violent crime rate and region? — 478
  - Scatterplot — 478
  - Boxplot: A simpler view of the data — 479
- 11.3 Question 2. Strength of the relationship? — 481
  - Region variable — 481
  - Residual variable — 483
  - Effect of both region and residual variable: Total sum of squares — 484
  - Measuring the strength of the relationship — 485
  - Explained amounts of variation — 486
- 11.4 Question 3. Could the relationship have occurred by chance alone? — 490
  - The null hypothesis — 490
  - p-value from F — 491
  - Going beyond the F-test: Making mean comparisons — 494
- 11.5 Question 4. Causal relationship? — 496
- 11.6 Analysis of variance: A bird's-eye review — 496
- 11.7 Matched pair analysis: Two observations per unit — 498
  - A t-test — 498
  - The sign test: A simple yes or no — 501
- 11.8 Summary — 502
- Additional Readings — 504
- Formulas — 504
- Exercises — 507

12 Rank Methods for Two Rank Variables — 525

- 12.1 Two rank variables with words as the values — 527
  - Question 1. Relationship between identification and interest? — 528
  - Question 2. Strength of the relationship? — 530
  - Question 3. Relationship in the population? — 532
  - Question 4. Causal relationship? — 533
- 12.2 Ranking numbers as values: How are the Phillies doing? — 534
  - Question 1. Relationship in the data? — 534
  - Question 2. Strength of the relationship? — 535
  - Question 3. Did the relationship occur by chance? — 536
  - Question 4. Causal relationship? — 537
- 12.3 Summary — 537
- Additional Readings — 538
- Formulas — 538
- Exercises — 542

xvi Contents

13	Multivariate analysis	559
13.1	Partial phis: Three categorical variables	561
	Control for a third variable: The neutralizing game 562	
	Partial phi 563	
13.2	Multiple regression with metric variables	567
	Question 1. Relationship in the data? 567	
	Question 2b. Form of the relationship? Partial regression coefficients 568	
	Question 2a. Strength of the relationship? Partial correlation coefficients 571	
	Question 3. Relationship in the population? 574	
13.3	Multiple regression with a dummy variable	576
13.4	Two-way analysis of variance	581
	One-way analysis with time of day only 582	
	One-way analysis with route only 583	
	Two-way analysis with time of day and route 584	
	A second study with interaction effects 589	
13.5	Establishing causality	592
13.6	Summary	593
	Additional Readings	595
	Formulas	596
	Exercises	598

14	Statistics in Everyday Life	617
14.1	Stepping stones to statistical sophistication	618
14.2	Approaching numbers with care	622
14.3	Data and statistical methods	624
14.4	How things can go wrong	626
	Dangers in the collection of data 626	
	Special problems of survey research 628	
	Misuses of analysis methods 631	
	Misuses of statistical inference 632	
	Misuses in interpretation of numbers 633	
14.5	Statistics and Big Brother	634
14.6	Ending on the upbeat	636
	Additional Readings	637
	Exercises	638

Glossary	641
Statistical Tables	648
Answers to Odd-Numbered Exercises	673
Index	731

Statistics

THE CONCEPTUAL APPROACH

CHAPTER 1

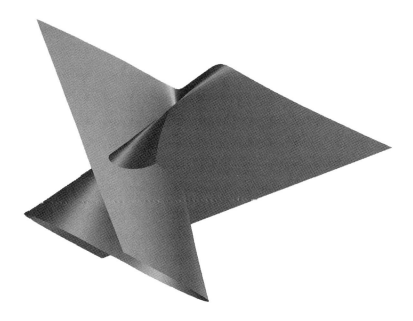

1.1 Statistics: What's in a word?

1.2 Knowing how statistics is used: Goals for the reader

1.3 Central ideas in statistics

1.4 Users of statistics

1.5 Relationship of statistics to mathematics, pencils, and computers

1.6 Summary

STATISTICS: RANDOMNESS AND REGULARITY

You probably are reading this book because you think it is important to know something about the subject of statistics. At the same time, you may suspect that studying statistics won't be the pleasantest task you have ever undertaken. We have seen too many reluctant students to think statistics courses are automatically crowd pleasers. We know some of you would prefer to analyze a poem, sing a ballad, or dissect a frog. But we think we have enough knowledge of student temperament to speak to all of you, eager and less than eager.

Some of you recognize that knowing how statistics is used in solving problems is critical in some parts of daily life; some of you may be looking forward to the challenge of statistics as a mental sport; others may see statistics as a means of solving mysteries that intrigue you. We think statistics can be intellectually stimulating and even fun. Our goal is not to introduce you to the inner sanctums of the profession of statistics. As the title suggests, this book is designed to help you understand statistics, to be comfortable with statistical language, and to know how to evaluate statistical results. If you wish to pursue statistics, this text will be just the beginning of a long and exciting road ahead.

To help you become oriented to the realm of statistics, we start each chapter with a few practical problems appropriate to the chapter content. We hope these problems will whet your appetite for the food for thought that follows. Here are some problems to start off Chapter 1.

1. As a prospective college student, you look at *Barron's Profiles of American Colleges*. Under Boston University, it indicates that the average SAT Verbal score of applicants is 550; SAT Mathematics is 600. What do these numbers mean? What is an average score? If your score is below the average score, should you not apply to BU? It is clear that you have to know something about statistics to select a college that will select you.

2. Imagine that you are a new manager in a marketing department. The statistical results of an advertising campaign are submitted to you for your comments. Among other things, the report declares certain results as "statistically significant." How do you interpret the report without exposing your ignorance of the terminology? Making sense of statistics suddenly is important to you and your career.

In requiring car manufacturers to produce electric cars as a percentage of their total output, to cut down on air pollution from travel by internal combustion automobiles, California state laws are beginning a national trend. Statistical information is crucial in creating the arguments that legislators believe in and in testing the effectiveness of electric car use on the quality of air. *(Peter A. Simon, Phototake NYC.)*

3. As a potential car buyer and a conscientious citizen, you would like to do your part for ecological preservation of the planet. What does the latest research indicate about the effects of consumer actions on natural resources? Should you buy a car with a diesel engine or purchase an electric car or maybe go all the way and ride a bicycle? Should you not use aerosol sprays? Should you not use chemical fertilizers on your lawn? Statistical studies presented in newspapers, magazines, and consumer reports become crucial to your decision making. What do all these studies really advise you to do about your consumer habits?

4. As a newspaper reader, you see headlines such as "Eat raw yogurt and live to be 100." Are there statistics that support this claim? What if you hate raw yogurt?

1.1 STATISTICS: WHAT'S IN A WORD?

Statistics is a word with many meanings, some of them better defined than others. The word *statistics* itself seems to have been coined by a German named Hermann Conring when he used the term *Statistik* in print in 1660. The first part of the word is an adaptation of the word *state* (*Staat* in German), and the term was first used to name the practice of states collecting information on births and deaths more than three hundred years ago. To this day, statistics remains a mainstay of bureaucratic organizations at all levels of government worldwide. Global statistics have become of vital concern to many international organizations, such as multinational corporations, the United Nations, and organizations concerned with such questions as population density, ecological disasters, and the prevalence of disease.

Beyond its origins in state policy, the word *statistics* has two important meanings. First, statistics can be thought of as numbers in one form or another: average rainfall in Texas, weekly temperature in Arizona, batting average of the Boston Red Sox, size of the national debt, or price of coffee in Brazil. Modern society seems to have an insatiable hunger for statistics, and in response to this need statisticians collect more and more of them. "This is a country run on numbers," said Janet Norwood, then Commissioner of the Bureau of Labor Statistics, in her presidential address to the American Statistical Association in 1989.

Statistical numbers are known as *data,* and one simple meaning of the word *statistics* is numerical data. In this text, we go beyond this meaning. We are interested in how data are obtained and what is done with the information data contain. In the end, we hope to show you that with the help of statistics, information in data can be turned into real knowledge.

> The word *statistics* can be either singular, as in "statistics *is* a fun course," or plural, as in "these statistics *indicate* an upswing in the economy," but the word "data" is always plural. Datum, not data, is the singular form. (Contemporary English professors have relented on this point, and so the once capital crime of saying "The data is impressive" has been reduced to a misdemeanor.)

Statistics is a set of concepts, rules, and methods for (1) collecting data, (2) analyzing data, and (3) drawing conclusions from data.

In its singular meaning, statistics can be defined as a discipline of study. You are taking a course in statistics, and your instructor may have a graduate degree in statistics. Within their discipline, statisticians explore and invent ways to obtain data and ways to work with the information contained in data in order to draw conclusions. They design new applications of statistics from mathematical equations, and they test theoretical models in practical settings.

By the end of this book, we hope that with the help of statistics you will be able to appreciate how data can be turned into useable knowledge that is more sophisticated than the numbers themselves. Unlike chemistry, sociology, or psychology, which are disciplines that study well-defined phenomena, statistics does not have its own empirical subject matter based on experiments or observations. Instead, statistics provides a set of methods that are used by the chemists, sociologists, and psychologists, among others.

1.2 KNOWING HOW STATISTICS IS USED: GOALS FOR THE READER

Because statistics is used in so many disciplines, the results of statistical analyses are all around us. Academic research journals, for example, depend on statistical results. In many disciplines, whether or not an article is published in a major journal depends heavily on whether or not statistical methods have been correctly applied.

Even cartoons are statistics-saturated. *(Reprinted with permission of the artist, Carol Cable.)*

Statistics is also heavily used outside the academic community. We cannot read a newspaper or a weekly news magazine without being exposed to articles based on statistics. Statistics is heavily used in industry, especially in research, quality control, and marketing. Statistics also forms the bases of stories in other print media. In *Playboy* and *Cosmopolitan, Vanity Fair* and *The New Yorker,* we read about percentages of people who are unfaithful to their spouses, percentages of people who contribute to charity, percentages of people who lose money on Broadway flops. The programs available on television, the particular anchor person we watch, and the kinds of advertisements we view depend on statistics; only TV shows, anchor persons, and advertisements with a high rating survive.

Opinion polls and surveys make use of statistics, and these days it is hard to imagine an election without polls on what the voters think about the issues and the candidates. The "image making" of presidents and party platforms depends on voter feedback obtained by statistics. Statistics also provide the basis for knowing who won an election and by how much. Even more dramatic is the power of statistics to predict

Are these people still on the air? That depends on the statistical results of surveys constantly being made in the race to be number one in television news broadcasting. *(UPI/Bettmann; Corbi-Bettmann.)*

with great accuracy the result of an election before the polls close or even before the election takes place. The fact that candidates either claim victory or admit defeat before all the votes are in and counted is a tribute to the confidence people have in statistics.

Now that you're thinking about statistics, the extent to which our culture is statistically indebted is probably dawning on you. Two examples to tickle your imagination:

When flights are overbooked because more tickets were sold than there are seats available, it is not an unfortunate oversight. The airline is relying on statistical analyses that indicate how many "no-shows" normally can be expected for any given flight. If they win, the flight is fully booked. If they lose, they have to give out a few free tickets.

Retirement communities depend on elaborate pricing schemes to attract clientele. When condominium costs are set for a retirement center, a factor in the estimate is the anticipated life expectancy of the residents of the complex. The longer people live, the higher the costs. Statistical analyses assist managers in setting competitive and at the same time profitable fees.

STOP AND PONDER 1.1

"Stop and Ponder" signals exercises scattered throughout the text to encourage you to bring your own creative juices to the process of understanding statistics. Instead of providing all the examples, applications, and incidentals, we invite you to add some to the mix. We think that if you take a moment to relate topics in the text to your own experiences, you will be better able to recognize how well you are comprehending the information, where trouble spots lie, and how you can tie together life outside the course with life within. This kind of activity leads to longer-lasting learning.

The first exercise is to think of a few examples from your personal experience where statistical analyses have been the basis for a decision that was either positive or negative, in your opinion. Also, try to come up with some reasons that the statistical results came out as they did. The cancellation of a recent TV show might be a good choice.

Understanding what can go wrong

If as consumers we are to fully understand the extensive applications of statistics, we need to know something about the rules and methods that were applied to get the results we read about. A knowledge of statistics helps us evaluate the results. It also helps us to be critical and to be aware of some of the things that could have gone wrong along

Predicting elections is not always easy! *(UPI/Bettmann.)*

the way from the time when the problem was first formulated to the writing of the final report.

The following story about a famous statistical survey that failed is part of statistical folklore by now. A highly respected magazine called *Literary Digest* conducted a poll of the electorate before the 1936 Presidential election. The burning question, of course, was who would be the next President—the challenger, Governor Alf Landon of Kansas, or the incumbent, President Franklin Delano Roosevelt. In order to assess voter preferences, the magazine pollsters had sent out sample ballots to a large number of people who were listed in telephone directories and car registries. (Telephones and cars were not as common in 1936 as they are today, but the lists were easy to obtain.) Although about 10 million sample ballots were sent out, not a very high percentage of people returned their ballots. Among those who did reply, however, Alf Landon was the hands-down favorite. The magazine predicted a Landon victory.

If readers had known something about statistics, they would have been skeptical about the claim that Alf Landon would win the election. As you might expect, polling people who owned telephones and automobiles during the middle of a great economic depression was not a very good way to get an accurate assessment of the spectrum of voter opinions. Furthermore, the low percentage of ballots that were returned was suspect. As it was, the readers had to wait until after the election to see how wrong the results of the poll were: Franklin Roosevelt, not Alf Landon, was elected President. Most current usage of statistics is not as wrongheaded as it was in this example, but even today we do not have to look far to find questionable uses of statistics, especially where choosing a correct sample is concerned. *(Source: Jeffrey Witmer, DATA Analysis: An Introduction. Englewood Cliffs, NJ: Prentice Hall, 1992, p. 97.)*

Understanding statistical terms

The results of statistical analyses do not help us much if we do not understand the terms that are used. For example, a typical statistical expression used to report findings is "statistically significant." In reporting the percentage of voters that favor a candidate, the terms "sampling error equals $\pm 3\%$" or "margin of error equals $\pm 3\%$" might be used. Two variables may have a "high correlation." These are three common statistical terms, and for people who know what they mean, the terms are informative and useful. People who do not know the meanings of the terms, however, may not understand what is being said or come to erroneous conclusions about the findings.

1.3 CENTRAL IDEAS IN STATISTICS

Randomness and regularity: Twins in tension

When we cannot predict the outcome of an event, randomness is associated with the event. For example, when tossing a coin we cannot tell whether the coin will land heads or tails. Similarly, when we take a trip, we cannot tell whether we will have an accident or not.

At the same time, when we put random events together, they display amazing regularities. Patterns and trends become evident, even when we examine something as random as coin tosses. If you toss the same coin 100 times, you know it will land approximately 50 times heads and 50 times tails. Similarly, while a single car accident is a unique concurrence of several unlikely events, if you could examine all accidents, you would find disturbing regularities among them. Year in and year out, around 40,000 people in the United States die in car accidents. This is an amazingly stable number despite the small probabilities surrounding a particular event. On a more personal level, each year when Mary Gergen surveys her Introductory Psychology class of 100 students at her campus, she finds that about 50% of them have been involved in automobile accidents in the past year. She has discovered that random events make up a regular rate of accidents.

Using statistical analyses of seemingly random phenomena, we can begin to make sense of the world. A basic knowledge of statistical ideas helps put randomness into the perspective of regularity. Statistical ideas help us realize the importance of randomness and regularity both in how we observe and in how events actually occur in the world. Thus, statistics can be seen as a search for regularities in randomness.

Randomness in regularity

But even the regularities display some randomness. If you toss the coin another 100 times, you would almost never get exactly the same number of heads and tails as in the first 100 tosses. In one round of 100 tosses you might get 48 heads, and in the next 100 tosses you might get 53 heads. This illustrates an important and central feature of statistics.

> Whether we take a single new observation or a new set of many observations, most of the time we do not get exactly the same result we did the first time.

This kind of variation happens not only with coin tosses but also with surveys, experiments, and every other means of data collection. If people in a survey are asked how they stand on an important issue of the day, a certain percentage of the respondents will have a particular opinion. If the same survey is done with a new sample of respondents, a different percentage of respondents will have the opinion. The variation in the two percentages is attributed to the randomness that is inherent in data. This way statistics becomes the study of the variation in the data.

With the mathematical theory that underlies statistics, we can find how much randomness is attached to a percentage from a survey and how much the percentage can be expected to vary from one repetition of the survey to another. We can even tell if the difference between two percentages is larger than can be explained by randomness alone. These ideas are expanded on and discussed in much greater detail in later chapters.

In regularities trends of change sometimes appear. The rate of car accidents is going down with increased seatbelt use and airbag deployment. Statistics puts the single, random events into regularities, and statistics reveals trends of change. If the numbers of accidents in separate periods (two patterns of regularities) differ by more than can be explained by randomness alone, a change is occurring.

Two examples in the study of randomness and regularity

As an example of whether the difference between two numbers is a result attributable to more than randomness, consider the introduction of the polio vaccine in the 1950s. Polio was a dreaded disease that struck in mysterious ways, often leaving its victims, many of them children, paralyzed or dead. After many years of epidemics, a vaccine was finally developed that scientists hoped would provide protection against the disease. But it was not clear whether the vaccine would actually work as the researchers hoped. Although laboratory and animal tests looked promising, the only way to find out was to test the vaccine on humans. Because polio was a rare disease, the vaccine had to be tried on a fairly large number of children to see if it had any effect, so the researchers decided to use a group of 200,000 children. They also decided to have a control group of the same size, in which the children received a placebo—a substitute that looked like the real vaccine—to see if the vaccine had any effect.

After the children received their vaccine or placebo, the researchers watched and waited to see what the outcome would be after the next "polio season." In the control group, 138 children contracted the disease. The researchers were not exactly sure what this number of cases meant. There is a certain amount of randomness in that number. If another group of 200,000 children had also received the placebo, the same number of children would not have contracted polio. Depending on how large the random component was, perhaps 130 or maybe 140 or some other number of children would have come down with polio.

In the group that received the vaccine, 56 children got polio, a number that also has a random component. The important question was whether 56 differed from 138 by more than could be explained by randomness alone. If that were the case, then the researchers could be confident that the vaccine had an effect. By methods explained in Chapter 7, it turned out that the difference between 138 and 56 indeed was larger than could be expected by randomness alone, and the vaccine was pronounced a success. In the years since then, the vaccine has essentially eradicated polio in many countries. Renewed efforts by health organizations around the world make it likely that even children in lesser developed countries will not have to suffer from polio in the near future. In an important sense, statistical reasoning provided support to the medical researchers who developed and tested this vaccine.

Another famous instance of randomness—or the lack of randomness, as is the case in this particular example—took place in the military. During the war in Vietnam, the United States government instituted a draft lottery to get enough soldiers to fight in the war. The plan was to assign a number between 1 and 366 randomly to each date in the year. The military was then to draft young men in the order of the numbers assigned to their birthdays. This method was designed to equitably distribute the risk of entering this unpopular war; the possibility of getting drafted was supposed to be determined randomly.

The draft lottery the first year assigned the number 1 to September 14 by the drawing of the appropriate ball from a large container of 366 Ping-Pong balls on which the dates were written. All eligible 18-year-olds born on September 14 were drafted first. The men born on the date assigned draft number 2 were drafted second, and so on. It was known that not all the draft numbers were needed and that therefore men born on dates receiving high draft numbers would probably never serve in the military.

The lottery seemed as good a method as any to decide who would get drafted. However, the day after the drawing, when all the dates and their numbers were published, statisticians began to investigate the data. After some looking and counting, the statisticians found certain patterns. For example, we would expect that about half of the low draft numbers—1 to 183—would be assigned to dates in the first half of the year, in the months January through June, and about half of them to dates in the second half of the year, in the months July through December. Because of the randomness of the drawing, there would not be exactly half the draft numbers in each half of the year, but it should be close to half. As it turned out, 73 of the low draft numbers were

Vietnam casualties were determined by a lottery. *(FPG International.)*

assigned to birthdays in the first half of the year, while 110 low numbers were assigned to birthdays in the second half of the year. In other words, if you had a birthday in the second half of the year, your chances of having a low draft number and therefore being drafted were considerably higher than if you had a birthday in the first half of the year.

In this case, where there should have been only a random difference, the difference between 73 and 110 was larger than what could be expected by randomness alone. The absence of randomness has been attributed to not stirring the Ping-Pong balls adequately before they were selected. The Selective Service consulted statisticians before they conducted the draft lottery the following year. (This was small comfort to those born in the second half of the year whose birthdays were overselected.)

Probability: What are the chances?

What this discussion of randomness says is that much of statistics is based on the very important concept of *probability*. Probabilities provide a building block for the third aspect of statistics, namely, how to draw

> A **probability** is a number between 0 and 1 that tells us how often a particular event will occur.

conclusions from data. We may never be quite certain whether two numbers differ by more than can be expected by randomness alone, but we can find out whether the probability that they do is small or not. From this basic idea emerges many interesting opportunities for drawing important conclusions about the world around us. How this is done is fleshed out in Chapter 5 and later chapters.

Variables: The names we give things

> A **variable** is a characteristic, a trait, or an attribute that can take on two or more possible values.

A second major building block in statistics is the concept of a *variable*. The human characteristic of gender is a variable with two values, because a person is either female or male. Religious affiliation is a variable, which in the western hemisphere might have the values Catholic, Jewish, Muslim, Protestant, and other; in India the values might be Hindu, Muslim, Buddhist, Sikh, and other. Other examples of variables are miles per gallon for cars, with a range of values from 8 to 50, weight of children in kilograms, with a scale from 10 to 70, dosage of a medication, and so on. Usually researchers begin their projects by defining the variables they are interested in and the possible values of the variables. We can think of the values of a variable as points stretched out along a line that represents the variable itself (Figure 1.1).

Variables, values, and elements

The *value* of a variable always is a measure of a specific unit, often thought of as an *element*. An element can be a person, a group of people, a plot of land, a plant, an animal, or a country, as long as the element is agreed on, obvious to the users, and does not change in the middle of the analysis. Table 1.1 lists some examples of variables and their values together with the element on which the variables usually are measured. Thus, the gender variable is observed in a person as the element. Number of children is a variable observed on a family as the element. In the case of the family, the element is an aggregation of single individuals.

Theoretical variables and empirical variables

The variables we have discussed so far are familiar to most of us as everyday kinds of items and events. These variables are called *empirical variables* because they deal with objects in the observable physical world surrounding us. In addition to empirical variables, we also use variables

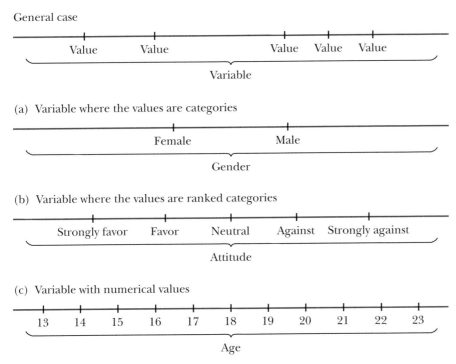

Figure 1.1 Variables and their values

created by statisticians. These variables, which are mathematically derived, are called *theoretical variables*. Several examples of theoretical variables are introduced in later chapters. Four of these variables are known as the z-, t-, chi-square (pronounced kī-square) and F-variables.

Table 1.1 Variables, values, and elements

Variable	Values of the variable	Element
Gender	Female, male	Person
Attitude	Oppose, neutral, favor	Person
Unemployment	Employed, not employed	Person
Unemployment	0.0%, . . . , 4.6%, . . .	County
Yield of corn	. . . , 5678 lb., . . .	Acre
Number of children	0, 1, 2, 3, . . .	Family
Poverty level	Severe, moderate, borderline, none	Precinct
Placement in a race	1st, 2nd, 3rd, . . .	Team

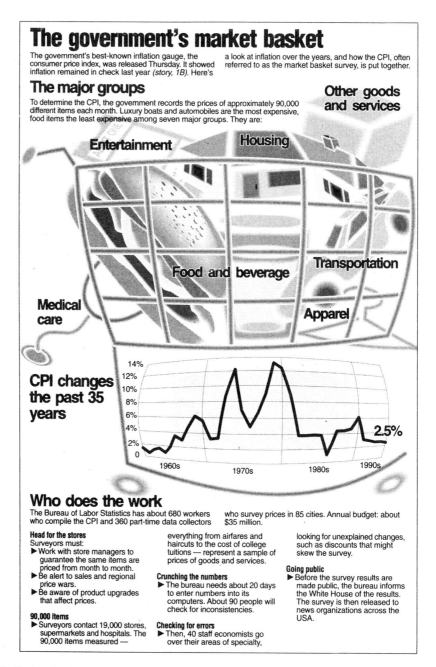

Inflation is difficult to measure. (© 1996, USA Today. Reprinted with permission.)

Constants

A *constant* is the opposite of a variable. Imagine that we survey all the students in a statistics course and find the percentage who think it is an interesting course. Assuming that nobody changes their mind, we would find the same percentage if we repeated this little study. A number such as this percentage is a constant: it does not change when the study is repeated. Obviously, it did not change because each time we asked all the students in the course. In statistics we make use of a type of constant called a *parameter* (Chapter 7 and later chapters).

> A **constant** always has one fixed value.

1.4 USERS OF STATISTICS

Let's take a closer look at some of the fields affected by statistics: government agencies, sciences, medicine, industry, even the law. In this country, the federal government is the largest collector of data and user of statistics through the Bureau of the Census and other federal statistical offices like the Bureau of Labor Statistics. The federal statistical system is well known for its excellence, although it has suffered from budget cutbacks in recent years.

Two of the best-known activities of the federal statistical system are the consumer price index and the unemployment figures. These results are published monthly and play a very important role in the economic life of the nation. The consumer price index dates back to the early 1900s. Many activities, such as labor contracts and Social Security payments, are tied to its value at any given time. The unemployment figures were developed during the Depression of the 1930s, when New Deal reformers realized just how important it is to know how many unemployed people there are in the United States. Both reports are based on large sample surveys conducted according to complex statistical principles.

Much of the data government agencies collect are analyzed in order to create public policy on a variety of issues. For example, to determine tax policies it is important to know how existing tax laws affect people in various income categories and to be able to predict the impact of changes. For a social welfare program to be successful, it is necessary to know the conditions in society that create the need for the program and to know in what ways the program affects the people for whom it was designed. The Head Start program for preschool children, for example, has been the object of intense scrutiny as experts attempt to determine if children who are enrolled in it enjoy long-term benefits. For a farm subsidy program, it is necessary to know about the magnitudes of current agricultural production and try to determine the consequences of the subsidies on future production.

People from most academic disciplines use statistics in their research: biology, economics, and psychology are three disciplines with such heavy usage that they have developed their own sets of statistical methods: biometrics, econometrics, and psychometrics, respectively. In the humanities, groups of historians, geographers, linguists, and classicists make use of statistics to draw conclusions as diverse as the number of deaths due to the Black Plague and the popularity of the French language in the English-speaking world. This means that almost all empirical academic research—reports, presentations at professional meetings, journal articles, and books—is based on statistics in one way or another. Academic research enriches the life of a society in manifold ways, and statistics plays a unique role in this process. No other discipline contributes so much across so many scientific fields.

A vivid example of the growing role of statistics in social life involves the practice of law. Many lawyers have found themselves in new territory when confronted with statistical issues in addition to legal ones. One major area where statistics has been required is in class action suits concerned with discrimination based on age, gender, and race. Lawyers must persuade judges and juries that differences in age, gender, or race in any given setting are either by design or are random. Statisticians have been challenged to act as expert witnesses to explain topics such as "confidence intervals" and "significance levels" to juries and judges. Without the expert testimony of statisticians, it would not be possible to conduct these cases in the courtroom in a fair and rational manner.

The field of medicine has been altered by the introduction of new statistical ways of evaluating treatment effects. For example, in cost containment measures imposed by managed health care organizations, physicians must follow the organizations' guidelines for care in order to be reimbursed. These guidelines are developed through careful statistical analyses of large numbers of medical practices and outcomes. If the probability of a satisfactory result is the same for both an expensive and a cheap intervention, the HMOs do not reimburse for the more costly one. The use of statistical methods to bolster the medical

DNA TESTING: BY THIS SPIRAL HANGS A TALE

During the famous O. J. Simpson murder trial, which concluded in 1995, much testimony was given regarding DNA samples and how they were collected, analyzed, and identified. The public learned a great deal about statistical measurement from the witnesses who gave evidence about the blood samples taken from various surfaces. The crux of the matter was how probable it was that the DNA samples collected matched the blood of the murder victims and the accused. Defense attorneys opposed the conclusions of the prosecutors that the chances that the blood samples were not Simpson's were minuscule at best.

In general, the process of DNA testing has been to look at a pattern of indicators in a DNA strand and to calculate the likelihood that the pattern could be shared by two individuals. Once this method of testing was made available, the public quickly came to appreciate its uses in many types of cases. In another trial, a man who was in prison for seven years on a rape conviction was released when his lawyers obtained evidence that the convict's DNA did not match the rapist's.

guidelines of insurance companies has led to concerned arguments from medical advocates that not enough attention is paid to the individual patient. In the New Jersey legislature, for example, the policy of allowing newborns and mothers only 24 hours of hospital coverage after birth was overturned by a state law, even though 95% of all mothers and infants had no serious complications. While the statistical methods are not themselves under scrutiny, the definition of "acceptable risk" is in question.

Major corporations are also heavy users of statistics. For example, to get a new drug approved by the Federal Drug Administration, a pharmaceutical company must prove that the drug is safe. Companies invest heavily in measuring the effectiveness of their new products through experiments on animals and humans. As a result, such companies employ large numbers of statisticians. They are responsible for setting up experiments properly, analyzing the resulting data for the impact of the experiments, and checking the validity of marketing claims to avoid lawsuits and costly retrenchments in drug development.

Many industries use statistics in their quality-control operations. Items rolling off a production line are not identical, partly because of random variation and partly because things can go wrong in the production processes. This variation can be studied using statistical methods that can help pinpoint what went wrong and where it went wrong. Good quality-control programs ensure that consumers will be satisfied with their purchases and not turn to competitors for their next purchases. The American statistician Edward Deming was a leader in the development of statistical methods for quality control. Ironically, many of his methods were first adopted by companies outside the United States, especially in Japan. One reason Japanese industry experienced such impressive growth after World War II is that its business leaders took up Deming's ideas early on. *(Source: W. Edward Deming,* Out of the Crises, *Cambridge, MA: MIT Center for Advanced Engineering Study, 1986.)*

1.5 RELATIONSHIP OF STATISTICS TO MATHEMATICS, PENCILS, AND COMPUTERS

The field of statistics is founded in mathematics. Today, independent departments of statistics in leading universities train statisticians, but statistics formerly was a part of mathematics departments. Statistical reasoning rests firmly on mathematical foundations. As a result, it is easy to find statistics texts that look like mathematics books with theorems and proofs. But it is possible to learn about statistics without knowing all the mathematical underpinnings, and that is how we present statistics in this book. Today most statistical analyses are done on the computer, so it is more important to understand what goes into and comes out of the computer than how the computer software computes.

The emphasis here, as we have mentioned, is on learning the basic statistical ideas—some of the specialized vocabulary, how data are collected, displayed, and analyzed, what results mean, and when they should and should not be used in everyday life—without getting bogged down in formulas and technical discussions of how computations are made. For most people today, an understanding of statistical ideas is critical to being a literate and well-rounded citizen; being able to do competent statistical analysis on one's own is part of a highly professional career path.

There are other reasons for doing statistical analysis, such as for the sheer fun of it or to get a "gut" feeling for the craft. For those who wish to do analyses of statistical problems using traditional paper-and-pencil techniques or the more advanced computer programs, the exercises at the end of each chapter contain many opportunities. The exercises are divided into three parts. Those in the first part test your general conceptual knowledge, those in the second test your abilities to interpret data and apply statistical results to daily events, and those in the third part require you to use formulas and mathematical techniques. The formulas for various statistical calculations are available in their full glory at the end of each chapter. In addition, statistical tables for the theoretical variables are found at the back of the book.

1.6 SUMMARY

1.1 What's in a word?

Statistics was first coined as a word for state-related indicators. Later, statistics came to mean a summary of individual data points. As a field of study, statistics can be defined as a set of concepts, rules, and methods for (1) collecting data, (2) analyzing data, and (3) drawing conclusions from data.

1.2 Knowing how statistics is used: Goals for the reader

Statistics does not have its own subject matter but is applied to data from other fields of study. Because of the prevalence and power of statistics in today's society, we cannot avoid the consequences of these analyses.

1.3 Central ideas in statistics

Randomness and regularity are two important statistical concepts. Randomness is the inability to predict the outcome of a particular event.

Regularity is the pattern we find when we collect data on many events. Regularities themselves contain randomness. Statistics can be defined as the search for regularities in the face of randomness. Trends of change occur when the difference between two patterns of regularity exceed the effects of randomness alone.

Probabilities provide the foundation for drawing conclusions from our data. A probability is a number between 0 and 1 that tells us how often an event happens. Statisticians judge whether numbers differ by more than can be expected from randomness alone by using probabilities.

A variable is defined as a characteristic or attribute, such as a person's age, that can take on two or more possible values (e.g., 0 to 100+ years). The value of a variable always refers to a specific element, such as a person, a group of people, a plot of land, a plant, an animal, or a country. Many variables are familiar to most of us as everyday items and events. These variables are called empirical variables. We also use variables created by statisticians, called theoretical variables, which are mathematically derived. Four of these variables are known as the z-, t-, chi-square, and F-variables.

The opposite of a variable is a constant. Constants are numerical values that do not change. In statistics a certain kind of constant is known as a parameter.

1.4 Users of statistics

Statistical methods are crucial to government agencies in the formulation and evaluation of policies. They are also necessary for development of knowledge in all fields of scientific scholarship. Statistical methods are also gaining in importance in professional fields, such as law and medicine, and in diverse business enterprises.

1.5 Relationship of statistics to mathematics, pencils, and computers

Statistics is founded in mathematics, but the thrust of this book is to acquaint you with basic statistical ideas, not to turn you into a statistical analyst.

ADDITIONAL READINGS

Gani, J. (ed.). *The Making of Statisticians.* New York: Springer-Verlag, 1982. Sixteen statisticians tell why and how they became statisticians.

Gonick, Larry, and Woollcott Smith. *The Cartoon Guide to Statistics.* New York: HarperPerennials, 1993. If ordinary textbooks do not appeal.

Huff, Darrell. *How to Lie with Statistics.* New York: W. W. Norton, 1954. A classic book on possible misuses of statistics.

Peters, William S. *Counting for Something: Statistical Principles and Personalities.* New York: Springer-Verlag, 1987. Teaches statistics in a historical context.

Tufte, Edward R. *Data Analysis for Politics and Policy.* Englewood Cliffs, NJ: Prentice-Hall, 1974. Chapters 1 and 2 give good introductions to various statistical issues.

EXERCISES

REVIEW (EXERCISES 1.1–1.14)

1.1 Why is the root of the word *statistics* derived from the word *state*?

1.2 Why might statistics be called a "helper" science?

1.3 a. Define randomness.
   b. Define regularity.
   c. What roles do randomness and regularities play in a statistical study?

1.4 Give three examples from daily life of random events that contain regularities.

1.5 How does the notion of probability help a researcher decide if the data indicate a difference in the variables under study greater than random fluctuation?

1.6 Define "trends of change" and give an example in which it might be found.

1.7 a. Define the term *variable*.

b. What is the difference between an empirical and a theoretical variable?

c. What are the names of the four theoretical variables mentioned in this chapter?

1.8 You are interested in studying hurricanes worldwide. Name five empirical variables you might want to use in such a study.

1.9 Create a list of the values for each variable in Exercise 1.8.

1.10 a. Which is the (more) correct sentence: "The data is interesting" or "The data are interesting"?

b. Explain your choice.

1.11 How is a constant different from a variable?

1.12 Describe how the practices of law and medicine have been influenced by statistics.

1.13 Name two federal statistical systems that are central to national economic management.

1.14 Describe a way in which statistical analyses lead to better manufacturing processes.

INTERPRETATION (EXERCISES 1.15–1.18)

1.15 Discuss the following: Whether or not an individual property owner on Cape Hatteras loses a roof during a hurricane seems to be a matter of chance, that is, a random event. Yet there seems to be some regularity in hurricanes hitting Cape Hatteras. How is property damage during a hurricane, such as roof losses, related to the notions of randomness and regularity?

1.16 Name one area of life described in this chapter in which decisions based on statistical analyses have affected your own life. Briefly describe it.

1.17 Sports broadcasting today depends heavily on computer-generated statistics, which can be calculated for everything from the amount of prize money won by each professional tennis player in a season's play to the total number of triple plays by a single player in the history of baseball.

    a. Why do you think televised sports coverage has become so statistically oriented in recent years?

    b. How do you think statistical orientation has affected viewer appreciation of sports?

    c. Do you think other aspects of culture, e.g., music, films, politics, amateur sports, have been (or will be) affected in the same way as professional sports, in terms of the "invasion" of computer-generated statistics? Give several examples to support your claim.

1.18 Comment on the following in light of the goal of most statisticians to go beyond the actual data collected: "Our samples are like the shadows at the entrance to a cave we may not enter."

ANALYSIS (EXERCISES 1.19–1.23)

1.19 This exercise is intended to illustrate the notion of variation and randomness. Close your eyes and open this book to a random page. Place your finger on a random spot on the page and select the nearest complete sentence of text below your finger.

    a. How many words do you find in the chosen sentence?

    b. Select another sentence the same way and count the words in it.

    c. Why are the numbers of words in the two sentences not the same?

    d. If everyone in the class counted the lengths of two sentences, you could estimate the average length of the sentences in this book. How do you think this average would compare with the average length of sentences in Shakespeare's *Hamlet*?

1.20 Turn on a water faucet until it just drips. Count the number of drips per 20-second interval for 5 minutes. Keep a record of the number of drips in each interval. Using your data, how would you describe the drips? That is, in what respect were they random and in what respect were they regular?

Table 1.2 Infant and maternal mortality rates 1915–1945* (Exercise 1.21)

	Infant mortality rates		Maternal mortality rates	
Year	White	Nonwhite	White	Nonwhite
1915	98.6	181.2	6.0	10.6
1920	82.1	131.7	7.6	12.8
1925	68.3	110.8	6.0	11.6
1930	60.1	99.9	6.1	11.7
1935	51.9	83.2	5.3	9.5
1940	43.2	73.8	3.2	7.7
1945	35.6	57.0	1.7	4.5

*The rates are all numbers of deaths in first year of life per 1,000 births.
Source: Data compiled by U.S. Bureau of the Census.

1.21 The following exercise is derived from the data in Table 1.2.

a. Looking at the trends in infant mortality over the 30 years, what two major conclusions can you draw?

b. Looking at the trends in maternal mortality rates over this period, what two major conclusions can you draw?

c. Which set of data seems to be simpler to describe, infant mortality rates or maternal mortality rates?

d. What major conclusion do the data suggest about childbirth death and race?

e. If there were problems with the accuracy of collection of these data, what might they be and which data are more likely to be inaccurate?

1.22 Find an article in a newspaper or a news magazine that includes statistical information.

a. Identify the variables used in the article.

b. Determine what the values are for each variable.

c. What readers would be particularly interested in the article?

d. Does the article describe change of any kind?

1.23 a. Would the article you selected in Exercise 1.22 have been more precise, interesting, or valuable if the variables had been reported differently?

b. Are there ways in which you think the article could have been improved in light of the material presented in this chapter?

CHAPTER 2

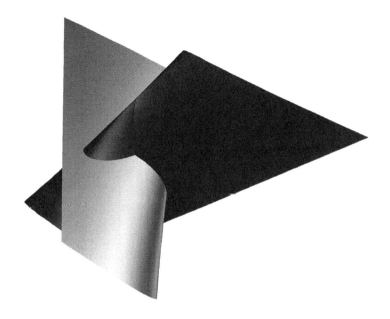

2.1 Defining the variables

2.2 Observational data: Problems and possibilities

2.3 Errors and "errors" in collecting observational data

2.4 Experimental data: Looking for the causes of outcomes

2.5 Data matrix/Data file

2.6 Summary

COLLECTION OF DATA

How many people in Los Angeles were infected with AIDS by a sexual partner last year? How much garbage was recycled in New York City last year? What caused scurvy to attack seventeenth-century sailors on long voyages? Does class size affect school performance? Is the President doing a good job?

To answer these questions and an enormous number of other ones, information must be gathered. In these instances, we need to know many things, from sexual habits to recycling practices. At first glance it seems easy to get this information. One needs only to go out and ask people or do an experiment to see how things work. But then the quandaries begin: Who should do the asking—you, me, unemployed college students, retired executives? And who should be asked? Can we afford to ask everyone concerned with the problem? For the first question, that would be the entire population of Los Angeles! Well, if not everyone, how about people who walk by a certain store at the mall on Saturday afternoon? Or those buying beer at the baseball stadium? Or do you think a presumably fairer way should be found?

Once these issues are settled, what should be asked? Some of the topics suggest "delicate" phrasing, to say the least. Will we get a straight answer if we ask people how many sexual partners they have had? Should we expect one? Should we even ask? How many people will tell us what they think we want to hear or what they think will make them look good? Will it make a difference if the asker is perceived as a medical worker, a police officer, a trash picker, or a bookie? What does "doing a good job" mean? Each of these questions deserves a thoughtful answer. Yet no answer seems to be exactly the right one.

A wise statistician says there are two kinds of data: good data and bad data. There are other ways of characterizing data, but this is as good a start as any. Good data are data that have been collected according to sound and proper statistical principles. Bad data are data that have been collected in other ways. This chapter describes some of the solutions statisticians and others have come up with to improve the quality of the data collected.

Data depend on many factors. (*"Sally Forth"* reprinted with special permission of King Features Syndicate.)

2.1 DEFINING THE VARIABLES

Data are collected in a variety of ways and in a multitude of settings. (At the moment, one author is drinking a sample of instant coffee provided by a company for a market survey.) At the most general level, collecting data involves measuring variables. Researchers ask people, for example, about their sleeping habits, count the number of dollars in gambling casino revenues, weigh how much trash is recycled, and give a plant a specific amount of water and measure how much it grows. Researchers weigh, measure, interrogate, and count their subjects in a multitude of ways.

The first rule of data collection is clarity about what is being measured. In other words, the variable must have a well-thought-out definition. Sometimes this sounds simpler than it turns out to be.

Suppose we are interested in family life and ask in a survey the following question: "How many children are in this family?" We may think we know what we want to find out, but there is no reason to expect that the person answering (commonly call the *respondent*) shares our view. We may rather thoughtlessly assume that a child is defined as a person who is under 18 years of age and who lives in a residence

How many children are in this family? *(Bruce Coleman, Inc.)*

with his or her biological parents. But what if the household includes biological children over 18, stepchildren, foster children, adopted children, or other young relatives? What about children who live elsewhere than with their biological parents? What if the parents are divorced and share custody of a child? The possibilities for confusion are many. If we, as the researchers, have not thought these issues through, we have no reason to expect the respondent to figure them out. If our ideas are muddled and the respondents' reports are inconsistent, our data will be extremely uneven in meaning. The lesson here is that before we can conduct a research, it is essential that we develop a clear, detailed definition of the variables. In the example, we must clarify our definition of "child."

2.2 OBSERVATIONAL DATA: PROBLEMS AND POSSIBILITIES

Observational data are data collected from observations of the world without manipulating or controlling it.

There are two major approaches to data collection. One method is collecting data on the world as we observe it, for example, the average number of pounds of aluminum cans recycled in different cities. Observational data arise when we simply observe the world around us. Researchers collecting observational data try not to intervene in ongoing patterns of behavior. Counting how many people in Los Angeles were diagnosed with the AIDS virus is an example of gathering observational data. Tabulating the results from a political survey is another example.

Observational studies are diverse. They examine the operations of local organizations and businesses, the behavior of humans and animals in their normal habitats, historical evidence found in libraries, interactions on the Internet, physiological, psychological, social, or environmental data, such as in blood samples, "inkblot" tests, stock market price indicators, quality control studies, or level of carbon monoxide pollution readings or any other phenomenon you can imagine! Statistics play an important role in all observational studies both in the planning of how the data should be collected and in the actual analysis of the data.

Population versus sample

Data are collected for the purpose of drawing conclusions from a collection of *elements*. Social scientists collect data on people to gain an understanding of human behavior. Botanists collect data on plants to gain an understanding of how they grow. Engineers collect data on ball bearings to make sure that they are of the right size for the engine for which they were made. All the elements we are interested in make up the *population*. All the inhabitants in Canada on January 1, 2000, is an example of a population; so are all the champagne corks in Times Square on New Year's Eve.

Sometimes we are able to collect data on all the elements in the population; in that case, we have conducted a *census* of the population, similar to the census conducted on the inhabitants of this country every ten years. In the harsh world of limited budgets, time constraints, and changing environmental conditions, however, it is usually impossible to conduct a census. Instead, we limit ourselves to collecting data on a *sample* of the elements in the population.

> A **population** consists of all the elements under study.
> A **census** is the process of collecting data on an entire population.
> A **sample** is a selected part of a population.

Let us look at how samples are selected, what makes a sample good or bad, and why a good sample is better than a mediocre census.

Selection of the sample: Making sure the pot is stirred

A critical issue facing all statistical researchers is how a sample should be selected. A researcher wants to be certain that the conclusions drawn from the study's sample can be applied to the larger population from which the sample was drawn. Without a "good" sample, this will not be the case.

An analogy from cooking assists in explaining why getting a good sample is so important. When we taste a spoonful of soup that we have been cooking, we are interested not in how that particular spoonful tastes but in how the entire pot of soup tastes. If the pot has been stirred adequately, we need to taste only a spoonful to find out how the entire pot tastes. We get the taste of the full pot from a spoonful whether the pot is a small one in a family kitchen or a large one in a soup factory. This is also the case when we choose a sample from a population—a sample, in a sense, from a population that has been properly stirred. If the population has been properly stirred, a sample of 1,000 respondents could tell us as much about a very large group, such as the

> **STOP AND PONDER 2.1**
>
> There are many ways to collect observational data. What other ways can you think of?

Mitch Reardon, Tony Stone Images.

entire population of the country, as about the population of a town or rural county.

We can apply this soup sample example to sample surveys. An opinion poll before an election finds that 57% of the people in the sample favor a candidate. If the sample is properly selected, the percentage will be approximately the same as in the entire electorate. Similarly, in a quality-control study, a sample of light bulbs is inspected not to see if the particular bulbs burn as they should but to see whether the manufacturing process is producing a general population of lightbulbs that function properly. The sample should be selected as a good indicator of the total production run and therefore a good indicator of the production process itself.

If a sample is not properly selected, misleading conclusions can be drawn about the "soup." If pollsters questioned only their families and friends, poor sample results would occur. If checkers inspected only the top layer of bulbs in "fragile" boxes and did not see that the bottom layers were crushed by insufficient padding, for example, the sample would be misleading. Because of the importance of sample selection on the trustworthiness of the results, it is imperative that samples be selected according to proper statistical principles. The failure of the draft lottery in selecting soldiers during the Vietnam war, mentioned in Chapter 1, was an example of poor sample selection.

Random sample: What is it?

A proper statistical sample that can be used for generalizations to a larger population is called a *random sample*. Drawing names from a hat is the simplest example of choosing a random sample. The slips of paper are the elements that make up an entire population, and all have an equal chance of being drawn. In this way, it is possible for all groups in a population to be represented in a sample in approximately the same magnitudes as in the population. Thus, if there are 10,000 Serbs and 100,000 Croatians in Dubrovnik, then a random sample from the city would have approximately 10 Serbs for every 100 Croatians.

> A **random sample** is a sample drawn from a population in which every element has a known (sometimes equal) chance of being included in the sample.

Convenience sample: How to produce a "bad" sample

Researchers are often tempted to study elements of a population that are easily at hand. For example, many studies reported in psychological journals use subjects who have been required to sign up for experiments, often, introductory psychology students. Medical researchers and therapists often do studies on their own patients; market researchers study shoppers they can urge to cooperate. Samples that are easy and economical to acquire are known as *convenience samples*. While in some cases a convenience sample might be perfectly adequate for the research study design, this is usually not true. The extent to which one can generalize the results from the subjects in a convenience sample to others in the population is limited.

The principle of random sampling casts into doubt the kinds of samples magazines get when they invite readers to fill out questionnaires and mail them back. Those who do not buy the magazine obviously have no chance of being included in the survey. Those who return the questionnaire become a self-selected group, and the data collected from them cannot be used for generalizations to any population larger than the group who returned the questionnaire; they aren't necessarily typical of even the population of the magazine's readers. The data provide a fine description of those who took the time and effort to return the questionnaire, but that is all the data can tell us.

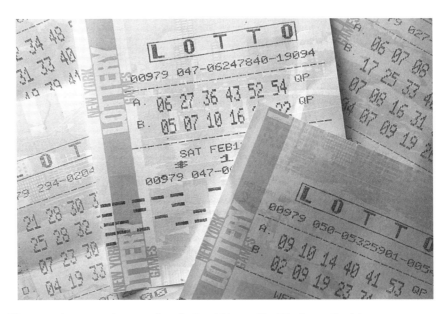

How random are the numbers? *(Patti Mcconville, The Image Bank.)*

The same kind of criticism applies to the conclusions reported in social surveys and self-help books. Shere Hite, a freelance writer who has become known as a specialist in women's love lives, has described the massive discontentment with marriage, sex, and husbands reported by thousands of women. Perhaps the most famous of her statistical claims was that 70% of women married more than five years have sex outside marriage. *(Source: Shere Hite,* The Hite Report: Women and Love: A Cultural Revolution in Progress, *New York: Knopf, 1987, p. 360.)* Evidently those were the results obtained from Hite's convenience sample. The argument here is not that there is no validity in this claim, but that the sample is not representative of the entire population of women in America because it was not randomly drawn. It is therefore incorrect to generalize to the population of all women married more than five years.

Selecting proper samples

Simple random sample When names or telephone numbers of a population are "put into a hat," well stirred, and drawn at random, the result is a *simple random sample*. All the formulas at the ends of the chapters in this book are based on the use of simple random samples.

One way to get a simple random sample is by using random computer-generated dialing of telephones across a population. Unlisted numbers have the same chance as listed numbers to be included, an advantage over random selections from telephone directories. However, this system of collecting data means that business phones also get called. Thus, a person with a business phone and a home phone has twice the chance of being included in the sample as someone who has only a home phone. Telephone interviewing also leaves out the small percentage of people who do not have telephones, another well-recognized drawback of telephone surveys.

Other forms of sampling It is possible to draw samples that are more complicated than simple random samples. One sampling method involves randomly selecting several small geographical areas drawn from voting unit lists and then personally interviewing a random selection of the people living in the areas. This is an efficient way to gather a sample. By interviewing several neighbors living in each geographic area, researchers avoid having to travel miles and miles from one dwelling to the next.

A common difficulty with any type of sampling procedure is that very few complete lists of everyone who belongs to a particular population exist. There exists, for example, no complete list of cocaine addicts, petty criminals, husbands on their third marriage, or children with overbites. Even if these lists existed, they could never be considered complete; an individual could enter or leave a list even as it was being obtained. (Even a list of living U.S. ex-Presidents could change within a heartbeat.) There is also no central population register in this country. Although this is inconvenient for survey studies, it is considered a way to safeguard civil liberties. A list of people with social security numbers does exist, but it is not available to anyone for sampling purposes.

Selection of variables on which to collect observational data

Researchers must ask themselves which variables should be measured in order to draw conclusions about their research questions. It is often difficult, if not impossible, using observational data alone to know which variables have causal effects on other variables and which ones do not. Sometimes researchers who are observing various phenomena may attribute causal power to one variable and overlook a more influential variable. For example, electoral research has shown that there is a tendency for women to vote for the Democratic party and for men to vote for the Republican party. Does that mean that being a woman *causes* one to vote for the Democrats? More formally, is there a causal influence of the gender variable on the vote variable? Or are other variables involved?

To the extent that a person's gender is defined by a certain pattern of chromosomes, it is hard to imagine that the chromosomes could in any way affect which lever a person would pull in a voting booth. Perhaps certain economic variables play a role. If women are less well paid than men, for example, and the Democratic party is more concerned than the Republicans with this type of issue, no doubt women will be influenced to vote Democratic. Researchers may not notice that it is really the more economically disadvantaged—not women—who vote Democratic and the economically advantaged who vote Republican.

It is much more difficult to disentangle effects like these in observational data than it is in experimental data. In properly obtained experimental data, the effects of other variables cancel out in the random assignment of subjects to experimental and control groups. Unfortunately for statistical purity, experimental data cannot always be collected because the requirements of the research designs would violate customs, laws, and sometimes ethical standards. (For example, randomly assigning newborn infants to families to study child-rearing differences would not be socially acceptable, and certainly no contemporary scientist would seriously entertain such an idea.)

In a sense, it is never possible to decide the best way to identify causal variables. If, for example, income level is more important than gender in causing people to vote for a political party, one might ask what it is about income level that produces the behavior of interest. Is it fear of losing what one has, the ability to buy what one wants, pride in one's social position, or any of a number of other attributes that are associated with one's income level that might be influencing voting choices? In an important way, the selection of variables to be studied is a function of the researcher's interests and goals, who is paying for the research, and what utility, in general, some explanations of the results have over others.

2.3 ERRORS AND "ERRORS" IN COLLECTING OBSERVATIONAL DATA

Studying sampling techniques makes us aware of the many things that can go wrong with data from samples and how they skew the results. Just because 60% of a sample approves of the way in which the President handles the job, we cannot conclude that 60% of the entire population approves of the President. Any number of things may have gone wrong from the time it was first decided to do the survey to the time the final results were reported. Most surveys do go wrong in one way or another.

To evaluate the results of a survey, we must know

- whether the sample is a *proper statistical sample* of data.
- the *response rate*.
- the *actual wording* of the question being asked.
- *where* the question was *placed* in the interview schedule.
- *who* the interviewers were.

Sampling error: The "error" that is not a mistake

Some of the errors made in surveys are purely statistical, while others go beyond the statistical aspects of the study. The main statistical error is the so-called *sampling error*. This is not an error in the sense that something is wrong. It refers to the fact that if a study were to be done over again, the results would not be exactly the same. For example, instead of 60% approving of the President, 59% of the next sample—or 62% or some other nearby percentage—might approve.

But even though different samples yield different answers, most of the answers lie within a certain range of the true percentage in the population. For example, with many repeated samples of around 1,000 respondents each, most sample percentages (95 of 100) lie within three percentage points of the true population percentage. Thus, the sampling error equals plus or minus three percentage points ($\pm 3\%$).

This finding is only a reflection of the randomness that is inherently part of every study. After all, the percentages come from different sam-

> The **sampling error** tells us how far from the true population value 19 of 20 different sample results will fall if many different samples have been selected.

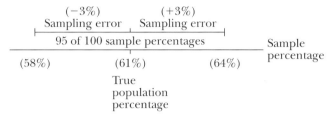

Figure 2.1 Example of a true population percentage and a sampling error of ±3%

ples, and there is very little reason to believe that the result from one sample will be identical to the result from another sample. Additionally, there is no reason to believe that the result from a particular sample is exactly equal to the result data that could have been obtained from the entire population.

Figure 2.1 illustrates the point. It shows a computer-generated case where the true population percentage equals 61%. Furthermore, the size of the sample is such that 95 out of 100 different samples will have a sample percentage that lies somewhere between 58% and 64%. In that case we say that we have a sampling error of ±3%—64% is 3 percentage points more (+) than 61%, and 58% is 3 percentage points fewer (−).

The example is based on a true population percentage of 61. In reality, we almost never know this number; indeed, the reason we did the survey in the first place was to get an idea of the population percentage. Still, from the sample we can compute how large the sampling error is. This remarkable result occurs because mathematical statisticians have been able to derive formulas for the computation of sampling errors. Some of these formulas are given in Chapters 6–13.

The size of a sampling error depends on the way the sample was drawn and the number of observations in the sample. The larger the sample, the smaller the sampling error. If the sample equals the entire population, then the sample percentage is exactly equal to the population percentage. If the study of an entire population is repeated before the population has changed, the result will be the same. In this case, the sampling error is zero.

Any presentation of results from a sample survey should state the size of the sampling error, whether for percentages or averages or anything else. The sampling error conveys a sense of how far away the sample value possibly could be from the true population value. We return to the issue of sampling errors in Chapters 6 and 7 on estimation and hypothesis testing.

Nonresponse error: Result of rude, rushed, and reticent respondents

A different type of error that affects the results of a sample survey is the *nonresponse error*. It may be that in spite of several callbacks, nobody ever answered the phone at a selected telephone number. Or it may be that somebody answered the phone but refused to be interviewed. Mail surveys typically suffer from larger response errors than telephone surveys; it is easier to ignore a sealed envelope than a ringing telephone. Also, the possibility of error in addressing an envelope is greater than dialing an unused telephone number. With follow-ups, a good telephone survey can have an 85–90% response rate, while a mail survey rarely reaches a 50% response rate.

> The **nonresponse error** is the error in the results that occurs because not everyone in the sample responded to the survey.

The percentage of people who refuse to participate in all types of surveys has been increasing. People may have become more reluctant to answer questions because they suspect that a survey is a thin disguise for selling a product or a service. Reputable survey organizations now often do not achieve more than a 60% response rate.

High refusal rate is a big problem for researchers because usually not much is known about the people who were selected but did not participate in the survey. Many unanswerable questions arise. Is there anything about the nonrespondents that makes them different from people who did respond? Are they richer or poorer, more conservative or more liberal, more influential or less powerful than the respondents? How much would their answers have affected the results of the study if they had responded?

A worst-case scenario shows what the effect of nonresponse error can be. Suppose we plan a study with 1,200 potential respondents and obtain 1,000 interviews. This means that data on 200 people are missing. Of the 1,000 respondents we did interview, we find that 600 (or 60%) are in favor of something while the rest are opposed. If we were to assume that all the missing 200 were also in favor, 800 out of 1,200 were in favor, for a 67% rate. On the other hand, if all the missing 200 were not in favor, 600 out of 1,200 were in favor, for a 50% rate. Thus, the observed sample percentage of 60% in favor could really have been anywhere from 50% to 67% due to the nonresponse error alone. That could make a big difference to the outcome of our study.

Some empirical evidence indicates that on most issues the nonrespondents are not very different from those who do respond. If we have a high response rate to begin with, then we can assume that the nonresponders would have answered in the same percentages. But with a low response rate, such as 50%, the impact of nonresponse can be quite large.

How do researchers deal with situations in which nobody answers the phone? It is tempting to substitute another phone number, but this changes things more than you might expect. In a telephone survey, substitution means that people who are seldom at home have much smaller chances of being included in the sample than people who are at home all the time. This violates the principle that everyone in the population should have a fixed chance of being included, and there is good reason to think that people who spend much of their time outside the home are different from those who are home most of the time. The only way to interview the people at numbers where there is no answer is to call back again later. But this takes time; it may take several days and several callbacks before an answer is obtained.

The data from overnight polls, when there is no time for callbacks, are therefore not as good as the data from surveys where callbacks can be made. A poll report on people's attitudes right after an event takes place is interesting, but these polls suffer large nonresponse errors, and we should be wary of the results. Overnight polls taken right after Presidential campaign debates are good examples of this type of situation.

STOP AND PONDER 2.2

Can you think of an example in which political opinions held by those who are mostly at home could sway the results of a telephone survey if the opinions of those who are seldom at home or never answer their phones are not taken into account?

Response errors

The data that result from surveys can be infected with *response errors* that are escapable, if the researchers are careful. We discuss some (but not all) of them here. And even after all these issues have been addressed, *all we know is what people surveyed actually tell the interviewer,* not what they actually do, or feel, or think. When we read in the newspaper that in a recent survey, 60% of the respondents approved of how the President currently does the job, then we should mentally qualify the statement to read that 60% of those people surveyed and who answered this question *said to* the interviewer on the occasion of the interview that they approved of the President's handling of the job.

> **STOP AND PONDER 2.3**
>
> Many of us have been involved in a market survey in a shopping center, on the telephone, or by mail. Can you recall instances when you tried to shorten or curtail an interview, answered in a careless manner, or tried to say what the interviewer wanted to hear, regardless of how you might actually have felt? How would you rate yourself as a respondent? Should market researchers depend on people like you when it comes to gathering good survey data?

Response errors are errors in the responses people give due to factors in the survey context, such as the formulation of the questions, the placement of the questions, and the effect of the interviewer on the respondent.

Wording of questions   The wording of questions in surveys influences the answers people give. On a subtle level, questions frame the issue to which the respondents must give an answer. Sometimes questions confuse the respondent, leading to unintended outcomes. For example, a 1992 survey done by the Roper organization found that a disturbing 22%—1 in 5—of the respondents reported that they doubted that the Holocaust had happened. After the initial reaction to the statistical result, readers of the report turned their attention to the question itself: "Does it seem possible or does it seem impossible to you that the Nazi extermination of the Jews never happened?" The question contains a double negative, a potential source of confusion to respondents. A new survey was done two years later, and this time the wording of the question was "Does it seem possible to you that the Nazi extermination of Jews never happened, or do you feel certain that it happened?" Worded this way, only 1% of the respondents thought the Holocaust never happened, quite a change from the original 22% from the first survey.

Despite wording changes, statisticians often raise the question of whether the respondent has any opinion on the issue in the first place or whether the wording gives the person an opinion by the word choices offered. If a couple asked you what name to give to their baby, you might be rather befuddled. But if the couple added, "We are debating three choices: Maria, Gertrude, or Maud," you might find that you have an opinion. In the Holocaust question, the possible choices that the event never happened and being certain that it happened allow for only two options. People who had not thought about the issue or who otherwise were unopinionated were given no appropriate choice. The neutral position probably went underrepresented, and as this group was sorted into either of the two options, the two options were probably overrepresented. *(Source:* The New York Times, *July 8, 1994, p. A10.)*

One way around the problem of response options creating opinions is to ask a screening question first. The question "Do you have any opinion on the issue of whether the Holocaust happened?" might have been asked in the Holocaust poll. Those who answer no to the screening question are then not asked the next question about their opinion. In general, unless the actual questions are reported along with the results, it is difficult to assess the results of surveys that propose to measure people's attitudes.

Placement of questions To add to the complexities of questionnaire design, the placement of a question in a survey can affect the responses. Early in the interview, the contact between the interviewer and the respondent is not well established. The respondent may be hesitant about expressing opinions. Well into an interview, the respondent may feel more comfortable with the interviewer and as a result speak more frankly and less formally. The respondent may make more prejudicial remarks and "politically incorrect" comments and may state personal opinions. By the end of the interview, the respondent may be experiencing fatigue or boredom. If the respondent wishes to terminate the session quickly, answers may be shorter, less precise, and more careless than answers given in the middle of the interview. Researchers try to accommodate respondents' comfort needs by asking fairly easy and impersonal questions at the beginning of an interview and more difficult and personal ones when rapport is higher. Questions on income, for example, are asked far along in most U.S. surveys. Closing questions are often short and simple.

Respondents may also want to remain consistent from one area of questioning to another. If they support a particular point of view in one question, they may feel the necessity of supporting it in another one, despite a lack of commitment to what they are saying. For example, someone who supports capital punishment in the answer to one question may be hesitant to declare herself or himself a pacifist when it comes to warfare. Throughout an interview, representing oneself positively is a constant need of the respondent, and surveyors try to

> **RESPONDENT BIAS OVER THE TELEPHONE**
>
> The National Black Politics Study is a telephone survey of 1,204 African-American respondents. African-American interviewers were used to make respondents feel comfortable in answering questions. But, since this was a telephone survey, the respondents could not see the interviewers. A political scientist, Lynn Sanders, studied the effect of the perceived race of the interviewer on respondents' answers to survey questions. In response to a question asking what race the interviewer was, 14% of the respondents said they thought the interviewer was white.
>
> The respondents were also asked if they agreed with the statement "American society is fair to everyone." Of those who thought the interviewer was African-American, 14% agreed. Of those who thought the interviewer was white, 31% agreed. *Source:* Chance, *vol. 8 (1995), no. 4, p. 5.)*
>
> **STOP AND PONDER 2.4**
>
> The effects of the perceived race of the interviewer on respondents' answers is clear. How many other undetected interviewer effects might have influenced survey results?

place questions to permit people to give opinions that they believe will reflect well on themselves.

Interviewer effect   Respondents' answers are influenced by their perceptions of who the interviewer is and what the interviewer believes. Survey designers often try to match interviewer and respondent as closely as possible on demographic features such as age, gender, and race. Especially with sensitive issues, such as attitudes toward other groups, ethical or legal behaviors, or sexual activities, talking to someone who may share one's views is preferable for both parties in the interview.

2.4 EXPERIMENTAL DATA: LOOKING FOR THE CAUSES OF OUTCOMES

The other method of data collection involves manipulating one or more variables in an experiment and measuring the results of the ma-

> **Experimental data** are data collected on variables resulting from the manipulation of subjects in experiments.

nipulations. For example, if we give one group of plants a fertilizer and another group no fertilizer, then we are manipulating the plants' soil content. We can then measure variables such as growth or vitality. An experiment is a way of studying causal relationships between variables. In an experiment, researchers try to control every relevant aspect of a situation, manipulate a small number of variables of interest, and then observe the results of the manipulations.

An early example of an experiment occurred at the beginning of the 1600s when the British navy tried to discover the causes of scurvy, an illness characterized by swollen and bleeding gums and livid spots on the skin, which often attacked sailors on long voyages. The Admiralty suspected that lack of citrus fruits might cause the disease. At the time this idea was suggested, four naval ships set out from England on a long journey. To investigate whether a lack of citrus fruit caused scurvy, the Admiralty arranged that on one of the ships each sailor would be given citrus juice to drink every day, while the sailors on the other three ships would not get citrus juice.

Before the voyage was over, there were so many sailors sick with scurvy on the three "juiceless" ships that sailors who had received citrus juice had to be transferred to these ships to help sail them to harbor. This experiment was obviously successful in proving a point, even though the actual experimental plan could have been improved in various ways.

Experimental group and control group

> A **control group** is a randomly selected subset of the subjects in an experiment that is not manipulated.

In the scurvy example, the sailors who drank citrus juice formed the *experimental group,* and the sailors who were not given juice formed the so-called *control group*. An experimental group is a randomly selected subset of the subjects in an experiment that receives a particular treatment that the control group does not receive. Almost all well-designed experiments (and some observational studies) have a control group and one or more experimental groups.

The reason a control group is needed is that without one there would be no way of determining whether the manipulation or some other variable (or several variables in conjunction) had an effect. If the sailors on all four ships of the scurvy experiment had been given the citrus juice, the lack of scurvy could have been attributed, for example, to the exceptionally good rum rations or some other treatment the sailors received on the voyage. But the *only* difference between the experimental group and the control group was that one group drank juice and the other did not. Therefore, it is logical to conclude that it was the citrus juice that kept the sailors from getting scurvy. This point is also illustrated in the experimental example in Chapter 1 of the testing of the polio vaccine in the 1950s; without the presence of a control group, there would not have been a baseline with which to compare the effect of the vaccine.

Selecting the experimental and control groups

Another issue in the setting up of an experiment is the question of who should be in the experimental group and who should be in the control group. The scurvy example is not a perfect experiment because we can think of alternative explanations for why the men on one ship did not get the disease. Perhaps there was something about the three ships themselves—but not the fourth—that produced scurvy. Although unlikely, such a phenomenon was a possibility, so it would have been better if the decision of who should get and not get the citrus juice had been made randomly for each sailor, without regard to ship. By randomly assigning the treatment, the effects of other variables related to the ships would have canceled each other out and not affected the results.

One might wonder if volunteers could be used, rather than randomly assigning people to the treatment and control groups. For example, what if the sailors who liked citrus juice had been the experimental group and those who preferred rum the control group? The problem with this method is absence of certainty that the men in both groups were equally healthy before the study began. If the subject assignment is random, then healthy and unhealthy sailors would be equally likely to be in each group. Health could then be eliminated as a cause of scurvy.

The principle of random selection of subjects was one of the major contributions of the great English statistician Sir Ronald Fisher, who worked with agricultural experiments in the 1920s. It is a principle that has been followed in all good experiments ever since.

Problems with experimenting on people

In experiments on human beings, the goal is still to assign people randomly to experimental and control groups, but this is difficult and sometimes even impossible to achieve. It is much less complicated to assign a potato plant to a poor dirt patch than a person to substandard living conditions.

Data should always be taken with a grain of salt. *"Calvin and Hobbes" copyright 1995 Watterson. Dist. by Universal Press Syndicate. Reprinted with permission. All rights reserved.*

Logistical issues We can all come up with reasons why it is more difficult to study people than potato plants. First and foremost, people have their own plans and interests and are not necessarily willing to oblige the research interests of the scientist. They may also have difficulty in meeting the conditions of the research, keeping appointments, following directions, and fulfilling their part of the arrangement. We have already mentioned the problems with getting good data from people in telephone and personal interviews, and the same types of limitations apply to experiments.

Psychological issues In an experimental study, people are highly sensitive to being studied. This makes them self-conscious, which can create many constraints on their behaviors. One of the first times this effect was documented was in a series of investigations of worker productivity at a General Electric factory from 1924 to 1933. In one investigation, a team of social scientists and company personnel studied the effects of various levels of illumination on the productivity of workers making light bulbs. The researchers increased the illumination level and found an increase in productivity. But strangely, when they reduced the lighting levels, productivity also increased. It seemed that no matter what the researchers did, the workers produced more. The workers seemed to respond to the attention of the researchers, not the light level.

Over time, the phenomenon of workers responding to the attention of researchers and not specifically to the intended manipulation was called the Hawthorne effect, taken from the name of the factory where the lightbulb study was done. Precautions against such effects can be taken, for example, ensuring that the control group receives as

Many factors affect workers' productivity. *(Michael Rosenfeld, Tony Stone Images.)*

much attention from the researchers as the experimental group. *(Source: See, for example, Robert K. Merton,* Social Theory and Social Structure, *New York: The Free Press, 1957, p. 66.)*

Ethical issues Ethical issues complicate the process of doing experiments on people and animals. While certain ethical dilemmas are associated with collecting observational data, such as standing by as negative events occur, the experimenter who manipulates and controls events is more likely to face ethical dilemmas. Is it right, for example, to expose people to drug treatments where the outcomes cannot be predicted? Suppose people suffer from unexpected, negative side effects? Thinking of side effects might lead one to be conservative about testing and introducing new drugs. Yet, on the other hand, what if a new treatment is beneficial? How long should people with fatal diseases have to wait to try a new drug? How long can they wait?

What about the absence of benefits from the treatments for the control group subjects who only received the placebo? In the polio vaccine experiment, many more children in the control group got polio than in the treatment group. If the children in the control group had received the vaccine also, there is every reason to believe that as many as 100 more children would not have gotten polio.

A similar dilemma occurred for researchers in a study of the effects of aspirin on heart attacks. The design of the study allowed for one group of male doctors to take an aspirin a day to see if this treatment would cut down on their risk of heart attacks. After the experiment had run for about five years, fewer heart attacks were recorded in the treatment group taking aspirin than in the control group taking a placebo. The results were so clear that the experiment was stopped long before the planned termination, and people in the control group were encouraged to start taking aspirin. In other projects results are not so clear, and long-term side effects can nullify short-term gains. This was the case with thalidomide, a drug that pregnant women took in the 1950s to suppress miscarriages; mothers who took the drug delivered an unusually high number of babies with deformed limbs.

Almost all research done in the United States, particularly research with health consequences, is screened by experts who specifically look for ethical problems. Imagine, for example, the possibility of testing a promising new drug to cure AIDS. If it is effective, people in the control group may risk death if they do not get the drug. However, if it is found that the drug has side effects that result in a higher mortality rate two years after the test is begun, the control group may have escaped a lethal dose. What is the ethical thing to do? No easy answers can be given. Ethical issues must be constantly considered and reconsidered. Fortunately, most studies are less dramatic and the consequences are less severe.

STOP AND PONDER 2.5

Can you think of any recent research you have read about in which companies have been sued for treating customers in ways that have violated their rights? Have there been ethical dilemmas involved in these cases? Given the nature of the case, have the companies, in your opinion, been ethically responsible or not?

Role of statistics in experimentation

Most researchers who do experiments receive statistical advice. The contribution of statistics to the running of experiments centers on three practical issues: obtaining the proper number of observations to make it possible to find if there are any effects; planning the experiment so that the standards for statistical analysis are met; creating a method for studying the impact of several variables simultaneously in the most efficient way possible.

How many observations? Statisticians can give advice on how many observations are needed to get results with the desired accuracy. More data are usually better than less data. But it costs more and takes a longer time to collect many observations as opposed to a few observations. A researcher may ask: Would it be enough to plant a new variety of corn on 10 plots of land or do I need data from 100 plots to discover if the new variety gives a higher yield? This question exemplifies the general quandary for many cost-conscious, result-oriented researchers.

Plan of analysis: Be safe rather than sorry Statisticians often help researchers set up a plan of statistical analysis. The best time to create this plan is before the data are collected. If the study is done without a careful plan of analysis, faults are built into the experiment that cannot be corrected later. Often statisticians are called in to clean up the mess, and it is often too late. A researcher cannot supply missing data once the data are collected, so the statistical analysis is less helpful than it would have been had the data been appropriately collected in the first place. Often statisticians are able to expand a statistical plan as further questions become relevant to a researcher. Initial results may suggest new avenues of inquiry, and statisticians may offer additional suggestions once the preliminary data analyses are run. In today's complex world of statistical analyses, even the most sophisticated researchers prefer to have statisticians at their sides when working up a major design for a study.

Studying the impact of several variables at the same time Planning becomes particularly important when we want to study the effects of several experimental variables on an outcome variable. One could do several experiments, each using one variable. For example, if a sport psychologist wanted to study the effects of diet, exercise, and self-confidence on bodily weight, he or she could do three experiments to study the effect of each variable separately.

Sir Ronald Fisher developed a way to use all the variables at the same time in one experiment, thus reducing the randomness in the data and allowing for comparisons of the relative strengths of each variable. Thus, in the example, the sport psychologist can measure how

SIR RONALD FISHER

One of the many contributions to statistical methods by the great British statistician Ronald Fisher was his work on how to conduct experiments. He realized that when several variables influence an outcome variable, it is better to study the effects of all the variables taken together than it is to study the effect of each variable separately. One of the plans for experimentation that Fisher came up with for studying several variables at the same time is known as a Latin-square design.

During part of his life Fisher was a Fellow at Caius College, Cambridge, England. As a memorial to Fisher, the college installed in its dining hall a stained glass window representing a Latin-square design. The window consists of a square about 3 feet by 3 feet. The square is divided into 7 rows and 7 columns giving a total of 49 small squares, or cells. Each of these cells is a square of colored glass in one of 7 different colors. The colors are laid out so that each color occurs only once in each row and once in each column. For example, yellow glass is used for the cell that lies in row 1 and column 6, the cell in row 2 and column 2, the cell in row 3 and column 1, the cell in row 4 and column 3, the cell in row 5 and column 4, the cell in row 6 and column 7, and the cell in row 7 and column 5.

The window represents three different variables, each with 7 different values. The rows designate one variable, the columns the second variable, and the colors the third variable. Thus, there are $7 \cdot 7 \cdot 7 = 343$ different combinations of values, the number of observations required for one observation for each combination of values. The window shows that we need only the 49 combinations of values shown in the window if we want to do a study with three variables where each has 7 values. *(Photo courtesy of UPI/Bettmann.)*

STOP AND PONDER 2.6

Can you create a stained glass window design that has properties similar to that designed by Fisher as the Latin square? You may use different patterns of color, but use yellow according to the Fisher window. Or try a similar type of design with a smaller number of rows and columns, for example, four.

relatively important diet, exercise, and self-confidence are in affecting a person's weight.

This approach led to the development of a variety of multivariate statistical designs for analyzing experiments. In Chapter 13 we start a discussion of multivariate analyses, which are needed to find answers to the sport psychologist's queries.

Putting it all together: Does class size affect school performance?

To conclude this discussion on experiments, let us look at an example of an actual experimental study. With this example, you can take on the role of the experimenter who must confront a series of difficulties in order to answer a question of interest.

The example is an educational experiment that took place in Tennessee, as reported in a news magazine. *(Source:* The Economist, *August 31, 1991, p. 23.)* Here we use only the information given in the news story to see what we can conclude about this experiment. (If we wanted more information about the study, we would need to consult more detailed descriptions.)

For a long time people have had the sense that smaller rather than larger classes result in better learning, but it is hard to find empirical evidence for this proposition. In the mid-1980s school officials in Tennessee decided to perform an experiment to test whether class size affected school performance. The main design of the study was a simple one, concerned with the effect of one single variable, class size, with only two values, small and regular size.

Operational definitions   Before the experiment could be performed, several important, nonstatistical issues had to be decided. Perhaps the most important one was to decide what was meant by a "small" class. In the study, the Tennessee officials decided that a class containing between 13 and 17 students would be called "small," and a class consisting of 22 to 25 students would be thought of as "regular" class size. Also, the researchers had to decide what "improved performance" meant. They decided to measure performance with standardized educational tests.

Thus, the researchers started the experiment with the research hypothesis that elementary school students in small classes do better than students in regular classes and then made it specific: 5-year-olds starting in classes consisting of 13 to 17 students who stay in "small" classes for four years will do better on standardized tests at the end of the four-year period than students who spend the same time in classes consisting of 22 to 25 students. It was important to specify the four-year period to eliminate the students who would inevitably move into and out of the experimental classrooms during the course of the study.

> **STOP AND PONDER 2.7**
> How would you feel if you lived in Tennessee and had a child who was put in a regular-size class instead of a small class?

Selection of sample schools   The next thing to be decided was which schools should be used for the experiment. The magazine story simply says that the study was done in 76 different elementary schools. There are many more than 76 elementary schools in Tennessee, and we can only hope that random selection was involved in choosing the schools. We also trust that students were placed randomly in small and regular classes.

> **STOP AND PONDER 2.8**
> Why is it so important that the schools were chosen randomly and that students were randomly assigned to the two types of classes?

Experimental design   Now the researchers were faced with the question of what to do with all the 5-year-olds in 76 elementary schools across the state. They could have created only small classes in some schools and only regular classes in other schools. But then whatever differences found between small and large classes could have been due to other variables. For example, if all the children in a school in a university town were put in small classes and all the children in an inner-city school were put in regular-size classes and it was found that students in small classes did better, the difference could have been due not to class size but to the fact that children from more academic families tend to do better on standardized tests. Instead, within each school the children were randomly assigned to the different types of classes. In this way, possible effects of the backgrounds of the different students were canceled out and did not affect the overall test scores.

Other variables also had to be taken into account. For example, teachers had to be randomly assigned, as well. It would not have been fair to put all the new teachers in larger classes and all the experienced ones in small classes. Other physical variations, such as classroom resources, also had to be balanced out so that neither group had an advantage over the other.

Results After four years, it was found that students in the small classes were performing "significantly" better than the students in the regular-size classes. (The term "significantly" is discussed in Chapters 6 and 7.) The test results showed that after only one year the students in small classes were 1.5 months ahead in reading and 2.5 months ahead in mathematics. The small-class advantage was also present after four years, when the experiment ended.

2.5 DATA MATRIX/DATA FILE

After data have been collected from a study, whether experimental or observational, they are commonly entered into a computer file in typical spreadsheet form. This means that each column refers to a variable, such as gender. Each row refers to an element, for example, person, plant, animal, group, or whatever units we have collected data on. Such a table of data is often referred to as a data matrix or a data file. Table 2.1 shows an example of a small data matrix for data from a sample survey.

Table 2.1 Data matrix for a sample survey

Person	Age	Gender	Vote	Attitude
1	20	Female	Democrat	Neutral
2	27	Female	Democrat	Against
3	19	Male	Republican	Against
4	38	Male	Democrat	Favor
5	38	Male	Republican	Favor
6	53	Female	Democrat	Favor
7	24	Male	Republican	Favor
8	41	Female	Republican	Against
9	35	Female	Democrat	Neutral
10	30	Male	Republican	Favor

For convenience in computer analyses of the data in the file, we often change the words in a data file to numbers. Each person is given an ID number as a name. The age variable is already measured using numbers, so no change is needed there. The gender variable has the two categories female and male, so female is replaced by the number 0, male by the number 1. Any two numbers could be used, say −17 for female and 23 for male, but for practical reasons, which are explained in Chapter 9, the numbers 0 and 1 are better to use. The values of the vote variable can similarly be changed to 0 or 1, and for the attitude variable we can use three ranked numbers, say 1, 2, and 3.

The way the data matrix appears on a printout from the computer is shown in Table 2.2. While the table is easy to read, a typical national survey could have 1,000 respondents instead of the 10 shown here, and there could easily be 100 variables instead of just 4. With 1,000 rows and 100 columns, there would be 100,000 numbers in the data file. This would not be so easy to read! The information would all be there, but the trends and patterns in the data would be obscured. A researcher could not extract what is of interest without simplification and condensation—analysis—of the data by statistical methods.

Table 2.2 Data matrix for a sample survey

Person	Age	Gender	Vote	Attitude
1	20	0	0	2
2	27	0	0	1
3	19	1	1	1
4	38	1	0	3
5	38	1	1	3
6	53	0	0	3
7	24	1	1	3
8	41	0	1	1
9	35	0	0	2
10	30	1	1	3

2.6 SUMMARY

2.1 Defining the variables

The first step in proper data collection involves carefully specifying the variables to be studied.

2.2 Observational data: Problems and possibilities

Observational data are data collected through observations of the world, without manipulating or controlling it.

A population consists of all the elements under study. A census is the process of collecting data on an entire population. A proper statistical sample that can be used for generalizations to a population is called a random sample, a sample in which every element in the population has a known (often equal) chance of being selected for the sample. Drawing a sample "out of a hat" produces a simple random sample.

In observational studies, it is often difficult to determine whether or not a variable is causally affected by another variable. The potential of other unknown variables to have a more direct impact on a variable than the one under study must be acknowledged in any observational study.

2.3 Errors and "errors" in collecting observational data

Sampling error tells us how far from the true population value 19 of 20 different sample results will fall if many different samples had been selected. This variation in the results from one sample to another is due to the randomness of sample selections. The size of the sampling error depends on how many observations there are in the sample and how it was drawn. The larger the sample, the smaller the sampling error becomes. Sampling errors should always be reported.

Nonresponse error is the error in the results that occurs when data are missing from the sample. Missing data may result from such causes as unwillingness of respondents to answer all queries and the inability to locate certain sample members. The effects of a worst-case scenario, in which all nonrespondents would have answered a survey question identically, could be enormous. Fortunately, studies have shown that on most issues nonrespondents are not very different from those who do respond.

Data collection response errors can be made by wording questions in ways that confuse respondents or suggest certain answers, by arranging questions in a nonpropitious order, and by using interviewers who bias respondents' answers, among other things.

2.4 Experimental data: Looking for the causes of outcomes

Data collection can also be done by manipulating one or more variables in an experiment and measuring the results of the manipulations. An experiment is a way of studying causal relationships between variables. In an experiment researchers try to control every relevant aspect of a situation, manipulate a small number of variables of interest, and then measure the results of the manipulations. Of critical importance to a good experimental design is the control group, a subgroup of subjects that is not manipulated but is in all other respects like the experimental group(s), which does receive the experimental manipulation. Experimental objects are randomly assigned to treatment and control groups. One major reason for random assignment is so that the effects of extraneous variables cancel out and do not affect the end results.

It is often difficult to study people experimentally because they may resist the efforts of the scientist to manipulate and control them, they may become self-conscious, and/or they may become bored or fatigued by the experimental situation. Subjects may also be manipulated in their behaviors by the nature of the experiment itself, in some cases becoming more cooperative than normal. Ethical dilemmas, such as weighing the pros and cons of giving or withholding treatment, are also important constraints on experiments.

The contribution of statistics to the production of successful experiments centers on three practical issues: obtaining the proper number of observations in order to get significant results, planning the experiment so that the standards for statistical analysis are met, and creating methods for studying simultaneously the impact of several variables.

2.5 Data matrix/Data file

After the data have been collected in a study, whether experimental or observational, they are commonly entered into a computer file in typical spreadsheet form. Such a collection of data is called a data matrix or a data file.

ADDITIONAL READINGS

Bartlett, M. S. "R. A. Fisher." In William H. Kruskal and Judith M. Tanur (eds.), *International Encyclopedia of STATISTICS*. New York: The Free Press, 1978. On the life of one of the best-known statisticians of the twentieth century.

Cochran, William G. "The design of experiments." In William H. Kruskal and Judith M. Tanur (eds.), *International Encyclopedia of STATISTICS*. New York: The Free Press, 1978. How statisticians help in the planning of experiments.

Converse, Jean M., and Stanley Presser. *Survey Questions: Handcrafting the Standardized Questionnaire* (Sage University Paper Series on Quantitative Applications in the Social Sciences, no. 07-063). Beverly Hills, CA: Sage Publications, 1986. On the art of asking questions in surveys.

Deming, W. Edwards. "Sample surveys: The field." In William H. Kruskal and Judith M. Tanur (eds.), *International Encyclopedia of STATISTICS*. New York: The Free Press, 1978. Deming was a sampling statistician before he became a famous quality control person.

Hoaglin, David C., et al. *Data for Decisions*. Cambridge MA: Abt Books, 1982. Chapter 2, Experiments (pp. 18–46), Chapter 4, Comparative Observational Studies (pp. 55–57), Chapter 5, Sample Surveys (pp. 78–106), and Chapter 9, Official Statistics (pp. 166–178) present discussions of different types of data.

Hobbs, Nicholas. "Ethical issues in the social sciences." In William H. Kruskal and Judith M. Tanur (eds.), *International Encyclopedia of STATISTICS*. New York: The Free Press, 1978. A discussion of ethical issues we run into when doing social research.

Kahn, Robert L., and Charles F. Cannell. "Interviewing in social research." In William H. Kruskal and Judith M. Tanur (eds.), *International Encyclopedia of STATISTICS*. New York: The Free Press, 1978. Two authors from the Institute of Social Research at the University of Michigan discuss issues that come up in interviewing.

Kalton, Graham. *Introduction to Survey Sampling* (Sage University Paper Series on Quantitative Applications in the Social Sciences, no. 07-035). Beverly Hills, CA: Sage Publications, 1983. An overview of how to draw random samples.

Mosteller, Frederick. "Nonsampling errors." In William H. Kruskal and Judith M. Tanur (eds.), *International Encyclopedia of STATISTICS*. New York: The Free Press, 1978. A discussion of errors that can affect the conclusions we draw from the data, by one of this country's leading statisticians.

Spector, Paul E. *Research Designs* (Sage University Paper Series on Quantitative Applications in the Social Sciences, no. 07-023). Beverly Hills, CA: Sage Publications, 1981. An overview of different plans for how to collect experimental data.

Witmer, Jeffrey. *DATA Analysis: An Introduction*. Englewood Cliffs, NJ: Prentice Hall, 1992. Examples of interesting data sets collected in a variety of different ways.

EXERCISES

REVIEW (EXERCISES 2.1–2.22)

2.1 A teen magazine wants to do a survey using college students to increase reader appeal for 17–19-year-olds. Name several decisions the survey team should make about the definition of the target population before the sample is drawn.

2.2 Define *experimental group*.

2.3 Define *control group*.

2.4 a. What is a random sample?

b. Name three difficulties in creating a random sample.

2.5 In what sense is the sampling error the best kind of error we can have in a statistical analysis?

2.6 Select an example of sample data, census data, or other data from a journal or book.

a. Does the author satisfactorily explain how the data were collected? Explain.

b. Is the population to which the findings are generalized well specified? Explain.

2.7 The student council wishes to survey the senior class regarding graduation ceremonies. You volunteer to draw a sample of 10% of the seniors in your school.

a. How would you arrange to draw the sample to assure randomness?

b. What possible problems might you encounter in drawing the sample?

c. What might be some possible impacts of these problems?

d. How would you attempt to solve these problems?

2.8 Several of your friends wish to help you complete the student council survey in Exercise 2.7. They volunteer to poll 12 friends each concerning the graduation ceremonies. This will then constitute a sample of 10% of the class. They also volunteer to buy the beverages if the class votes to have an all-night barbecue at a nearby lake.

a. For what reasons do you decline your friends' offer of assistance?

b. If you yourself want the barbecue as well, what safeguards might you suggest to prevent your views from influencing your classmates?

2.9 Explain whether a sampling error indicates a poor job of statistical analysis.

2.10 What factors are important in determining the size of the sampling error?

2.11 a. What is the true population value?

b. Where is it supposed to be located?

2.12 What is a response error?

2.13 Make up a survey question that would have the same property of unacceptability as the following question: Have you stopped beating your dog yet? Why is your question a bad question?

2.14 Construct a survey question that you would find acceptable if the goal were to establish the level of financial well-being of the respondent.

2.15 A poll found that 56% of the respondents favored the *Roe* v. *Wade* Supreme Court decision of some years ago. The sampling error was reported to be $\pm 2\%$.

    a. What can you say about how the other 44% of the respondents felt about *Roe* v. *Wade*?

    b. What are some other things you should know about how these data were collected before you can make anything of this poll?

    c. Show how to use the sampling error percentage and interpret the result.

2.16 Would you permit your child to participate in an experiment with a medication, like the experiment with the polio vaccine, where it is not clear whether the medication will have bad side effects, have no effect at all, or be beneficial to all humans? Explain.

2.17 Why is it difficult to interpret the results of an experiment that did not include a control group?

2.18   a. Under what conditions would you volunteer for an experimental study of toothpastes without knowing the possible side effects?

    b. Under what conditions would you volunteer for an experimental study of a drug designed to alter mental states without knowing the possible side effects? Consider the well-being of other people as well as yourself.

    c. Do you think your answers are similar to those of most other people, some other people, or a few other people? Why?

2.19   a. Is it possible to construct a survey question concerning favorable or unfavorable attitudes toward abortion that would appear to be totally value-neutral? Why?

    b. What major effects does the wording of questions have on the answers given?

2.20 You are directing a housing survey in a conflict-ridden community consisting of families of Korean, Pakistani, Filipino, Armenian, and Icelandic origins.

    a. What considerations might be important in your decisions about who should do the interviewing?

    b. Describe who you would hire, including any gender, age, ethnic, educational, or other distinctions.

c. What biases would you accept, and which would you try to avoid?

2.21 What is a data file?

2.22 a. Do the columns in a data file usually refer to a variable or to an element?

b. Do the rows in a data file usually refer to a variable or an element?

INTERPRETATION (EXERCISES 2.23–2.36)

2.23 The wise statistician declares: "There are two kinds of data: Good data and bad data." The difference between good data and bad data depends on whether or not proper statistical principles were adhered to during the collection process. Given the difficulties in collecting good data, do you think the statistician should have said: "There are two kinds of data: Bad data and worse data?" Explain.

2.24 a. What are some circumstances under which you would be (or have been) unwilling to participate in a survey?

b. What do you think the results of your refusal to participate might be on the outcome of the survey (assuming that there were others who refused for the same reasons you did)?

c. In what ways might this affect the ways in which the survey helped or hindered those who commissioned the study?

2.25 During the spring of 1991, 7-Eleven stores around the country conducted a poll in which customers who bought drinks could "vote" on an issue by choosing a beverage cup marked either Yes or No. According to the results of this poll, 50.9% of the respondents in the Philadelphia/Trenton area "voted" that they would marry for money, while nationally the percentage was 53.6. (*Source:* The Philadelphia Inquirer, *April 16, 1991, p. B3.*)

a. Did the results imply that people in Philadelphia and Trenton were less inclined to marry for money than the people in the rest of the country?

b. What might explain the difference in percentages?

2.26 In the local mall, customers of all shapes and sizes are being stopped by a group of interviewers of all shapes and sizes who are asking the shoppers about their recent purchases of diet beverages and diet foods.

    a. What effects could the interaction of pollsters and respondents have on the results of this poll?

    b. More generally, is it ever possible for surveys such as these to be done in a totally neutral and unbiased fashion? Explain.

2.27 City planners are interested in the level of fire prevention awareness among the volunteer firefighters of Delaware County. A survey of the Garden City firefighters contains a sampling error of $\pm 7\%$.

    a. Will it be helpful to the planners if a sample of the other firefighters of Delaware County—in the town of Media and the boroughs of Swarthmore and Rutledge—are included in the report? Why?

    b. If all the volunteer firefighters in the entire county were given the survey, would the sampling error be greater or smaller? Why?

2.28 Suppose you want to ask students to rate how favorable their overall academic experience has been.

    a. Comment on the difficulties you might encounter in defining this variable.

    b. How might the difficulties result in the favoring of some types of educational institutions over others? (Hint: The results may favor small schools devoted almost entirely to undergraduate teaching.)

2.29 Quoting a study on commercial matchmaking enterprises, "A matchmaker who uses video technology claims that 40% of her first-time matches result in committed relationships." Given the information, would you spend the money for this service if you were eager to get married soon? (Are there any problems with this claim in terms of the definition of the agency's success at matchmaking?) *(Source: Mara B. Adelman and Aaron C. Ahuria, "Mediated channels for mate seeking: A solution to involuntary singlehood?" Critical Studies in Mass Communication, vol. 8 (1991), pp. 273–289.)*

2.30 According to the National Institute on Alcohol Abuse and Alcoholism, alcoholic fathers and sons are less creative than nonalcoholic fathers and sons. "Creative people may be alcoholic, but alcoholics are rarely creative, the head of the study concludes." How is the variable creative/noncreative redefined in the description of the outcome of this study, according to this news item? *(Source:* The Philadelphia Inquirer, *October 17, 1993, p. F1.)*

2.31 Surveys indicate that cheating in college is a serious problem. Yet it is not always possible to know who has cheated and who might be falsely accused. You are interested in studying the effects of cheating on college exams on physiological indicators of lying. As the professor of an introductory psychology class, it is possible for you on the midterm exam to arrange for half the students to receive answers to the multiple-choice test questions (apparently by mistake) in their answer booklets. You will be able to tell which students received the "cheater" booklets and which did not. Later, you will be able to secretly videotape the class as you confront them as a group with the cheating episode. You will be able to observe physiological indicators and check to see how the students who cheated react to your charges. Assuming that this study has scientific merit and can be conducted (from a logistical standpoint), do you see any problems with going ahead with it? What are they?

2.32 Researchers led by Dr. Arthur Kellermann of Emory University, Atlanta, compared people murdered in their homes to nonvictims of the same age, sex, race, and neighborhood. The homes of the 388 murder victims were different from the homes of the nonvictims in several ways. The people who lived in the homes of the victims were more likely to have guns, especially loaded guns, to use illicit drugs, to have arrest records, and to have a history of domestic violence. The study reported that "gun ownership increased risk 2.7 times, regardless of other risks."

a. What was the control group in this study?

b. What was the most important factor in whether or not one would be murdered?

c. Were there any significant variables that were not controlled for, according to this report, that might have been important in understanding factors that could lead to murder in the home?

2.33 In a study of curriculum development in mathematics, researchers included a random selection of ten schools in the district. Among them was the Wallingford Elementary School, where teachers were asked to volunteer for a workshop that was to be conducted over the Thanksgiving vacation. Of the 38 eligible teachers, 14 agreed to go; from this group 9 names were chosen at random (the researchers had asked that the selection process be done "at random"). Despite the care in trying to pick a random sample, 8 of the 9 teachers selected were men, although 65% of the teaching staff were women. After the workshop, the teachers came back with glowing reports about the new mathematics program, especially how it challenged them to "really brush up on their own math skills." They strongly recommended that the program be instituted the following year. The researchers concluded that teacher reception to the math program was very warm and recommended that the program be given a high priority in the following year's budget. The research results were especially important because previous attempts to introduce new mathematics programs in these schools had not been very successful, due to teacher resistance and sometimes their lack of preparation in mathematics.

a. Do you think the teacher sample for the research was a random selection of the teachers in the district?

b. What factors interfered with the sample being statistically ideal?

c. Was it chance that led to so many men being selected from Wallingford? Explain.

d. Do you think the new math program will be well received next year? Explain your answer.

e. Do you think it is primarily laziness on the part of researchers that keeps samples from being randomly selected? Explain.

2.34 A salesperson for the class ring company wants to do a survey at your high school to determine how much money the average student plans to spend on the class ring. The principal suggests that the question be asked of the students at a pep rally for the championship game on Friday afternoon. As the nosy statistics student, you would like to put in your two-cents worth about this planned survey.

a. What will you say about the data collection procedure?

b. What consequences do you think it will have on the survey results if the principal's plan is carried out?

2.35 Whenever we collect data and then begin to summarize them in graphs or tables, in numbers or in statements, we lose information. From the following statements, taken from various accounts in newspapers and scientific reports, give an opinion on what important information has been lost.

a. "In this survey, students at the Illinois Institute of Technology were most likely to say they were 'unhappy.'" (The Philadelphia Inquirer, *October 10, 1993, p. B5.*)

b. "According to the Alan Guttmacher Institute, more Catholic women seek abortions than women of any other faith." (The Philadelphia Inquirer, *December 8, 1992.*)

ANALYSIS (EXERCISES 2.36–2.41)

2.36 A hometown newspaper reports on the eve of a local election that a survey of the electorate has found that Rainwater is leading Goldthorp in the city council race by 53% to 47%, with a sampling error of ±4%. The editor wants to know if she should begin the headline for tomorrow's paper "Rainwater Tromps Goldthorp." What would you advise, and why?

2.37 Following are questions from a survey designed to gather data from moviegoers on the popularity of recent films shown at local theaters. List at least ten problems with these questions.

Name _____ Age _____
Address _____ Telephone _____
Salary _____ Job title _____
Movie you saw tonight _____
Name of theater _____
How good was the movie? Very good _____ Good _____ Bad _____
Rate the movie on a ten-point scale: 1 2 3 4 5 6 7 8 9 10
How good was it compared to the last movie you saw? 1 2 3 4 5
How did it compare to *Some Like It Hot?* 1 2 3 4 5
What did you like best about the movie? _____
Was it the actors? _____ yes _____ no
Did you buy popcorn? _____ soda? _____ candy? _____
What model car do you drive? _____

2.38 Adoption is a very important personal and societal issue in the 1990s. For the past decade, the social agencies in charge of arranging for adoptions have adhered to a policy goal of racial similarity in matching parents and child. This policy stands in contrast to that of twenty years ago, which encouraged interracial adoptions. Social attitudes surrounding the family form the base of both policies. Imagine that a social agency wants to develop a survey to assess community attitudes toward whether or not interracial adoptions should be resumed. With particular attention to the issue of interviewer-interviewee relationship, answer the following questions.

    a. How would you design a research study to address this question in order to minimize racially motivated responses?

    b. How do you think racial similarities and differences (between interviewers and interviewees, for example) might play a role in influencing the responses to the study?

    c. Assuming adequate resources, authority, and time, how might this study best be carried out?

2.39 Assume that, at an institution with which you have an affiliation, you are going to create a survey (either for yourself, your group, or your supervisor). You are interested in asking a few questions (perhaps no more than ten) to determine how satisfied the people in the institution are with a particular policy, boss, activity, or recent change that has affected them.

    a. After you determine the goal of your survey, consider how you would like to carry it out: personal interview, telephone, anonymous survey, voice mail or e-mail, and so on.

    b. Create a written mock-up of your research design, including the mode of delivery, the questions to be asked, and the way you would analyze the results.

    c. What dilemmas did you face in creating the design? How did you decide to handle them? What limitations do you think still exist in the design?

    d. If you had many resources, more authority, and much time, what might you do differently?

2.40 Create a data matrix from the following information. A Boy Scout troop is on an overnight camping trip, and they begin talking about their families. Chris, age 9, has three brothers and three sisters, lives with his mother and father, and has a pet gerbil; Andy, age 10, has no brothers and sisters and lives with his mother and a dog; Carl, age 9, has a stepbrother, Sam, and lives with his father and stepmother and a cat also named Sam; Greg, age 10, has a sister, a stepsister, and a halfbrother, lives with his mom and dad and a dog named Rex; Alex, age 8, lives with his grandmother and grandfather; Paul, age 11, has four brothers and a stepsister and lives with his mother and stepfather and a fish named Wanda.

a. What decisions did you make about creating variables for siblings? For parental figures? For pets? How could you have made them differently?

b. If you had a large number of Scouts to summarize, which variable choices for siblings, parental figures, and pets would you prefer?

c. If you were interested only in issues of divorce, remarriage, and single parenthood, how would you design the data matrix?

d. What data did you discard in creating your data matrix? Why did you discard it?

e. Was there any information missing that you think would have been useful if you had wanted to study the likelihood that boys with no siblings and firstborn boys were more likely to belong to the Boy Scouts than other boys?

2.41 In 1789, in Massachusetts, the average male at birth could be expected to live 34.5 years. The average female could expect to live to be 36.5 years old. In 1850, male life expectancy was 38.3 years and female life expectancy 40.5. In 1890, male life expectancy was 42.5 and female life expectancy was 46.6. In 1910, male life expectancy was 54.0 and female life expectancy was 56.6. In 1930, the numbers were 59.3 and 62.6 years.

a. Create a data matrix and put the numbers in the proper rows and columns to make them understandable and ready for statistical analysis.

b. Name two findings that are evident from this data file.

CHAPTER 3

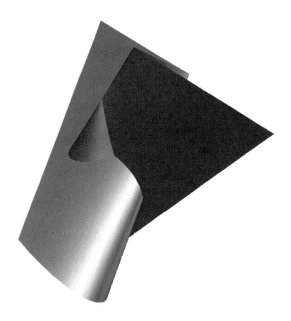

3.1 Graphs: Picturing data

3.2 Categorical variables: Pie charts and bar graphs

3.3 Metric variables: Plots and histograms

3.4 Creating maps from data

3.5 Graphing: Standards for excellence

3.6 Tables: Turning can be timely

3.7 Summary

DESCRIPTION OF DATA:
GRAPHS AND TABLES

What is the fastest growing group in the American work force today? Does taking a literature course reduce crime among convicted criminals? How old are women and men when they marry these days?

In Chapter 2 we discussed ways of collecting data. Once the data are gathered, we must search them for the information they contain. The data are available in the data file, but with so many numbers there, we cannot comprehend them all. Some way or other we must extract information from the data and put it into usable form. This means we need to *analyze* the data by graphing, tabulating, and computing.

Data analysis usually consists of one or more of three activities:

1. Make a **graph** of the data.

2. Make a **table** of the data.

3. **Compute** something from the data.

All three methods involve some degree of simplification. Computing an average, for example, simplifies a collection of numbers. If we compute an average age for ten girls, then the ten numbers are reduced to one. Similarly, graphs and tables involve simplifications and reductions of data. Simplifications make it much easier to understand and to extract information of new kinds from data.

Data simplification has an important drawback. From simplified data we cannot recover the original observations. Thus, there is almost always a loss of some kind of information when we analyze the data.

In analyzing statistical data, we are torn between two conflicting goals—to simplify and to be complete. First, we want to simplify a body of data enough to discover the patterns it contains. We want to highlight important information and to suppress "noise." But at the same time, we do not want to lose interesting details. A football game can be summarized by the final score, but that datum does not describe how the game was played and won. This conflict between simplicity and loss of detailed information is often difficult to resolve. Fortunately, practical considerations provide guidelines for producing useful forms of information. How we describe our data usually depends on what we have in mind for the analysis—where it will be seen, by whom, and for what purpose. In addition, we must satisfy our own judgment and those of our colleagues about what is the best statistical picture we can offer.

> A gain in simplicity involves a loss of information, and a good statistician tries to strike a balance in the tension between these two competing concerns.

3.1 GRAPHS: PICTURING DATA

One way to analyze data is to graph them. A graph is extremely informative because a great deal of data can be summarized in it and understood at a glance. To put a new twist on the old saying, a graph is worth a thousand numbers.

Graphs are made for two main purposes: to help the researcher extract information from the data and to help communicate the information to others.

A graph is essentially a rhetorical device; it is a form of persuasion, first to the researcher and then to others. A graph is constructed to illustrate particular patterns found in data. Many other graphs could be drawn from a particular data file, but they rarely are. Only those graphs are produced that seem important to the analyst in order to understand and to communicate what the data mean. As with so many other statistical methods, it is possible (intentionally as well as unintentionally) to misuse graphs in making an argument. We want you to be able to distinguish between a good graph and a bad graph. Knowing this difference may help protect you from making poor choices and drawing bad conclusions.

Creating statistical graphs

Statistical graphs have been in existence for more than two hundred years. But graphs were invented long after many other important mathematical discoveries. Relatively rare, at first, they were originally drawn by hand and were often extremely imprecise. Today, computer software has taken much of the drudgery and inaccuracies out of constructing

Graphs can be very important. (*"Calvin and Hobbes" copyright 1992 Watterson. Dist. by Universal Press Syndicate. Reprinted with permission. All rights reserved.*)

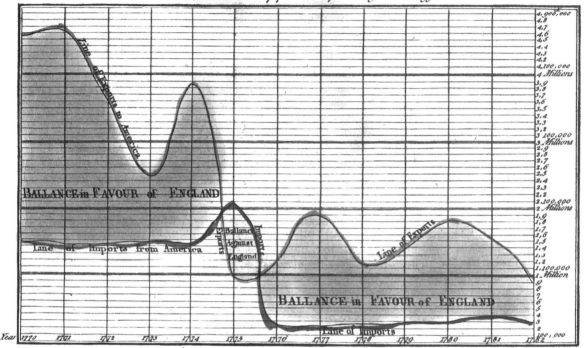

One of the first graphs created—such a rarity. *(Source: Edward R. Tufte,* The Visual Display of Quantitative Information, *Cheshire, CT: Graphics Press, 1983.)*

graphs, and it is very seldom now that professional researchers draw a graph by hand.

There are advantages and disadvantages to the computerization of graphic design. With computer software constructing graphs, many facets of the graphic form are automatically shaped by those who wrote the software, and researchers find it easy to rely on them. But if the computer program is not good, bad graphs result. What is meant by a "bad" graph is described in more detail throughout the chapter.

Statistical graphs have become increasingly commonplace in the media. Graphs taken from computer screens are shown in newspapers, news magazines, and on television. As the media have become increasingly saturated with graphic representations of information, consumers have been required to be more knowledgeable about their construction. Graphic literacy is a must for the twenty-first-century adult.

Types of graphs

In Sections 3.2, 3.3, and 3.4 we discuss some of the more common types of graphs, and we introduce you to some of their respective advantages and disadvantages. In Section 3.5 we take up the principles that underlie the construction of graphs. These principles can be used to judge whether a graph is good or bad.

The simplest type of graph summarizes the data on one variable only, for example, gender, age, or IQ. Such a graph involves the data from only one column in the data file. More elaborate graphs summarize data on two variables, from two data columns, for example, gender and age. Making graphs from data on three or more variables is more difficult but not impossible.

Many graphs are used to show a count of the observations of each value of a variable. For example, a graph could illustrate how many rainy days and how many sunny days occurred last month. This graph would compare the two observed values (rain, sun) by showing which occurred more often and which was more unusual. Other graphs show values of variables measured on a scale. Age in years and income in thousands of dollars are simple variables of that kind.

3.2 CATEGORICAL VARIABLES: PIE CHARTS AND BAR GRAPHS

For the gender variable, the values are female and male. The only thing we can say about two observations on such a variable is that either they are the same or they are different. Such a variable is called a *categorical variable*.

Graphing one categorical variable

In analysis of data on one categorical variable, the first step usually consists of counting the observations of each value. As an example, we focus on data about 72 criminals convicted in the New Bedford District Court in Massachusetts. We want to know whether or not they were convicted of new crimes within from one to two and a half years after they had served their sentences. *(Source:* The New York Times, *October 6, 1993, p. B10.)*

> A **categorical variable** is a variable where two observations are either the same or different. The observations cannot be ordered; one observation is not more of something than another observation.

76 Chapter 3 • Description of Data: Graphs and Tables

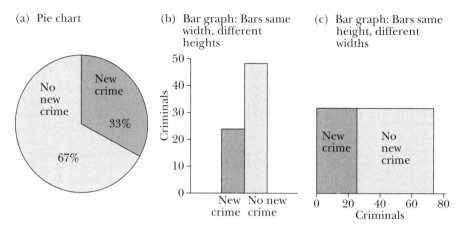

Figure 3.1 Pie chart and bar graphs for a variable (criminals) with two categories (whether or not they were convicted of new crimes within from one to two and a half years after they had served their sentences)

When the observations for this group of criminals are counted, we find that 24 were convicted of new crimes and the remaining 48 were not, at the time the data were collected. Figure 3.1 shows a pie chart and two different bar graphs for the data on the 72 criminals.

Pie chart The pie chart (Figure 3.1a) indicates that about one third of the convicts were convicted of new crimes and two thirds were not. While it may be hard to see that the pie is divided exactly into 1/3 and 2/3 parts, it rapidly conveys that one group is about twice as large as the other.

Pie charts are good for showing the relative sizes of groups. Several different groups can be represented and compared in a pie chart. Pie charts are particularly good for categorical variables because they do not order values. One piece of a pie can move to another location in the pie without changing the meaning of the chart. Also, nearby groups can be easily combined into larger units in a pie chart.

Pie charts are not good for showing how many observations there are in each group. If 240 criminals were convicted of a new crime and 480 criminals were not convicted, the pie would be divided the same way. Also, pie charts are not useful in representing a large number of groups: the "slices" become so tiny and so numerous that they lose their usual impact.

> **STOP AND PONDER 3.1**
>
> Draw a pie chart composed of observations of a set of values from a categorical variable with which you are familiar, for example, the number of phone calls you and several of your friends received in a day. Does the pie chart you create represent your variable in a convenient way or not? What would make the pie chart better or worse?

Bar graph The two bar graphs (Figure 3.1b and c) tell the same story about the criminals. The bar graph in Figure 3.1b—where the bars are the same width and the height of each bar represents the number of observations of the corresponding value of the variable—is the most common. But a bar graph like Figure 3.1c, where the bars are the same height and the width of each bar represents the number of observations of the corresponding value of the variable, can also be used. Note that in each of the bar graphs the bars start at the value of 0. Sometimes this is not the case, and the bar graph then usually conveys a very different story (see Stop and Ponder 3.2).

The bar graph in Figure 3.1b is good for showing the number of observations of each value of the variable but not for showing the total number of observations; it is awkward to mentally place one bar on top of the other to visualize the total. The bar graph in Figure 3.1c is good for showing the total number of observations and the number of observations of the first category of the variable (criminals who committed new crimes) but not for showing the number of observations of the other category (criminals who did not commit new crimes). The more values of a categorical variable, the more complex and difficult a same-height, different-width bar graph becomes.

> **STOP AND PONDER 3.2**
>
> How could a bar graph that did not start with zero on the vertical axis be used by a skillful politician who wants to exaggerate the tax increases proposed by a rival party?

Graphing two categorical variables

There is more to the story about the 72 criminals. Judge Robert Kane of New Bedford District Court in Massachusetts, with the encouragement of Professor Robert P. Waxler of the Dartmouth campus of the University of Massachusetts, gave some of the criminals found guilty in his court the choice of going to jail or taking a literature course taught by Professor Waxler. Professor G. Roger Jarjoura of Indiana University followed the 32 men who took the course and found that 6 were later convicted of new crimes. Among the 40 criminals who went to jail, 18 were convicted of new crimes after release. *(Source:* The New York Times. *October 6, 1993, p. B10.)*

Now we know more about the 24 criminals who committed new crimes and the 48 who did not. We have data on a second categorical variable, namely, whether they took a literature course or went to jail. Figure 3.2 shows three different ways bar graphs can be used to tell the story about the two variables.

In the graph in Figure 3.2a, the two bars represent the criminals who took the literature course and the criminals who went to jail. Each bar is divided into two groups, those who committed new crimes and those who did not. The bars clearly show that a much smaller group of literature-course takers committed new crimes, even though it is hard to read from the scale how many they were because that part of the bar does not start at zero. Among those who went to jail, about one half committed new crimes and the other half did not.

In Figure 3.2b, the tops of the bars in Figure 3.2a have been moved to the horizontal axis. In this graph it is easier to see how many committed new crimes, since all four bars now start at zero. But in this graph it is harder to see the totals of criminals who took the course and criminals who went to jail.

In Figure 3.2c, the total bars are of the same height and different widths. The widths of the bars represent the number of criminals who took the course and the number who went to jail. The vertical divisions of the bars show the proportions of each group who committed and did not commit new crimes. This is an unusual bar graph, but it contains more information than the other two.

All three graphs show that fewer criminals who took the literature course committed new crimes than criminals who went to jail. While the results of this small study are intriguing, there are too many unanswered questions to know whether or not giving literature courses to criminals is the way to reduce the number of crimes. Tuition is probably cheaper than prison stays, at any rate.

(a) Bars same width, different heights, stacked

(b) Bars same width, different heights, next to each other

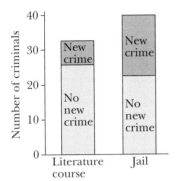
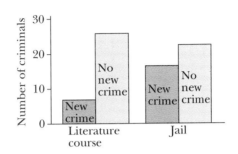

(c) Bars different widths, same heights, stacked

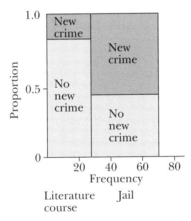

Figure 3.2 Three types of bar graphs for two variables (criminals who did/did not commit new crimes after either taking a literature course or serving a jail term)

3.3 METRIC VARIABLES: PLOTS AND HISTOGRAMS

On certain variables we can *measure* the value of an observation on a scale; for example, we can measure the height of a plant with a ruler marked off in inches. The *unit of measurement* we use to measure the height of the plant is the inch. Similarly, using dollars we can measure the income of a household; using years we can measure how old a person is. Measurable variables such as height, income, and age are called *metric variables*. A metric variable is not metric in the sense of the metric system but in the sense that its values can be numerically measured.

Because meaningful numerical values of a metric variable can be collected, arithmetic operations can be performed on the values of the variable, something that cannot be done with categorical variables. The values of a metric variable can be added, subtracted, multiplied, and divided.

Metric variables are sometimes known as interval or ratio variables. The distinction between interval and ratio need not concern us in this book.

> A **metric variable** is a variable on which we can determine whether one observation is different from another. We can also determine if one observation is more (or less) of something than another observation and how much more (or less) one observation is than another.

STOP AND PONDER 3.3

Give an example of a metric variable and list some of the values of the variable. Why is the variable a metric variable?

Graphing one metric variable

How old are women when they marry? Following is a list of the ages of women who applied for a marriage license in one week, according to the local newspaper (note that not all were necessarily first marriages):

30 27 56 40 30 26 31 24 23 25 29 33 29 22 33 29 46 25
34 19 23 23 44 29 30 25 23 60 25 27 37 24 22 27 31 24 26

What do these numbers tell us? It is easy to spot that the youngest woman is 19 years old, the oldest is 60 years old, and several seem to be in their twenties, but beyond that it is difficult to get much sense of the overall age of the 37 women. With a larger number of observations, it would be even more difficult to understand the data without further analysis. With a metric variable such as age, several different graphs can aid in understanding the data better. Four types of graphs are shown in Figure 3.3.

Lineplot A small number of observations, as we have here, can be organized in a lineplot to get a better understanding of the data. In Figure 3.3a, a line represents the variable; the values of the variable label the line. Each observation is marked as a point above the line.

3.3 Metric Variables: Plots and Histograms

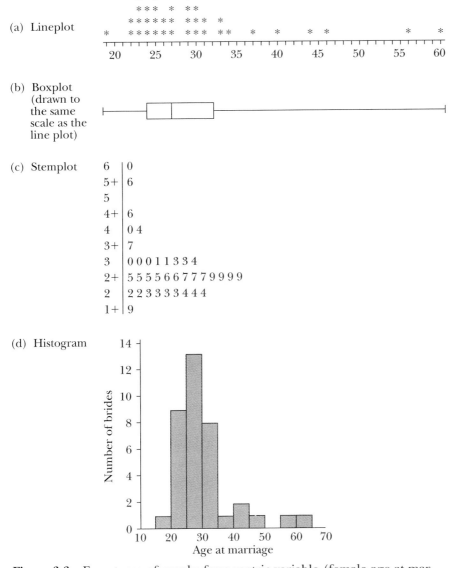

Figure 3.3 Four types of graphs for a metric variable (female age at marriage)

The lineplot clearly shows that most of the women were in their middle to late twenties and early thirties, with a scattering between 35 and 60.

An advantage of a lineplot is that it shows directly how the observations distribute themselves across the variable. We can see where many observations cluster and where few observations scatter; the pattern of the ages is clear. And the original values of the variable are right there: none of the information contained in the original data has been lost, in spite of the simplification the graph offers.

With a large number of different observations of the variable, the lineplot gets messy. For example, a scale of hourly wages would have a large number of values, from the wage of a baby sitter who earns $5 an hour to the wage of a rock star who gets $500,000 an hour for performing. Similarly, a large total number of observations—say, the ages of brides in an entire year—would make a large and messy collection of points above the line. The lineplot would look more like an ant colony than a statistical aid. Larger data sets are better served with graphical methods other than lineplots.

Boxplot Figure 3.3b is a boxplot of the female age at marriage data, drawn to the same scale as the lineplot. Boxplots are not common in the popular press, but they are making inroads in professional journals. They require a bit more work to understand and create than other graphs.

The boxplot in Figure 3.3b shows a line that starts at 19, the age of the youngest bride. The line stops at a rectangular box, from which the plot takes its name. In the box is a vertical line. The line picks up again after the box and extends to 60, the age of the oldest bride. The boxplot is constructed to show one quarter of the observations (37 ÷ 4 ≈ 9) as the line between the minimum value and the beginning of the box. Another quarter of the observations lie between the beginning of the box (at 24) and the vertical line in the box (at 27). Another quarter of the data lie between the line and the end of the box (at 32). The last quarter of the data lie from the end of the box to the largest observation (at 60). Thus, one half of the data lie in the range spanned by the box.

Sometimes the line is not drawn all the way from the box to the smallest observation and/or all the way from the box to the largest observation. This is done when the smallest and/or largest observation lies more than a certain multiple of the length of the box away from the box. In that case, one or more of the extreme observations are marked only as points.

Boxplots are informative graphs. They show the two extreme values as well as the range of the middle values. In Figure 3.3b, the middle half of the brides are between 24 and 32 years old, and the other half

3.3 Metric Variables: Plots and Histograms

> **STOP AND PONDER 3.4**
>
> What would a boxplot of the following data on the yields on thirteen of the nation's biggest money market funds look like?
>
> | Money market fund | Yield (%) |
> |---|---|
> | Vanguard MMR/Prime Port | 5.69 |
> | Schwab Value Advantage MF | 5.66 |
> | Dean Witter/Active Assets MT | 5.59 |
> | Fidelity Spartan MMF | 5.50 |
> | Fidelity Cash Reserves | 5.45 |
> | Dean Witter/Liquid Asset Fund | 5.44 |
> | Kemper MMF/Money Market Port | 5.40 |
> | Smith Barney Cash Port/Class A | 5.29 |
> | Merrill Lynch Retirement Res. MF | 5.28 |
> | Merrill Lynch CMA Money Fund | 5.26 |
> | Merrill Lynch Ready Assets | 5.18 |
> | Dreyfus Liquid Assets | 5.17 |
> | Prudential MoneyMart Assets | 5.16 |
>
> *Source: Money Fund Report,* USA Today, *January 26, 1995, p. 3B.*
>
> How might seeing the data in a boxplot help you to assess in which fund you might wish to invest?

are scattered over the rest of the age range—a good picture of the common ages at which women marry.

Boxplots are particularly useful with data from several groups. A boxplot can be made for each group and compared with one another to see how the groups differ. A collection of boxplots for rates of violent crime in seven regions of the country appears in Figure 3.19. How the center lines compare for the different regions is directly apparent, and so are longer boxes in some regions than others. With the boxplots it is easy to see which regions are most violent (by comparing the center lines), which are most diverse (by comparing the length of the boxes), and which are most likely to include some very peaceful or very violent states (by comparing the locations of the lines on either side of the boxes).

With a boxplot, the original data are lost and cannot be recovered from the plot. At the same time, a boxplot provides a powerful and simplified view of the data.

Stemplot The third graph of the female age at marriage data is a *stemplot* (Figure 3.3c). As the name implies, the graph has a stem, which is drawn as a vertical line. From the stem, branches grow out on both sides. The branches at the left are the first age digits and the branches at the right are the second digits, listed as many times as needed for all the brides of that age. For clarity, each decade is shown in two parts; 2, for example, stands for the ages 20–24, 2+ stands for ages 25–29.

In the stemplot in Figure 3.3c, the youngest bride is 19 years old, two brides are 22 years old, four brides are 23 years old, and so on. Note that the original data are saved in a stemplot. At the same time, the distribution of the observations across the range of values of the variable is clear. Most brides fall in the range 25–29 years of age, and in this group of women most were marrying before reaching 30.

A stemplot does not work well for a large number of observations of a variable because each observation takes up a space in the graph. You can imagine the length of the branches if many observations were listed.

STOP AND PONDER 3.5

Draw a stemplot of the ages of twenty of your family members and friends. To the left of the stem, use a 0 for ages 0–9, 1 for ages 10–19, 2 for ages 20–29, and so forth. What does your stemplot tell you about the ways in which the ages of people you know are clustered?

Histogram The histogram is the most commonly used graph to display the number of observations across a range of values of a metric variable. To create a histogram, the range of values is divided into intervals, usually but not always of the same length, and then the observations in each interval are shown in a rectangle whose area represents the number of observations. (When all the intervals have the same width, as in the histogram in Figure 3.3d, the height of the rectangle shows the number of observations. But it is still important to realize that it is the area of the rectangle that represents the number of observations.)

ORIGIN OF THE WORD HISTOGRAM

The term *histogram* seems to have been first used in print in 1895 by the great English statistician Karl Pearson and defined in a footnote. In a talk he gave to the Royal Society in London, Pearson refers to some data on valuations of house properties in England and Wales for the year 1885–1886:

> It will be observed that so far as the observations can be plotted to the theoretical curve, it leaves little to be desired. The histogram* shows, however, the amount of deviations at the extremes of the curve. Footnote: *Introduced by the writer in his lectures on statistics as a term for a common form of graphical representation, i.e., by columns marking as areas the frequency corresponding to the range of their base. *(Source: K. Pearson, "Contributions to the Mathematical Theory of Evolution. II. Skewed Variations in Homogeneous Material,"* Philosophical Transactions of the Royal Society of London (A), *vol. 186 (1895), part I, p. 399.)*

Pearson gives no explanation for why he uses this particular term.

A modern scholarly interpretation of the word histogram comes from comparative linguist (and author's son) Eric Iversen:

> Re: histogram, starting from the end, which presents an easier explanation, "gram," of course, refers to a picture or representation of something, as in pictogram—a painted image; telegram—an image from far away; epigram—an image attached to something. You can see that "gram" can denote either words or pictures, depending on the usage. But "hist" is more interesting, because of the various things one might mistake it for, like "history," in which the root is the Greek "historia," or "histology," the study of body tissue, in which the Latin root "hist" suggests connectivity or tissue. But "histo" in "histogram" is Greek and means mast or beam, and I should think the word became "histogram" simply because the columns look like masts on a ship or beams used in construction.

In the histogram in Figure 3.3d, the greatest number of brides falls in the interval 25–30, since that rectangle is the largest. There are also quite a few brides in the intervals 20–25 and 30–35. The other brides distribute themselves fairly evenly and sparsely across the remaining age intervals.

The histogram in Figure 3.3d looks very much like the stemplot in Figure 3.3c laid on its side. Each row in the stemplot corresponds to a rectangle in the histogram. Since a histogram shows the shape of a distribution, there is a gain in simplicity and we see a pattern that is not apparent in a list. But at the same time, a histogram loses information. From a histogram the original observed values of the variable cannot be recovered as they can be from a stemplot. In a stemplot an original observation is represented by its actual value in the plot; in a histogram an observation is represented only by a part of a rectangle.

A histogram is therefore useful in simplifying a large number of observations; each observation occupies only a small part of a rectangle. For example, a histogram of ten times as many brides, for a total 370 brides, with 10 in the interval 15–20 years, 90 in the interval 20–25 years, and so on, would look just like the histogram in Figure 3.3d. The difference would be that the frequencies on the vertical scale would be 20, 40, 60, and so on instead of 2, 4, 6, and so on. Whereas it is hard to imagine a stemplot for as many as 370 observations, histograms display large data sets with ease.

The main interest in histograms is their shapes, which can be quite varied. The histogram in Figure 3.3d is *unimodal,* so called because it has one peak. This shape tells us that there is one main group of observations. Histograms can also be *symmetric;* that is, the left half of the distribution is a mirror image of the right half. The histogram of brides' ages does not show a symmetric distribution since there is a longer tail of observations to the right than to the left. This histogram is *skewed.*

Many variables, for example physical characteristics and test scores, often show both unimodal and symmetric distributions. A unimodal and symmetric histogram tells us that most of the observations are in the middle of the distribution of values and that fewer observations are very large or very small. Unlike in Lake Wobegon, most children are average, not above the average. "Averageness" is often an interesting characteristic of a variable, and it becomes apparent only in a graph such as a histogram.

A *bimodal* histogram shows two peaks. To illustrate this shape, imagine a histogram showing the distribution of income in a community that consists mainly of rich people and poor people, with not too many people who have a middle-range income. The shape of the histogram would show the two peaks, telling us that this is a polarized community.

A histogram can change shape, depending on how it is constructed and give a very different impression of the data. First, the number of intervals used on the horizontal axis to divide the variable into intervals affects the shape of the histogram. If the variable is divided into a large number of intervals, each interval will contain only a few observations, and the histogram will look ragged and uneven (Figure 3.4a). If all the observations are shown in one interval, then the histogram will simply consist of one big bar (Figure 3.4b)—not a very useful histogram. The histogram in Figure 3.3d lies somewhere between the two extremes and is more informative and attractive than the two in Figure 3.4. But we have to be on our guard: if a histogram has only a few intervals and

(a) Too many intervals

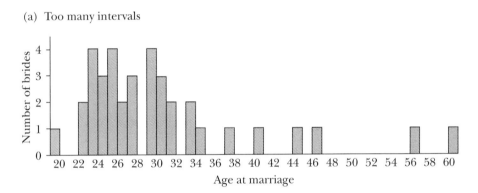

(b) Too few intervals

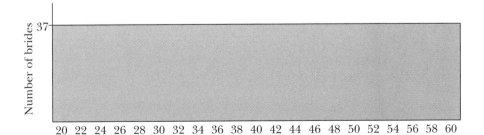

Figure 3.4 Histograms with too many and too few intervals (female age at marriage)

a unimodal shape, the graph may hide the fact that more intervals would have revealed a bimodal shape.

Second, the shape of the histogram changes if the bars are tall and thin versus short and wide. The histograms in Figure 3.5 show the fe-

(a) Wide and short

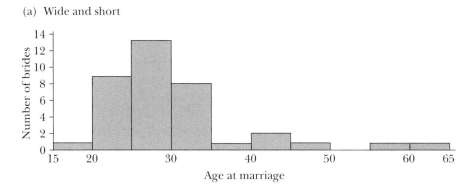

(b) Tall and thin

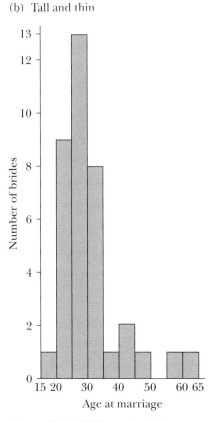

Figure 3.5 Wide and tall histograms (female age at marriage)

male age at marriage divided into the same number of intervals. But the histogram in Figure 3.5a is short and wide, while the one in Figure 3.5b is tall and thin. Because the differences in height of the rectangles in Figure 3.5a are so small, the differences between the number of observations in the intervals also appear small. The opposite perception holds for Figure 3.5b.

No matter their shape, of course: the two histograms are identical; they tell exactly the same story about the age variable. But we should be wary of histograms in which the rectangles are tall and thin or short and wide. Their designers may be trying to create an impression of something that really is not present in the data.

STOP AND PONDER 3.6

Imagine drawing a histogram based on the following data on the population of Mexico from 1930 to 1990. The variable is the average number of children per woman over 12 years old at five-year intervals, starting with 1930.

5.0 4.8 4.6 4.6 4.6 4.6 4.5 4.5 4.5 4.0 3.4 2.8 2.5

Source: Adapted from Zavala de Cosio (1992) in Matthew C. Gutmann, "The meanings of macho: Changing Mexican male identities," Masculinities, vol. 2 (1994), p. 29.

How might you draw the histogram if you wished to emphasize the differences between the high and the low numbers? How might you draw it if you wished to deemphasize the differences? Is either histogram skewed? If so, in what way?

Graphing two metric variables

Statisticians often need to display data on two metric variables—the ages of brides and the ages of grooms, for example, or height and weight of individuals, age and income, SAT scores and grade point average, national literacy and gross domestic product. A common way to display data on two variables is a scatterplot.

A scatterplot consists of two axes, a horizontal axis and a vertical axis. The horizontal axis (the mathematical x-axis) is used for one variable (e.g., age of grooms), and the vertical axis (the mathematical y-axis) is used for the other variable (e.g., age of brides). A pair of observations on the two variables is shown as a point in the graph. For example, if a groom is 37 years old and the bride is 30 years old, a point is drawn in the graph where an imaginary vertical line from point 37

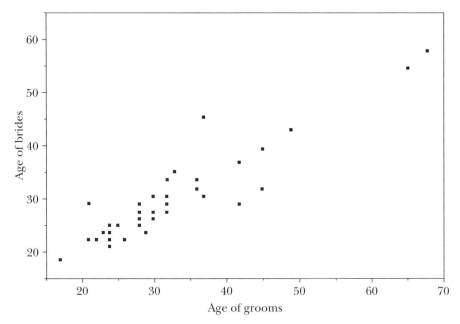

Figure 3.6 Scatterplot showing ages of brides and grooms

of the *x*-axis intersects an imaginary horizontal line from point 30 on the *y*-axis.

Figure 3.6 is a scatterplot of data on the ages of the 37 couples. (Note that 37 points do not appear in the graph because the ages of the brides and grooms in some couples were the same; those couples are represented by a single point.) Looking at the original data, it is difficult to see patterns in them beyond the tendency for older grooms and older brides to marry. When the ages are displayed as points in a scatterplot, the relationship of the two variables is clearer. The points start in the lower left corner of the scatterplot and continue roughly to the upper right corner: by and large, younger grooms marry younger brides and older grooms marry older brides. The path of the points from the lower left corner to the upper right corner has, in mathematical terms, a positive slope; it indicates a positive relationship between the two variables. The points also show that in some couples the groom is older than the bride and in other couples the bride is older than the groom.

In a scatterplot, no numerical information is lost and simplification of the data is gained. A scatterplot is easy to create and to interpret.

3.3 Metric Variables: Plots and Histograms

> **STOP AND PONDER 3.7**
>
> How would you picture a scatterplot of the variables on years of marriage and number of serious quarrels per year for the 24 couples below?
>
> | Years married | Quarrels | Years married | Quarrels |
> |---|---|---|---|
> | 5 | 10 | 10 | 5 |
> | 2 | 20 | 15 | 3 |
> | 4 | 16 | 13 | 4 |
> | 1 | 15 | 20 | 2 |
> | 3 | 9 | 16 | 4 |
> | 6 | 6 | 25 | 1 |
> | 5 | 8 | 22 | 3 |
> | 8 | 5 | 14 | 3 |
> | 3 | 10 | 15 | 4 |
> | 7 | 7 | 19 | 3 |
> | 3 | 8 | 17 | 3 |
> | 9 | 6 | 20 | 2 |
>
> What conclusions might you draw about the nature of married life? What missing information might prevent you from making generalizations from the data about the road to marital bliss?

Time series plot

Variables often consist of data that have been collected over a period of time. The consumer price index for the last forty years is a variable with time series data, as are value of annual imports from Japan since World War II, average length of a baseball game since 1940, or average length of a skirt hem since 1932. Graphs of time series data are special scatterplots. Time as a variable is plotted along the horizontal axis, the other variable along the vertical axis. The points do not look as scattered as they do in the age at marriage scatterplot because time values on the horizontal axis are usually evenly spaced. Also, only one value of the variable on the vertical axis is plotted for each value on the horizontal axis.

Using the data file in Table 3.1, the scatterplot in Figure 3.7 shows the height in inches jumped by the male Olympic gold medalists in the

Table 3.1 Gold-medal-winning male Olympic high jumps 1900–1936

Year	Height of jump (inches)
1900	74.8
1904	71.0
1908	75.0
1912	76.0
1920	76.2
1924	78.0
1928	76.4
1932	77.6
1936	79.9

high jump competition in the years from 1900 to World War II. In the data file, the two columns of numbers show that we are dealing with two variables. It is evident from the table that the heights increased, and this is clearly illustrated in the figure. A new Olympic record was set each time except for the years 1904 and 1932; the line dips as it connects 1900 with 1904 and again as it connects 1928 and 1932. The line then extends upward after each of these years. (Imagine continuing the graph to the present. Not only would the year axis have to be

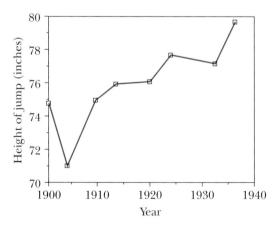

Figure 3.7 Time series graph showing gold-medal-winning male Olympic high jumps 1900–1936

extended, so would the winning jumps axis: champion jumpers now think nothing of 8-foot—96-inch—jumps.)

As with many other graphs, it is possible to change the shape of a time series plot and give entirely different impressions of the data. Figure 3.7 seems to indicate considerable change in the heights of winning jumps from one Olympic games to another. But note that the vertical scale starts at 70, not at 0, and extends only 10 inches. There is a reason for starting at 70. Imagine extending the vertical line downward to 0 while keeping the graphed line where it is and maintaining the scale of the 1-inch intervals. The differences from one year to the next would not seem as large as they do in Figure 3.7, but the graph would be too deep for practical purposes. It would be eight times the depth of Figure 3.7, and the bottom seven eighths would be empty!

If the graph is kept the same size as Figure 3.7 and the vertical axis marked with a scale from 0 to 80 inches instead of from 70 to 80 inches, the immediate visual impact is much different (Figure 3.8). The lines connecting the points do not go up and down quite as steeply as they do in Figure 3.7. Figure 3.8 shows a general increase in the height of the jumps, but the change from one Olympics to another seems much less dramatic.

Obviously, we should examine a time series plot carefully, imagining what it would look like with changes such as the ones discussed

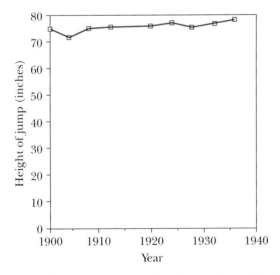

Figure 3.8 Redrawn time series graph showing gold-medal-winning male Olympic high jumps 1900–1936

here, before we draw conclusions about the data. In general, the time series graph has the major advantages and disadvantages of scatterplots: numerical information is retained and data are simplified, but the shape of the graph can be misleading.

> **STOP AND PONDER 3.8**
>
> Rate your mood states, with 0 being extremely negative and 7 being extremely positive, for each day of this week, starting with Monday. Then create a graph in which you place the days on the horizontal or *x*-axis and your ratings on the *y*-axis. You have just produced a time series plot! By comparing your ratings, you can get a better picture of the entire week. You probably did not use the entire range of numbers. The usual finding for this kind of scale is that the upper end of the scale is used more than the lower end. (See Exercise 3.33.)

3.4 CREATING MAPS FROM DATA

Maps can represent not only geographical features, such as rivers and mountains, but also statistical information. For example, in a map of the United States the states can be colored according to some variable, such as Presidential election voting patterns. Figure 3.9 shows a map in which the states are colored by the magnitudes of their divorce rates.

> **STOP AND PONDER 3.9**
>
> What does the map in Figure 3.9 tell us about variations in divorce rates across the country? To what extent does using a map to display these data create false impressions about the nature of the data? That is, is this map a good or a bad representation of the data? What could some of the reasons be for variations in divorce rates across the country? Can you think of other cases in which you have seen a map such as this do a good job representing the data? Under what conditions could maps be an excellent way of presenting data?

Maps can be useful in identifying conditions in which viewers are interested. (Maps on the weather channel are a case in point!) If you were considering moving to California, you would be interested in knowing where the air pollution is greatest, for example. Maps also

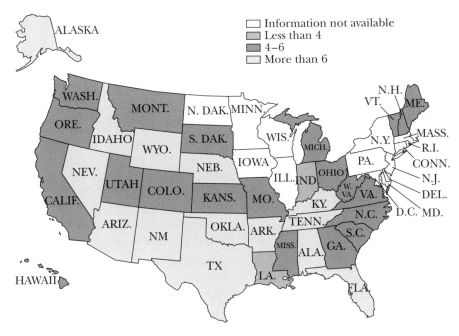

Figure 3.9 Map showing the distribution of divorce rates in the United States (per 1,000 people) *(Source: Adapted from a map based on data from the National Center for Health Statistics by Paul O. Pugliese in* Time, *December 6, 1993, p. 23.)*

help in studying regional trends, such as the prevalence of certain insects in one area of the country rather than in another.

The approximately 3,000 counties of the United States could be outlined and colored to show cancer rates (the number of reported cancer patients in each county divided by the population of the county). The categories low, medium, and high cancer rate could be established and the counties colored in three shades of a color according to their level of cancer rate. Such a map would show regional patterns.

Useful as maps are, they can be misleading in a major way. It is the geographical area that is shaded, and geographical areas vary a great deal in size. A small eastern county with a high cancer rate will not show up on a map as much as a large western county with a low cancer rate. A high cancer rate in a geographically small county in New Jersey affects many people, whereas a low cancer rate in a large county in Nevada affects only a few.

3.5 GRAPHING: STANDARDS FOR EXCELLENCE

This chapter introduces you to a range of standard statistical graphs. Most of them were made using statistical software on a computer. With a click of the mouse, graphs that used to take a long time to design can now be almost instantaneously created and revised. Graph makers can experiment with multiple forms, many of which are unfamiliar to the public. Each form may be useful in revealing some facet of the data and suppressing others. But each new breakthrough in visual imagery also produces new pitfalls. To evaluate a graph, we have to have some idea of what constitutes a "good" graph.

An excellent introduction to "good" and "bad" graphs is found in Edward R. Tufte's book *The Visual Display of Quantitative Information* (Cheshire, CT: Graphics Press, 1983). Tufte, an expert in the field of visual displays of data, uses the term *graphical excellence* to describe a "good" graph. In his view, an excellent graph is one in which complex ideas are communicated with clarity, precision, and efficiency (page 51).

> **Graphical excellence** is that which gives to the viewer
>
> the greatest number of ideas
>
> in the shortest time
>
> with the least ink
>
> in the smallest space.

"The least ink": Is the simplest graph best?

Figures 3.10 and 3.11 show the same data on the relationship between the size of the U.S. population and a price index for 20-year intervals in the nineteenth century, but Figure 3.11 uses less ink than does Figure 3.10; it does away with parts of the graph in Figure 3.10 that are not necessary to convey the information in the data. It uses the dates as points, rather than dates plus points. (Even the points in Figure 3.10 are redundant; they don't need to be marked with with a dot and an open square.) It removes the parentheses around the years. And, since

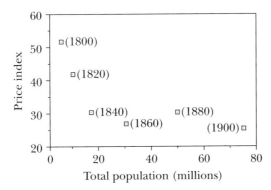

Figure 3.10 Scatterplot of population and price index 1800–1900 *(Source: U.S. Bureau of the Census,* Historical Statistics of the United States, Colonial Times to 1970, *Bicentennial Edition, Part 1 (Washington, D.C.: U.S. Bureau of the Census, 1975). Population: Series A57–72, pp. 11–12; consumer price index: Series E135–166, p. 211.)*

the title states that the data apply to the nineteenth century, it abandons the first two digits in the years (the sequence of the years makes it quite clear that the upper left '00 refers to the year 1800 and the lower right '00 refers to 1900). Thus, Figure 3.11 is a better graph than Figure 3.10, at least for Tufte!

Do the histograms on age of brides in Figure 3.4 use too much ink? Yes, according to Tufte. The height of each rectangle is indicated by the length of the left side of the rectangle, the length of the right side,

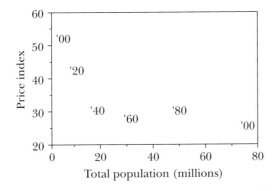

Figure 3.11 Simplified scatterplot of the data in Figure 3.10 *(Source: U.S. Bureau of the Census,* Historical Statistics of the United States, Colonial Times to 1970, *Bicentennial Edition, Part 1 (Washington, D.C.: U.S. Bureau of the Census, 1975). Population: Series A57–72, pp. 11–12; consumer price index: Series E135–166, p. 211.)*

the location of the top line, and the shading—for redundant clues. Assuming that Tufte would subscribe to the rectangle idea at all, he would probably argue that the shading does not add anything to the graphs. No shading would certainly cut down on ink, but it would also make the graphs less clear. The shading makes the rectangles stand out against the background. Without shading, the inside and outside of a rectangle would have the same color, and the point of the graphs—that the shapes of the two histograms are different—would be less obvious. The extra ink makes the figure readable.

"Chartjunk": A new name for garbage

Graphs sometimes include features that have nothing to do with the data presented in the graph—features that the graph maker includes in an attempt to make the graph more attractive or interesting. Tufte refers to unnecessary features as "chartjunk." Chartjunk includes shadings on rectangles, grids on scatterplots, figurative symbols to represent quantities, and illustrations that decorate the margins or the graph itself. Tufte's view is based on the premise that "less is more" in proper graph design. We may or may not agree; one viewer's chartjunk may be what makes the graph comprehensible to another viewer. A graph that is attractive to the eye, displays a touch of humor, or stirs a reaction such as curiosity or dismay may not suit strict statistical standards of simplicity and order, but it might attract the viewer's eye.

Data density

The purpose of a graph is to transmit information to the viewer. Figure 3.8 on winning Olympic high jumps shows 9 jumps and the years of 9 Olympic games, for a total of 18 numbers. The graph itself is fairly large, so it has only a few numbers per square inch. The more numbers per square inch in a graph, the higher the data density and the more informative the graph.

An example of a graph with high data density is the daily national weather map in the newspaper. The map shows the outline of each of the 50 states, temperature, barometric pressure, and precipitation. Another example of graphs with high data densities are the graphs in *Consumer Reports* on repair records of cars.

3.5 Graphing: Standards for Excellence 99

> **STOP AND PONDER 3.10**
>
> Follow Tufte's advice and eliminate some of the ink in the accompanying graph. Which parts of the graph are not necessary? Which ones seem to serve an important function in conveying the data at hand? Imagine defending your actions to Tufte.
>
>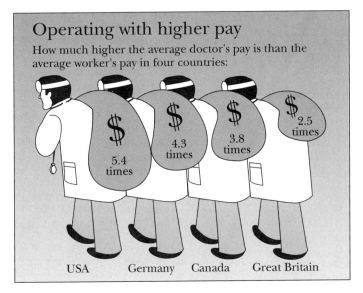
>
> Source: *Graph by Marcy E. Mullins from data of Stuart Altman,* USA Today, *November 13, 1993, p. 1.*

"Revelation of the complex"

Tufte concludes his book by discussing a famous graph showing how Napoleon's army suffered a grand defeat in Russia in 1812 (page 40; Figure 3.12 reproduces the graph):

> [This] is the classic [graph] of Charles Joseph Minard (1781–1870), the French engineer, which shows the terrible fate of Napoleon's army in Russia.... Seeming to defy the pen of the historian by its brutal eloquence, this combination of data, map and time series, drawn in 1861, portrays the devastating losses suffered in Napoleon's Russian campaign of 1812. Beginning at the left on the Polish-Russian border near the Niemen River, the thick band shows the size of the army (422,000 men) as it in-

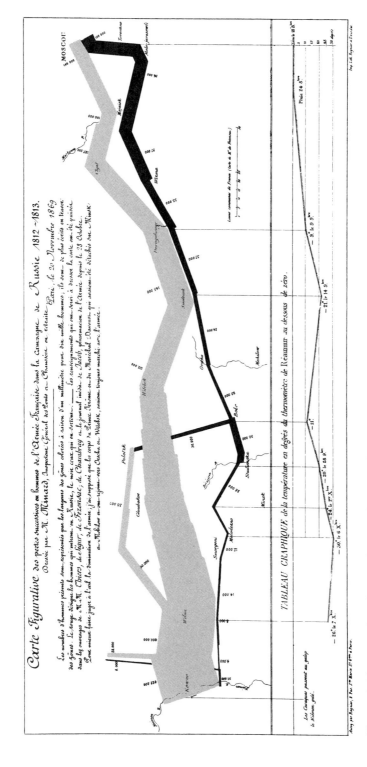

Figure 3.12 Napoleon's march to and retreat from Moscow. (*Source:* Edward R. Tufte, The Visual Display of Quantitative Information, Cheshire, CT: Graphic Press, 1983.)

vaded Russia in June 1812. The width of the band [invasion is gray, retreat is black] indicates the size of the army at each place on the map. . . . The crossing of the Berezina River was a disaster, and the army finally struggled back into Poland with only 10,000 men remaining. . . . *Six* variables are plotted, the size of the army, its location on a two-dimensional surface, direction of the army's movement, and temperature on various dates during the retreat from Moscow.

It may well be the best statistical graphic ever drawn.

3.6 TABLES: TURNING CAN BE TIMELY

Tables are another way to summarize data in compact form. Usually tables are composed of numbers organized in rows and columns. Tables often show how many or what percentage of observations fall in different categories, for example, children of different ages in an educational study.

Tables are used for two broad purposes. One purpose is to support arguments in accompanying text; the other is to organize data. Tables in newspapers, journal articles, and books are usually of the first kind, and tables presented by official statistical agencies such as the Bureau of the Census are usually of the second kind. A table for the purpose of supporting an argument must make a point. A table that simply presents data must be easy to read and interpret.

Table 3.2 contains the same data as those in Figure 3.1 on the criminals who did and did not commit new crimes. The visual experience the table provides is very different from that of the corresponding graphs. When we examine Figure 3.1 and Table 3.2 together, it is much easier to compare the numbers of people in the different categories in

Table 3.2 Number of criminals convicted of new crimes and not convicted of new crimes within from one to two and a half years after they had served their sentences

Convicted of new crimes	24
Not convicted of new crimes	48
Total	72

the graphs than in the table. The different sizes of the two pie slices or bars immediately convey the differences in the two categories. In the table, the numbers have to be first read and then mentally compared. To see how many more criminals not convicted of new crimes there are than criminals reconvicted, we have either to subtract 48 from 24 or to divide 48 by 24.

What the table does show very directly are the actual frequencies. The table states how many criminals were convicted of new crimes, while in the graph we have to draw a mental line from the top of the bar left to the vertical axis. And depending on how detailed the numbers are on the axis and how good our eyesight is, we still might have difficulty judging whether the number is 24 or 26. Thus, if exact numbers are important, a table is better than a graph. For a quick impression of the data, a graph is better than a table.

A table always has a title and rows and columns are clearly headed. Totals for rows and columns should be included, where appropriate. The totals provide a context for the details in the columns and rows. If the table contains only one collection of numbers, as in Table 3.2, the numbers should run vertically. The table could be arranged with the numbers running horizontally, but it is not as easy to get a sense of the total frequency. Even though the total is given and we do not have to do any addition ourselves, we are accustomed to seeing numbers that are added arranged in a column. It is also easier to compare numbers arranged in a column than numbers arranged in a row. The differences between two numbers are more apparent when the numbers are stacked.

In her book *Plain Figures,* Myra Chapman (in collaboration with Basil Mahon; London: Her Majesty's Stationery Office, 1986) gives an instructive example of how a table can be improved by rearrangement (Tables 3.3 and 3.4). The data are from England and Wales.

Table 3.3 is typical of statistical summaries. It shows data on two variables for students who had finished compulsory education. One variable is time, running in seven academically annual increments from 1973 to 1980. The other variable is destination of pupils after compulsory school was over, that is, what each group of students did after finishing school. The numbers are all percentages except for those in the last row, which are total numbers of pupils in thousands. For example, the 1973–1974 column shows what percentages of 701,000 pupils ended up in the five different destination categories. The percentages in each column add up to 100%, except perhaps for

Table 3.3 Destination of pupils attaining the statutory school-leaving age

Destination (%)	School year						
	'73–'74	'74–'75	'75–'76	'76–'77	'77–'78	'78–'79	'79–'80
Staying on at school	25.9	26.1	27.5	28.3	27.6	27.4	27.8
In full-time or nonadvanced further education	9.7	11.5	13.6	13.6	14.1	14.3	14.1
In employment							
with part-time day study	17.4	16.4	12.1	10.2	14.1	12.1	12.2
with no day study	44.1	41.7	38.0	37.7	34.4	38.7	38.7
Unemployed	3.0	4.2	8.8	10.1	10.0	7.5	7.2
Total pupils (=100%; thousands)	701	723	744	746	773	801	814

rounding-off errors in the computations of the percentages, but these totals are missing in the table. (The percentages across the rows do not and should not add up to 100%.)

The main purpose of Table 3.3 is to show how the percentages for the various categories changed over time. Among other difficulties in clarity of presentation, the table requires comparison of percentages across rows rather than down columns to see what happens from one year to the next.

Table 3.4 Table 3.3 transposed and simplified

School year	Destination					Total pupils	
	In employment		Staying on at school	In full-time or nonadvanced further education	Unemployed	Percent	Thousands
	With no day study	With part-time day study					
1973–1974	44	17	26	10	3	100	701
1974–1975	42	16	26	12	4	100	720
1975–1976	38	12	28	14	9	101	740
1976–1977	38	10	28	14	10	100	750
1977–1978	34	14	28	14	10	100	770
1978–1979	39	12	27	14	8	100	800
1979–1980	39	12	28	14	7	100	810

In Table 3.4, the data are rearranged to assist the viewer in understanding the changes in the percentages over time. Perhaps the most striking improvement is simplification: all the percentages have been rounded to two figures without decimals. This removes 35 numbers from the table, together with 35 periods. (The principle of less ink in a graph also seems to hold for a table.) The percentages are now not completely accurate, but for the purpose of this table, complete accuracy is not important. The total numbers of pupils themselves have been rounded; the purpose of the table is to show the trends of change in pupils' destinations, not the precise numbers of pupils for each destination.

Another large difference between the two tables is the order of the destination categories. In Table 3.3, the category "Staying on at school" comes first, in Table 3.4 the category "In employment with no day study." The reason for rearranging the categories is to put the higher numbers in the upper left corner of the table and the lower numbers (more or less) in the lower right corner. The reason for such an arrangement is ease in comparing numbers, particularly the ones in the same column. Table 3.4 also shows that the percentages in each row add up to 100%.

3.7 SUMMARY

3.1 Graphs: Picturing data

A graph is an extremely informative way of analyzing data because an entire data set can be summarized in a graph and understood in one glance. Graphs help the investigator to extract important findings from the data and help to communicate these findings to others.

3.2 Categorical variables: Pie charts and bar graphs

A categorical variable is a variable where two observations are either the same or different. The major graphs used to display categorical variables are the pie graph and the bar graph. The pie graph is easy to understand if there are not too many categories, but the number of observations in each category is usually lost. The bar graph is easy to read, but the details of different categories are difficult to observe if the bars are composed of different groups.

3.3 Metric variables: Plots and histograms

A metric variable requires a unit of measurement that assesses how much bigger or smaller one value is than another. Lineplots, box plots, stemplots, and histograms can be used to display single metric variables. The lineplot displays a small data set along one continuous line; the original values of the variable appear in the plot.

The stemplot is well suited for a small data set; it is less useful for data sets with a small range of numbers. Boxplots show the two extreme values and the range of the middle values of a variable. Boxplots are helpful in comparing data from several groups on the same variable. The histogram indicates by the areas of its rectangles the relative numbers of observations of each value for the variable. A unimodal histogram has one peak while a bimodal histogram has two peaks. A histogram is useful for showing a large number of observations. A drawback of a histogram is that the original values of the observations are lost.

Graphs often used to plot two metric variables at a time are the scatterplot and the time series plot. A scatterplot has two axes on which each point—composed of the observed values of two variables—can be graphed for each element. Scatterplots show patterns of relationships between the two variables. Time series scatterplots show time values, which are usually evenly spaced, on the horizontal axis and another variable on the vertical axis.

Bar graphs and time series graphs can be made for several variables. While multiple-variable graphs allow the comparison of a large amount of information, they can be difficult to read if they contain too many different variables.

3.4 Creating maps from data

Maps colored or shaded in symbolic ways convey statistical data and/or help in the generation of hypotheses regarding regional trends. Maps can be misleading because shaded areas represent land masses, not population densities.

3.5 Graphing: Standards for excellence

Edward Tufte, an expert in the field of visual displays of data, uses the term graphical excellence to describe a "good" graph. In his view, a good graph communicates complex ideas with clarity, precision, and efficiency. Tufte argues that the purpose of graphics is to create a "revelation of the complex."

3.6 Tables: Turning can be timely

Tables consisting of numbers organized in rows and columns summarize data in a compact form. The manner in which tables are constructed can strongly influence the way in which the data are interpreted by the viewer.

ADDITIONAL READINGS

Cleveland, William S. *Elements of Graphing Data.* New York: Chapman & Hall, 1993. Contains many interesting graphs.

Monmonier, M. *How to Lie with Maps.* Chicago: University of Chicago Press, 1991. How not to use maps.

Tufte, Edward. *The Visual Display of Quantitative Information.* Cheshire, CT: Graphics Press, 1983. A classic on how to display data in graphs, with good historical background.

Wainer, Howard. "How to display data badly." *The American Statistician*, vol. 38, no. 2 (May 1984), pp. 137–147. Entertaining article on how *not* to display data.

Witmer, Jeffrey. *DATA Analysis: An Introduction.* Englewood Cliffs, NJ: Prentice Hall, 1992. Interesting graphs of different data sets.

EXERCISES

REVIEW (EXERCISES 3.1–3.15)

3.1 What are the two conflicting goals in analyzing statistical data?

3.2 a. When are pie charts useful in displaying data?
b. What is a major disadvantage of pie charts?

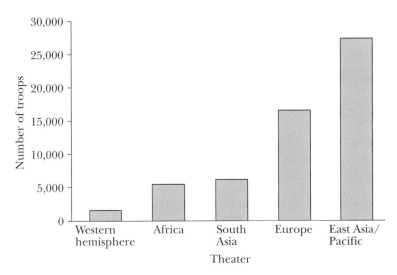

Figure 3.13 U.S. troops stationed on ships around the world, June 30, 1993 (Exercise 3.5) *(Source: Data of the U.S. Department of Defense.)*

3.3 a. What is a lineplot?

b. What is an advantage of using a lineplot to display data?

c. What is a limitation of a lineplot?

3.4 a. What is a stemplot?

b. Why do you think it is called a stemplot?

c. What is an advantage of using a stemplot to display data?

d. What is a limitation of a stemplot?

3.5 Figure 3.13 is a bar graph of the number of U.S. troops stationed on ships around the world in 1993.

a. Copy the histogram and indicate the following parts: A. Horizontal axis. B. Vertical axis. C. Variable being measured. D. Number of troops in Europe.

b. How is the number of observations for each area of the world displayed in the graph?

c. What would be incorrect about indicating the different numbers of troops with figures of sailors instead of bars?

d. What are some conclusions that can be drawn from the information provided in this histogram?

e. Are there any aspects of this histogram that are confusing or that could use further clarification in order to be comprehended more easily?

3.6 In a histogram is it the base, the height, or the area of a bar that corresponds to the number of observations that fall into each interval?

3.7 Why is the histogram one of the best graphs for very large samples?

3.8 When a histogram has two high points, or peaks, what is the name of the shape of the distribution?

3.9 Would the distribution of heights of 100 boys in the eleventh grade have a unimodal or a bimodal distribution?

3.10 What five numbers in data are necessary to display the data in a boxplot?

3.11 What is a skewed distribution?

3.12 What type of data is a scatterplot useful for?

3.13 Figure 3.14 shows total U.S. population and consumer price index for the nineteenth century.

a. Without reviewing the text, interpret this graph in your own words.

b. Why is the graph useful?

c. How would it be easy, with this kind of figure, to convince someone of a strong relationship between the two variables? (Hint: Two scales are used on the vertical axes.)

3.14 What are the key characteristics graphical excellence in a statistical graph?

3.15 What are the major characteristics of a well-produced statistical table?

INTERPRETATION (EXERCISES 3.16–3.33)

3.16 Discuss the statement "A single graph is better than a thousand words."

3.17 In what respects are graphs a type of persuasion?

3.18 Figure 3.15 is a pie chart showing the number of crime victims per 1,000 population for different income groups.

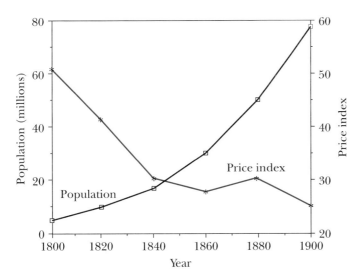

Figure 3.14 Total U.S. population and consumer price index at six census years in the nineteenth century (Exercise 3.13) *(Source: U.S. Bureau of the Census,* Historical Statistics of the United States, Colonial Times to 1970, *Bicentennial Edition, Part 1 (Washington, D.C.: U.S. Bureau of the Census, 1975). Population: Series A57–72, pp. 11–12; consumer price index: Series E135–166, p. 211.)*

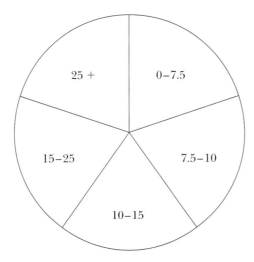

Figure 3.15 Number of crime victims per 1,000 population for different income groups (thousands of dollars) (Exercise 3.18)

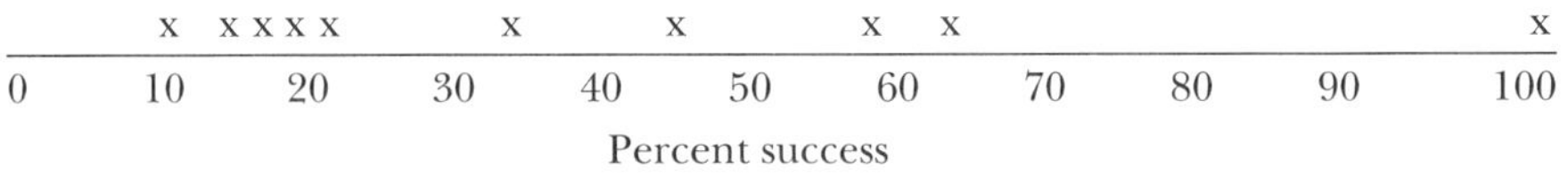

110   Chapter 3 • Description of Data: Graphs and Tables

```
        x  x x x x        x         x            x   x                              x
     _____
     0      10     20     30     40     50     60     70     80     90     100
                                     Percent success
```

Figure 3.16 Success rates of matchmaking services (Exercise 3.19) *(Source: Mara B. Adelman, and Aaron C. Ahuvia, "Mediated channels for mate seeking: A solution to involuntary singlehood?"* Critical Studies in Mass Communication, *vol. 8 (1991), pp. 273–289.)*

   a. What conclusions would you draw about crime victims and income levels?

   b. What questions are unanswerable because a pie chart is used for displaying the data?

   c. What would be a better graph for analyzing these data? Why?

3.19 Figure 3.16 is a lineplot for success rates reported for matchmaking services, with success being defined as "long-term romantic relationship" resulting from arrangements made by the agency.

   a. What is one advantage of using a lineplot to display these data?

   b. When is a lineplot a poor method for displaying data?

   c. What conclusion might you draw from Figure 3.16 about investing in a matchmaking service if you were eager to find a long-term romantic relationship?

   d. Because there is such a wide range of percentages of success rates, what more might a potential customer want to know about matchmaking services that is not presented in Figure 3.16?

3.20 Figure 3.17 is a double stemplot of marriage ages of 37 couples listed in the Sunday issue of a local newspaper.

   a. What patterns in the data does the stemplot help us observe?

   b. Do you find any aspects of the stemplot disadvantageous in terms of visual appeal or convenience?

   c. Would any details of the data not presented in the figure be useful to know?

3.21 Find a statistical graph in a newspaper, news weekly, or scientific journal, for example, and copy the graph.

   a. Describe what the graph tells you about the data it displays.

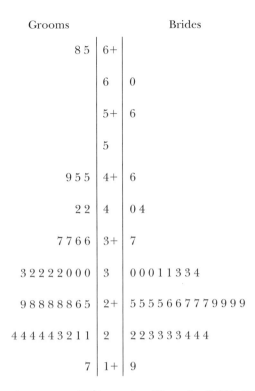

Figure 3.17 Marriage ages of 37 couples (Exercise 3.20) *(Source:* The Philadelphia Inquirer, *September 10, 1995, p. MD12-d.)*

b. Discuss the quality of the graph in terms of principles of graphical excellence.

c. Could the graph be redrawn in any way to improve it? Explain.

d. Can you think of a way in which information available in the data matrix might have been suppressed in order to produce the graph in the figure?

e. Can you suggest another type of graph that could have been made from the original data matrix? Or is this the only graph the data matrix would allow?

3.22 Figure 3.18 compares the heights of several defense secretaries with national defense spending.

a. What does the graph tell us?

b. Does the graph meet the criteria for a good graph? Explain.

112   Chapter 3 • Description of Data: Graphs and Tables

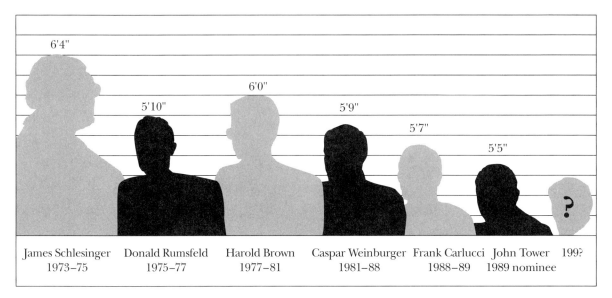

Figure 3.18   Heights of Secretaries of Defense 1973–1989 (Exercise 3.22)
(Source: Data provided by the Secretaries; adapted from the graph in The Economist, *February 11, 1989, p. 20.)*

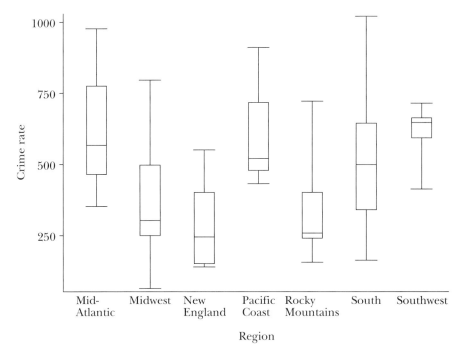

Figure 3.19   Violent crime rates in the 48 contiguous United States in 1986 (per 1,000 population) (Exercise 3.23) *(Source: F.B.I. Uniform Crime Report for the United States)*

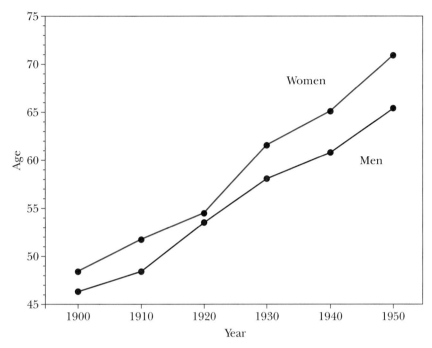

Figure 3.20 Life expectancy in the United States 1900–1950 (Exercise 3.24) *(Source: U.S. Bureau of the Census.)*

3.23 Figure 3.19 shows a boxplot of the violent crime rate in each of seven regions in the United States.

a. From the graph, how do the regions differ in crime rates?

b. Does the graph meet the criteria for a good graph? Explain.

3.24 Figure 3.20 shows the life expectancy for men and women at ten-year intervals in the first half of the twentieth century.

a. Why are the differences between men and women as large as they are?

b. How might such a graph be used politically by certain interest groups?

c. Redraw the graph to meet criteria of good graph making.

3.25 Compare Figures 3.7 and 3.8 on winning Olympic high jumps.

a. Which do you think is the better graph?

b. What did you take into account in your answer to part a?

114 Chapter 3 • Description of Data: Graphs and Tables

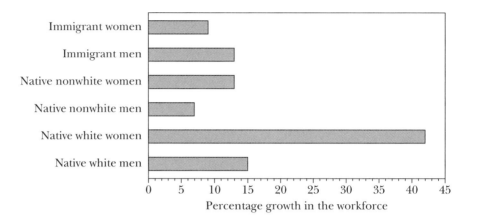

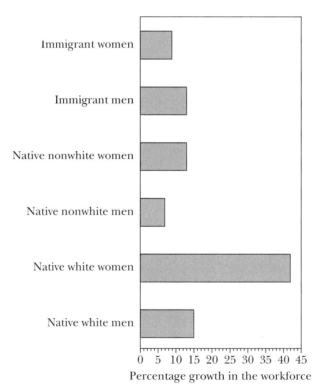

Figure 3.21 Two bar graphs showing the same data (Exercise 3.26) *(Source: Data from* Workforce 2000, *produced by the Hudson Institute, 1987.)*

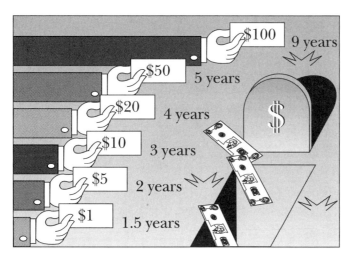

Figure 3.22 Life span of paper currency (Exercise 3.30) *(Source: Data of U.S. Bureau of Engraving and Printing, adapted from the graph by Marty Baumann in* USA Today, *August 19, 1991, p. 1.)*

3.26 The two bar graphs on growth in the workforce for a given time period in Figure 3.21 display the same data, yet they look quite different from one another.

    a. Why is it possible to say that the two bar graphs display the same data?

    b. What effect on the casual reader might each graph have?

    c. How would you redraw these bar graphs if you were trying to be as neutral as possible? (Draw or describe your changes.)

3.27 Explain whether stemplots can be used for categorical and metric variables.

3.28 Explain whether boxplots can be used for categorical and metric variables.

3.29 Compare the strengths and weakness of stemplots and boxplots.

3.30 Money actually wears out, and Figure 3.22 shows how long the average life span is for paper currency of different denominations.

Table 3.5 Data for Exercise 3.31

Dish (number of cups)	Calories	Fat (grams)	Percent of calories from fat	Sodium (milligrams)
Egg roll (1 roll)	190	11	52	463
Moo shu pork (4 pancakes)	1,228	64	47	2,593
Kung pao chicken (5)	1,620	76	42	2,608
Sweet and sour pork (4)	1,613	71	39	818
Beef with broccoli (4)	1,175	46	35	3,146
General Tso's chicken (5)	1,597	59	33	3,148
Orange (crispy) beef (4)	1,766	66	33	3,135
Hot and sour soup (1)	112	4	32	1,088
House lo mein (5)	1,059	36	31	3,460
House fried rice (4)	1,484	50	30	2,682
Chicken chow mein (5)	1,005	32	28	2,446
Hunan tofu (4)	907	28	27	2,316
Shrimp in garlic sauce (3)	945	27	25	2,951
Stir-fried vegetables (4)	746	19	22	2,153
Szechuan shrimp (4)	927	19	18	2,457

Source: Data from Center for Science in the Public Interest, tabulated by the Philadelphia Inquirer, *September 2, 1993, page D1.*

    a. What type of average do you think the figure refers to? Give reasons for your answer.

    b. Does the graph meet Tufte's criteria for graphical excellence?

    c. Redraw the graph another way and explain why your graph may be better.

3.31 Table 3.5 shows the fat content of several Chinese foods.

    a. Comment on how the sentence "Dishes are ranked from worst (highest percent of calories from fat) to best (lowest percent)" could be misleading to someone who wanted to order a Chinese meal and eat the "best" food possible. (What, if anything, is problematic about this table?)

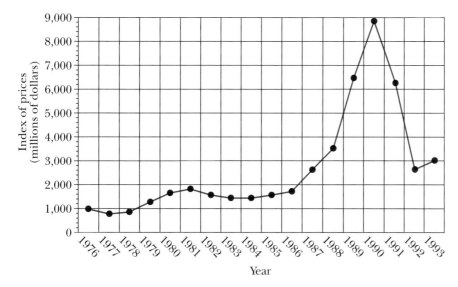

Figure 3.23 Worth of 20 Cézannes, 16 Renoirs, and 15 Matisses in the Barnes Foundation Collection 1976–1993 (Exercise 3.32) *(Source: Robin Duthy, "The boom for Barnes," Connoisseur's World, 1994, p. 108.)*

b. How could you reorganize this table for another purpose than showing fat content? Redesign the table and state the alternative purpose.

3.32 Figure 3.23 was copied from a report on the value of Cézanne. Matisse, and Renoir paintings held by the Barnes Foundation in Philadelphia, produced by a consultant at Art Market Research in London.

a. How would you describe the monetary history of this collection in the last fifteen years?

b. Overall, what would you say the *trend* in the values of the paintings has been over the years as shown in the graph?

c. How else could the graph have been drawn to show more detailed information about the collection?

d. Are there any ways in which the graph could have been made more helpful to a reader who wanted to quickly scan it and not read the accompanying article?

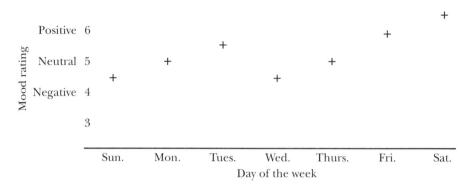

Figure 3.24 Retrospective accounting of pleasantness of week (Exercise 3.33) *(Source: Jessica McFarlane, Carol Lynn Martin, and Tannis MacBeth Williams, "Mood fluctuations, women versus men and menstrual versus other cycles,"* Psychology of Women Quarterly, *vol. 12 (1988), p. 214.)*

3.33 Figure 3.24 is a time series graph that charts a subject's mood ratings for a week as the subject remembered them later.

a. According to the data in the graph, what were the best and worst days in the week for the subject, in terms of mood?

b. How are the differences among the mood levels for different days in the week emphasized in the graph?

c. What would be the effect of including 0–3 on the mood rating axis?

d. Why do you think 5 on a 7-point scale is called neutral instead of 4, which is the middle of a 7-point scale?

e. The authors of the article from which this graph is taken superimposed another time series graph over it showing how the subject rated mood at the time each day in the week. (Thus, the time series graph can be used to present two different variables over the same time period.) The line for the concurrent report of moods was flatter than the retrospective one. That is, the concurrent report indicated that weekends were not so wonderful and Sundays and Wednesdays were not so bad. Why do you think the retrospective report differed from the concurrent one?

ANALYSIS (EXERCISES 3.34–3.54)

3.34 A sample of socioeconomic scale scores follows:

$$42\ 35\ 48\ 26\ 52\ 47$$
$$29\ 65\ 42\ 51\ 47\ 35$$

a. What could we expect to learn about this sample from a histogram of these data?

b. Use an appropriately small number of intervals and make a histogram of the data.

c. What do you conclude about the variable on the basis of what the histogram shows?

d. What information is missing about these data that would help you to make better sense of it?

3.35 How people choose to spend their time can reveal a great deal about our society. In an extensive time-use study, it was found that during weekdays employed men spend 8.1 hours on work-related activities, 1.0 hours doing housework, 9.9 hours on personal care such as eating, sleeping, and grooming, 1.2 hours traveling, and 3.8 hours in free-time activities such as sports and television viewing. For employed women the corresponding figures were 6.5, 3.4, 9.8, 1.1, and 3.2 hours, and for homemakers the figures were 0.0, 7.8, 10.3, 0.7, and 5.2 hours. These values are all means. *(Source: J. P. Robinson,* How Americans Use Time: A Social-Psychological Analysis of Everyday Behavior, *New York: Praeger, 1977, p. 90.)*

a. Display these data graphically. Use two different types of graphs and discuss which type displays the data better.

b. What information do your graphs convey about the time use of each of the three groups of people?

3.36 The signers of the Declaration of Independence were a select group of people, and we are interested in whether they lived longer than the average man did in this period. For example, George Wythe was 50 years old when he signed. Having reached 50, Wythe could have been expected to live another 21 years, but he lived 30 more years. He therefore lived 9 years longer than he was expected to. The differences for all the signers follow:

24	−3	−24	2	−4	−19	21	16	−4	7	−11	−1
8	9	−6	−14	−6	2	−4	−18	14	−8	13	1
−4	22	−9	−1	13	−14	−6	1	−16	−1	−1	9
−4	19	−6	−12	−13	−1	13	4	−3	13	−14	
29	4	−9	−4	−6	−12	−13	−19	−14	−19	11	
7	9	−19	21	−9	−4	−28	−14	−21	−18	−7	

A negative difference means that the person fell short of living his life expectancy at his age of signing by that many years. *(Source: U.S. Bureau of the Census, Bicentennial Statistics. Quoted in Pocket Data Book, USA 1976, Washington, DC: U.S. Government Printing Office, 1976, p. 370.)*

a. Make a table and a histogram showing the distribution of this variable.

b. What do you conclude about the longevity of the signers on the basis of the shape of the histogram?

3.37 Down syndrome is a genetic disorder that shows up in some newborn babies. The following datafile shows the age of the mothers of Down syndrome children at birth of the children and the number of children with Down syndrome born in Sweden in 1971.

Age of mother	15–19	20–24	25–29	30–34	35–39	40–44	45–49
Number of babies	18	87	96	72	73	73	19

Source: E. B. Hook, and A. Lindsjö, "Down syndrome in live births by single-year maternal age interval in a Swedish Study," American Journal of Human Genetics, vol. 30 (1978), pp. 10–27, as reported in C. J. Geyer, "Constrained maximum likelihood exemplified by isotonic convex logistic regression," Journal of the American Statistical Association, vol. 86 (1991), pp. 717–724.

a. Make a histogram of the data.

b. What information does the histogram convey about the distribution of the number of babies with Down's syndrome across the ages of the mothers?

c. The fewest number of babies with Down syndrome are born when mothers are very young or very old. Does that mean that a woman should have babies when she is either very young or very old?

d. What additional data, if any, do you need to recommend the age at which a woman ought to have children to lower the risk of Down syndrome?

3.38 The calorie values of sixteen different snack foods are as follows:

110 120 120 164 430 192 175 236
429 318 249 281 160 147 210 120

Source: USDA data and manufacturer's data in an advertisement in The New York Times Magazine, *April 20, 1990, p. 20.*

a. Make a histogram of the data, using 50 as the width of each bar.

b. Make a stemplot of the data, using two digits on the left side of the line.

c. Make a boxplot of the data.

d. What are some of the strengths and weaknesses of each graph?

e. Which graph do you prefer? Give your reasons.

3.39 The band uniform hat is being changed at the high school. The band leader has used a tape measure to collect the head circumference of each of the 150 band members. The band president has asked everyone to order a hat in size small, medium, or large. The hat store sells hats in ten sizes (from $6\frac{7}{8}$ to $8\frac{1}{4}$).

a. Describe in general how you would organize the data for maximal effectiveness in buying the hats.

b. What errors have the band leader and the band president made in measuring the band members for hats?

3.40 Select a question you would like to have answered. Example: What were the most popular CDs being sold in the music stores last week: country/western, rap, rock, heavy metal, ballads?

a. Find data to answer your question.

b. Create a graph that best illustrates your data.

3.41 Draw a stemplot that illustrates the amount of money you spent for each of 20 items you purchased in the last month.

3.42 a. Draw a histogram representing your estimate of the average number of days each month of the year the temperature drops below the freezing point (use either 32 °F or 0 °C) in your present community.

b. Describe the shape of the histogram.

3.43 Write down the names of twenty friends and relatives, along with the number of years you have known each one and the number of intense arguments you have had with each one.

a. Make a scatterplot showing the data on the two variables.

b. Does there seem to be any relationship between the two variables in your sample?

c. What problems do you see in this analysis?

d. What type of data might be better for studying the variables?

3.44 a. Draw a bar graph with horizontal bars going to the left for men and to the right for women using the following information. The median incomes in a sample of full-time workers with four or more years of college are as follows: White men earned $42,000; white women earned $29,000; Asian American men earned $37,000; Asian American women earned $29,000. *(Source: U.S. Census data for 1990.)* You may drop the 000s for the purpose of the graph.

b. What major conclusions can you draw from your graph?

3.45 According to a survey of 60,000 households by the U.S. Bureau of the Census in October 1990, the age distribution in percentages of 5,644,000 full-time college students in four-year colleges that year was as follows:

Age	15–17	18–19	20–21	22–24	25–29	30–34	35–39	40–44	45–59
%	1.9	34.7	34.1	16.6	6.4	2.7	1.8	1.2	0.6

Source: The Chronicle of Higher Education, vol. XXXIX, no. 1 (August 26, 1992), p. 11.

a. Make a histogram of this age distribution. Note that the intervals will be of different lengths. This means you will have to adjust the heights of the rectangles so that the area of each rectangle shows the magnitude of the corresponding percentage. Also, for any age group the age starts at the lower limit and goes up to but does not include the lower limit of the next older group. For example, 20–21 years means an interval that starts at 20 and goes up to 22.

b. What does the shape of the histogram convey about the age distribution of college students?

c. Does the shape of the distribution surprise you? Why or why not?

3.46 Swordfish absorb mercury in their bodies, and it is thought that a mercury concentration of more than 1.00 ppm (parts per million) is not good for human consumption. In a sample of 28 swordfish the following concentrations of mercury in ppm were found:

0.07 0.24 0.39 0.54 0.61 0.72 0.81 0.82 0.84 0.91
0.95 0.98 1.02 1.08 1.14 1.20 1.20 1.26 1.29 1.31
1.37 1.40 1.44 1.58 1.62 1.68 1.85 2.10

Source: Larry Lee and R. G. Krutchkoff, "Mean and variance of partially-truncated distributions," Biometrics, *vol. 36 (1980), pp. 531–536.*

a. Make a stemplot of the data, using the first two digits as the stem.

b. Describe the shape of the distribution of mercury concentration.

c. The reason many of the swordfish were found to have a concentration of more than 1.00 ppm is that not all swordfish are tested before they are brought to the market. Does it seem as if the overall, average level of mercury concentration is larger than 1.00?

3.47 Most people seem to drive above the speed limit on interstate highways. The chances of getting a speeding ticket may therefore mainly depend on how many police officers are out on patrol. The states differ in their numbers of state police officers, and one way to measure police coverage is to look at the number of miles of interstate highways per police officer in each state. This number ranges from a low of 0.1 mile per officer in Delaware to 7.0 miles per officer in Wyoming. Thus, if every officer in Delaware were out on the road, there would be one officer every tenth of a mile, while in Wyoming there would be one officer every seven miles.

Figure 3.25 is a stemplot for the 48 continental states showing the number of miles per state police officer, with miles to the left of the stem and tenths of miles to the right.

    a. Describe the shape of the distribution shown in the stemplot.

    b. Construct a boxplot from the stemplot.

    c. What are the advantages and disadvantages of the stemplot compared to the boxplot for these data?

3.48 Discuss how the graph in Figure 3.26 on rescue missions in national parks meets Tufte's criteria for graphical excellence.

3.49 When criminal attacks are committed, in the great majority of cases the attacker and the victim are of the same race. In 1991, according to the FBI, 85% of black victims were attacked by blacks, 75% of white victims were attacked by whites, 8% of black victims were attacked by whites, 17% of white victims were attacked by blacks, and the rest were the result of other combinations of victims and attackers.

    a. Draw two pie charts, one for black victims and one for white victims.

    b. Imagine a pie chart that combines these data. In what respect might a single pie chart be somewhat misleading (that is, what would a single pie chart assume)?

3.50 Create a data set for yourself that could be displayed to good advantage in a pie chart.

    a. Draw the chart.

    b. Summarize the findings presented in the chart.

    c. Are there any problems with the pie chart as you have designed it?

3.51 Create a data set for yourself that could be displayed to good advantage in a stemplot.

    a. Draw the stemplot.

    b. Summarize the findings presented in the stemplot.

    c. Are there any problems with the stemplot as you have designed it?

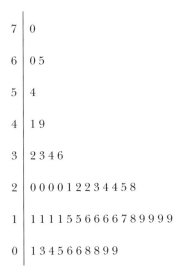

Figure 3.25 Miles per state police officer, 48 contiguous states (Exercise 3.47) *(Source:* Autoweek, *July 9, 1990, p. 37.)*

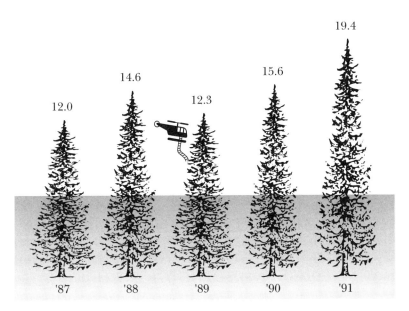

Figure 3.26 Search and rescue operations per million visits to the 367 national park system areas (Exercise 3.48) *(Source: Data of National Park Service; adapted from the graph in* The New York Times, *March 25, 1993, p. A18.)*

Table 3.6 Data for Exercise 3.53

Country	Total deaths	Type of event			
		Transportation	Natural	Homicide	Other
Austria	75.2	34.8	29.7	1.6	9.1
France	77.8	23.8	31.0	0.9	22.1
Italy	47.2	22.8	19.2	1.1	4.1
Netherlands	40.3	17.8	18.2	0.7	3.6
Norway	48.4	17.3	25.1	0.7	5.3
United States	60.6	23.4	15.8	10.0	11.4

Source: Social Indicators III, *U.S. Census Bureau, December 1980, p.252, reprinted in Howard Wainer, "Tabular Presentation,"* Chance, *vol. 6 (1993), no. 3, p. 53.*

3.52 As in Exercise 1.20, turn on a water faucet until it just drips. Count the number of drips per 20-second interval for 3 minutes. Keep a record of the number of drips in each interval.

    a. Draw a graph illustrating your data.

    b. Would you say the drips were randomly or regularly distributed over the 3-minute period? In what respects were they random, and in what respects were they regular?

3.53 Table 3.6 is an abbreviated version of a table of death rates from various causes in selected countries in the mid-1970s per 100,000 population. Redo this table so that it is more readable, and justify your changes.

3.54 a. Make a graph of your choice for the following data. When pollsters interviewed 1,000 adults employed by private sector companies about issues of privacy, 61% said their employers respected after-hours privacy "very well," 29% said "somewhat well," 8% "not very well," and 3% "not well at all."

    b. Make a graph of your choice for the following data: Respondents believed employers had the right to verify information provided by job applicants to various degrees. Eight in 10 thought it appropriate for employers to check on a job applicant's claims regarding educational background and criminal record; tests for nicotine use away from work were opposed by 93%; 69% objected to urine tests for alcohol use; 69% thought psychological tests for attitudes and social preferences were inappropriate; 59% opposed using blood samples to test for AIDs virus. *(Source: "U.S. workers are concerned about privacy on the job, survey finds," The Philadelphia Inquirer, August 23, 1994, p. F6.)*

CHAPTER 4

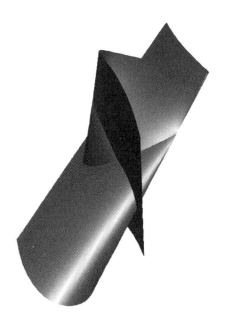

4.1 Averages: Let us count the ways

4.2 Variety: Measuring the spice of life

4.3 Standard error of the means

4.4 Standard scores: Comparing apples and oranges

4.5 Gain in simplicity, loss of information

4.6 Real estate data: Out-of-sight prices

4.7 Summary

DESCRIPTION OF DATA: COMPUTING SUMMARY STATISTICS

Did Shakespeare write Shakespeare? How many children does the average American family have? Do men make more money than women? How many times per minute does the "normal" heart beat? How old are brides and grooms at the time they marry? How many men die of cirrhosis per year? Is it feasible to raise taxes in a community to provide more money for schools?

As noted in Chapter 2, the original observations in a data file contain all the information there is in a set of data, but it is almost impossible simply to look at a data file and extract the information. All the information is there, but it is hidden by the randomness in the data.

Chapter 3 discussed the use of graphs and tables to organize data. Graphs and tables often need to be augmented by summary statistics—new numbers computed from the data. From the values on one or several variables we can compute a few new numbers that represent the variables, thus summarizing the data into just a few values.

This chapter is concerned with two problems.

1. How to summarize many observations of a variable into a single number that gives us a central tendency, or average value. Is it possible to find a single value that illustrates what all the observations are like?

2. How to summarize how different the values of a variable are from one another. Are the observations much alike or are they very different? That is, is there much variation in the data?

As with graphical representations (Chapter 2), computing summary numbers has one major advantage and one major disadvantage:

Advantage: A summary number gives a great simplification of the data.

Disadvantage: Any simplification means a loss of information.

Before the 1960 Presidential election, the Survey Research Center at the Institute for Social Research at The University of Michigan asked people in a survey who they intended to vote for. Of the 1,396 respondents who planned to vote, 634 planned to vote for John Kennedy. Thus, of the total number of respondents who intended to vote, 45% planned to vote for Kennedy (who did win in a very close election later that year). The representation of 634 of 1396 separate answers by a single percentage is an enormous simplification of the original data. At the same time, however, the values of the original variable are irrecoverably lost. If we know only a single number, we cannot recover the original data, and many different data sets could yield the same average value.

As with visual representations of data, summary computations of data need to strike a balance between gain in simplicity and the loss of information. Doing so is not always easy and requires knowledge about the strengths and weaknesses of the most commonly used summary numbers.

4.1 AVERAGES: LET US COUNT THE WAYS

The most common number computed from data is an *average* or *central value* of some kind. Most of us were introduced to the notion of averages in elementary school. Today, we read about the average salary of MBAs, average house prices, Dow Jones average stock prices, average homicide rates, and so on. But how aware are we of the various forms of averaging possible to us, or how simply calculating a particular average can create false impressions?

> An **average** is a single number computed from the observed values of a variable.

There are many kinds of averages, not just one. To explore this variety, take a close look at the following sentence:

> The average person in this country today is a woman who has 2.1 children and lives in a house worth $80,000.

Three common kinds of average are referred to in that sentence. Can you distinguish how the three differ?

Mode: The hostess with the mostes'

The gender variable has two values, man and woman; in this country, there are more women than men. The statement that the average person in this country is a woman uses a statistical average called the *mode*.

The mode is commonly used to describe categorical variables, especially those with many values such as religion, race, or social class. One might find, for example, that within a particular neighborhood the modal religion is Muslim, the modal race is Asian, and the modal social class is "upper middle."

> The **mode** for a variable is the value that occurs most often.

The mode can be used for other types of variables as well. Figure 4.1 shows a histogram of the age at marriage for 37 women, the same women in the graphs in Figure 3.3. The major peak in the histogram in Figure 4.1 occurs in the age range from 25 to 30 years. We take the midpoint in that range, 27.5 years, as the modal value of age at marriage.

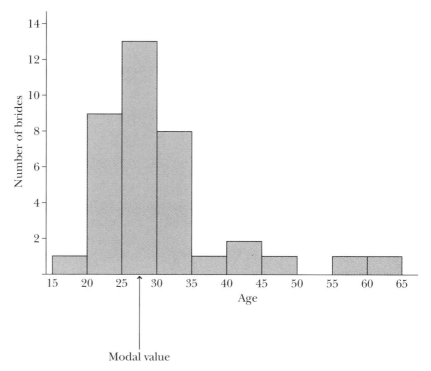

Figure 4.1 Modal value of the age variable: Midpoint of the tallest rectangle

Sometimes a variable has two values that occur most frequently. Thus it has two modes and what is called *bimodal distribution.* When a variable has two modes, the observations of it often consist of a mix of data from two groups of elements. A histogram of the heights of students in a statistics class would be bimodal, for example, when the class contains a mix of men and women.

The mode tells us that there are more of that value than of any of the other values of the variable but not whether there are many more of this value or only a few more. If there are 100 people in a group, the modal gender value would be woman if there were 51 women (and 49 men) or 99 women (and 1 man). The two cases are very different, but the mode does not distinguish between them. Thus, the mode may at times mask more information than it reveals.

For a metric variable the mode does not make use of all the actual, observed values of the variable. In addition, by choosing different widths of the intervals in a histogram, as shown in Chapter 3, different modes can be obtained depending on how the histogram is drawn.

SHAKESPEARE À LA MODE

For centuries scholars doubted that William Shakespeare actually wrote the many wonderful plays and poems attributed to him. Doubt arose in part because Shakespeare, the historical figure, was thought to be too provincial, illiterate, and unknown to have accomplished such remarkable feats. No one during his lifetime referred to him as an author, and he left not one personal letter, manuscript, or literary record when he died. Indeed, these clues lead to well-grounded suspicions that the real William Shakespeare was someone else. Other authors, more famous, better educated, and better documented, were said to have written these classics, among them Francis Bacon, John Donne, Christopher Marlowe, Walter Raleigh, Edmund Spenser, and even Queen Elizabeth I herself.

Some expert statistical sleuthing has helped to give Shakespeare his due. The method used depended on the application of the mode.

For three years, a Shakespeare Clinic composed of Claremont College undergraduates used statistical analyses to see which of 58 contemporary authors had writing styles that most closely matched that of works attributed to Shakespeare. Blocks of the 58 authors' writing were selected and divided into 500-word passages. Counts of several variables were made in the blocks. For example, the students explored the occurrences of 52 key words, looking for the modal uses for each of the authors. Using various statistical strategies, they created profiles of authors. When all was finished,

> None of the tested poems of 27 claimants passed the modal test. Thomas Heywood, the closest author test, was 2.2 standard errors distant from Shakespeare. By this measure, . . . were Heywood actually Shakespeare, poems as different as the ones he wrote under his own name would have occurred less than 5% of the time. John Donne, the most distant claimant tested, was 36.9 standard errors distant from Shakespeare.

Pretty strong evidence that Shakespeare wrote Shakespeare's verse! *(Source: Ward Elliot and Robert Valenza, "Who Was Shakespeare?" Chance, vol. 4 (1991), pp. 8–14.)*

Advantages of the mode: The mode for a variable is easy to find from a graph or a table of the data. For categorical variables, it is typically the best way to describe the average value. For a variable with a bimodal distribution and few observations in the middle of the range, the two modes tell us more than a single value located somewhere in the middle of the variable where there are not very many observations. The mode requires very little actual computation, since the value can be seen directly from a bar graph.

Disadvantages of the mode: The mode is not used often, and many statistical computer software programs do not even calculate the mode. The modal value of a variable does not convey much about the entire data set, and conversely, the information in the data set is not well used by finding only the mode.

> **STOP AND PONDER 4.1**
>
> Think about any job you have had in which you were paid weekly. Which day of the week were you paid? If you created a bar graph of the days in the week for 52 weeks of pay, which would be the modal day on which you were paid? Is this why TGIF is such a popular phrase?

Median: Counting to the middle

Our "average" woman lives in a house valued at $80,000. "Average" house prices and many other economic variables are most often described by the *median* value. Because price is a metric variable and has higher and lower values, unlike a categorical value such as religion, the values can be ranked from lowest to highest. The middle value of the variable in the ranking, is the median value. When the median house price is $80,000, half the houses in the data cost less than this value and the other half cost more than this value.

The median is found by arranging the observations by size, from smallest to the largest, then counting halfway through them to the middle. For a very small data set it makes a difference whether the total number of observations is even or odd, but for larger data sets this distinction is not important. It is also possible to find the median from data in a table or a histogram that does not show the original observations.

*The **median** value of a variable is the value of the variable that divides the observations into two equal groups so that half the observations are smaller than the median and half are larger than the median.*

It does not matter for the median how much more or how much less the other houses in the example cost. For example, suppose the data set includes two other houses besides the one that costs $80,000— a house that costs $79,000 and one that costs $500,000. The median price for the three houses is $80,000, even though one house costs only $1,000 less than the median and the other costs $420,000 more. All that matters for the median is that it is the middle number of the data set.

Finding the median   Imagine a family with five children, ages 17, 14, 12, 9, and 5 years. The data set is an odd number of observations, already arranged by size. The middle number is the third observation; two observations are less than and two observations larger than this value. Thus, the median age of the five children is 12 years. (Formula 4.1 at the end of the chapter shows how to find the median in an odd number of observations.)

Suppose this family has twins who are 5 years old, for a total of six children, ages 17, 14, 12, 9, 5 and 5 years. This data set contains no actual observed value that divides the data into two equal parts. But for any age between 9 and 12 years, three children are older and three children are younger. By convention, the midpoint between the two middle values is the median. The midpoint between 12 and 9 is 10.5, and that is the median age of the six children. (Formula 4.2 shows how to find the median in an even number of observations.)

Median and other percentiles   The median is also known as the 50th percentile, since 50% of the observations are smaller than the median. The 25th percentile is the value of a variable such that 25% of our observations are smaller than this value. In the example on the brides' ages, the 25th percentile equals 24 years, the 50th percentile or median equals 27 years, and the 75th percentile equals 32 years.

STOP AND PONDER 4.2

You have collected a group of recipes for chicken curry. The number of ingredients varies from one recipe to another: 12, 16, 8, 9, 15, 10, 11, 14, 20, 12, 18. Determine the median number of ingredients in the typical chicken curry recipe. Is there a modal number? Which dish would you like to make?

Median from a stemplot Medians are particularly simple to find from data arranged in a stemplot. In the stemplot, the values of the variable are already arranged by size from the smallest to the largest, and the median or any other percentile is found by simply counting up to the desired number. Refer to the stemplot of the brides' ages in Figure 3.3. With 37 observations, the median age is the age of the 19th bride when the ages are arranged from youngest to oldest; 18 brides are younger and 18 are older. Starting from the bottom and counting from youngest to oldest, the 19th bride is 27 years old; 27 is the median age of this group.

Median from a histogram Can the median be found from a histogram, without the original observations? The answer is a modified yes, if we are willing to assume that the observations are evenly distributed within the middle interval. In the histogram in Figure 4.1, the total number of brides can be computed by adding the values represented by the height of the rectangles: 37 brides were observed. So the median age is the age of the 19th-oldest bride. The histogram shows that 10 brides are included in the first two age intervals; nine more are needed. Since the next interval (25–30) includes 13 brides, the median must lie within that interval. Assuming that the 13 brides in that interval are evenly distributed across the ages, the 9th bride is found by going 9/13 into the interval. The interval is 5 years wide; 9/13 of that is $5(9/13) = 3.5$. Adding 3.5 to the lower value of 25, the median is $25.0 + 3.5 = 28.5$.

This median can also be shown in a histogram. Figure 4.2 repeats the histogram in Figure 4.1 on the ages of the brides. Chapter 3 emphasized that the *area* of a rectangle indicates how many observations the bar represents. So to find the median, the total area shown in the histogram is divided into two equal parts. The dashed line in Figure 4.2 is the median: the area in the histogram to the left of the dotted line represents 18.5 units, and the area to the right also represents 18.5 units. The value of the variable at the dashed line—the median value—is 28.5 years. The estimated value of the median at 28.5 years is not quite the true median of 27 years because of our assumption that the observations in the middle interval are distributed evenly; the brides in the range from 25 to 30 years are not, in fact, distributed evenly within the interval.

Use of the median The median is used most often when a histogram of the data shows a skewed distribution. House prices typically show a

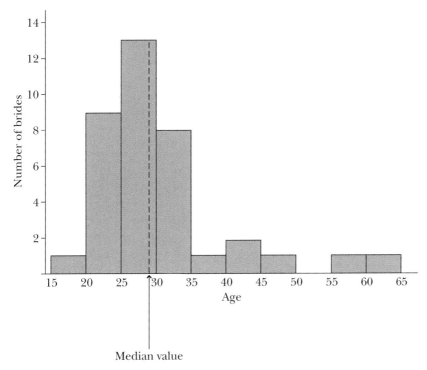

Figure 4.2 Median value of the age variable: Midpoint of the total area of the bars of the histogram.

skewed distribution. Most house prices fall in a middle range, but usually a few houses are expensive. Thus, the histogram has a "tail" on the right side.

The histogram in Figure 4.2 shows that the two oldest brides skew the data. The median is useful for these data because the median is not much affected by a few extreme scores. The median would be the same value whether the two oldest brides were 30, 40, 50, or 60 years old.

Advantages of the median: The median gives a good indication of the midpoint of a set of observations, particularly if the histogram shows a skewed distribution. The median requires little computation. The observations need only be ranked from the smallest to the largest, and the median is simply found by counting up to the middle observation. The median is not sensitive to extreme observations, and this can be an advantage for certain purposes.

Disadvantages of the median: Aside from the middle value, the median does not make use of the actual values of the other observations. Thus, it does not make use of all the information in the data. The median is not sensitive to extreme observations, and this can be a disadvantage for certain purposes.

Mean: Balancing the seesaw

When we say that the average American family has 2.1 children, we are saying that the *mean* number of children per family in the United States equals 2.1. The mean is the most commonly used type of average. Just like the median, the mean gives a value of the variable somewhere in the middle of the observed data. The difference is that the mean is a value of the variable that can be viewed as the center of gravity of the data. If we placed the observations on a seesaw according to their values, the seesaw would balance right at the mean. For the age data on the 37 brides, the mean works out to be 30.0 years (Figure 4.3). If we imagine that each bride weighs the same and stands on a horizontal seesaw at her particular value, the seesaw will balance at 30.0.

To find the value of the mean, the values of all the observations are added and then the sum is divided by the number of observations. This statement is written mathematically in Formula 4.3 at the end of the chapter. Finding the mean according to Formula 4.3 is the same as finding the value where the seesaw will balance, that is, the center of gravity of the distribution of the data.

The mean is commonly used for metric variables to find a central value of the observations in a set of data. As with other averages, a good deal of information is lost when the original data are replaced with the mean, but the exact value of each observation is used to find the mean. If any data points are changed, then the mean changes. This is not necessarily the case with either the median or the mode.

Figure 4.3 illustrates an important weakness of the mean. The two oldest brides have a large effect on the mean because they are so far away from the mean. If they step off the seesaw, the balance point moves from 30.0 years down to 28.4 years. Those two brides cause the mean to be larger than the median for these data. If two of the brides close to the mean age step off the seesaw, the effect on the balance point is very small.

Because the mean is so sensitive to isolated extreme observations, we prefer not to use the mean when a data set has extreme observations. The mean works well with roughly equal numbers of small ob-

> The **mean** is the value of the variable obtained when the values of all the observations are added and the sum is divided by the number of observations.

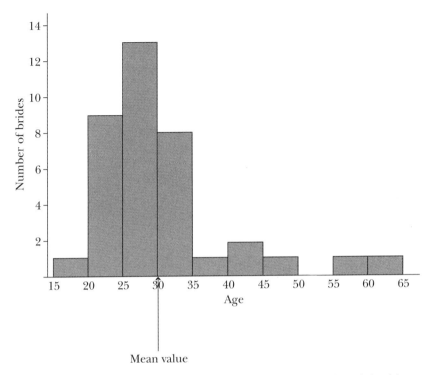

Figure 4.3 Mean value of the age variable: Center of gravity of the histogram

servations and large observations; the small observations balance the large observations. When the distribution of the data is skewed, as in Figure 4.3, we prefer to use the median to the mean because the median is not sensitive to extreme observations. To decide between the mean and the median for a set of data, first compute both. If they are

Computing the mean is not always easy. *"Calvin and Hobbes" copyright 1995 Watterson. Dist. by Universal Press Syndicate. Reprinted with permission. All rights reserved.*

approximately equal to each other, then use the mean. If they are very different, use the median.

The mean is the most frequently used average value, and because it is so common, it has its own symbol: $\bar{x}$, read as "x bar." An x is used because it is the most common symbol for a variable, and the bar symbolizes the mean. The mean for a variable denoted by some other letter, for example by y, is denoted similarly: $\bar{y}$.

Advantages of the mean: The strength of the mean is that it uses the numerical value of each observation, implying that it uses more of the information in the data than do the mode or the median. As shown in later chapters, conclusions about data can more easily be drawn from the mean than from either the mode or the median.

Disadvantages of the mean: Since the mean makes use of the actual, numerical value of each observation, it can be cumbersome to compute. The mean is sensitive to extreme observations. This can be particularly bad if there is an error in the measurement of an observation and that is the reason the observation is extreme in the first place.

Mode, Median, or Mean?

We should get into the habit of asking ourselves which kind of average is being used in a data analysis and whether it is the right kind. Occasionally people use the wrong type of average on purpose to create an impression from the data that may not be fully warranted. When a distribution is skewed with many small observations and only a few large observations (the distribution of household income is an example), then the mean will be larger than the median. Anyone who wanted to summarize this distribution with as large a value as possible would then use the mean, even though the median would be a more appropriate choice of average.

This kind of twisting is particularly tempting in comparing two or more groups. Suppose a headline says that men make more money than women. What is the implication of such a sentence? That every man makes more money than all the women? Of course not. The headline probably comes from a comparison of averages for the two groups. If so, then it ought to say so. Maybe the median income is higher for men than women; or maybe the mean income for men is higher. Thus, how groups differ is related to the *particular statistical method* used to compare them.

> **STOP AND PONDER 4.3**
>
> Imagine that you are working for the Department of Transportation in your state and you wish to inform the governor of the average amount of federal aid the state has received for 26 highway projects. One new highway received the largest amount of aid ($22 million), while the other 25 projects received between $200,000 and $1 million each. The median amount was $250,000, the mean was $1,000,000, and the mode was $200,000. How would you choose to represent the average amount of money received by the state per highway project? What would be the drawbacks of any average you compute?

4.2 VARIETY: MEASURING THE SPICE OF LIFE

Usually, an average is a useful way of summarizing data, but sometimes an average can be misleading. There is an old joke about the statistician who puts her head into the oven and her feet into the refrigerator and says, "On average, I feel just fine." In the computation of the statistician's "average," two extreme temperature values, the heat in the oven and the coolness in the refrigerator, cancel out to produce a comfortable average temperature. Thus, any average masks the extreme values in a set of data, and extreme values are sometimes of particular interest. The mean household income in a community may be a comfortable $100,000 a year, but if this mean is computed from the incomes of 200 extremely poor families and 20 extremely rich families, it does not represent the incomes of any of them. Sometimes we need to go beyond averages to summarize data.

Imagine two different data sets that have the same average value but are still very different. In one data set, the observations are all close to each other, while in the other data set the observations are spread out. No average—mode, mean or median—would catch this crucial difference. In this case, the spread of the data needs to be taken into account.

Range: Lassoing the two extreme values

One easy way to measure the spread of a set of data is to find the *range*, which is simply the difference between the values of the largest and the smallest observations:

$$\text{range} = \text{value of largest observation} - \text{value of smallest observation}$$

For the data on the brides' ages,

$$\text{range} = 60 \text{ years} - 19 \text{ years} = 41 \text{ years}$$

> The **range** is the difference between the value of the largest observation and the value of the smallest observation.

The range is easy to compute, and it can often be a very useful number to know. An average value and the range of a data set tell us quite a lot about the values of the observed variable. This is particularly true if the data include no extreme observations. One drawback is that the range is sensitive to extreme observations. If the two largest observation, 56 and 60 years, are dropped, then the largest observation is 46 years and the new range is 46 − 19 = 27. A mere 2 observations out of 37 added 50% to the range! Dropping some of the extreme observations and finding the range of the remaining values is indeed a statistical strategy—as long as the number of observations to drop is agreed on.

STOP AND PONDER 4.4

College handbooks that list the characteristics of various schools often list the interquartile ranges of the incoming classes' SAT scores for verbal and mathematical tests. At Swarthmore College, for example, for a recent entering class the interquartile range was 690 − 580 = 110 for the verbal test and 720 − 630 = 90 for the mathematics test. (This range is helpful in showing not only that the median score is quite high but that most students at the school had scores of 580 or better on their verbal SATs, 630 or better on the math test.) At a less selective private university nearby, the median score was 530 verbal and 597 mathematics, the interquartile range was 579 − 483 = 96 for verbal and 653 − 552 = 101 for math. Where would a median score in the verbal SAT at the university place you in terms of the Swarthmore SAT quartile divisions?

When the smallest 25% and the largest 25% of the data are dropped, the range is the middle half of the observations. This is the so-called *interquartile range*. Both the full range and the interquartile range are well illustrated in a boxplot. In Figure 3.3, a boxplot for the ages of the brides, the interquartile range covers the length of the box, so the interquartile range is 32 years − 24 years = 8 years.

Standard deviation: The crucial deviant

The standard deviation is the most commonly used statistic designed to show how the observations on a variable differ from one another. The *standard deviation* describes how far away the observations are from the mean (Figure 4.4). The farther from the mean, and therefore from each other, the observations are, the larger is the standard deviation. If we know, for example, that the standard deviation equals 6.9 heartbeats, then we know that a typical observation lies 6.9 beats away from the mean and that it is either larger or smaller than the mean. The smallest value for the standard deviation is 0.00, the value for a set of observations that are all alike. But no variation in the data is rare indeed. More common are distributions that are somewhat dispersed. There is no limit to how large the standard deviation can be.

> The **standard deviation** is an average distance from the mean of the observations in a data set.

The standard deviation is typically denoted by the letter *s*. Standard deviation is a somewhat strange name; how can something at the same time be both a standard and a deviation? The name will become clearer as you learn more about the standard deviation.

Comparing data spreads   Histograms for the four following data sets are shown in Figure 4.5.

(a) 6 6 6 6 6 6
(b) 5 5 6 6 6 7 7
(c) 3 3 4 6 8 9 9
(d) 3 3 3 6 9 9 9

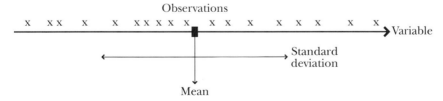

Figure 4.4   Mean as center and standard deviation as spread in data

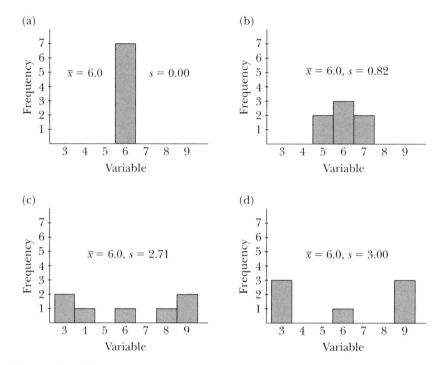

Figure 4.5 Histograms of four data sets with the same mean and different standard deviations

The histograms show that the farther the data are from the mean $\bar{x}$, the larger is the standard deviation s.

> **STOP AND PONDER 4.5**
> Think of some instances from ordinary life where the standard deviation for a set of observations would be quite small. Consider other instances where the standard deviation for a set of observations would be quite large. (Your mind may wander as far as the race track or the grocery store.)

In Figure 4.5a, all the observations are equal to the common value of 6, and since they are all equal, the standard deviation for those numbers is 0.00. In Figure 4.5b, the observations are somewhat spaced out between 5 and 7, and the standard deviation increases to 0.82. In Figure 4.5c, the observations are spaced out farther, and the standard deviation is 2.71. In Figure 4.5d, most of the observations are located at two extremes, and the standard deviation is 3.00. The four data sets all have the same mean, and if we knew only the mean, then we could not tell the four examples apart. But there are differences between the data sets in the spread of the observations around the mean and the variation in the ranges of the data; thus, they have different standard deviations.

Average distance to the mean: Dissecting the numbers   The standard deviation s is found by computing the square root of the average squared deviation of the observations from the mean. To see how it works, let us apply the computation one step at a time to the data set for Figure 4.4b—a variable with observations 5, 5, 6, 6, 6, 7, and 7. How do we arrive at a standard deviation of 0.82 for these data? As we know, the mean of the observations equals 6. The definition asks for the deviations of the observations from the mean. The deviation (distance) from the first observation to the mean is $5 - 6 = -1$, the second is $5 - 6 = -1$, the third is $6 - 6 = 0$, the fourth is $6 - 6 = 0$, the fifth is $6 - 6 = 0$, the sixth is $7 - 6 = 1$, and the seventh is $7 - 6 = 1$. These are the numbers in the second column of Table 4.1.

Next we need the squares of the deviations, and they are shown in the third column of the table: 1, 1, 0, 0, 0, 1, and 1. Their sum equals 4. We then average the sum and arrive at 0.67. Finally, the standard

Table 4.1   Computation of the standard deviation s as the square root of the average squared deviation from the mean

Observation	Deviation from mean	Deviation squared
5	$5 - 6 = -1$	$(-1)^2 = 1$
5	$5 - 6 = -1$	$(-1)^2 = 1$
6	$6 - 6 = 0$	$0^2 = 0$
6	$6 - 6 = 0$	$0^2 = 0$
6	$6 - 6 = 0$	$0^2 = 0$
7	$7 - 6 = 1$	$1^2 = 1$
7	$7 - 6 = 1$	$1^2 = 1$
Sum	0	4
Average	0	$\frac{4}{6} = 0.67$
Square root		$s = \sqrt{0.67} = 0.82$

deviation s equals the square root of this average, or 0.82. Formulas 4.4 and 4.5 at the end of the chapter show the computation of the standard deviation.

The deviations from the mean in the second column of Table 4.1 range in value from -1 to 1. The standard deviation $s = 0.82$ is somewhere in the middle of these distance. It is smaller than the largest deviation of 1, and it is larger than the smallest deviation of 0. Thus, a value of 0.82 does not seem unreasonable for an average deviation.

The reason we first square the distances to the mean is to get rid of the minus signs. The unit of the squares is the square of the unit of the original observations. For example, if the original numbers were dollars, then the squares would have the unit (dollar)^2. (Dollar)^2 would also be the unit of the average square 0.67. But it is hard to interpret such a number: What is a square dollar? By taking the square root at the end, the unit is restored to its original form.

We have not addressed the minor fact that while there are 7 observations in the figure and table, we divide by 6 to get the average square. This is not a mistake. It is simply better to divide by one less than the number of observations than it is to divide by the number of

MEAN ABSOLUTE DEVIATION: THE LOW-CAL CHOICE

The mean of the original deviations in Table 4.1 is 0 because the sum of the deviations equals 0; negative deviations always cancel out positive deviations. Thus, the mean of the original deviations does not tell us anything. You can check this out yourself by looking at any example on the previous pages. We could, perhaps, take the absolute values of all the deviations and find the mean of those numbers. That way we also get rid of the minus sign for some of the deviations. The sum of the absolute values is 4, and the mean of the absolute values is $4/7 = 0.57$. The mean absolute deviation is typically less than the standard deviation.

Because the standard deviation is useful for estimation and hypothesis testing, statisticians usually prefer to use the standard deviation instead of the mean deviation. However, as a quick, low-calculation alternative to the standard deviation, the mean absolute deviation is fine. It is an easy "guesstimate" of the differences among the scores.

observations itself. This issue is discussed in more detail in Chapter 7 on using s^2 for estimation.

Most of the time, we stay away from lengthy interpretations and simply think of the standard deviation as a number that conveys how different, on average, a set of observations are from each other. If the standard deviation is small, then the observations are much alike. If the standard deviation is large, then the observations are different from each other.

Subtracting and adding standard deviations to the mean The standard deviation can be put to use for another interesting interpretation. Figure 4.6 shows a histogram for 27 values of the number of human heartbeats per 30 seconds. The mean pulse rate equals 34.4 heartbeats and the standard deviation equals 6.9 heartbeats. As expected, the mean value falls in the middle of the histogram, since this is the value where the histogram would balance.

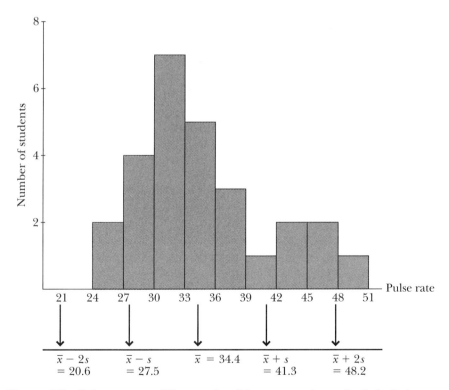

Figure 4.6 Pulse rate per 30 seconds, with mean and standard deviation ($\bar{x} = 34.4$) *(Source: Data collected from Students in Statistics 1: Statistical Thinking, Swarthmore College, spring 1995.)*

The standard deviation added to the mean is 34.4 + 6.9 = 41.3, and this value is shown on the line below the histogram. The line also shows the mean plus two standard deviations, or 34.4 + 2(6.9) = 48.2. Similarly, the mean minus one standard deviation is 34.4 − 6.9 = 27.5, and the mean minus two standard deviations is 34.4 − 2(6.9) = 20.6. The graph shows that the interval from the mean minus two standard deviations to the mean plus two standard deviations, in this case from 20.6 to 48.2, contains almost all the data. Only 1 of the 27 observations lies outside this range. The mean plus and minus one standard deviation, here from 27.5 to 41.3, contains about two thirds of the data. For most unimodal and reasonably symmetric distributions, we can expect the same kinds of results. Thus, if we know the values of the mean and the standard deviation, we can almost recreate the histogram. It follows that the range of the observations is often approximately equal to four standard deviations. Finding the range and dividing by 4 results in an estimate of the standard deviation. This little rule can often provide a quick sense of the size of the standard deviation.

STOP AND PONDER 4.6

You know that the length of the ice fishing season on Lake Milles Lacs has ranged between 25 days and 73 days for the past 50 years. Estimate the standard deviation of the days of the ice fishing season. Most years it would be fairly safe to estimate that you would spend how many days in the ice hut if you wanted to fish every day? You may give a range of the possible number of days.

Variance: Squaring the standard deviation For mathematical reasons, statisticians sometimes prefer to use the *variance* instead of the standard deviation as a way of measuring the difference in a set of observations. The variance is the square of the standard deviation, s^2. In the example with the ages of the brides, the standard deviation is 9.0 years, and the square of this number, the variance, is 81.0:

$$s^2 = (9.0 \text{ years})^2 = 81.0 \text{ years}^2$$

> The **variance** is the square of the standard deviation and is a measure of variability.

The variance does not tell us anything more than the standard deviation. Also, the variance is harder to interpret because the unit of the variance is the square of the unit of the variable we are working with; the standard deviation as well as the mean are in the same unit as the variable itself. What are 81 square years?

4.3 STANDARD ERROR OF THE MEAN

One of the major principles underlying statistical analyses is that if we measure something over again, we usually get a different result. In the female age at marriage data, one bride was 19 years old, another bride was 22, and so on. When we look at all the observed values of the variable, we find that most of the observations are different from each other. The standard deviation tells us how different the observations are from each other.

The 37 women in the data on the brides represent one sample, with a mean of 30.0 years. Suppose we selected another random sample of 37 brides and observed their ages. Doing the study over again would yield another value of the mean number of years. Repeating the study many times would yield many different values of the mean. Thus, just as individual observations in a study are usually different, sample means are usually different across different samples.

How different from each other are the means in repeated studies? Are they as different or less different from each other as the individual observations are?

STOP AND PONDER 4.7

Can you answer this question before we answer it for you? What is your guess?

One way to answer the question is to find the standard deviation of all the means. The means are simply a string of numbers, just as the original 37 observations were a string of numbers, so finding the standard deviation of a set of means from different samples for the same variable is almost no different from finding the standard deviation of a set of observations on a variable. The only difference is that to find the standard deviation of the means, we have to first compute the mean in each sample. Thus, sometimes we work with a standard deviation for a set of observations in a sample, and sometimes we work with a standard deviation for a set of numbers that have been computed from the observations in a sample, like a mean. To distinguish between the two kinds of standard deviations, the one for a set of observations is called a *standard deviation* and the one for a set of means a *standard error*. By implication, standard error can also be computed for a set of medians or a set of standard deviations!

> The **standard error of the mean** is the standard deviation of means from many different samples.

> **STOP AND PONDER 4.8**
>
> Why is it not surprising that the standard error of the mean is smaller across large samples than across small samples?

The standard error of the means is smaller than the standard deviations of the observations; that is, the means do not vary as much among themselves as the values of the variable itself do. This is not surprising. A particular sample contains some large and some small observations that tend to cancel out, when we compute the mean, leaving a mean somewhere near the middle. The same thing happens in each sample, so the sample means cannot differ among themselves as much as the values of the variable do. And the larger the samples, the less variation from one sample mean to another, making the standard error of the mean even smaller.

The biggest difference between standard deviation and standard error is that finding a standard deviation requires data from only one sample, while finding a standard error requires data from many samples. However, it is often possible to *estimate* the value of a standard error from the data in just one sample (see Formula 4.6 at the end of the chapter). The standard error of the mean in a large number of samples of 37 brides can be estimated to be equal to 1.5 years. The standard deviation of the variable in the example is 9.0 years. Obviously, the standard error of the mean is considerably smaller than the standard deviation of the observations.

The standard error of the mean is a very useful number. For one thing, two standard errors of the mean equals 3.0 years. Plus and minus two standard errors of the mean gives us an interval of length 6.0 years. If we did have many samples and many sample means, most of those means would therefore lie within an interval that is 6.0 years long.

4.4 STANDARD SCORES: COMPARING APPLES AND ORANGES

Different variables generally have different means and standard deviations. Values of one variable cannot statistically be compared with values of another variable when the means and standard deviations are different. In the age at marriage example, the bride's ages have a mean of 30.0 years and a standard deviation of 9.0 years, while the groom's ages have a mean of 32.4 years and a standard deviation of 11.1 years. In the youngest couple in this group, the groom is 17 years old and the bride is 19.

4.4 Standard Scores: Comparing Apples and Oranges

Which brides and grooms are most typical in today's world of weddings? Statistical techniques help us to identify them. *(Source: Kaluzny/Thatcher, Tony Stone Images; Telegraph Colour Library, FPG International; Bruce Stoddard, FP6 International.)*

How do we compare the two ages for this couple? The groom is obviously younger than the bride, but is he a younger groom than the bride is as a bride? Who is the more statistically unconventional, bride or groom? How does this couple stack up against the other brides and grooms? One handy solution is to change both bride's and groom's ages to a common scale: we convert raw scores to *standard scores* (Formula 4.7). The bride's and groom's ages—raw scores—are changed into scores that tell how far from the mean the raw scores fall, in stan-

> A **standard score** equals the value of an observation minus the mean, and this difference divided by the standard deviation.

dard deviation units. Using standard scores, any value of one variable can be compared with any value of another variable because we know the relative position of any score from the mean.

In converting the age data into standard scores, the goal is to construct a new scale of standard scores to replace the old scale of raw scores. For the bride who is 19 years old, the raw score is 19, the mean of the sample is 30, and the standard deviation is 9.0. The standard score is

$$\frac{19.0 - 30.0}{9.0} = -1.22$$

Similarly, the groom at 17 years old has a standard score of $(17 - 32.4)/11.1 = -1.39$. Through the means and standard deviation, we find that this groom is farther away from the male mean than the bride is from the female mean. The groom's age at marriage is more unusual than his bride's.

Figure 4.7 shows the brides' ages converted to standard scores. In this example, the mean plus one standard deviation equals 39 on the original scale, and the standard score for that value becomes 1.00. The mean plus two standard deviations equals 48 on the age scale, and the standard score for that value becomes 2.00. The mean minus one standard deviation equals 21, and that corresponds to -1.00 on the standard score variable. The mean minus two standard deviations is 12, and that corresponds to -2.00 on the standard score variable.

Most standard scores for any variable range in values from about -2.00 to -2.00. If the standard score for an observation is larger than $+2.00$ or smaller than -2.00, the value of the observation is unusually large or small observation. Unusual values help in drawing conclusions from samples and applying them to the real world from which the samples were drawn. Standard scores are often known as *t*-values.

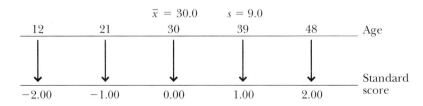

Figure 4.7 Conversion of brides' ages to standard scores

4.5 GAIN IN SIMPLICITY, LOSS OF INFORMATION

Replacing the data with a graph

The purpose of making graphs, creating tables, and computing summary numbers is to understand data better. Each of these techniques simplifies data and brings out patterns that are not directly obvious from the data themselves. At the same time, some of the detail in the original data is lost. We close this chapter with some thoughts on the conflict between gain in simplicity and loss of information.

Look at Figure 4.8. The data in the box are death rates per 100,000 men in 30 different countries for a liver disease called cirrhosis. What can we learn from the 30 values of the variable? Beyond the smallest value, 1.5, and the largest value, 50.1, it is hard to see how the values distribute themselves.

When we replace the data at the left by the histogram at the right, it is much easier to understand the data. The 30 observations have been simplified to six rectangles in a histogram. The histogram shows a unimodal and skewed distribution, with more than half of the data lying between 10 and 30.

At the same time information about the data—the values of the individual observations—has been lost. For example, the histogram shows that one observation occurs somewhere between 50 and 60, but

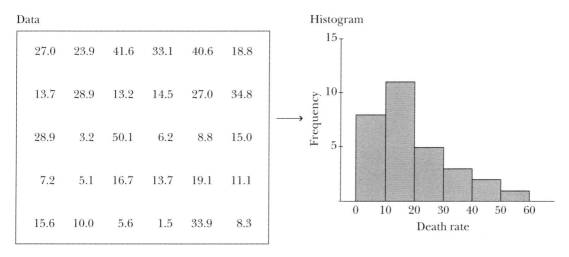

Figure 4.8 Data on cirrhosis deaths per 100,000 men in selected countries simplified to a histogram *(Source: Ann Cronin, "The Tipplers and the Temperate: Drinking Around the World," The New York Times, January 1, 1995, p. E4.)*

it does not specify what that value is. Histograms also destroy the ability to know *when* something occurred. For example, if we collected data on the price of a quart of milk at the end of every month, we would lose the ordering of the observations by putting them into a histogram. The histogram would only show the number of times the price fell in a particular interval.

Replacing the data with a summary value

Figure 4.9 shows the data on cirrhosis death rates reduced to a single number, the mean. The mean death rate of 19.2 conveys an immediate overview of the magnitudes of the death rates; it is the center of the data. It is much easier to understand a single number like 19.2 than it is to comprehend 30 different values of variable. Knowing the mean, we immediately know where the center of the data is located.

Still, a considerable amount of information is lost in reducing 30 observations to the value of the mean, and the original data cannot be reconstructed from the mean. Balancing the loss of information and the gain in simplicity that take place in computing the mean depends on the purposes for the data. The data were collected in the first place because the researchers had certain questions about this disease.

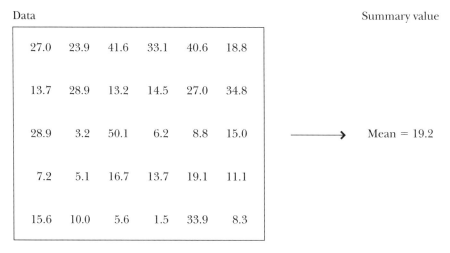

Figure 4.9 Data on cirrhosis deaths per 100,000 men in selected countries simplified to a mean *(Source: Ann Cronin, "The Tipplers and the Temperate: Drinking Around the World,"* The New York Times, *January 1, 1995, p. E4.)*

4.6 REAL ESTATE DATA: OUT-OF-SIGHT PRICES

A school board is looking into raising real estate taxes in order to have more money for the schools the following year. Before taxes can be raised, the board needs information on the taxes paid in their school district as well as in surrounding districts. Board members could ask the town offices to provide them with tax figures for the preceding years and examine page after page of figures giving the property assessments and the taxes for each tax payer. But instead they get from county records summaries of the things they are interested in, such as the average property values, average assessments, and average taxes each unit pays. Table 4.2 shows a small excerpt of the data for their own district.

Table 4.2 Sales prices, tax assessments, and taxes for a few residential sales in Swarthmore, PA, in 1995

Address	Sales price	Assessment	Taxes
520 Cedar	$335,000	$ 6,400	$4,752
326 Cornell	220,000	3,300	2,700
9 Cresson	183,750	6,500	5,260
609 Elm	237,000	6,000	4,620
60 Forest	246,000	6,000	4,456
9 Guernsey	370,000	9,500	7,055
624 North Chester	249,000	5,000	3,849
513 Ogden	290,500	7,000	5,774
310 Park	195,000	4,200	2,800
529 Rutgers	176,000	5,600	4,696
633 Strath Haven	272,500	8,000	6,001
621 University	265,000	6,300	5,132
10 Wellesley	340,000	10,000	7,501
Mean	$259,981	$ 6,446	$4,969
Median	249,000	6,300	4,752
Standard deviation	61,086	1,890	1,420
Interquartile range	105,250	2,200	1,735

Source: We are grateful to David Welsh, D. Patrick Welsh Realtors, who obtained these data for us from the Office of Registry of Deeds, Delaware County, PA.

In addition to the real estate data on sales prices, tax assessments, and taxes paid, Table 4.2 also shows some summary data. What do these summary numbers tell us about the three variables now that we have come to the end of this chapter on the computation of summaries?

The mean and the median convey that the average house in the list costs around $250,000. Since the mean is larger than the median, the list must contain a few very expensive houses that pull the mean above the median. How the assessed values of the houses correspond to the sales prices cannot be determined from the four summary numbers at the bottom of the table. For that we have to turn to the statistical methods discussed in Chapter 10. But, again, the median is larger than the mean, so some houses must have particularly large assessments that pull the mean up.

The taxes for these houses run almost $5,000, on the average, with the mean higher than the median. Thus, on the average, the taxes are about 1/50th of the sales prices. This may be a useful figure for the school board members to know when they compare these figures with similar figures from other parts of the school district and try to determine how the taxes could be raised.

The magnitudes of the means and the medians indicate that the three distributions are skewed, but the standard deviations still can be used to get some sense of the variations in the prices, assessments and taxes. Subtracting one standard deviation from each of the means shows that not many of the houses sold for less than $200,000 and that not many house owners pay less than about $3,500 in real estate taxes. Going up one standard deviation shows that more expensive houses sell for $320,000 and up, and the owners pay about $6,500 and up in taxes.

Using the statistical methods from this chapter and similar data from other communities in the school district, the school board can now begin to get a sense of the tax base in the district and whether taxes can be increased, perhaps by reassessing properties that have been sold at high prices.

4.7 SUMMARY

To find patterns in a set of data, the observed numerical values can be summarized. As with graphs and tables, a summary number greatly simplifies the data, while at the same time information is lost.

4.1 Averages: Let us count the ways

The three most common averages are mode, median, and mean. The mode is equal to the value of a variable that occurs most often. A bimodal distribution has two values that occur most often. It is essential to use a mode when describing categorical variables.

The median value is the value of the variable that divides the observations into two equal groups in such a way that half the observations are smaller than the median and half are larger than the median. The median, is the most common average used when a histogram of the data shows a skewed distribution. This is because the median is not greatly affected by extreme scores that are isolated from the majority of the values. The median is also the 50th percentile.

The mean—an average value of a variable that takes into account all the actual observed values—is the most commonly used type of average. The mean is found by adding up all the observations and dividing this sum by the number of observations. The symbol for mean is $\bar{x}$. If the mean and the median are approximately equal, the mean is the preferred average. If they are very different, then the median is preferred. For skewed data, the median gives a more realistic sense of where the middle of the data is located.

4.2 Variety: Measuring the spice of life

In addition to knowing about the central value of a set of data, it is important to understand how spread out the data are. One way to measure the spread is to find the range, the difference between the values of the largest and the smallest observations. One drawback of the range is that it is very sensitive to extreme observations. Occasionally we drop the smallest 25% and the largest 25% of the data and find the range of the remaining values. This range of the middle 50% of the data is the interquartile range.

The standard deviation s is the square root of the average squared deviation from the mean. It describes how far, on average, an observation is located from the mean. It is the most frequently used and most statistically sophisticated way of measuring the spread of data. Usually, about two thirds of the observations are within one standard deviation on either side of the mean, and almost all the observations are within two standard deviations of the mean. The square of the standard deviation is known as the variance, and it is denoted s^2.

4.3 Standard error of the means

The standard error of the mean is the standard deviation of a large set of means from many different samples. The standard error of the means is smaller than the standard deviations of the observations themselves because the means do not vary as much as the individual observed values of the variable.

4.4 Standard scores: Comparing apples and oranges

All observations on a variable can be converted to standard scores. A standard score equals an observation minus the mean, and the difference divided by the standard deviation. Its function is to judge how large an observation is in relation to the mean and standard deviation of all the observations. Most standard scores lie between -2 and $+2$; standard scores outside that range are unusual.

4.5 Gain in simplicity, loss of information

Simplifying information in a graph or a summary number means a gain in comprehensibility, but the detail of the original data is lost.

4.6 Real estate data: Out-of-sight prices

Concepts from the chapter are used to solve a real-life problem.

ADDITIONAL READINGS

Weisberg, Herbert F. *Central Tendency and Variability* (Sage University Paper Series on Quantitative Applications in the Social Sciences, no. 07-083). Newbury Park, CA: Sage, 1992. This book discusses different ways of computing measures of central tendency and variation.

Witmer, Jeffrey. *DATA Analysis: An Introduction*. Englewood Cliffs, NJ: Prentice Hall, 1992. This book gives many different quantities computed from the data.

FORMULAS

A variable is denoted by x, and the number n of observed values of this variable are denoted.

$$x_1, x_2, x_3, \ldots, x_n$$

The observations are then ranked from the smallest to the largest. To show that we have ranked the observations, we put parentheses around the subscripts, so that $x_{(1)}$ is the smallest observation, $x_{(2)}$ is the second smallest, and $x_{(n)}$ is the largest observation. With this notation, the ranked observations are denoted

$$x_{(1)}, x_{(2)}, x_{(3)}, \ldots, x_{(n)}$$

MEDIAN

When n is an odd number, then the median is found as the middle observation in the listing of the ranked observations. In symbols this can be written

$$\text{median} = x_{((n+1)/2)} \tag{4.1}$$

For example, if a data set has $n = 11$ observations, then $(n + 1)/2 = (11 + 1)/2 = 12/2 = 6$, and the median is equal to the sixth largest observation, $x_{(6)}$. Five observations are smaller than the median and five observations are larger than the median; the median is the value of the middle observation.

When n is an even number, the median is found by calculating the midpoint between the two middle observations;

$$\text{median} = \frac{x_{(n/2)} + x_{(n/2+1)}}{2} \tag{4.2}$$

If a data set has $n = 12$ observations, then the median becomes

$$\frac{x_{(12/2)} + x_{(12/2+1)}}{2} = \frac{x_{(6)} + x_{(7)}}{2}$$

This is the midpoint between the sixth and the seventh observation. Half the observations are smaller than this number and the other half is larger.

MEAN

The mean $\bar{x}$ is the sum of the observations divided by number of observations:

$$\bar{x} = \frac{x_1 + x_2 + \cdots + x_n}{n} = \frac{\Sigma x}{n} \tag{4.3}$$

where the symbol Σx (Σ is the Greek capital letter sigma) stands for the sum of all the observed values of the x variable.

STANDARD DEVIATION AND VARIANCE

The standard deviation is the distance from the mean of a typical observation in a data set. To find the standard deviation s, the variance s^2 is found and then the square root of the variance. The variance is found by subtracting the mean from each observation, squaring each difference, adding the squares, and dividing the sum by $n - 1$:

Observation	Difference	Square
x_1	$x_1 - \bar{x}$	$(x_1 - \bar{x})^2$
x_2	$x_2 - \bar{x}$	$(x_2 - \bar{x})^2$
x_3	$x_3 - \bar{x}$	$(x_3 - \bar{x})^2$
.	.	.
.	.	.
.	.	.
x_n	$x_n - \bar{x}$	$(x_n - \bar{x})^2$
Sum	0	$\Sigma(x - \bar{x})^2$

The sum of the differences themselves is always equal to 0, and making certain that this sum equals zero provides a check that the differences have been computed correctly. The variance s^2 is found by dividing the sum of the squared differences by $n - 1$:

$$s^2 = \frac{\Sigma(x - \bar{x})^2}{n - 1} \tag{4.4}$$

The variance is an average squared difference from the mean. The reason for dividing by $n - 1$ instead of n is discussed in Chapter 6 on estimation.

The standard deviation s is found as the (positive) square root of the variance, that is,

$$s = \sqrt{s^2} \qquad (4.5)$$

The three steps in computing the standard deviation are cumbersome, and any rounding error in the mean is introduced in every square computed. But even though there are other formulas for the computation of the variance that are easier and more exact, this is the definitional formula. Using it with a calculator or a computer makes the procedure less taxing.

STANDARD ERROR OF THE MEANS

The standard error of the means of two or more data sets can be found from a single random sample of n observations from a large population. The standard deviation s of the sample is found first, and the standard error of the mean s.e.$(\bar{x})$ is found by dividing the standard deviation s by the square root of the number of observations n:

$$\text{s.e.}(\bar{x}) = \frac{s}{\sqrt{n}} \qquad (4.6)$$

Sometimes the standard error of the mean is denoted by the symbol $s_{(\bar{x})}$.

STANDARD SCORES

A standard score for an observation is found by subtracting the mean from the value of an observation and dividing the difference by the standard deviation:

$$\text{standard score} = \frac{x - \bar{x}}{s} \qquad (4.7)$$

EXERCISES

REVIEW (EXERCISES 4.1–4.30)

4.1 What are the two major goals for summarizing data discussed in this chapter?

4.2 a. Why does too much information in the data make it difficult to understand a data file?

b. What is the major drawback with summarizing a data set?

4.3 Define mode, median, and mean.

4.4 Give an example of a situation in which each type of average (mode, median, mean) would be preferable over the other two types.

4.5 Give an example of a variable where you would expect to find a bimodal distribution of the data.

4.6 Find a newspaper article that makes use of an average of some type to make its point.

a. Describe the kind of average the journalist uses.

b. Describe what the journalist tries to achieve by using this average.

c. Does the journalist use the proper type of average?

4.7 You are interested in getting a clear picture of the economic well-being of your community.

a. Discuss whether it would be better to take the mean or the median average of the incomes of the people in your community and why.

b. Can you imagine a situation where a mode or modes might be a fairer way of describing a group's income level than either median or mean? Describe such a situation, if you can envision one.

4.8 Explain this statement: It is better to summarize skewed data with the median than the mean.

4.9 Create an example in which a summary statistic would greatly enhance understanding of a variable that has many values (e.g., number of words per page in this book).

4.10 a. Define *range*.

b. Name one positive quality about the range.

c. Is the range a measure of central tendency or variability? Why?

d. Are you at home with the range?

4.11 a. To what factor in a distribution of scores is the range insensitive?

b. To what aspect of the data is the range extremely sensitive?

4.12 a. The farther apart the observations tend to be from the mean, the [greater, smaller] the standard deviation is. Choose the correct adjective and explain the statement.

4.13 What letter of the alphabet do we use to designate the standard deviation?

4.14 Shaquille takes 6 trials of 5 shots of free throws. These are his scores: 5, 5, 5, 5, 5, 5.

a. What is the standard deviation of his shots?

b. Why?

4.15 For most unimodal and reasonably symmetric distributions, what proportion of the data would you expect to find within one standard deviation on either side of the mean?

4.16 Almost all of the data in a unimodal and reasonably symmetric distribution is found within how many standard deviation units on either side of the mean?

4.17 What is the result of squaring the standard deviation?

4.18 a. A handy rule suggests that the range of most distributions can be estimated as approximately how many times the standard deviation?

b. On the other hand, the standard deviation can be figured roughly as how many parts of the range?

4.19 Suppose several people are evaluating different pizzas on a scale from 0 to 10, with 10 being the best. Why might you prefer to purchase a pizza that has a high mean score and low standard deviation?

4.20 What is the *standard error of the mean*?

4.21 Why is the standard error of the mean smaller than the standard deviation of the observations in a sample?

4.22 What information is needed to estimate the standard error of the mean?

4.23 a. Explain why it is often useful to change raw scores to standard scores.

b. Give an example from your own experiences where it would have been helpful to be able to do this.

4.24 What is the method for changing a raw score to a standard score?

4.25 Normally, standard scores range between what two numbers?

4.26 If a fortune teller told you that your standard score on an IQ test was +15.55, would you immediately go out and celebrate your genius? Or would you decide the fortune teller was using hallucinogenic tea leaves? Explain.

4.27 Standard scores are often called _____-values. (See pun in Exercise 4.26.)

4.28 If you were defending the practice of changing raw scores to standard scores to a coworker, what would you say the greatest advantage of standard scores over raw scores is?

4.29 What does the following statement mean? The standard deviation measures the randomness of the data.

4.30 What is the difference between a standard deviation and a standard error?

INTERPRETATION (EXERCISES 4.31–4.52)

4.31 The observations of a variable have a standard deviation equal to zero. What does that tell you about the observations?

4.32 Until 1992, members of Congress could write checks against their accounts in an internal bank without incurring penalties for writing checks for amounts larger than the balances in their accounts. Newspapers published the number of overdrawn checks for each member of the House of Representatives. The median number of overdrawn checks was 3 and the mean number of checks was 47. What do these two numbers tell you about the distribution of the number of overdrawn checks?

4.33 A newspaper story on typical Americans reported the household income in 1989 to be $35,225. *(Source:* The New York Times, *July 26, 1992,* *p. E5.*) Why is this figure probably the median and not the mean household income?

4.34 The same story as in Exercise 4.33 reported that the typical American person is a woman who weighs 144 pounds, lives in a house with 2.6 bedrooms, watches television 28 hours and 13 minutes each week, and has a household income of $35,225.

    a. Which of these characteristics is a mode?

    b. Which characteristic is a median?

    c. Which characteristic is a mean?

4.35 A newspaper article claims that the average woman has 2.1 children. "How is that possible?" your 10-year-old brother asks. "Babies don't come in parts." What would you tell him?

4.36 The modal value of the gender variable is female. Name one strength and one limitation to the mode as a summary statistic.

4.37 Name a chief strength of the median as a summary statistic.

4.38 If the median score of students acceptable to Slippery Rock State is 550 on the verbal SAT, and your friend has a score of 500, should you tell your friend to not bother applying to Slippery Rock State, or should you first look for more information? Explain your answer.

4.39 A survey of workers indicated that in productivity, on a scale of 1 to 100, U.S. workers were rated 100, French workers 95, West German workers 89, Japanese workers 77, and British workers 75. The headlines indicated that American workers topped the French, Germans, Japanese, and British. Later in the article, economic indicators were reported on the productivity of the groups. "In 1990, each full-time U.S. worker produced $49,600, compared with $44,200 for West German workers, $38,200 for Japanese workers, and $37,100 for British workers." The study excluded statistics about workers in government, education, health, and real estate. *(Source: Alex Dominquez, "Study says US workers are the world's top producers,"* The Philadelphia Inquirer, *October 14, 1992, p. D-1.)*

    a. The headline says, "U.S. workers are the world's top producers." In what respects is this headline accurate, and in what respects is it misleading?

    b. Did you find any error or omission in the report? (There is at least one.) How would you correct it, if you were the business page editor?

c. Why do you suppose workers in government, education, health, and real estate were excluded from the study? Could you make any assertions about the effects of excluding these workers on the results?

4.40 How would you describe the differences in the way two distributions of scores are arranged where the numbers of scores are equal, the means are equal, and the standard deviation of one distribution is twice as large as that of the other?

4.41 The Atlanta Braves commit a mean of 1.3 errors a game. The standard deviation of the number of errors for games over the entire season is 1.0. The Philadelphia Phillies have a mean of 2.0 errors a game. Their standard deviation is 0.3. Which of the following statements would you feel confident about making and why?

1) The Braves play more errorless ball than the Phillies.

2) The Phillies are more consistent in making errors than the Braves.

3) The Braves sometimes play very sloppily and sometimes very well.

4) The Phillies seldom play errorless ball.

4.42 You read in the newspaper that at small four-year colleges, students under the age of 24 drink, on average, 7.0 alcoholic drinks a week, versus 4.6 drinks at campuses with over 20,000 students. Assume that the standard deviation for each sample was 2.0, and discuss the following, using your knowledge about standard deviations.

a. At small schools, about 66% of the students reported drinking between _____ and _____ drinks a week.

b. At large schools, about 66% of the students reported drinking between _____ and _____ drinks a week.

c. If a student says she drinks 6 drinks a week, can you predict with confidence that she attends a small college?

d. How would you describe the drinking behavior of the 33% of the students at the large schools not described in part b? How would you describe the drinking behavior of the 33% of the students at the small schools not described in part a?

e. Are there many students who do not drink at all on these campuses?

4.43 You are told that your child has a standard score of $+1.80$ in reading and $+2.00$ in mathematics. You are also told that your child has a standard score of 0.00 in musical understanding.

    a. What are the chances that your child is achieving at a fairly high level in academic work, assuming that the class includes a broad cross section of children?

    b. Should the music score confirm your suspicions that your family is not very musically inclined or not? What does the musical understanding score mean?

4.44 From data on the first 19 modern Olympic summer games, the mean for the winning distances in the men's long jump equals 308 inches, the median equals 310 inches, and the standard deviation equals 19 inches. What do these three numbers tell you about the original data?

4.45 One year, the modal temperature in Hibbing, Minnesota, was 32 degrees Fahrenheit (0 degrees Celsius). In Duluth, Minnesota, that year, there was a bimodal distribution of 33 degrees and 61 degrees Fahrenheit. What can we say about the difference in temperature in Hibbing and Duluth from these data?

4.46 You are applying for a sales job with an encyclopedia company. The recruiter explains to you that the field is very lucrative; in fact, the previous year, the top salesperson of 50 salespeople earned a million dollars, and the mean salary for all the salespeople was $35,000.

    a. Are you convinced you too can be a successful salesperson in this company?

    b. What more information would you like to have?

4.47 The recruiter from the encyclopedia company in Exercise 4.46 senses that you would like more information. She tells you that, in fact, not all the salespeople are great successes, and that the range of salaries was between $5,000 and $1,000,000. Does this information satisfy your curiosity about the salary prospects at the company? Explain what other information you might want.

4.48 The accountant from the encyclopedia company in Exercise 4.46 tells you that the interquartile range of salaries for the salespeople is from $10,000 to $30,000.

    a. How would you use this information in deciding whether or not to take the sales position?

b. Why was the mean salary so much higher than the interquartile range?

c. Can you hazard a guess about what the median salary might be?

4.49 You receive the following information about three areas of town in which condominiums are being built. You are interested in buying a condo to live in and from which you will receive an assured return on your investment when you sell.

Rose Valley: Mean price increase for all condominiums resold last year was $7,000, with a standard deviation of $4,000.

Garden City: Mean price increase for all condominiums resold last year was $5,000, with a standard deviation of $1,000.

Media: Mean price increase for all condominiums resold last year was $6,000, with a standard deviation of $800.

a. In which area will you be most certain of making a profit? In which area will you be least certain?

b. In which area will you make the most money, if all goes well?

4.50 Standby travelers are waiting at La Guardia to catch various New York-to-Boston flights and New York-to-Washington flights. The mean waiting time for all standby passengers is 1 hour. For the Boston passengers, the standard deviation is 30 minutes. For the Washington passengers, the standard deviation is 10 minutes. How would you describe to a nonstatistical friend what these facts mean in terms of the transit of passengers and their moods at the ticket counter?

4.51 During the baseball strike of 1994, reports revealed that the mean salary of the players was approximately $1,200,000 and the median salary was $500,000. What do these numbers tell you about the distribution of salaries for baseball players?

4.52 Consider the mean income in two different states. Suppose a person moves from one state to the other. How can it be that as the result of this move the mean income increases in both states?

ANALYSIS (EXERCISES 4.53–4.72)

4.53 Go to the Springer Web site (htt://www.springer-ny.com/supplements/iversen/) to find files relating to this book. Open the data file called Baseball Individual Scores.

a. For each column, find the mean, median, standard deviation, and range.

b. Obtain a histogram for each variable, using statistical software.

c. On the basis of the histograms, for which variables is the mean the better measure of central tendency and for which variables is the median the better measure?

d. Why is the range approximately equal to 4 times the standard deviation for only some of the variables?

4.54 Exercise 3.36 gives data on the longevity of the signers of the Declaration of Independence.

a. From looking at the data, do you think the signers as a group lived a longer or a shorter time than they were expected to?

b. The mean of the observations equals -1.8 years. What does the mean tell you about the how long the signers lived?

c. The standard deviation equals 13.2 years. How large is the range in these data compared to the standard deviation?

d. How many observations lie more than two standard deviations away from the mean?

e. What can you say about these men?

f. Judging from the histogram of all the data, would you expect the median to be very different from the mean? Explain.

g. These observations are found as the *difference* between how long a man actually lived and how long he was expected to live after signing the Declaration. Would there be any reason to analyze the *ratio* of those numbers instead? Explain.

4.55 Exercise 3.34 gives a sample of values of socioeconomic scores. Another group of people have the following values of the same variable: 55, 36, 70, 66, 75, and 49. You are interested in how long the two groups can expect to exist; you think that the more homogeneous a group is the longer the members will remain a group.

a. Explain how to measure the homogeneity of each group.

b. Compute the measure of homogeneity for each group.

c. Compute a comparison of the groups. Is there a great difference between the groups? Explain.

4.56 Refer to the socioeconomic scores given in Exercise 3.34 and Exercise 4.55.

a. Find the medians for the two sets of data.

b. What do the medians tell you about the two groups?

c. What is the combined median for all 18 observations?

d. How many observations in each of the two samples are smaller than the combined median and how many are larger?

e. What do the answers to part d tell you about how different the two groups are?

4.57 As a rule for good nutrition, no more than 30 percent of our daily calorie intake should come from fat. In a group of frozen chocolate desserts, the mean percentage of the calories that come from the fat equals 18.9, with a standard deviation s equal to 9.2. For comparison, the data also include information on regular chocolate ice cream, in which 39 percent of the calories come from fat. (Source: "Low-fat frozen desserts: Better for you than ice cream?" Consumer Reports, vol. 57, no. 8 (August 1992), pp. 483–487.)

a. Change the percentage for the chocolate ice cream to a standard score.

b. Does chocolate ice cream seem different from the other desserts?

4.58 The calorie values of 16 different snack foods follow (you made graphs for these data in Exercise 3.38).

$$110 \quad 120 \quad 120 \quad 164 \quad 430 \quad 192 \quad 175 \quad 236$$
$$429 \quad 318 \quad 249 \quad 281 \quad 160 \quad 147 \quad 210 \quad 120$$

(Source: ASDA data and manufacturer's data shown as an advertisement in The New York Times Magazine, April 20, 1990, p. 20.)

a. Find the mean and the median of the data.

b. Which of these two averages seems more appropriate for these data?

c. Find the range for the observations.

d. Use the range to find an estimate of the standard deviation of the data.

4.59 In the school year 1995–1996, the members of the Department of Mathematics and Statistics at Swarthmore College had the following numbers of children: Eugene 2, Don 0, Gudmund 4, Helene 0, Charles 2, Aimee 0, Stephen 2, Michael 0, Janet 0.

   a. Draw a histogram illustrating these findings.

   b. What was the modal number of children?

   c. What is the mode for men and the mode for women?

   d. What do these modes tell you?

4.60 To find the average numbers of pages in the textbooks for his courses, Clark first listed the books by course as follows: Biology 657, 189; History 348, 237, 181; English 104, 201, 298, 87; Math 302, 99; Psychology 607, 139.

   a. Organize the items in the list so that it is possible to find a median by "eyeballing" the numbers.

   b. Find the mean of the pages.

   c. Compare the two scores. What accounts for the discrepancy between the scores? Which one would you think is the fairer answer, given Clark's question?

4.61 According to the U.S. Bureau of the Census, the following were the number of medical schools in the country in each year between 1915 and 1945.

1915	1916	1917	1918	1919	1920	1921	1922	1923
96	95	96	90	85	85	83	81	80

1924	1925	1926	1927	1928	1929	1930	1931	1932
79	80	79	80	80	76	76	76	76

1933	1934	1935	1936	1937	1938	1939	1940	1941
77	77	77	77	77	77	77	77	77

1942	1943	1944	1945
77	76	77	77

Source: Historical Statistics of the United States 1789–1945, *p. 50.*

a. Draw a stemplot illustrating the data.

b. Compute the mode, median, and mean number of medical schools over the years.

c. What special insight that the others do not reveal does each summary statistic give?

d. Is there anything about these data that surprises you?

d. How would you account for the trends historically? (Additional information: the number of medical school graduates went from 3,500 in 1915 to 4,000 in 1925 to 4,500 in 1930 and leveled off at slightly more than 5,000 from 1930 to 1945. The number of physicians rose from approximately 143,000 in 1915 to approximately 181,000 in 1945.)

4.62 Recall the hourly wage you made in each job you have had over your entire life. Calculate the range.

4.63 Draw a histogram of Shaquille's free throws in Exercise 4.14. (You will waste some paper on this exercise!) What does the histogram tell you?

4.64 The following data come from a sample of high school students' reports of smoking cigarettes and marijuana and drinking alcohol. To simplify the task, first draw a histogram of each distribution.

Number of days smoked cigarettes during month: 0 0 30 29 30 0 0 10 0 30 29 30 0 0 0 0 1 30 28 10 0 0 0 30 30 29 0 0 30 0 0 30 0 0 1 0 0 30 30

Number of days smoked marijuana during month: 0 0 0 0 0 0 1 0 0 0 0 1 2 2 1 0 0 3 0 0 2 0 0 1 0 0 1 0 0 1 0 0 4 0 0 0 0 0 1 1

Number of days used alcohol during month: 0 1 0 5 0 4 0 0 3 0 0 2 2 0 0 0 1 0 0 4 0 0 3 0 0 2 0 0 0 1 2 0 0 1 0 0 1 0 3 0

a. Estimate (or calculate) the mean of each distribution.

b. Which distribution will have the highest standard deviation?

c. Which distribution will have the smallest standard deviation?

d. Would it be possible to estimate a standard deviation that would be appropriate for each distribution knowing what you do about the percentage of the distribution within one standard deviation of the mean?

4.65 Refer to the information in Exercise 4.64.

a. What percentages of high school students in this sample smoked cigarettes, smoked marijuana, or drank alcohol at least once in the past month?

b. How do your findings agree with the following results of a survey of a sample of all U.S. students: 46% had drunk alcohol at least once; 24% had smoked cigarettes; 11% had used marijuana at least once. *(Source: "Teen-age drug use high,"* The New York Times, *September 20, 1992, p. 33.)*

4.66 Refer to the data on mercury concentrations in swordfish in Exercise 3.46.

a. Find the mean concentration of mercury in the sample of 28 swordfish.

b. Find the standard deviation of the mercury concentration.

c. How many of the swordfish have a mercury concentration within plus or minus two standard deviations from the mean?

d. Why is it that the mean concentration is larger than 1.00 when those swordfish that are tested and found to have a concentration larger than 1.00 are not even brought to the market?

4.67 A small company employs 9 people who earn the following hourly wages:

$6.50 $6.20 $6.50 $7.00 $10.00 $10.00 $11.00 $15.00 $21.00

a. How large is the median wage?
b. How large is the mean wage?
c. It was decided that the four lowest wages should each be increased by $4.00 per hour. What is the median wage of the new wages?
d. What is the mean wage of the new wages?
e. Why do the median wage and the mean wage not change by the same amount when the low wages are increased?

4.68 The observed values of one variable are 1, 3, 3, 3, 3, 3, 3, and 5. The observed values of the another variable are 2, 2, 2, 2, 4, 4, 4, and 4.

a. Make histograms of the data on the two variables.

b. From the histograms, does it appear that the two variables have the same or different means?

c. From the histograms, does it appear that the two variables have the same or different standard deviations?

d. Find the means and standard deviations for the two sets of data.

e. What do you learn about the two data sets from the results in part d?

4.69 Refer to Exercise 10.34 for data on the percentage of calories that come from the fat in ten different ice creams as well as a score measuring the flavor of the ice creams as determined by a group of testers.

a. Find the mean and the standard deviation for the percentage of calories from fat variable.

b. How many of the observations lie in the range from the mean minus two standard deviations to the mean plus two standard deviations?

c. Find the mean and standard deviation for the flavor variable.

d. How many of the flavor observations lie in the range from the mean minus one standard deviation to the mean plus one standard deviation?

4.70 a. Find the standard deviation of the data in Figure 4.6.

b. What does the data tell you when you know both the mean and the standard deviation?

4.71 Find the median ages at time of marriage for the data in Exercise 3.20. What do the two medians tell you?

4.72 Following are the results of two well-known tests of physical strength taken by 10 college swimmers.

Test	Adam	Bob	Emil	Juan	Sam	Lou	Ken	Paul	Mike	Lee
A	20	23	24	18	17	16	25	24	21	19
B	31	39	39	29	28	31	40	30	31	30

a. On which test did each person do better? To answer this question it is necessary to convert the raw scores to standard scores. The mean of test A for the national sample is 20 and the standard deviation is 2; the mean of test B for the national sample is 35 and the standard deviation is 3.

b. If you were the coach and wanted your team to feel good about themselves, which test would you prefer to use?

c. Which team member seems to be the weakest?

d. Which team member seems to be the strongest?

e. Which team member(s) seem most inconsistent from one test to the other?

f. Which team member(s) seem most consistent from one test to the other?

CHAPTER 5

5.1 How to find probabilities

5.2 Computations with probabilities

5.3 Odds: The opposite of probabilities

5.4 Probability distributions for discrete variables

5.5 Probability distributions for continuous variables

5.6 Using probabilities to check on assumptions

5.7 Decision analysis: Using probabilities to make decisions

5.8 Summary

PROBABILITY

5

What is the probability that all four children in a family are girls? What is the chance that there will be two no-hitter baseball games in the same day? What is the likelihood that the lucky number in the daily double will be 71? How certain can Libby's parents be that she will be accepted at Carleton College? What is the probability that the mean number of children in a family is 2.0 or less in a sample that comes from a population where the mean equals 4.0 children? If a voting population is split evenly between two parties, what is the probability of a sample percentage of 55% or more voting for one of the two candidates?

> A **probability** is a number between zero and one that tells us how often an event happens.

Questions about probability occur regularly in everyday conversations as well as in statistics classes. In this chapter we discuss what statisticians mean by the term *probability* and how we use it in statistical analyses. From the questions on page 177, you can see that the word *probability* has to do with the chance, or likelihood, or degree of certainty that some event will happen, and that the term is used well beyond the scope of statistics.

A probability is simply a *number*. More specifically, it is a number between zero and one that describes how often an event happens. An event with a small probability (near zero) happens seldom, while an event with a large probability (near one) happens often. For example, the probability of two no-hitter baseball games on the same day is small; the probability that at least one hurricane will hit the United States somewhere during a year is large, since more than one such storm occurs in most years.

The idea of probability goes back a long time. References to chance and probability are even found in the Old Testament: "And Saul said, Cast lots between me and Jonathan my son. And Jonathan was taken" (Samuel 1:42). About a thousand years ago, legend has it that the Norwegian king (Olav the Holy) and the Swedish king threw a pair of dice to determine the ownership of a disputed piece of land.

There were sporadic writings on chance and probability up until the 1600s. At that time, interest in probability was stimulated when gentlemen gamblers tried to determine what the payoffs should be in certain card and dice games. Because events with small probabilities do not occur very often, the gamblers thought these events should have high payoffs. On the other hand, events with large probabilities should have smaller payoffs because these events happen frequently. Also, the probabilities—and thereby the payoffs—ought to be such that the winnings were fair, meaning that people putting on the games should neither go broke nor make excessive profits. Problems of these kinds were presented to mathematicians of the time, and they began to develop probability theory as we know it today.

Today, statistical interest in probabilities is somewhat different from that of gamblers. To stress a basic notion of statistics, if we measure something several times, a different result will occur most of the times. For example, the measure of the length of one leaf on a tree is a certain number of inches, and the measure of the length of another leaf on the same tree is a different value. This is because of the randomness in the variable (the length of a leaf). Similarly, the percentage of people in favor of the current president's policies in one sample is

Probabilities can be used for many purposes. (*Source:* Peanuts® reprinted by permission of United Feature Syndicate, Inc.)

different from the percentage of people who favor them in another sample. This too is because of the randomness we experience from the random drawing of the people in one sample to the next.

Variability between measurements of the same variable raises the question of how often a specific result would occur if the measurement were repeated many times over, regardless of whether we measure a single object (a leaf) or a whole sample of objects (people). The question can be answered by probabilities, which are designed to show how often something happens over a long series of observations. For example, if we draw many, many samples of voters, and in three quarters of the samples the percentage of people in favor of the president's action is larger than 60%, then we can say that the probability that the observed sample percentage is larger than 60% equals 0.75: in 75 out of 100 samples, the percentage approving of the president is 60% or more. The other 25 samples would have a sample percentage less than 60%. For the leaf example, the probability may equal 0.10 that the leaf length is more than 2.34 inches: in only 1 of 10 measurements is the leaf longer than 2.34 inches; 9 of the 10 leaves are shorter than 2.34 inches.

Probability statements are made throughout this book: the probability is 0.023 that a sample mean of a variable is larger than a certain value, the probability is 0.15 that the sample standard deviation is less than 5.67, and so on. In Chapter 4, in the age at marriage example, the mean age of the brides is 30.0 years and the mean age of the grooms is 32.3 years, for a difference of 2.3 years. The probability of getting a mean difference of 2.3 or more in a sample coming from two populations where the means are the same is only 0.002. That is, in only 2 of 1,000 different samples coming from populations where the brides and grooms are the same age would the mean for the grooms be 2.3 or more years larger than the mean for the brides.

Such probabilities tell us on the basis of sample data how often these kinds of results would occur if the study were repeated for many different samples. Probabilities are useful for applying to the real world findings on an observed sample.

5.1 HOW TO FIND PROBABILITIES

So far we have established that a probability is a number between zero and one. How do we find these numbers?

Using equally likely events

The early method of finding probabilities came from card and dice games. If a die (one of a pair of dice, for noncrapshooting readers) has six sides and all sides are equally likely to show when the die is tossed, then the probability of any one side showing is 1/6. Similarly, if a deck of cards has 52 cards and 13 of them are clubs, then the probability that a randomly chosen card is a club equals 13/52 or 1/4.

This way of thinking about probabilities suggests that if there is a specific number n of possible outcomes when an experiment is performed and a subset k of them is considered favorable, then the probability of a favorable outcome is k/n. For the die, $k = 1$ for one side and $n = 6$ for the sides for the probability 1/6. For a deck of cards, $k = 13$ clubs and $n = 52$ cards, so the probability of drawing a club (or a card from one of the other three suits) becomes $13/52 = 1/4 = 0.25$.

This system for finding probabilities works for cards or dice because the possible number of outcomes is known and, because of their symmetry, all are equally likely. However, often it is not known whether all possible outcomes are equally likely (for example, all the horses in a race are not equally likely to win). Sometimes the possible outcomes are not known (for example, the number of gamblers choosing numbers in a football pool). Under such conditions, the "equally likely" way of finding a probability is impossible.

Using relative frequency

In the second and most common way of finding simple probabilities, the probability of an event occurring is based on the proportion of times an event actually occurs in a great number of cases. Take child-

Poker players' paradise—a royal flush and a full house. What is the probability of drawing either hand? Of drawing both? *(Source: First Image West, Inc.)*

WATCHING OUR LANGUAGE: WHAT IS THE *n* AND WHAT IS THE *k*?

The probability that a woman having an abortion is Catholic is not the same as the probability that a Catholic woman will have an abortion. Excerpt from a letter to the editor from Charles F. McLaughlin of Philadelphia to *The Philadelphia Inquirer,* December 8, 1992:

> I am writing in response to a statement on Catholic women in Victoria A. Brownworth's Commentary Page article. She stated: "According to the Alan Guttmacher Institute, more Catholic women seek abortions than women of any other faith."
> That statement misleads. It tends to leave one with the impression that, as individuals, a Catholic woman is more likely than a woman of any other faith to seek or have an abortion, which is not true. The Roman Catholic church has more members than any other religious denomination in the United States. Catholic women outnumber women in the next largest denomination by more than 2 to 1. In numbers, it may be true that Catholic women seek more abortions than women of another faith. However, it is unlikely that a Catholic woman is more likely than a woman of another faith to seek an abortion.

birth as an example. Each birth results in one or more babies who are either boys or girls. Whether the two possibilities are equally likely is not known.

Over many years of record keeping, the proportion of girls among newborn children was found to equal 0.49. The proportion is found by dividing the number of girl babies by the total number of babies born. A probabilist (a person who studies probabilities) would say that as the number of births approaches infinity (by this is meant a very large number of observations), the observed proportion of girls approaches the true value of the probability of a girl.

In this example, a probability is a long-run proportion, the result of investigating a large number of events over the long run. The exact numerical value of such a probability never is identified, but many observations bring the estimate close to the actual value. The problem with the long-run way of finding probabilities is that, as the famous economist Lord Keynes said, "in the long run we will all be dead." No statistician can hope to stay around long enough to find the true values of a probability. Instead, the statistician relies on the observed proportion as an estimate of the true probability.

THE SPIN OF A COIN

Tossing a penny can be used to illustrate long-run probability. A variation is spinning the coin instead of tossing it. Stand a penny on its edge and support it with a finger on the top. Use your other hand to give the penny a good spin with the snap of a finger. Is the penny likely to show heads and tails about the same number of times as if it were tossed?

To answer this question, we tried spinning pennies for a while in one of our statistics classes. There were 25 people in the class, and each person spun a penny 10 times, for a total of 250 spins. Of these, 97 showed heads and 153 showed tails. Thus, the proportion of heads was $97/250 = 0.396$ (or 39.6%) instead of about 0.5 if the two sides were equally likely.

Did the true probability still equal 0.5, and was the result within the range of possible results expected from the randomness of coin spinning alone? This question is answered in Exercise 7.58.

Using subjective probabilities

Even the relative frequency approach to finding probabilities does not always work. What is the probability that Mr. Kaye will arrive safely at his destination after taking a planned trip tomorrow? Only one specific trip is occurring tomorrow, a *unique* event. He cannot take the trip and then roll back time and take the trip again and again and again to see how many times he arrives safely out of the total number of times he takes the trip. When there is no measurement that can be repeated, there is no way to find an observed proportion of how many times a specific event occurs. But it is still useful to think in terms of probabilities. Mr. Kaye cannot be certain that he will arrive safely, but from what he knows of travels like these, he judges the probability of a safe arrival to be large enough that he should take the trip.

Probability for a unique event is called *subjective probability*. In the example, personal probability is simply an expression of the uncertainty Mr. Kaye feels about traveling, based on all the information available at the time. We can all have different values of the probability for a safe arrival, so there is no right or wrong value of a personal probability. That makes personal probabilities subjective.

STOP AND PONDER 5.1

Which method of finding a probability would be most suitable for the following problems?

Method
a. Equally likely events
b. Relative frequency
c. Subjective probabilities

Problems
1. A 10-year-old commuter airline will continue to have an accident-free record.
2. The poker player will draw an ace from the deck.
3. The snowfall in March in Minneapolis will exceed 5 inches.
4. It will rain at the picnic tomorrow.
5. A family with 6 children includes twins, triplets, quadruplets, quintuplets, or sextuplets.
6. A probability problem that has been of special interest to you lately.

Personal or subjective probabilities form the basis for what is known as Bayesian statistical inference, which we do not pursue in this text. This book most often uses the long-run proportion as a probability.

5.2 COMPUTATIONS WITH PROBABILITIES

Probabilities are simply numbers, and as numbers they can be added to each other, subtracted from each other, and multiplied and divided by each other. Computations can help us find the probabilities of more complicated events from the probabilities of simpler events.

For example, the probability that a randomly chosen new baby is a girl is 0.49. Given the probability for that simple event, what is the probability of having a boy? Subtracting 0.49 from 1.00 yields 0.51, and that is the probability of a boy, since girl and boy are the only two possible outcomes of a birth. A more complex problem that the simple probability helps solve is, What is the probability that there are 3 girls and 1 boy in a family with 4 children? That is, how often are the children in families with 4 children composed of 3 girls and 1 boy?

One way to find this out would be to actually locate many, many families with 4 children and count how many of them have 3 girls and 1 boy and how many do not. We would find that about 0.24 (or 24 out of 100) of them have 3 girls and 1 boy. But this empirical method to find a long-run proportion would be costly and time consuming—and unnecessary. Instead, we can use the rules for how to multiply and add probabilities to compute the answer from the original female birth probability of 0.49. The answer to the question is also 0.24, and the computations are shown in Section 5.4 in the subsection on binomial distribution.

Adding probabilities

When we want to find the probability of one event *or* another event that both cannot happen at the same time, then we simply *add* the probabilities for the two events. For example, to find the probability that a family of 4 children has 3 girls *or* 4 girls, we assume that particular family cannot have both 3 and 4 girls at the same time, so we add the two probabilities. The probability of 3 girls is 0.24 and the probability of 4 girls is 0.06, so the probability of 3 *or* 4 girls is 0.24 + 0.06 = 0.30. If we want to find the probability that something is either large or small—for example, that a sample mean is smaller than 5.6 or larger than 17.8—since both of these events cannot happen at the same time, we add the probability that the mean is smaller than 5.6 and the probability that the mean is larger than 17.8.

Multiplying probabilities

To find the probability of one event *and* another event happening at the same time, we *multiply* the probabilities of the two events. The probability of both events happening at the same time is smaller than the probabilities of either of the two events happening by itself; two specific events happening at the same time occurs less often than either of them happening alone. This piece of common sense is borne out mathematically: multiplying two numbers that are each less than one yields a product that is less than either one of them. For example, 0.3 times 0.4 equals 0.12, and 0.12 is smaller than either 0.3 or 0.4.

Returning to the family of 4 children, what is the probability that a family had a girl, then a boy, then a boy, and finally a girl? Just multiply the probabilities for each child: $0.49 * 0.51 * 0.51 * 0.49 = 0.062$. Thus, only 62 out of 1,000 families with 4 children would have a girl, a boy, a boy, and a girl, from oldest to youngest.

In many situations probabilities cannot be multiplied directly. So-called conditional probabilities have to be taken into account when the multiplication is done.

5.3 ODDS: THE OPPOSITE OF PROBABILITIES

In 1993, before the International Olympic Committee had decided where to hold the summer Olympic games in the year 2000, bookmakers in London gave odds on where they thought the games would be held. The bookmakers thought that some cities had a higher probability of getting the games than other cities, and they offered *odds* against where they thought the games would be held:

Sydney, Australia	4 to 9
Beijing, China	5 to 2
Manchester, England	10 to 3
Berlin, Germany	16 to 1
Istanbul, Turkey	66 to 1
Brasilia, Brazil	200 to 1

Odds against an event are expressed in a ratio of whole numbers showing how often an event fails to take place versus how often it does take place.

These numbers look just like the odds for a horse race. Since in odds *against* an event occurring the number of times the event does *not* occur is always given first, it is clear that the bookmakers felt that Sydney had a pretty good chance of being the Olympic site and that Brasilia was a long shot.

Odds rather than probabilities are commonly used when money is being wagered on an outcome. The odds of 200 to 1 on Brasilia tell us that if we had paid a bookmaker $1 and Brasilia actually got the games, then we would have received the dollar back plus an additional $200. The odds therefore describe how much money we have to pay the bookmaker and how much money we get back if we win.

Odds are given in whole numbers, like 4 to 9, for ease in expression. Odds of 4 to 9 are the same as odds of 2 to 4.5, but decimal numbers are cumbersome. This means that odds take some getting used to before they can be compared.

STOP AND PONDER 5.2

In the list of Olympic sites odds, which are better, the odds for Beijing or the odds for Manchester?

Brasilia as a long shot in the race for the Olympic games shows that the bookmakers thought Brasilia had a very small probability of getting the games. Odds of 200 to 1 translate into a probability for Brasilia getting the Olympic games of $1/(200 + 1) = 1/201 = 0.005$. Sydney's odds of 4 to 9 mean that if we paid the bookmaker $9 and Sydney got the games, then we got our $9 back plus another $4. We would not have received much money because the bookmakers thought that because Sydney had such a large probability of getting the games, many people would pick Sydney as the winner.

The probability of Sydney getting the games, in the eyes of the London bookmakers, was $9/(4 + 9) = 9/13 = 0.69$. The probability of Beijing getting the games was 0.29, Manchester 0.24, Berlin 0.06, Istanbul 0.015, and Brasilia 0.005. Formulas 5.1, 5.2, and 5.3 at the end of the chapter take us from odds to probabilities and back to odds again. Rather than giving the odds *against* each of the cities, the bookmakers could have given the odds *in favor* of each city.

On September 23, 1993, the International Olympic Committee awarded Sydney the summer Olympic games in the year 2000. The bookies were content.

5.4 PROBABILITY DISTRIBUTIONS FOR DISCRETE VARIABLES

Often there is an easy way and a hard way to find something out. For example, a hard way to find the distance from New Orleans to Chicago would be to drive the distance and measure the miles. An easy way would be to look in the back of a road atlas where cities and distances are listed. Statisticians have an easy way to find the probabilities for complex events once certain probabilities for simple events are known.

Probabilities for simple events can be used to compute probabilities of more complex events when it is too difficult to find the probabilities of the complex events directly. In the example family with 4 children, the simple probability is the probability of .49 that a randomly chosen baby is a girl. The simple event is the birth, resulting in a boy or a girl. The complex event is the occurrence of 3 girls and 1 boy in the family.

By creating preformulated solutions to various problems of probability, statisticians save themselves a great deal of time and trouble. Two examples of these energy-saving opportunities are the binomial distribution and the Poisson distribution (Poisson was the French mathematician who introduced the method).

Binomial distribution

Imagine that you would like to know the probability of a coin landing heads twice in a row. Do you have to sit in a room all day tossing coins to find the probability of tossing heads twice in a row? Maybe not, if you know that (1) the probability of tossing heads once is 0.5; (2) there are only two options (heads or tails); (3) each toss is independent of the other. To find the probability of tossing heads twice in a row, you multiply 0.5 times 0.5 to get 0.25. Thus, there is a 25% chance of tossing heads twice in two tosses. This you can do without a calculator, a day tossing coins, or high-level mathematics.

Consider the more difficult problem of calculating the probability of a family with 4 children having 3 girls and 1 boy from the simple probability of 0.49, the probability that a baby is a girl. Mathematicians realized as long as 300 years ago that it does not matter whether the probabilities being sought are for girls or boys, heads or tails, or dead or alive goldfish. From the correct probability for a simple event, formulas, printed tables, and now computer software have been created that help us find the correct probability for more complicated events. The most common of these formulas is called the *formula for binomial distribution,* and it charts the distribution of numbers of successes (such

What is the probability that all of these randomly selected babies are girls? *(Source: Penny Gentieu, Tony Stone Images.)*

as a girl birth) in *n* trials (baby births). By using this formula with paper and pencil, computer software, or printed tables and plugging in the simple event information, the probability of any given outcome can be obtained. Formula 5.4 at the end of the chapter is the formula for binomial distribution.

A variable that has only two possible values, such as the gender of a newborn child, forms the basis for binomial distribution. (The word *binomial* means "two numbers or names.") Suppose we know the probability for one of the two values. For example, for the gender variable the probability of a newborn child being a girl is 0.49; the probability of the child being a boy is therefore 0.51. The two probabilities must add up to 1.00, since the child must be either a boy or a girl.

The next step in creating the binomial distribution consists of making several independent observations of the base variable, gender of child. If a family has 4 children, then a certain number of them are girls and 4 minus that number are boys. This is a new variable: number of girls among 4 children. This variable has the possible values 0, 1, 2, 3, or 4. Such a variable is known as a *binomial variable.* A binomial variable indicates the number of occurrences of one of the two values under study.

The next step is to find the probability for each value of the binomial variable (number of girls) in families with 4 children. The probabilities can be computed using Formula 5.4, and they are shown in Table 5.1. Such a collection of values of the binomial variable and their corresponding probabilities is known as a *binomial distribution.*

Let us look at a family with 3 girls and 1 boy as an example. One order in which the children could have been born is the 3 girls first and then the boy. The sequence can be represented GGGB, where G

Table 5.1 Binomial probabilities for the number of girls in a family of 4 children

Number of girls	0	1	2	3	4	Total
Probability	0.07	0.26	0.37	0.24	0.06	1.00

stands for girl and B for boy. The probability that the first child was a girl equals 0.49. The probability that the second child was a girl equals 0.49. The probability that the first *and* the second children are girls is therefore the product of 0.49 * 0.49. Generalizing from here, the probability that the first child is a girl *and* the second is a girl *and* the third is a girl *and* the fourth is a boy is 0.49 * 0.49 * 0.49 * 0.51 = 0.06. This is the probability that a family will have three girls and a boy in exactly that order of births.

A family can have three girls and a boy in four different orders:

3 girls and 1 boy	Probability
GGGB	0.49 * 0.49 * 0.49 * 0.51 = 0.06
GGBG	0.49 * 0.49 * 0.51 * 0.49 = 0.06
GBGG	0.49 * 0.51 * 0.49 * 0.49 = 0.06
BGGG	0.51 * 0.49 * 0.49 * 0.49 = 0.06
	Sum = 0.24

Each of these possible sequences has a probability of 0.06. Adding the probabilities results in the overall probability 0.24 of three girls and a boy.

It can be tricky to find the number of possible sequences when the sample is larger than 4, but the first term in Formula 5.4, given at the end of the chapter, makes it easier. And published tables of binomial probabilities as well as computer programs for finding binomial probabilities eliminate doing the computations altogether.

The binomial distribution is used only for a small sample, such as the 4 children in the example family. If the product of the sample size and the original probability is larger than about 5, then there are simpler ways of analyzing the data. (In the example family, 4(0.49) = 1.96, which obviously is less than 5.) To find probabilities in a sample of 1,200 respondents, with 720 people in favor of a proposition and 480 against, a better method is the so-called *normal approximation to the binomial distribution*, discussed in Chapters 6 and 7. As long as the basic probability is around 0.5, as is the case in the example family, this approximation can be used even with samples of as few as 10 or 15 or so observations.

Poisson distribution

On June 3, 1990, the sports pages were full of discussions of the unlikely phenomenon that had happened the day before: *two* no-hit baseball games had been pitched, one by Mark Langston and Michael Witt together for the California Angels and the other by Randy Johnson for the Seattle Mariners. No-hitters do not occur often in baseball, so even one gets considerable press. Two no-hitters on the same day had not occurred since 1898.

To find just how unlikely this event was, the Poisson distribution can be used. Siméon Denis Poisson was particularly intrigued by problems with small probabilities and potentially many occasions that the event could occur. Poisson developed his approach with data on the number of Prussian army soldiers killed by horse kicks in the days when the cavalry rode horses instead of tanks. His work was published in 1837.

A no-hitter is a dichotomous situation. Any baseball game is either a no-hitter or it is not, so there are only two possibilities. But unlike the probability of a baby being a girl, the probability of a no-hitter is very small, a no-hitter is very unlikely, especially given the potentially large number of times (every game) a no-hitter could occur. The Poisson variable here is the number of no-hitters in a day; possible values of the variable are 0, 1, 2, 3, and so on.

In such a case, when the occurrence of an event has a small probability and many possibilities, the probabilities of the different values of the Poisson variable can be calculated with Formula 5.7, the formula for the Poisson distribution, at the end of the chapter. (The Poisson distribution is mathematically derived from the formula for the binomial distribution, but if you examine it, you will see why some people think it is not as intuitively obvious as the binomial formula.) Poisson probabilities can be computed from the formula or they can be looked up in tables. It is also possible to program a computer to find the probabilities.

Data on no-hitters starting with the year 1900, the year both the American and National Leagues as we know them came into being, show an average 1.9 no-hitter games pitched every year. Let us say that

Table 5.2 Poisson probabilities for the number of baseball no-hitters occurring in one day

Number of no-hitters	0	1	2	. . .	Total
Probability	0.989234	0.010708	0.000058	. . .	1.000000

the baseball season lasts 180 days. An average of 1.9/180 = 0.0108 no-hitters, then, were pitched every day. Applying the Poisson formula to that number yields the probability of 0, 1, 2, . . . no-hitters being pitched any day (Table 5.2). There is no upper bound on the number of no-hitters that could be observed beyond the number of games played on a given day.

Based on 1.9 no-hitters a year and 180 days of playing, in most games there is at least one hit, since the probability of a no-hitter is 0.989234. At the same time, the probability of two no-hitters on one day is 5.8 in 100,000, or 1 in 17,241. In 100 years, baseball is played a total of 18,000 days, so a day with two no-hitters can be expected about once every 100 years. The first one occurred almost on schedule, 90 years after records began to be kept; there may be quite a wait for the next one.

Cy Young. This famous pitcher threw a no-hitter on May 5, 1904. *(Source: UPI/Bettmann.)*

Hypergeometric distribution

A third statistical distribution, the so-called *hypergeometric distribution*, can be used in the analysis of two categorical variables when the samples are very small; see Chapter 9. Formula 5.10 shows how to find probabilities for the different values of the hypergeometric variable. The distribution is used in Section 5.6 in the example about the fair workplace.

Displaying probabilities in graphs and tables

We can do many of the same things with probabilities as with observed data. We can display probabilities in graphs and tables and we can use the probabilities for computations of quantities such as means and standard deviations.

Any kind of graph that can be produced for frequencies can also be produced for probabilities: pie charts, boxplots, and so on.

Figure 5.1 shows a histogram of the binomial probabilities for the number of girls in a family with 4 children. A histogram of data on many, many families with 4 children and different numbers of girls would look exactly the same. Computing the probabilities using the binomial distribution saves much time and effort that otherwise would have to go into data collection.

It is worth repeating that it is not the height but the *area* of each of the rectangles in Figure 5.1 that shows the corresponding probabil-

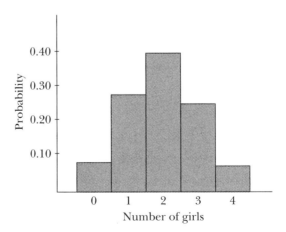

Figure 5.1 Binomial probabilities for number of girls in families with 4 children

ity. Since the base of each rectangle has length 1.0, the area of any rectangle is shown on the vertical scale. Also notice that the total area of all five rectangles in the graph equals the sum of the probabilities, which is 1.00.

Computations with probabilities

We do computations with probabilities to summarize a probability distribution just as we can do computations to summarize a frequency distribution. To find the mean number of girls from the distribution in Figure 5.1, we find the point on the horizontal scale at which the probability distribution balances. With observed data, we add the values of the variable and divide the sum by the number of observations. In this case, all possible values 0, 1, 2, 3, and 4 are present, and each value is accompanied by a probability or how often the value occurs. We then act as if the value of 0 had been observed 0.07 times, the value of 1 had been observed 0.26 times, and so on. Instead of adding 0 a total of 0.07 times, we multiply the value of 0 by 0.07 to get the contribution of that value to the mean and similarly for the other values of the variable:

$$\text{mean} = \mu = 0(0.07) + 1(0.26) + 2(0.37) + 3(0.24) + 4(0.06) = 1.96$$

The u with the tail is the Greek letter μ (mu), and it is used to distinguish a mean found from probabilities from the empirical mean $\bar{x}$ found from actual data. The number 1.96 tells us that in a very large number of families with 4 children, the mean number of girls would equal 1.96. Computing the mean from the original probability 0.49 of a girl and using of the binomial distribution eliminates spending time and money collecting from a large number of families. The mean for the binomial distribution can also be found with Formula 5.5 at the end of the chapter. The mean of the Poisson distribution can be found with Formula 5.8.

We can also find a standard deviation for the variable, and it is denoted by the Greek letter σ (sigma) to distinguish it from the standard deviation s computed from observed data. For this probability distribution, the standard deviation of the number of girls variable is $\sigma = 1.00$. The mean plus and minus two standard deviations equals $1.96 \pm 2(1.00) = -0.04$ to 3.96. This range of values takes in just about all the values of the variable and almost all the total probability of one. The standard deviation for the binomial distribution can be found with Formula 5.6 at the end of the chapter. The standard deviation for the Poisson distribution can be found with Formula 5.9.

5.5 PROBABILITY DISTRIBUTIONS FOR CONTINUOUS VARIABLES

Much of the data gathered for statistical analysis come from *continuous variables,* for which between any two values are other values. Examples of continuous variables include distance, dollar amount, weight, and time.

Four theoretical variables are useful in determining certain probabilities. They are known as the standard normal z variable, the t variable, the chi-square variable, and the F variable. Each has its own special distribution. Just as we compute a mean and a standard deviation from the data in a sample, we can compute a similar value of one of these four variables from a sample. Thus, a z or a t or a chi-square or an F is no different from any other sample statistic. As we see in later chapters, the values we compute of these four variables are useful for generalizing from the information in our sample to the larger population from which the sample came.

Standard normal distribution: The bell curve

There is nothing "normal" about the standard normal distribution, but perhaps this word is used in English to maintain a neutral stance between its German name (Gauss distribution) and its French name (DeMoivre distribution). Figure 5.2 shows a normal distribution, or bell curve. This distribution, the most easily recognized and aesthetically pleasing one, is famous for its shape, which resembles a bell in a bell tower. Among its characteristics is its symmetry, with equal areas under the curve on both sides of the midpoint.

One way to think of the normal or z-variable is to imagine a large number of observations of a variable, each written on a piece of paper that is put into a barrel. Each value is called a z-score. (Use of the letter z does not have special significance.) Most of the values of the z-variable are in the range from -2.00 to 2.00; more specifically, 95% of the z-values lie between -1.96 and 1.96. Very few of the values of z are smaller than -3.00 or larger than 3.00.

The mean of the z-values equals 0.00, and their standard deviation equals 1.00. (These numbers are arrived at using some fancy mathe-

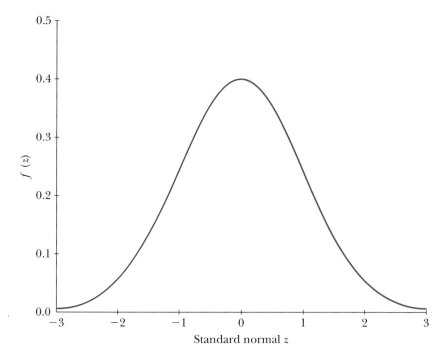

Figure 5.2 Distribution of the standard normal variable z

matics, but the mean and standard deviation can still be imagined as computed from a large number of observations, as discussed in Chapter 4.) A normal variable with mean 0 and standard deviation 1 is said to have the *standard normal distribution*.

A histogram might give a better sense of the distribution of these values. The range of values can be divided into small intervals and the number of z-scores in each interval represented by a rectangle whose area equals the proportion of observations that fall in the interval. But the histogram is messy. When the intervals are very small, the rectangles are very narrow, and the vertical lines for the rectangles are very close to each other. To clean up the graph, the vertical sides of the rectangles could be eliminated, leaving only the tops of the rectangles, which would look almost like a smooth curve, as shown in Figure 5.2.

A bell curve describes many phenomena in the real world, for example, height and weight. It also describes psychological test scores of many kinds; the curve has become a focus of contention regarding whether it measures the distribution of intelligence test scores for different ethnic groups.

Just as the total area of the rectangles in a histogram can be said to equal 1, so can the total area under the bell curve; the area of each thin rectangle is the proportion of z-values in the corresponding interval, and these proportions add up to 1. We see from Figure 5.2 that the shape of distribution is unimodal and symmetric around the value 0. Because of this symmetry, the area under the curve to the right of 0 equals 0.5 and the area to the left of 0 also equals 0.5.

Because of the design of the curve, the probability equals 0.95 that a randomly chosen value of z lies between -1.96 and 1.96. Because the curve is symmetric, the probability equals 0.025 that a randomly chosen value of z is equal to or larger than 1.96. It is also true that the probability equals 0.025 that a randomly chosen value of z is equal to or smaller than -1.96. This distribution has been extensively studied, and tables have been created to show various areas under the curve. The tables are useful for calculating probabilities from z-scores. One such table (Statistical Table 1) is shown at the end of the book. There also exists an equation that describes the curve.

The main use of the standard normal distribution is in finding the probability of any particular z-value and more extreme values. For example, suppose $z = 2.34$. Does that value belong to an unusual set of values or not? From looking at Statistical Table 2, the probability of z being equal to or larger than 2.34 is $p = 0.0096$. Only 96 of 10,000 z-values are larger than 2.34. Since this probability is very small, the observed value of z belongs to an unusual—even far out—group of z-values. Unusual z-values are discussed further in Chapter 7 on hypothesis testing.

The *t*-distribution

Around 1900, statisticians began to suspect that the standard normal distribution was not always the correct distribution to use for finding probabilities. William Gosset, a chemist who worked for Guinness Breweries in Dublin, Ireland, with a minor in mathematics, was one of these curious people. He decided to examine empirically whether the standard normal distribution was always the correct one to use in problems of probability.

Rather bizarrely, Gosset started his explorations by obtaining data on the height and left middle finger length of 3,000 criminals. From each of the two data sets (height and finger length), he selected samples of four observations of each variable, which gave him two groups of 750 different samples. For each sample he computed a value he called *t*. Then he made two histograms to see what the distribution looked like for all the *t*-values in each sample. How close did they match the standard normal distribution?

Gosset found that the shapes of his two histograms were close but not identical to the shape of the normal distribution. He called the new distribution the *t-distribution,* and the values he computed are still known as *t*-values. When he published his results, he signed his paper with the pseudonym "Student" because his employers did not like their people to publish papers for fear of giving away secrets about how to brew beer. Thus, the *t*-distribution is sometimes known as *Student's t.*

Later, Fisher did mathematically what Gosset had done empirically; he derived the mathematical function for the curves that display the *t*-distributions. Today this is by far the most common distribution in use.

Degrees of freedom: Different distributions for different degrees There is a whole family of *t*-distributions, and each member of the family is a little different from the other members. Envision not just one but a whole collection of barrels, each full of slips of paper with *t*-values on them. To distinguish between the *t*-distributions, they are numbered 1, 2, 3, . . . and the numbers are known as *degrees of freedom,* abbreviated d.f. or df. In dealing with the *t*-distribution with 10 degrees of freedom, we go to the barrel marked 10.

The statistical equivalent for the barrels is a table for probabilities for *t*-values. When statisticians use the *t*-table (Statistical Table 2), they go to the row labeled 10 degrees of freedom. The size of the sample partly determines which *t*-distribution they use, and readers of results are always told how many degrees of freedom were used in an analysis, since that is not an easy number to determine.

It is possible to find the graph of each *t*-distribution based on the idea of a histogram with small intervals described for the *z*-variable. The graph in Figure 5.3 shows the *t*-distribution for 10 degrees of freedom. The total area under the curve equals 1.00, just as for the normal distribution. The distribution is unimodal, and it is symmetric around the value $t = 0$. This sounds just like the description of the normal distribution, and it is hard to see any difference between Figures 5.2 and 5.3. But there are some differences.

The normal and the t-distribution One way to see the difference between the normal distribution and a *t*-distribution is to put the two curves together in one figure (Figure 5.4). The two curves have the same basic

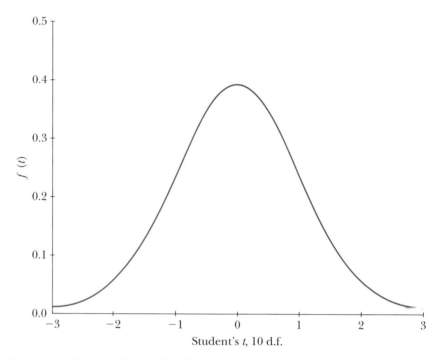

Figure 5.3 Graph of the *t*-distribution with 10 degrees of freedom

shape, but the normal distribution is higher in the middle and the *t*-distribution does not converge with the horizontal axis as quickly as the normal distribution. The differences indicate that the *t*-values are less concentrated around the mean than are the normal *z*-scores.

For example, the probability that *z* is larger than 2.5 equals 0.0062, while the probability that *t*, for 10 degrees of freedom, is larger than 2.5 equals 0.0152. In other words, only 62 of a thousand *z*-values are larger than 2.5, while 152 of a thousand *t*-values are larger than 2.5. Also, with 10 degrees of freedom, 95% of the *t*-values lie in the interval from −2.23 to 2.23. This means that we go farther away from the midpoint to take in 95% of all the *t*-values than we do in the case of the normal distribution. Recall that in the normal distribution, 95% of the values lie between −1.96 and 1.96.

As the number of degrees of freedom gets larger, the curve for the *t*-distribution gets closer to the curve for the normal distribution. After 30 degrees of freedom it is very difficult to tell the two curves apart, and by 50 degrees of freedom the two are almost identical. This is the reason why statistical tables for the *t*-distribution go up to only about

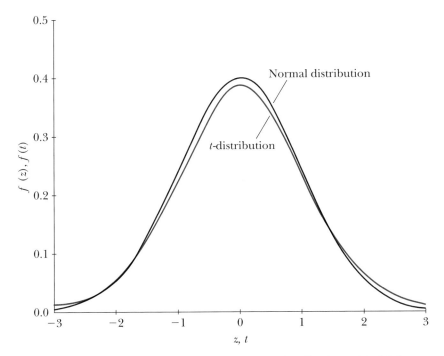

Figure 5.4 Standard normal distribution and *t*-distribution with 10 degrees of freedom

100 degrees of freedom; after that the table for the normal distribution can be used.

Chi-square distribution

The chi-square variable is named for the Greek letter χ (chi, pronounced kī). (Chapter 9 explains its place in statistical work.) The chi-square distribution, like the *t*-distribution, is also a family of distributions, not just a single distribution. Again, think of many barrels full of slips of paper, this time with chi-square values written on them. The chi-square distributions are numbered 1, 2, 3, . . . , and these numbers are also known as degrees of freedom. Thus, in dealing with the chi-square distribution with 3 degrees of freedom, we go to the barrel marked 3 degrees of freedom—in the chi-square statistical table (Statistical Table 3), the row marked 3 degrees of freedom.

Just as with the standard normal distribution and the *t*-distribution, it is possible to graph each chi-square distribution based on the idea of a histogram with small intervals. The graph in Figure 5.5 shows the

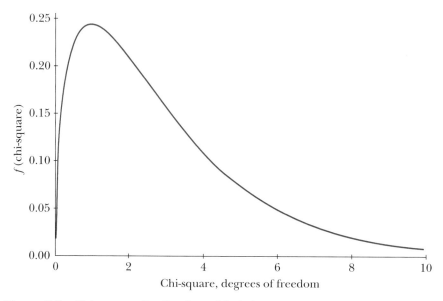

Figure 5.5 Chi-square distribution with 3 degrees of freedom

chi-square distribution for 3 degrees of freedom. The total area under the curve equals 1, just as it does for the normal and *t*-distributions. But the shape of the chi-square distribution looks very different from those of the *t*- and *z*-distributions. This is because the chi-square variable has no negative values; the graph starts at 0. The distribution is skewed (that is, asymmetrical), and most of the values of this variable are less than 8 or so. Only 5% of the chi-square values are larger than 7.82. Another way of saying this is that the probability equals 0.05 that a randomly chosen value of chi-square with 3 degrees of freedom is equal to or larger than 7.82. The mean of the chi-square values equals 3, the same as the degrees of freedom.

We use a chi-square distribution the same way we use the normal or a *t*-distribution. If a statistical problem requires us to compute a value of chi-square (with a certain number of degrees of freedom) from our data, then we use the chi-square distribution to find the probability of getting that or a larger value of chi-square. If the probability is small, then the value of chi-square is unusual; this means an unusual sample result. The result enables us to draw conclusions about our data and the larger population from which the sample was taken. This idea is pursued further in Chapter 7 on hypothesis testing.

F-distribution

The family of *F*-distributions is named in honor of the great English statistician Sir Ronald Fisher. Again, imagine a whole collection of barrels full of numbered slips. Each barrel represents an *F*-distribution and is marked by a pair of numbers, for example 4 and 40. This barrel represents the *F*-distribution with 4 and 40 degrees of freedom. A fairly detailed *F*-table would have information on as many as 1,000 different *F*-distributions.

Figure 5.6 shows the graph for the *F*-distribution with 4 and 40 degrees of freedom. The graph shows that, like the chi-square variable, the values of the *F*-variable are not negative, and the values of F for most *F*-distributions tend to lie in the range from 0 to about 5. With small numbers of degrees of freedom, the values of F are somewhat larger. For this particular *F*-distribution, most of the values of the *F*-variable seem to be less than 3.

According to the table of *F*-distributions (Statistical Table 4), 5% of *F*-values are larger than 2.45 and only 1% is larger than 3.51. Thus, the probability equals 0.05 that a randomly chosen value of F with 4 and 40 degrees of freedom is larger than 2.45. When we compute a value of F on 4 and 40 degrees of freedom from data and find it to be

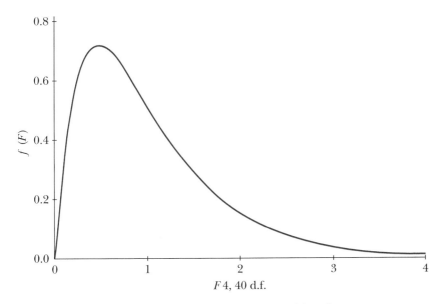

Figure 5.6 *F*-distribution with 4 and 40 degrees of freedom

larger than 2.45 (or maybe even larger than 3.51), then we have found an unusual *F*-value.

Need for normally distributed data

Everything we have said about these four theoretical statistical variables can be studied mathematically. The three variables t, chi-square, and F are all derived from the normal variable z, so every time a t-, chi-square, or F-variable is used, certain assumptions are made that the observed data follow normal distributions. For data that do not follow normal distributions, it is sometimes inappropriate to use any of these three distributions.

5.6 USING PROBABILITIES TO CHECK ON ASSUMPTIONS

Is it a fair coin?

Any probability is based on certain assumptions being true. If you tell me you have a coin in your hand and ask me for the probability that it will come up tails when tossed in the air, I would say that the probability is one half. I say this because I *assume* that your coin is a regulation U.S. Mint coin. But it could be a fake coin with both sides showing heads, and then the probability of tails would be zero.

Assume that you are a magician practicing coin tricks, and I do not know whether you hold a genuine coin or a fake coin. If you do not want to show me your coin, I could check my assumption that your coin is a fair coin by collecting data on the coin. Suppose you toss the coin 10 times and it comes up heads each time. The data consist of 10 heads, and I can now find the probability of these data using my assumption that you have a fair coin. Under the assumption of a fair coin, the probability of 10 heads and 0 tails can be found from the binomial distribution. The probability is equal to $(1/2)^{10} = 1/1{,}024 = 0.001$; only 1 in 1,024 times can we expect a fair coin to come up heads 10 times in 10 tosses. Under the assumption of a fake coin with heads on both sides, the probability of 10 heads in 10 tosses simply equals 1 because the coin would always land heads up when tossed.

Now there are two possibilities:

1. The assumption of a fair coin is *correct,* and the observed data have a very small probability of 0.001.

2. The assumption of a fair coin is *incorrect,* and the probability of the data is higher than 0.001.

One of these possibilities is true, even though I do not know which one it is. The first possibility is based on the assumption of a fair coin, and I computed the probability of the observed data using that assumption. In this possibility, the probability of the data is very small. The other possibility is that I do not think the observed data have that small a probability; after all, I observed the tosses of the coin, and it didn't seem to me that anything unusual actually took place. These are the data that did occur, and events that actually occur do not usually have small probabilities.

Now I must choose between the two explanations for what happened. I run the risk of choosing the wrong explanation, since I will never know which of the two explanations is the correct one. But because the probability of the data is so small for the first possibility, I choose the second possibility, where the probability of the observed data is higher. Indeed, events that have large probabilities occur much more often than events that have small probabilities. Having made the choice, I can now to say something about the coin based on the observed data: The coin is not fair!

To summarize, first we make certain assumptions about the world under study. Then data are collected and the probability of obtaining the data based on the assumptions is computed. Finally, if this probability is very small, say less than 0.05, the conclusion is that the assumptions must have been wrong in the first place. In the case of the coin, there is strong evidence that the assumption of a fair coin is incorrect.

This line of reasoning is very important to the basic rules of scientific investigations, and it is used in Chapter 7 on hypothesis testing and elsewhere.

Is it a fair workplace?

Consider another example where certain assumptions are made and then the probability of the observed data based on that assumption is computed. You work in a group of 10 people at an office; 5 of the employees are men and 5 are women. A committee of 4 employees is to be formed to study certain issues related to gender in the office environment. Some people assume that the management wants to have as many women on the committee as possible. The employees want the selection of people to the committee to be random. Management says they will choose the committee randomly, but when the committee is announced, its members are 4 women and 0 men. Did management really choose this committee randomly or did some other selection criterion influence the choices?

The possible explanations for what is going on are similar to the explanations for the coin toss:

1. The assumption of random selection of committee members is correct, and the observed data have a very small probability of 0.02.

2. Management used some other way of making the committee assignment, and the probability of the data is higher than 0.02.

Using the assumption that the assignment to the committee was truly random, you follow the general principle and compute the probability of getting the observed data or more extreme data. If this probability is small, you will have reason to doubt management's claim.

The data cannot be any more extreme, since all 4 committee members are women. If the claim by management of randomness is true, then the probability of getting the observed data can be computed to equal 0.02 (see Figure 5.7). This probability is very small, and it tells you that it is very unlikely that 4 women could be chosen if the selection was random. This kind of committee would occur very seldom if many committees were created at random; only 2 out of 100 committees would have this particular configuration. The fact of such an unlikely committee makes you question the assumption of random assignment

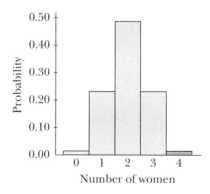

Figure 5.7 Probability distribution for number of women in a randomly selected committee of 4

to the committee. As with the coin example, you choose the second possibility, based on the small probability of the data under the assumption of random assignment. You go back to management and tell them the data are such that the employees do not believe the committee was selected randomly.

In addition to the probability of 4 women and 0 men, Figure 5.7 also shows the probabilities of 3 women and 1 man (0.24), 2 women and 2 men (0.48), 1 woman and 3 men (0.24), and 0 women and 4 men (0.02). The computations are shown at the end of the chapter as Formula 5.1. The actual observed value of number of women is 4, and the probability of this value is shaded in the graph. Such a shaded area is known as a tail probability, since it occurs in one of the tails of the distribution.

Is it an evenly split electorate?

Following is a larger and perhaps more realistic example of using probabilities to check assumptions. Before an election, we do a survey and ask people how they would vote if the election were held today. We find that 650 people would vote for the Democratic candidate and 550 would vote for the Republican candidate. We want to use the data to predict how the entire electorate will split between the two candidates.

First, we assume that the electorate is evenly split between the two parties: the probability equals 0.5 that a randomly chosen person plans to vote for either party. Using that assumption, we compute the probability of getting 650 or more Democrats in a sample of 1,200. We need the probability of 650 *or more* because the probability of any specific number of Democrats is very small. By computing the probability of 650 or more, we can find out whether 650 belongs to a set of values that all have small probabilities or to a set of values that could occur quite often.

To find the probability of 650 or more, assuming a probability of 0.5 that a person is a Democrat, we could use the binomial distribution. But that would be a great deal of work because we would have to use the binomial formula to find the probability of 650, of 651, of 652, and so on. Instead, we use the standard normal distribution to find the probability (discussed in greater detail in Chapter 7). Based on the assumption of an even split in the electorate, then 650 out of 1,200 can be translated to a z-value of the standard normal variable equal to 2.89. The probability that z is equal to or larger than 2.89 is 0.002, which means that the probability of 650 or more Democrats in 1,200 is also equal to 0.002. Thus, assuming an evenly divided electorate, we would get 650 or more Democrats in a sample of 1,200 votes in only 2 of 1,000 different samples.

Such a probability is known as the *p-value* for data. The *p*-value is shown as the shaded part under the curve in Figure 5.8. If the electorate is evenly split, then the *p*-value for our data is a small 0.002. This may mean some unusual data in the one sample, or it may mean that our basic assumption of an evenly split electorate is not correct. We prefer to think that the small *p*-value indicates that the assumption of an evenly split electorate is not correct. Thus, the sample tells us that the electorate is not evenly split and that the Democratic candidate would win if the election were held today.

This kind of conclusion is typical of conclusions of statistical analyses. First an assumption is made about the population from which the data were generated. Then a limited amount of data are collected and certain computations made on them. Depending on the results of the computations, the decision is made about whether the original as-

> A *p-value* is the probability of the observed data or more extreme data, given some basic assumptions about the data.

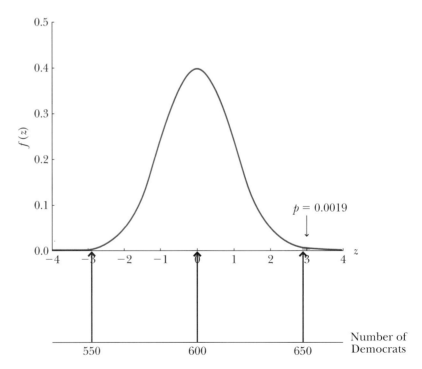

Figure 5.8 Normal distribution with *p*-value for $z = 2.89$

sumption is correct or incorrect. With only limited data, some uncertainty about whether the conclusion is correct or not remains.

5.7 DECISION ANALYSIS: USING PROBABILITIES TO MAKE DECISIONS

We are constantly faced with choices and decisions to be made, and in some very informal ways, we make most of our decisions on the basis of probabilities. When we need to travel a long distance, we are faced with the decision whether or not to fly. Airplanes crash from time to time, so there is a small probability the plane we take will crash with possible loss of lives. We could, of course, not go and avoid the chance of a crash. But the probability of a crash is so small that people do fly every day and do get to their destinations. We have to evaluate the probability of a crash against the benefit of getting where we want to go. This balance between risks and benefits forms the basis for a process known as *decision analysis*.

Decisions in the face of uncertainty are made on the group as well as the personal level. On the group level, decisions are made on setting public policy in the face of uncertainties about outcomes of new plans and laws. Because of the health risks from pollution in the air, should a community close down all sources of pollution such as cars and factories to eliminate pollution? Most do not seem willing to; they avoid this choice because the closings would also have other and perhaps less desirable consequences. Should a bridge be closed if there is a small probability that it will collapse? A closed bridge will send the traffic on detours and perhaps increase the chances of traffic accidents on smaller roads.

Scientists and statisticians often combine forces to develop forms of decision analysis based on probabilities. Decision analysis was used to form the decision to evacuate Americans from Lebanon in a time of trouble there; it was used to lift an embargo on the sale of advanced computers to the former Soviet Union; it was used in the Star Wars project.

How can we measure the amount of uncertainty? If the probability of an event equals 1, then the event is certain to happen all the time. With the probability of, say, 0.95, the event is somewhat less certain to happen; when the probability is that high, the event happens often but no longer always. Similarly, a probability of 0.05 means that the event will almost certainly not happen. The most uncertainty occurs when the probability of an event equals 0.5. The event happens and does not happen equally often, and it is impossible to predict what will happen next time. One way to measure the uncertainty in a situation is to create the product $p(1 - p)$. When the probability is 0 or 1 then there is no uncertainty, and the product equals 0. The product has its largest value of 0.25 when $p = 0.5$, the probability of most uncertainty.

Many of the probabilities that form the basis of daily decisions are not well known. We know that the probability of an airplane crash is small, but we do not know what the probability is. An engineer would have difficulty giving an exact probability that other engineers would agree on of a bridge collapse. When we are deeply in love and decide to get married, we do so with the sense that the probability of the marriage succeeding is very high. If we knew that the probability was 0.6, we might decide not to go ahead with the wedding. Of course, recent divorce data suggest that many people have an unrealistic sense of the probability of a successful marriage, while others may go ahead anyway, even when they know the exact probability.

Even when we know a probability, it is sometimes hard to comprehend what such a number means. What can we make of a statement that the probability of dying from drinking a can of soft drink with saccharin every day equals 0.00001? It helps to change that to 1 person in 100,000, but even that is hard to comprehend fully. Sometimes it is easier to understand small probabilities when they are compared with other probabilities. If we read that the average cigarette habit is 100 times as dangerous as drinking a diet soft drink, then it is a bit easier to understand the risks involved.

Some evidence indicates that people have excessive reactions to small probabilities. In the late 1980s, isolated cases of terrorism, in which tourists were killed, occurred in Europe; the reports had a great impact on tourism at the time. The number of acts of terrorism did not exceed the number of other kinds of individual accidents that normally take tourists lives, such as traffic accidents, suicides, heart attacks, drownings, and food poisonings. Still, tourism rates were greatly reduced that year, which suggests that many people perceived the probabilities of terrorism acts as high enough to cancel their travel plans.

Making decisions can be troublesome in part because probabilities are not static. Our personal probabilities of uncertain events are constantly upgraded in the face of new evidence. Our probability that it will rain next weekend firms up increasingly as we follow the weather developments during the week; after watching the forecast on TV Sat-

PERSONAL PROBABILITIES

Psychologists have long been fascinated with personal probabilities and have tried to assess some of the factors that cause people's personal probabilities to differ from those of the statistician. In one study, women were shown a variety of nightgowns and asked which one of them they would be most likely to buy. The gowns were randomly reordered for each subject so that they were constantly seen in different orders. The researchers assumed that the most beautiful nightgown would have the highest probability for selection and the least attractive gown the lowest probability for selection. The results of the study indicated that, in general, the nightgown seen first received the highest rating, regardless of which gown it was. The second most preferred gown was the one seen last. The subjects were not aware that the order in which the gowns were presented affected their choices. *(Source: R. E. Nisbett, E. Borgida, R. Crandall, and H. Reed, "Popular induction: Information is not necessarily informative," in J. S. Carroll and J. W. Payne (eds.),* Cognition and Social Behavior, *Hillsdale, NJ: Lawrence Erlbaum, 1976.)*

In other studies, the effects of stereotypes were clearly shown. Subjects were told that a panel of psychologists gave the following description of a person drawn at random from a sample of 70 engineers and 30 lawyers: "John is a 39-year-old man. He is married and has two children. He is active in local politics. The hobby that he most enjoys is a rare book collection. He is competitive, argumentative, and articulate." The subjects were asked: What is the probability that John is a lawyer rather than an engineer? Note that according to the numbers in the sample, the probabilities were 0.7 engineers and 0.3 lawyers; the chances were 7 in 10 that a randomly chosen person was an engineer and 3 in 10 that he was a lawyer. But 95% of the subjects chose lawyer over engineer. They tended to ignore the base rate of engineers to lawyers and over emphasize the information that seemed to fit cultural stereotypes of what lawyers are like. *(Source: D. Kahneman and A. Tversky, "On the psychology of prediction,"* Psychological Review, *vol. 80 (1973), pp. 237–251.)*

There are other reasons people rely on personal probabilities over statistical probabilities: the concreteness of an event—for example, developing a belief that you will acquire a rare disease because someone you know has it; the strong opinions of a close friend—"There's no future in majoring in mortuary science"; one's own personal experiences—"I don't care what the statistics say, smoking does not harm me"; or the word of authorities—"Marijuana causes senility."

urday night, our probability for rain the next day will be quite close to either zero or one.

The way in which people upgrade probabilities in the face of new information is often inconsistent with the mathematical theory for upgrading probabilities. People tend to be more conservative in their assessment of probabilities than they should be and do not move their probabilities quickly enough toward 0 or 1. If the probability of an event is 0.5 and new information is obtained, statisticians may recalculate the probability as 0.9. But when people are asked about the new probability, most will have changed their personal probability less. The many reasons for this conservatism include reluctance to change one's mind or to understand events in radically new ways. Other influential factors also affect how people use or misuse information in making decisions (see the box on page 209).

5.8 SUMMARY

A probability is a number between 0 and 1 that tells us how like event will happen.

5.1 How to find probabilities

There are three major ways of finding the numerical values of probabilities: equally likely events, relative frequency, and personal assessment. When events are equally likely, the probability of a desired outcome is found by dividing the number of favorable outcomes by the total number of outcomes. When records or an event have been kept over long periods of time or with very large samples, the proportion of times an event occurs is a good estimate of the probability of the event. For unique events, the probability is a personal evaluation of the likelihood of the event occurring, based on all available information.

5.2 Rules for computations with probabilities

As numbers, probabilities can be added and subtracted, as well as multiplied and divided. Such computations can help in finding the probabilities of more complicated events from the probabilities of simpler events. The probability of one event or another event, when both events cannot both occur at the same time, is the sum of the probabilities of the two events. The probability of one event and another event happening is the product of the probabilities of the two events, when the two events are independent of each other.

5.3 Odds: The opposite of probabilities

Odds express in whole numbers how many times an event fails to occur versus the number of times it occurs. Odds of 5 to 1 means that the event does not occur in 5 of 6 trials and does occur in 1 of 6.

5.4 Probability distributions for discrete variables

The binomial distribution finds the probability that one of two events occurs a certain number of times in a total of n trials; the binomial distribution is convenient to use only for small samples. For an event with a small probability and many possibilities, such as a no-hitter baseball game, the Poisson distribution is more useful than the binomial distribution for finding the probability. The Poisson distribution can be regarded as a specialized case of the binomial distribution.

5.5 Probability distributions for continuous variables

Four theoretical variables that can be used to find probabilities are the standard normal z, the t, the chi-square, and the F. Each variable has its own distribution and thus its own characteristically curved shape when graphed.

The standard normal curve is bell-shaped, with 50% of the observations on either side of the midpoint. The mean of the standard normal observations is set to 0 and the standard deviation to 1. 95% of the area under the curve is located between -1.96 and 1.96. Sample scores can be converted to special z-scores. From tables of the standard normal variable we can find probabilities that z-scores are larger or smaller than certain values.

The t-distributions, sometimes known as Student's t, are similar but not identical in shape to the normal distribution. They are the most commonly used statistical distributions. The number of degrees of freedom and indirectly the number of observations determine which of the t-distributions to use.

The distributions of the chi-square are skewed, with values starting at 0. The F-distributions are also skewed, with values starting at 0. The F-distribution depends on a pair of degrees of freedom.

By definition, the three variables t, chi-square, and F are all derived from the normal variable. To use these statistical variables, the data should be normally distributed.

5.6 Using probabilities to check on assumptions

A *p*-value is the probability of obtaining the observed data or more extreme data, given some basic assumptions about the data.

5.7 Decision analysis: Using probabilities to make decisions

Scientists and statisticians often combine forces to develop forms of decision analysis based on probabilities.

The way in which people upgrade personal probabilities in the face of new information is often inconsistent with mathematical theory.

ADDITIONAL READINGS

Chernoff, Herman. "Decision Theory." In William H. Kruskal and Judith M. Tanur (eds.), *International Encyclopedia of STATISTICS*. New York: The Free Press, 1978. Brief discussion of how statistics can be used to help decision making.

Fairley, William B., and Frederick Mosteller. *Statistics and Public Policy*. Reading, MA, 1977, "People v. Collins. The Supreme Court of California" and "A conversation about *Collins*." Interesting uses of probabilities to determine guilt or innocence.

Huff, Darrell, and Irving Geis. *How to Take a Chance*. New York: W. W. Norton, 1959.

Snell, F. Laurie. *Introduction to Probability*, New York: Random House, 1984.

FORMULAS

ODDS AND PROBABILITIES

If we are given odds of a to b of an event happening, then the probability p of the event happening is

$$p = \frac{b}{a+b} \tag{5.1}$$

This equation can also be written

$$p = \frac{b/a}{1+b/a} = \frac{\text{odds}}{1+\text{odds}} \tag{5.2}$$

It is also possible to go the other way, from probabilities to odds. If we solve the equation above for the odds b/a, we find

$$\frac{b}{a} = \frac{p}{1+p} \tag{5.3}$$

In the Olympics site example in Section 5.3, the probability of Sydney getting the 2000 summer games was 0.692:

$$\frac{b}{a} = \frac{0.692}{1-0.692} = 2.25$$

This means that the odds are 1 to 2.25, the same as odds of 4 to 9 when converted to whole integers. Odds are given as "against the event happening (a)" to "for the event happening (b)." Thus, the probability for the event happening is $p = b/(a+b)$, and the probability against the event happening is $a/(a+b)$.

BINOMIAL PROBABILITIES

The two values of the dichotomous variable are often called success and failure. Let π be the probability of success and $1-\pi$ the probability of failure. In a sample of n observations, the number of successes is x and the number of failures is therefore $n-x$. The probability of x successes and $n-x$ failures, the mean number of successes, and the standard deviation of the number of successes are

$$p(x \text{ successes in } n \text{ trials}) = \binom{n}{x} \pi^x (1-\pi)^{n-x}$$

$$= \frac{n!}{x!(n-x)!} \pi^x (1-\pi)^{n-x} \tag{5.4}$$

$$\text{mean of } x = \mu = n\pi \tag{5.5}$$

$$\text{standard deviation of } x = \sigma = \sqrt{n\pi(1-\pi)} \tag{5.6}$$

where the exclamation mark means a long (factorial) multiplication:

$$n! = n(n-1)(n-2)\cdots(3)(2)(1)$$

The parentheses enclosing n over x is the binomial coefficient known as "n choose x," and it tells us in how many ways we can choose x objects from a group of n objects.

As an example, let $n = 4$, $x = 3$ and $\pi = 0.49$. The probability of 3 successes and 1 failure is

$$\binom{4}{3}0.49^3(1-0.49)^{4-3} = \frac{4!}{3!(4-3)!}0.49^3(1-0.49)^{4-3}$$

$$= \frac{(4)(3)(2)(1)}{(3)(2)(1)(1)}(0.49)^3(0.51) = 4(0.49)^3(0.51)$$

$$= 0.24$$

The mean and standard deviation are

$$\text{mean } \mu = 4(0.49) = 1.96 \text{ girls per family}$$

$$\text{standard deviation } \sigma = \sqrt{4(0.49)(1-0.49)} = 1.00 \text{ girls per family}$$

POISSON PROBABILITIES

Let the mean number of occurrences be denoted μ. The probability of the event occurring x times is then found from the expression

$$p(x) = \frac{e^{-\mu}\mu^x}{x!} \tag{5.7}$$

The mean and standard deviation of the number of occurrences are

$$\text{mean} = \mu \tag{5.8}$$

$$\text{standard deviation } \sigma = \sqrt{\mu} \tag{5.9}$$

If the phone rings on the average 2.1 times an hour, what is the probability that there will be 5 phone calls in an hour? We substitute $\mu = 2.1$ in the formula for the Poisson probabilities and find

Table 5.3 Gender and committee selection

	Females	Males	Total
Selected	4	0	4
Not selected	1	5	6
Total	5	5	10

$$p(5) = \frac{e^{-2.1}2.1^5}{5!} = 0.042$$

In only about 4 of 100 hours will there be as many as five phone calls. The standard deviation of the number of calls is $\sigma = \sqrt{2.1} = 1.45$ calls.

HYPERGEOMETRIC PROBABILITIES

The data in the example about fairness in the workplace in Section 5.6 can be displayed as shown in Table 5.3. Each of the 10 people belongs in one of the four cells in the table. The two sets of totals are fixed, since we know the distribution of the gender variable and how many people are to be selected for the committee. The four numbers inside the table are random and would not necessarily be the same if another committee were formed.

Generalized data are shown in Table 5.4. The total of all the objects is n, b of one kind and r of the other kind. From the n objects, m are randomly chosen and are not replaced after they have been chosen. The probability of getting x of one kind is

Table 5.4 Selecting x objects from b objects of Type 1 and $m - x$ objects from $n - b$ objects of Type 2

	Type 1	Type 2	Total
Selected	x	$m - x$	m
Not selected	$b - x$	$n - b - m + x$	$n - m$
Total	b	$n - b$	n

$$p(x) = \frac{\binom{b}{x}\binom{n-b}{m-x}}{\binom{n}{m}} \tag{5.10}$$

The two numbers with their parentheses are a binomial coefficient, computed same as explained for binomial probabilities (Formula 5.4). For the committee example:

$$p(4) = \frac{\binom{5}{4}\binom{5}{0}}{\binom{10}{4}} = \frac{(5)(1)}{\frac{(10)(9)(8)(7)}{(4)(3)(2)(1)}} = \frac{(5)(24)}{5040} = 0.02$$

MEAN AND VARIANCE FROM PROBABILITY DISTRIBUTIONS

When a discrete random variable can take on the values $x_1, x_2, \ldots, x_k$ with probabilities $p(x_1), p(x_2), \ldots, p(x_k)$, the mean μ and the variance σ^2 can be computed as follows:

$$\mu = x_1 p(x_1) + x_2 p(x_2) + \cdots + x_n p(x_n)$$

$$\sigma^2 = (x_1 - \mu)^2 p(x_1) + (x_2 - \mu)^2 p(x_2) + \cdots + (x_n - \mu)^2 p(x_n)$$

EXERCISES

REVIEW (EXERCISES 5.1–5.29)

5.1 What are some synonyms for *probability* that we use in daily life?

5.2 Find an example of the use of probabilities in a newspaper article and describe how the probabilities are used.

5.3 How would you define *probability* according to this text?

5.4 Probabilities can range in values from what number to what number?

5.5 What is the meaning of the basic formula k/n for finding the value of a probability?

5.6 Describe a method for arriving at the conclusion that the probability of drawing a heart from a deck of cards is 0.25.

5.7 If you knew that the probability of a student getting expelled from school for cheating was 0.12, how would you find out what the probability of not getting expelled for cheating would be?

5.8 When the "real" probability of an event is not known, then the probability can be estimated. Describe a method for estimating the probability of an event.

5.9 a. When you tell one friend that the probability of your going to the holiday dance with a particularly engaging new friend is 0.90, what kind of probability are you giving?

b. What kind of event are you giving a probability for?

5.10 a. What is the difference between probabilities and odds?

b. How are they related to one another?

c. Why are odds preferred over probabilities in daily betting events?

5.11 A variable with two values, such as heads or tails for the outcome of a coin toss, forms the basis of a binomial distribution. What does the word *binomial* mean?

5.12 In a binomial distribution, what do the probabilities of the different values of the binomial variable add up to?

5.13 Revisit the family that has 4 children.

a. If you multiply by 4 the probability of a birth resulting in a girl, what does the product tell you?

b. Which Greek letter designates this product?

5.14 a. When do statisticians use a Poisson distribution to find particular probabilities?

b. Describe a situation in which you would think using a Poisson distribution of probabilities would be a good idea.

5.15 It is possible to find the standard deviation of a variable from the probability distribution of the variable. What Greek letter designates such a standard deviation?

5.16 What is a major difference between the binomial and the Poisson distribution?

5.17 a. Name the four major continuous theoretical variables used in statistics.

b. Which of these variables is most commonly used?

5.18 a. Name three important characteristics of the standard normal distribution.

b. What value is given for the total area under the curve of the standard normal distribution?

c. What is the mean of the z-variable of the standard normal distribution?

d. What is the mode of the standard normal variable?

e. What is an evenly balanced distribution called?

f. Most of the values of the z-scores of a standard normal variable are found between which two values?

5.19 a. Name the distribution developed by a brew master.

b. What was his pseudonym, and why did he use one?

5.20 We have a family of t-distributions, a family of chi-square distributions, and a family of F-distributions.

a. How do we distinguish between the members of a particular family of distributions (use t-distributions as an example)?

b. When do the distributions in a family begin to look very similar to one another?

5.21 When you use a statistical table to look up the probability for a chi-square value, what must you know in addition to the value of the chi-square?

5.22 a. In a chi-square distribution with 2 degrees of freedom, what is the total area under the curve equal to?

b. In a chi-square distribution with 3 degrees of freedom, does the total area under the curve change?

5.23 a. If someone told you they found a chi-square value for a given problem equal to -11.11, would your reaction be?

b. If the chi-square value was 11.11, would you be impressed at its size? Why or why not?

5.24 a. What does F stand for in the F-distribution?

b. Except for very low degrees of freedom, what generally is the range of F-values?

5.25 In a sample of voters, 700 people indicate that they are going to vote for a Republican candidate. Assuming an even split in the electorate, the *p*-value for these data is 0.002. Should you conclude that the voters are split 50/50 between the two parties or not?

5.26 If an economist indicates that she works in the area of decision analysis, what kinds of problems might she study?

5.27 "Statistics means never having to say you're certain" may be both funny and true; can you explain why?

5.28 In Section 5.3 on odds, probabilities are given for the possibilities of various cities getting the Olympic games in the year 2000. What kind of probabilities are they? That is, are they probabilities based on equally likely outcomes, probabilities based on long-run proportions, or personal probabilities? Explain.

5.29 Give an example of a probability based on equally likely outcomes, one based on a long-run proportion, and one based on personal opinions.

INTERPRETATION (EXERCISES 5.30–5.52)

5.30 When a political analyst says, "I believe the probability is 0.6 that the President will be reelected next year," what type of probability is it and what does the analyst mean by the statement?

5.31 At the stock car races, your friend tells you the odds for three racers on which you are interested in betting: Trudi 3 to 2, Andy 8 to 2, and Rod 20 to 1. If you want a "sure thing," who would you bet on? If you want to make a "big killing" and don't care if you lose small, how would you bet? Explain your strategies.

5.32 Suppose 32% of the adult males in the United States own at least one gun. It can then be shown that the probability equals 0.0015 that in a sample of 300 adult males, 120 of them or more own guns.

    a. How could you find the probability of 0.0015?

    b. What is such a probability called?

5.33   a. Give an example of an event with a probability near zero.

    b. Give an example of an event with a probability near 1.

5.34 A research report indicates that the probability equals 0.61 that a pregnant teenager who has an abortion has talked about the situation with both her parents. What would you say this probability statement means?

5.35 Explain why it is possible to use the binomial distribution to figure out how many girls or boys might be expected in families of various sizes, once one knows that boys are born 51% of the time, and why it is not necessary to go out and actually sample a large number of families of various sizes to collect the data.

5.36 a. The following indicates the calculation done to estimate the probability of having three girls and a boy in a family: p(GGGB) = 0.49 * 0.49 * 0.49 * 0.51 = 0.06. Explain the reasoning behind this formula as though you were talking to a friend who is not taking statistics.

b. What would the formula be to calculate the probability of having a family with 4 boys?

5.37 To estimate the number of U.S. families with 4 sons and no daughters, which would be more accurate: using a binomial distribution or sampling 100 families drawn at random from a national survey? Why?

5.38 If you have a very large sample of respondents, how do you find the binomial probabilities if you do not wish to do the computations yourself?

5.39 The mean number of girls in families of 4 is equal to 1.96. This mean is derived from adding the possible values of girls in a family of 4, from 0 to 4, which have been multiplied by the probability of each of them. Explain how you would go about finding the mean number of girls in families with 3 children.

5.40 In the 1992 Presidential campaign, incumbent President Bush, in his efforts to discredit the pro-ecology stance of vice-presidential candidate Gore, stated in his last campaigning days, "If we aren't careful, we'll be up to our necks in owls, without any jobs for the people." How would you describe Bush's complaint in terms of decision analysis?

5.41 How would you convince a sixth-grade class that the probability of their breaking a bone in the next six years is 0.0009? A psychology class of undergraduates that the probability of their being in an automobile accident in the next year is 0.50? A BMW owner that the chances of the car being carjacked are 0.00001 worldwide?

5.42 According to the text, "the probability that z is larger than 2.5 equals 0.0062, while the probability that t, for 10 degrees of freedom, is larger than 2.5 equals 0.0152."

a. For which variable is more likely one would find an amount larger than 2.5?

b. Why is there a difference between the z and the t in this regard?

c. What would make the difference between the two statistics very small?

5.43 If a statistician indicated that the probability of getting a particular chi-square or larger was 0.46, what would you say about the event measured by this chi-square?

5.44 a. In what way does the chi-square distribution look different from the t-distribution? Explain.

b. What would be a very rare value of a chi-square variable with 3 degrees of freedom?

c. What other distribution looks similar to the chi-square distribution?

5.45 A social club has 52 old members and 7 new members. A supposedly random drawing results in 5 new members and no old members selected to do party clean-up. (The statistics major said that the probability of such an outcome was 0.000004.)

a. Would you suspect that something was not fair about the drawing? Explain.

b. As a new member, how would you defend your argument to the club officers? Fill in the table to clarify your argument.

	Old member	New member	Total
Chosen			
Not chosen			
Total			

5.46 a. If you assume that the members of your campus organization would be equally divided as to whether or not they wished to participate in the annual blood drive, so that the probability of a person participating is 0.5, and later you discover that 10 out of the 100 members gave blood, what do you conclude about your original probability?

b. What two conclusions might you want to draw from your data about your earlier hunch regarding support for the blood drive?

c. What is the statistical decision rule that could help you decide which of the two conclusions to draw?

5.47 Say that the probability that you may get hit by lightning while playing golf in Minnesota in the summertime is 0.00002.

a. What is an easier way of making this statistic meaningful to people?

b. Why do you think this is so?

5.48 For your vacation, you are trying to decide whether to fly to Egypt to see the pyramids on a cruise down the Nile or to fly to Miami to explore the Florida Keys in a rental car. In each locale tourists have, at times, been endangered by unfriendly locals.

a. What are some probabilities you might want to take into account when making your decision?

b. How might you calculate which you might prefer doing?

c. How important would recent historical events be to you in making your choice?

5.49 You are interested in contributing to one charity this year. You want to select one that does research that could have an impact on the largest number of victims.

a. How would you make a choice among a fund for breast cancer research, an AIDS research fund, the March of Dimes for Crippled Children, or the American Heart Association?

b. In what respect do advertising campaigns create the potential for poor decision making on the part of charitable donors?

5.50 In 20 consecutive baseball World Series, the team that had more stolen bases during the regular season won the championship 13 times and lost 7 times. If you use 0.5 as the probability of either team winning the World Series any given year, then the probability of winning 13 or more times out of 20 equals 0.13. What can you say about the assumption that the probability of winning the World Series is an even 50/50 based on the data on stolen bases?

5.51 In the example in Chapter 4 on the age of brides, the mean age of the grooms is 32.3 years and the mean age of the brides is 30.0 years, for a difference of 2.3 years. By changing 2.3 years to a value of the *t*-variable, you find that, if there is no difference in mean age between brides and grooms in the population of all couples getting married, then the probability of getting a sample difference of 2.3 years or more equals 0.002. From the magnitude of this *p*-value, what do you conclude about the assumption of no difference in mean ages in the population of all couples?

5.52 The U.S. Postal Service claims that 83% of letters mailed in New York City are delivered overnight. A person checked this out by mailing 10 letters to himself, and 4 of them arrived the next day. Using a probability of 0.83 for overnight delivery and 10 cases, the binomial distribution gives a probability of 0.0027 of 4 or fewer overnight deliveries. *(Source: Daniel Seligman, "Ask Mr. Statistics,"* Fortune Magazine, *July 24, 1995, pp. 170–171.)*

    a. How do you interpret the *p*-value of 0.0027?

    b. What reservations might you have about this value?

ANALYSIS (EXERCISES 5.53–5.79)

5.53 a. What is the probability that a single die when rolled will come up 6?

    b. What is the probability that a single die when rolled will come up 1?

    c. What is the probability that a single die when rolled will come up either 1 or 6?

5.54 If you know that the probability of any student getting a B on the psychology final is 13% and the probability of getting an A is 5%, what is the probability of a student getting either an A or a B on the exam?

    b. In an English literature course, the probability of any student getting a B on the final is 20% and the probability of getting an A is 10%. What is the probability of a student getting a B in psychology and a B in English literature?

    c. What is the probability of a student getting either an A or a B in both subjects?

    e. What is the probability of a student being a straight A student?

5.55 f. Can you think of a practical limitation to this exercise? One of four poker players, named Chris, has been closely watching the cards for a game being dealt face up on the table. In this game, 16 cards have been dealt, and the last 4 are about to be dealt face down. On the table two aces are showing, one in each of two hands.

a. What is the probability that one ace will be dealt in the last round of cards?

b. What is the probability that two aces will be dealt in the last round of cards?

5.56 Suppose you know the probability of the Joneses having twins when Loretta Jones gives birth next month, and you know the probability of Lizzie Smith giving birth to triplets.

a. How would you calculate the probability of either Jones twins or Smith triplets?

b. How would you calculate the probability of both Jones twins and Smith triplets?

5.57 Mars, Inc., the company that makes M&Ms, claims that they use the following distribution of colors:

Brown	Red	Yellow	Green	Orange	Blue	Total
30%	20%	20%	10%	10%	10%	100%

a. Buy a small bag of plain M&Ms (not the peanut kind).

b. Count the M&Ms of each color.

c. Record your frequencies in a table.

d. Take the total number of M&Ms and multiply it by the various percentages above. Record the results in a table. This table is the *expected* frequencies; your original table was the *observed* frequencies. Find the expected frequencies to 1 decimal of accuracy.

e. The observed frequencies should not deviate too much from the expected frequencies. One way to measure how much they deviate is to compute for each color a fraction in which the numerator is the square of the difference between the observed and expected frequencies and the denominator is the expected frequency of M&Ms. Find the sum of the six fractions. This sum is a value of the chi-square variable with 5 degrees of freedom, and it would be surprising if your value lies outside the range from 1 to 10.

f. Find the p-value of your data, which is the probability of the deviation you observed and more extreme deviations of the observed from the expected frequencies. Was your bag unusual?

g. Eat the M&Ms.

5.58 The following data are 50 observations of the chi-square variable with 1 degree of freedom:

```
1.76  1.64  0.38  0.48  0.01  1.90  0.32  0.01  1.92  1.56
0.57  0.73  0.60  0.01  6.86  0.17  1.09  1.01  0.02  0.15
0.09  0.10  0.60  0.38  2.04  0.07  0.95  1.52  0.06  4.21
0.05  0.08  0.25  0.15  0.36  1.84  0.23  0.00  2.19  1.57
1.28  0.30  0.73  0.19  0.07  0.01  0.47  0.91  0.92  0.05
```

a. Use intervals of length 1.00 and make a histogram showing the distribution of these observations.

b. Describe the shape of the distribution. (Note that the chi-square distribution has a different shape for different degrees of freedom.)

c. How many observations are larger than 3.84 in this sample of chi-square values?

d. According to Statistical Tables of the chi-square distribution, what percentage of the values can be expected to be larger than 3.84 with chi-squares with 1 degree of freedom?

e. How do the answers to parts c and d compare?

f. Add the five values in the first column in the table above, then add the five values in the second column, and so on, until you have found ten sums. These sums are now ten observations from the chi-square distribution with 5 degrees of freedom.

g. Use intervals of length 2.00 and make a histogram showing the distribution of these ten new observations.

h. Describe the shape of the distribution and compare it with the distribution in part a.

i. Use the table of the chi-square distribution to find the value of chi-square with 5 degrees of freedom such that only 0.05 (5%) of the observations are larger than this value.

j. Are there any values that large in our sample of ten values?

5.59 a. Collect the M&M chi-squares (Exercise 5.58) from the other students and make a histogram of the chi-squares for the entire class of students.

b. According to the theoretical chi-square distribution with 5 degrees of freedom, half of the values are larger than 4.35 and one tenth are larger than 9.24. How do these numbers compare with the chi-squares observed by the class?

5.60 Use the binomial distribution to find the following probabilities:

a. In 10 tosses of a fair coin, what is the probability of getting 8 heads? 9 heads? 10 heads?

b. What is the probability of getting 8 or more heads?

5.61 A psychologist observes small groups and classifies each group as either competitive or cooperative. She assumes that the probability is 0.5 for each of the two classifications, and she classifies 7 groups as competitive and 1 as cooperative.

a. What is the probability of getting 7 or more competitive groups?

b. Do you think this probability is so small that it is a mistake to use 0.5 as the probability for competitive?

5.62 Use a statistical software program or a statistical table to find the following probabilities.

a. z is equal to or larger than 2.34.

b. t on 17 degrees of freedom is equal to or larger than 2.34.

c. t on 17 degrees of freedom is equal to or smaller than -2.34.

d. t on 17 degrees of freedom is equal to or smaller than -2.34 or equal to or larger than 2.34.

5.63 Use a statistical software program or a statistical table to find the following probabilities.

a. chi-square on 2 degrees of freedom is equal to or larger than 6.78.

b. chi-square on 20 degrees of freedom is equal to or larger than 27.8.

5.64 Use a statistical software program or a statistical table to find the probability that F on 2 and 46 degrees of freedom is equal to or larger than 3.45.

5.65 The Bureau of the Census reports that in 1989 the *median* family income was $35,225.

    a. What is the probability that a randomly chosen family had an income larger than $35,225?

    b. What is the probability that in a sample of 10 families all 10 had incomes larger than $35,225?

    c. Why is the median rather than the mean family income more informative in a problem of this kind?

5.66 The Bureau of the Census reports that in 1990 the median age was 32.7 years.

    a. What is the probability that a randomly chosen person was younger than 32.7 years?

    b. What is the probability that in a group of 5 people 4 of them were younger than 32.7 years?

    c. What is the probability that in this group 4 or more of them were younger than 32.7 years?

5.67 A Gallup Poll in February 1991 found that on any given day, 33% of Americans read a book for pleasure. (*Source:* The New York Times, *July 26, 1992, p. E5.*) What is the probability that all four people at a bridge table had read a book the day before?

5.68 If, in a family of 4 children, the probability of having 4 girls is 0.06, what is the probability of having fewer than 4 girls?

5.69   a. If you know that the probability of one event is 0.25 and that of another independent event is 0.08, how do you find out the probability of both occurring at the same time?

    b. This answer is smaller than each probability alone. Can you explain why?

    c. Could you apply this reasoning to the probability of thunder and lightning occurring together? Why or why not?

5.70 Once we have found a *z*-score (recall that we find a standard score *z* by subtracting the mean of a distribution from the raw score and divide by the standard deviation), we can use the standard normal distribution to figure out how unusual (or usual) the *z*-score is. If we had the following *z*-scores on a geometry test, how would you interpret them in terms of how usual or unusual they are?

a. $z = 0.22$
b. $z = 2.50$
c. $z = -1.96$
d. $z = -0.013$

5.71 Among car owners in Germany, the probability of owning a Porsche is 0.07 and the probability of owning a Mercedes is 0.29.

a. What is the probability of owning one Porsche or one Mercedes?

b. What is the probability of owning one Porsche and one Mercedes (assuming that ownership of the two cars are independent events)?

c. What is the probability of owning neither type of car?

5.72 In Bob's first-year law school class, the probability of a student having a GPA of 3.8 or more is 0.15. The probability of a student having a GPA of 2.5 to 3.8 is 0.80. What is the probability of a student having a GPA that is not 2.5 to 3.8?

5.73 Robin wants to leave for spring vacation on Thursday, but she does not want to cut mathematics or physics class on Friday. According to ancient lore, the probability that the mathematics teacher will cancel a class is 0.05; the probability that the physics teacher will cancel a class is 0.10. What is the probability that Robin's wish—that both classes be canceled on the same day—will come true?

5.74 In a continuous probability distribution, the resulting smooth curve outlines an area beneath it. This area under the curve becomes a probability. How large is the total area under the curve?

5.75 In a normal distribution, 95% of the z-scores can be found between -1.96 and 1.96.

a. What percentage can be found below -1.96 and above $+1.96$?

b. What percentage can be found only below -1.96?

c. What percentage can be found between -1.96 and 0?

5.76 In the t-distribution with 10 degrees of freedom, find a range of t-values that includes 95% of the distribution of the scores.

5.77 Looking at the information in Exercise 5.46, can you show how the probability of 0.000004 that only the new members were chosen for clean-up duty was computed?

5.78 The National Highway Traffic Safety Administration reports from past data that among 100,000 licensed drivers, about 6 females are involved in alcohol-related fatal crashes in a year.

   a. In a city of 100,000 licensed drivers, use the Poisson distribution to show that the probability equals 0.002 that no female will be involved in an alcohol-related fatal crash in a year.

   b. What is the probability that one or more women will die in an alcohol-related crash during the same period?

5.79 In a binomial variable, $n = 6$ and $\pi = 0.4$.

   a. Use the probabilities from Statistical Table 3 (binomial distribution) to make a histogram of the probability distribution.

   b. Use the binomial probabilities to compute the mean μ.

   c. Compare this value with the value you get from $\mu = n\pi$.

   d. Does it look as if the probability distribution might balance at the mean value?

CHAPTER 6

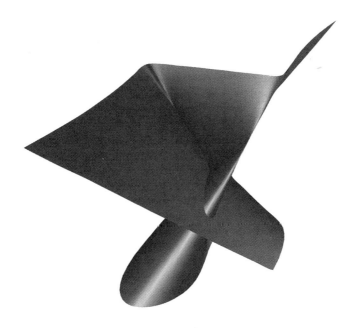

6.1 Sample statistic and population parameter

6.2 Point estimation

6.3 Interval estimation: More room to be correct

6.4 Summary

DRAWING CONCLUSIONS:

ESTIMATION

A Gallup poll surveying consumers in the United States, Germany, and Japan found that 55% of American consumers believe that U.S. products are of very high quality, compared with 26% of German consumers and 17% of Japanese consumers. "The sampling error is plus or minus 3 percentage points," according to the Associated Press news report. What does this plus or minus 3 percentage points contribute to the meaning of this report? (Source: The Philadelphia Inquirer, *Oct. 2, 1991, p. C-7.)*

Survey results and other statistical reports are often presented in newspapers and magazines and on television news. Statistical studies show, among many varied results, what percentages of African Americans in a sample prefer "African American" to "Black" as a name for their race (26%; 1989 telephone poll taken by Yankelovich Partners, Inc., for Time/CNN); what percentage of white Americans say they do not have enough money to buy food (13%; 1989 Gallup poll); what the mean age of female gymnasts equals (12.3 years; G. E. Theintz, et al., "Evidence for a reduction of growth potential in adolescent female gymnasts," *Journal of Pediatrics,* vol. 122 (1993), pp. 306–313); what percentage of their time people spend sleeping (30.9; *The New York Times,* Tuesday, September 6, 1995, p. C6).

While the results of particular studies are interesting in themselves, researchers go beyond the sample data to draw conclusions about the underlying population from which the sample came. They ask what the results would have been if they had done a complete census on all the elements (people, plants, etc.), in the population: if the Gallup Poll surveyors could have asked all Americans, what percentage would have thought that American products were of a high quality? The sample percentages tell us how only a few hundred people in the sample answered the interviewers' question about quality. At times there is no well-defined population to survey, but researchers still want to better understand the world that generated the data.

Going beyond the actual data is the part of statistics known as *statistical inference*, which is composed of *estimation* and *hypothesis testing*. In this chapter we discuss parameter estimation, and in the next chapter we present hypothesis testing. These methods are then used in the subsequent chapters.

Researchers use a sample instead of a population for pragmatic reasons: it would cost too much money and take too much time to collect data on an entire population, even if it were possible to do so. Researchers satisfy themselves with the data from a sample, even though the information from a sample is not complete and the reported results from the sample are not exactly equal to the true results in the entire population. To compensate for the inexact sample results, researchers calculate the sampling error, a number constructed such that 19 of 20 sample results will lie within plus or minus the sampling error of the true population values.

> **Statistical inference** is the process by which conclusions are drawn from sample data about values of the parameters in the population. It consists of two parts: **estimation** and **hypothesis testing.**

"Calvin and Hobbes" copyright 1993 Watterson. Dist. by Universal Press Syndicate. Reprinted with permission. All rights reserved.

6.1 SAMPLE STATISTIC AND POPULATION PARAMETER

Statistic is the singular form of the word *statistics*. Common examples of a *sample statistic* include the sample mean $\bar{x}$, the sample percentage *P*, and the sample standard deviation *s*. A sample statistic is typically denoted by one of the 26 Roman letters used in English. Because a statistic is computed from a sample of data, the value of the statistic can always be found.

A similar piece of information about a population is known as a *population parameter*. Population parameters are denoted by Greek letters. For example, the population mean is denoted by the Greek lower-case mu (μ), the population percentage by capital pi (Π), the population standard deviation by lower-case sigma (σ).

These ideas are illustrated in Figure 6.1, a graphic representation of a population, a sample, the parameters of the population, and the statistics from the sample. The population is represented by the large oval at the left. The elements of the population could be people, counties, plants, pigs, light bulbs, or anything else. From the population we draw a random subset of elements as a sample, and the sample is shown as the small oval at the right. Because the sample is a random sample, it is a fair representation of the population.

It is worth noting that there are two kinds of populations: finite and infinite. The finite population is an actual, large, delimited collection of elements. For example, the 100 million or so people who voted in the last presidential election is the population of all voters. The infinite population is a hypothetical collection of elements, such as all the lightbulbs that have been made and will be made by a certain machine, or all the tosses that could be made with a certain coin to test

> A **sample statistic** is a number computed from the data in a sample.

> A **population parameter** is a number that in principle could be computed from data for the entire population.

234 Chapter 6 • Drawing Conclusions: Estimation

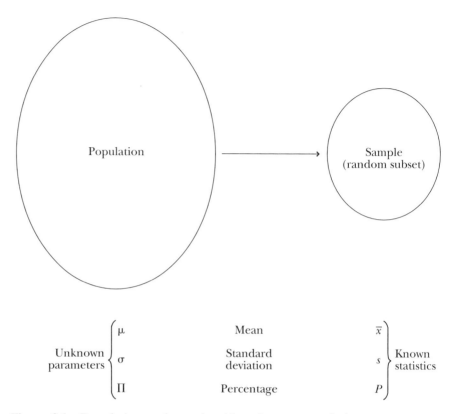

Figure 6.1 Population and sample with unknown population parameters and known sample statistics

the probability of heads or tails. All the bulbs the machine could make or all the tosses that could be made with a coin are impossible to actually count.

STOP AND PONDER 6.1

Think of some different populations you can identify in the world. What practical problems would be involved in drawing a sample of each one? What are the difficulties these problems pose for the surveyor? How do these problems interfere with finding out information about a population?

6.2 POINT ESTIMATION

Imagine that you are studying how many traffic tickets the average American collects in a lifetime. Consider the advantages and disadvantages of reporting the results in the following ways: "10" or "a range of values between 8 and 12." These two results represent two different approaches to estimating a population parameter. The simplest is a *point estimate.* In the example, the result of 10 tickets is a point estimate. If a sample survey finds that 58% of the respondents are in favor of raising taxes on gasoline, the percentage can be used as a point estimate of the percentage of people in the population who are in favor of such a tax increase. Similarly, for the population mean, the value of the sample mean can be used as the point estimate. It seems reasonable to use the sample statistic as the point estimate of the population parameter, even though we do not yet know how good this estimate is.

> A **point estimate** is a single number that serves as an estimate of a parameter.

What is a "good" point estimate?

Because a particular numerical estimate from a sample is never exactly equal to the true population parameter value, it does not make sense to ask if a particular numerical value is a good estimate or not. What can be asked is if the method for computing the estimate is a good one or not.

To determine whether a method is a good method or not, the results of the same study repeated many times over are compared. Let's do a thought experiment. Suppose we do many sample surveys and find a sample percentage from each. Assume that these are perfect samples and the only kind of error present is the sampling error. The following numbers, for example, represent the percentages of people in each of 10 surveys of 500 observations who are in favor of raising gasoline taxes:

      58.0 57.8 61.0 59.4 55.8 63.2 59.0 60.6 57.4 58.6

The percentages are all different from one other, even though all the samples came from the same population with one fixed—although unknown—percentage. This is because of the randomness of the drawing of the samples, a phenomenon that occurs in polls taken just before major elections. Several different survey organizations take sample surveys asking how people intend to vote, and the reported percentages of votes the candidates can expect to receive in the general election differ from one survey to the next. Most of the differences between the reported percentages can be explained by the variation and randomness inherent in drawing random samples, assuming that they are all well-designed surveys.

A good estimating method can be defined as one where the mean of the estimates from infinitely many samples equals the actual population parameter. The mean of these 10 estimates equals 59.1. The 10 different percentages were actually found by letting a computer draw 10 random samples from a population where the true percentage was known to equal 60%. If the computer had drawn many more than 10 samples, the mean of the many sample percentages would have equaled the true value of 60%. We therefore say that the sample percentage is an *unbiased* estimate of the population percentage. Each result may not be right, but the average of results across many repetitions of the study are right. If the mean of many sample statistics across many repeated samples is not equal to the true population value, then the estimate is *biased*.

Criteria for a good point estimate:

1. When the mean of a sample statistic from a large number of different random samples equals the true population parameter, then the sample statistic is an unbiased estimate of the parameter.

2. Across many repeated samples, the estimates should not be very far from the true parameter value.

The smallest sample value in the example is 55.8, a little more than 4 points fewer than the true value of 60.0. The largest sample value is 3.2 points greater than the true value. The standard deviation of the 10 percentages is approximately equal to the standard error of the percentages, in the example equal to 2.1%. This means that on the average, a sample percentage is 2.1% away from the mean of the 10 values. So the sample percentage in the example is a good estimate of the population percentage because (1) on the average across many samples the sample percentage is equal to the population percentage and (2) the sample percentages from many different samples are all close to the population percentage (how close usually depends on the number of observations in a sample).

It can be shown that any other way of estimating the population percentage from sample data gives worse results with larger sampling errors. Thus, even if the result is not good, this is as good as it gets. Similarly, it is possible to show that the sample mean is a good estimate of the population mean and is better than the sample median as an estimate of the population mean.

A strategic use of the point estimate: How many tanks did the Germans have?

Sometimes it is not clear how sample data should be used to estimate a population parameter. Statistical theory can be used to derive possible formulas for the computation of the point estimate. An unusual example is the method the Allies used to estimate the production of tanks in Germany in World War II.

For many strategic reasons, the total number of tanks produced in Germany during World War II was of great interest to the Allies. The Germans were methodical in their production of tanks, and they numbered the tanks consecutively starting with 1. As the war went on, the Allies captured enemy tanks and recorded the production numbers. How did they use the numbers of the captured tanks to estimate the total number of tanks? The population parameter consisted of an unknown number N, the total number of tanks produced, and the data consisted of the production numbers of the captured tanks.

A German World War II tank. Sometimes statisticians prove their worth underfire. *(Source: Culver Pictures.)*

Imagine that we are the Allied statisticians addressing the problem. The total number of tanks produced has to be equal to or larger than the largest number recorded. To find out how much larger, we could find the mean number of the captured tanks; by definition the mean is somewhere in the middle of all the production numbers. Therefore, doubling the sample mean produces an estimate of the total number, particularly assuming that the captured tanks represent a random sample of all the tanks. One drawback of this formula for the estimation of N is that there is no guarantee that 2 times the mean is a number larger than the largest captured number.

Another formula we could use for the point estimate of N is to multiply the largest observed number by the factor $1 + 1/n$, where n is the number of captured tanks. For example, if we have 10 captured tank numbers and the largest of them is 50, then an estimate of the total number of tanks is $(1 + 1/10)50 = 55$. That is, we think the actual number of tanks produced is a little larger than the largest number we observed in the sample.

Variations of this method of estimation were actually used during the war, and German records found after the war showed that the Allied estimates were close to the actual number of tanks produced. The records also showed that the statistical estimates were much closer to the true value than intelligence estimates obtained in other ways. The statisticians did better than the spies!

> **STOP AND PONDER 6.2**
>
> What assumptions about warfare and the deployment and capture of tanks are made in this problem as it is presented here?

6.3 INTERVAL ESTIMATION: MORE ROOM TO BE CORRECT

An **interval estimate,** called a **confidence interval,** is a range of values that serves as an estimate of a parameter.

The second approach to estimating a parameter is an *interval estimate*. "A range of values between 8 and 12" is an interval estimate. The statement that the backing for an increase in gasoline taxes is estimated to be between 52% and 64% is more informative than the statement that it is estimated to equal 58%.

For most population parameters, estimation intervals are found in the following way. First researchers find a sample statistic, for example, a mean or a proportion. Then the sampling error is computed from the data, and finally the sampling error is added and subtracted around the sample statistic. These three steps yield an interval known as a *confidence interval*—an interval that the statistician is confident contains the true population parameter value. A *confidence interval* for a population parameter is found by subtracting and adding sampling error to a sample statistic:

$$\text{statistic} - \text{sampling error} \quad \text{to} \quad \text{statistic} + \text{sampling error}$$

As an example, take a look at the sample survey at the beginning of the chapter. The percentage of people who believe U.S. products are of high quality in that sample equals 55% and the sampling error equals ±3%. The confidence interval for the unknown population percentage therefore is

$$55 - 3 = 52 \quad \text{to} \quad 55 + 3 = 58$$

Hopefully, this interval contains the unknown population percentage. Formula 6.1 at the end of the chapter can be used to find the confidence interval. Formula 6.2, a simplified version, works well in many cases.

Let's do another thought experiment. If another sample were drawn, it would yield a somewhat different sample percentage and a different confidence interval. Hopefully, this interval also contains the population percentage. With many different samples with different percentages and confidence intervals, we would still expect that the intervals would contain the parameter.

The reason for the name *confidence interval* is that statisticians have a certain degree of confidence that the interval actually contains the true and fixed value of the unknown parameter. The thinking goes as follows: If we collect many different samples, we can set up a confidence interval for each one. The confidence intervals are constructed wide enough that 95% of them contain the true population percentage, and 5% do not contain the true population percentage. The value 95% is known as the *confidence level*. This is a commonly used value, but other confidence levels can also be used.

What about the first confidence interval, from 52 to 58? Does that interval contain the unknown population percentage? For a particular interval, we will never know the answer to this question. Saying that we are 95% confident that the interval from 52 to 58 contains the unknown population percentage is saying that we do not know absolutely about this particular interval. What we do know is that across many repeated samples, 95% of them will give confidence intervals that contain the true parameter value.

Figure 6.2 illustrates this point. It shows the confidence intervals for each of the 10 gasoline tax samples. The interval of the first sample is from 54 to 62, the interval of the second from 53.7 to 61.9, and so on. The center points for the lines differ because the sample percentages differ.

Since these data were created by a computer, the true population percentage is known—it was set equal to 60—and is illustrated by the vertical line at 60. All 10 confidence intervals contain the true population percentage. With 100 instead of only 10 samples, 5 of the confidence intervals would be expected not to contain the value of 60, and 95 of the intervals would contain 60. This means:

> A 95% confidence interval is constructed according to a method such that 95% of all the confidence intervals contain the true value of the population parameter and 5% of all intervals do not contain the true value of the parameter.

In most settings, researchers collect data from only one sample. Whether the confidence interval from a single sample contains the true value of the parameter is not known. They hope that the interval is one of the great majority of possible intervals that do contain the parameter. But it could be one of the few flukey ones that do not contain the parameter.

The reason we have to express ourselves in such a roundabout fashion is that the value of the population parameter is a fixed but unknown number. The interval we construct from our sample is not fixed; it is a random interval in the sense that if we did the study over again, we would get a somewhat different interval. Thus, the intervals vary from sample to sample, and not all intervals contain the true parameter value. A confidence interval is like throwing out a net to capture the

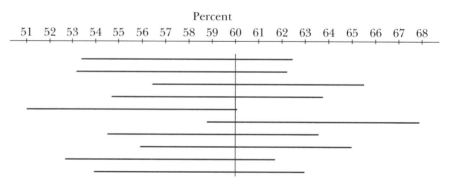

Figure 6.2 Confidence intervals from 10 different samples, all containing the true value of 60

> **Warning!**
>
> Remember, a probability tells us how often something occurs. Consequently, we *cannot* say that there is a 0.95 probability that the particular interval from 52 to 58 contains the unknown percentage of the population who think U.S. products are of high quality. The implication is that if we look at this interval 100 times, we would expect the interval to contain the true parameter value 95 times and not contain the parameter 5 times.
>
> But that, of course, makes no sense. Suppose the actual parameter value equals 57. Then the interval from 52 to 58 captures the parameter *all the time*, not just 95% of the time. Similarly, suppose the true value equals 50. Then the interval from 52 to 58 *never* captures the true value, no matter how many times we look.
>
> Probabilities cannot be used to say how often a *particular* confidence interval contains the value of the unknown parameter. A particular interval *always* contains the true value or it *never* contains the true value. Probability is used to refer to *how many* confidence intervals contain the unknown parameter value across many repeated samples.

unknown parameter; not all intervals land in such a way that they catch the parameter.

> **STOP AND PONDER 6.3**
>
> Why might a sample, taken with care and presumably random, be the 1 in 20 (5%) in which the confidence interval does not contain the true parameter?

Length of confidence interval

> **STOP AND PONDER 6.4**
>
> What might the connection be between the number of observations in a sample and the length of a confidence interval? Can you give an example that clarifies this relationship?

1. The length of a confidence interval depends on the number of observations in the sample. Large samples produce shorter confidence intervals and small samples produce longer intervals.

2. The length of a confidence interval depends on the confidence level. Low confidence levels (say 90%) produce shorter intervals and high confidence levels (say 99%) produce longer intervals.

A shorter confidence interval conveys more about the population parameter than a longer one. If you tell me that a confidence interval for an unknown population percentage goes from 0% to 100%, then you have essentially not told me anything: obviously it has to be somewhere between 0 and 100. If you tell me that the confidence interval goes from 30% to 70%, then I know something about the unknown parameter value. And if you tell me that the confidence interval goes from 49% to 52%, then you have told me a great deal about the value of the parameter.

"IN MY OPINION THE ANSWER IS ELEVEN. BUT THE MARGIN OF ERROR IS PLUS OR MINUS TWO."

Reprinted with permission of the artist, Carol Cable.

Effect of sample size on confidence interval One way of getting a short confidence interval is to have a large sample with many observations. With a larger sample comes more information, and with more information the interval is smaller. With large samples, the sample statistics are clustered more closely around a central value than they are with small sample. Mathematically, in formulas for sampling errors, the number of observations in the sample typically occurs in a denominator, and the larger a denominator the smaller the fraction.

Figure 6.3 shows confidence intervals for six samples of different sizes. It supposes that the sample percentage P equals 60% in each of the samples, and the intervals are computed using the formulas at the end of the chapter. The figure clearly shows that the larger the sample is the shorter the interval is. Also clear is that the lengths of the intervals do not decrease as fast as the numbers of observations in the samples increase: if the size of the sample is doubled, the corresponding interval is not half as long. Cutting the length of a confidence interval in half requires four times as many observations. But increasing the number of observations can be expensive; beyond a certain number it is not worth increasing the size of the sample. This is the reason most nationwide surveys ask 1,200 or so respondents for answers. The sample is large enough to get a 3% sampling error for a population percentage.

Effect of confidence level on confidence interval Another way to get a short confidence interval is to use a lower confidence level. Most often a 95%

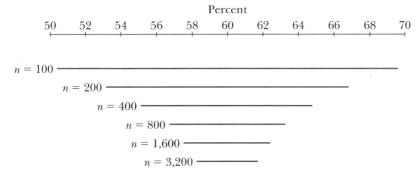

Figure 6.3 Confidence intervals for an unknown population percentage based on a sample percentage of 60 for samples with different numbers of observations.

confidence level is used. However, a 90% confidence level could be used. Its confidence interval is shorter than that of a 95% confidence level. But 90% confidence levels are not as desirable as 95% ones. With 90% confidence levels, only 90% of the confidence intervals from repeated studies contain the true value of the population parameter, and 10% of the intervals do not contain the true value.

Look at the difference between a 95% and a 90% confidence interval in a sample of 1,200 respondents. Suppose 60% of the people in the sample approve of the President's performance in office. The 95% confidence interval goes from 57.2 to 62.8, since the sampling error equals ±2.8. Hopefully, this interval is one of the many intervals that contain the true population value. Similarly, the 90% confidence interval goes from 57.7 to 62.3, with the sampling error in this case equal to ±2.3. The length of the first interval is 5.6, while the length of the second interval is 4.6. If it is important that the interval contains the true value, the slightly longer confidence interval is more likely to contain it.

Using confidence intervals   When the Associated Press reported that among a total of 3,500 people polled in three different Gallup surveys in three different countries, 55% of American consumers, 26% of German consumers, and 17% of Japanese consumers thought U.S. goods were of very high quality, it reported a sampling error of plus or minus 3. A sampling error of plus or minus 3 conveys that in 95 of 100 repeated samples, the sample percentage lay within 3 percentage points of the true population percentage. Of course the true value was not known. *(Source:* The Philadelphia Inquirer, *October 2, 1991 p. C-7.)*

Taking the reported percentage and then adding and subtracting 3 gives the 95 percent confidence interval for each consumer group. Thus, between 52% and 58% of Americans, 23% and 29% of Germans, and 14% and 20% of Japanese consumers thought U.S. products were of very high quality. Hopefully these confidence intervals actually contain the true population percentages.

The news article did not report that the confidence level was 95%, but it is commonly understood that reported sampling errors are based on a 95% confidence level if nothing else is said. All good surveys should report sampling errors and make it possible for the reader to construct confidence intervals.

> **STOP AND PONDER 6.5**
>
> Would you say that there are "real" differences between the views of the American, German, and Japanese consumers about the quality of American products? If 49% of Mexicans polled had rated the quality of American products as "very high," would you say they were more skeptical of American products than the Americans?

Confidence intervals for differences

A confidence interval for the difference between two parameters can also be created to study whether or not two groups are different or whether a group has changed over time, for example, whether or not a difference exists between the attitude of Republican and Democratic senators to the child welfare bill.

Difference between two proportions In February 1989 a telephone poll of 503 African Americans was taken by Yankelovich Partners, Inc., for Time/CNN, asking, "Which would you prefer as a name for your race, 'Black' or 'African American'?" In this poll, 26% of the respondents said they preferred "African American." The survey was repeated in February 1994; 53% of the respondents said they preferred "African American." Had there actually been a change in opinions in the five years between the two surveys? What does an inspection of the confidence intervals for these two studies suggest?

In the five-year period from 1989 to 1994, the percentage changed from 26 to 53, a difference of 27 percentage points. Can the difference of 27 percentage points simply be ascribed to the randomness that always exists when we compare two numbers, or is the difference between the two numbers larger than can be explained by randomness alone? Would the difference in the percentages be as large if we compared data on the populations of all African Americans in 1989 and 1994? One way to answer that question is to use the sample data and construct a confidence interval for the difference between the two percentages in the populations.

The difference between the two sample percentages has a sampling error of $\pm 9\%$, so the 95% confidence interval for the difference between the two percentages in the population goes from $27 - 9 = 18$ to $27 + 9 = 36$. The striking feature of the interval from 18 to 36 as an estimate of the true difference is that it does not contain 0. If a difference is 0, then two percentages are equal and there is no difference in what people like to be called. But the interval does not contain 0, so we conclude that this is not a possible value for the true difference. Rather, there seems to be somewhere between 18 and 36% more people who would prefer to be called African Americans. Based on the sample, there seems to have been a change in this five-year period in what African Americans like to be called. Formula 6.4 at the end of the chapter shows how to compute a confidence interval for the difference between two population percentages.

STOP AND PONDER 6.6

Can you construct a problem similar to this in which you could not be certain if there had been a significant change over the five-year interval?

Difference between two means In one of our classes, we asked the students to count their pulse for a minute. The mean rate for the men was 71.6 heartbeats and the mean for the women was 64.0 heartbeats for a difference of 7.6 heartbeats per minute. The difference between the two means shows how much faster the heartbeat is for one gender than the heartbeat for the other gender. Is the difference simply due to the random difference between any two sample means, or is it larger than can be explained by randomness alone?

The question can be answered by constructing a confidence interval for the difference between the two population means. If the confidence interval contains the value 0, then it may be that the observed difference between the two sample means is due to randomness alone. On the other hand, if the confidence interval does not contain the value 0, then evidence that the observed difference is larger than can be explained by randomness alone.

A 95% confidence interval for the difference between the two population means goes from −1.8 to 17.0 heartbeats. This interval includes the value 0. Therefore, the difference between the two population means may be equal to 0, and the difference between the two sample means can possibly be explained by randomness alone. The necessary computations are shown in Formula 6.5.

6.4 SUMMARY

Statistical inference is used to draw conclusions from sample data about the values of population parameters. It consists of two parts: estimation and hypothesis testing.

6.1 Sample statistic and population parameter

A sample statistic is a number computed from a sample, for example the sample mean $\bar{x}$, the sample percentage P, the sample standard deviation s. Because a sample statistic is computed from a sample of data, its value can always be computed. A population parameter is a number computed from a population, for example the population mean μ, the population percentage Π, the population standard deviation σ. Population parameter values are almost never known, so the available information in the sample data is used to estimate parameter values.

A finite population is an actual, large collection of elements, such as all the voters in a particular election. In an infinite population there is no limit to the number of elements. All possible coin tosses that could be made with one coin is an example of an infinite population.

6.2 Point estimation

A point estimate is a single number that serves as the estimate of the parameter.

A sample statistic is an unbiased estimate of a population parameter if, when a great many different random samples are drawn from the same population and the statistic is computed from each sample, the mean of the statistics equals the true population parameter. If the mean of many sample statistics across many repeated samples is not equal to the true population value, then the estimate is biased. Whether or not a particular statistic is biased or unbiased can usually be established mathematically.

In estimating a population mean, we prefer the sample mean to the sample median or other averages, because in most cases the sample means from many samples cluster more closely around the population mean than other estimates do.

6.3 Interval estimation: More room to be correct

An interval estimate is a range of values that serves as the estimate of the parameter. An interval is more informative than a single value, but the construction and interpretation of such an interval are also more difficult. Most intervals for the estimation of population parameters are found by (1) calculating a sample statistic, such as a mean; (2) computing the sampling error; (3) adding and subtracting the sampling error from the sample statistic. Such an interval is called a confidence interval for the population parameter.

Confidence levels describe what percentage of intervals from many different samples contain the unknown population parameter. A 95% confidence level indicates that 95% of the intervals in many samples contain the true value of the population parameter and 5% of all intervals do not contain the true value of the parameter. Whether the particular interval under study is one of the many intervals that contain the parameter value or one of the few that do not contain the population parameter is never known.

Shorter confidence intervals are more informative than longer ones. A short confidence interval can be obtained by having a large sample or by using a lower confidence level. When a survey has about 1,200 respondents, the sampling error for a percentage equals ±3%, which implies that in 95 of 100 different samples, the sample percentage lies within 3% of the true population percentage. News reports commonly use a sampling error that gives a 95% confidence level.

It is possible to construct a confidence interval for differences between two parameters, such as two percentages or two means, to study whether the two parameters could possibly be equal.

ADDITIONAL READINGS

Burkholder, Donald L. "Point estimation." In William H. Kruskal and Judith M. Tanur (eds.), *International Encyclopedia of STATISTICS*. New York: The Free Press, 1978. More on how to estimate the value of a parameter by a single number.

Pfanzagl, J. "Confidence intervals and regions." In William H. Kruskal and Judith M. Tanur (eds.), *International Encyclopedia of STATISTICS*. New York: The Free Press, 1978. More on how to estimate values of parameters.

FORMULAS

CONFIDENCE INTERVAL FOR A POPULATION PERCENTAGE

The simple random sample consists of n observations from a large population, and the sample percentage equals P. We want a 95% confidence interval for the population percentage Π. This interval goes from

$$P - 1.96\sqrt{\frac{P(100 - P)}{n}} \quad \text{to} \quad P + 1.96\sqrt{\frac{P(100 - P)}{n}} \quad (6.1)$$

The number 1.96 comes from the normal distribution. This is the value of the variable z such that 2.5% of the z-values are smaller than -1.96 and 2.5% of the z-values are larger than 1.96. This means that 95% of the z-values lie between -1.96 and 1.96, making the interval a 95% confidence interval. For another level of confidence interval, the corresponding value of z is found from a table of the normal distribution (Statistical Table 1).

A quick approximation for a 95% confidence interval involves using $P = 50$ and rounding the 1.96 to 2:

$$P - \frac{100}{\sqrt{n}} \quad \text{to} \quad P + \frac{100}{\sqrt{n}} \quad (6.2)$$

This is a slightly conservative confidence interval in the sense that most of the time it is a bit longer than the interval found by Formula 6.1. But it is much easier to compute, and the intervals from the two formulas are usually not very different. Formula 6.2 shows that with a sample of 900 observations, the error equals $100/30 = 3.3$. For a sample of 1,600 observations, the error is 2.5, and so on. For the error to equal 3, a sample of 1,111 observations is needed. Errors around 3 are commonly used, and this is the reason most samples have about 1,200 respondents.

CONFIDENCE INTERVAL FOR A POPULATION MEAN

The sample has n independent observations of a variable that follows the normal distribution, the sample mean is denoted $\bar{x}$, and the sample standard deviation is denoted s. The confidence interval for the population mean then is

$$\bar{x} - t^* \frac{s}{\sqrt{n}} \quad \text{to} \quad \bar{x} + t^* \frac{s}{\sqrt{n}} \qquad n - 1 \text{ d.f.} \qquad (6.3)$$

Here t^* is a value of the t-variable, and it is found in Statistical Table 2 of the t-distribution for $n - 1$ degrees of freedom. For a 95% confidence interval, the value of t is found such that 95% of all t-values lie between $-t$ and $+t^*$.

In the unusual case where the population standard deviation σ is known, then σ is used instead of the sample standard deviation s. This also means using a value z^* from the normal distribution instead of the value t^* from the t-distribution. In this case the confidence interval for the population mean is

$$\bar{x} - z^* \frac{\sigma}{\sqrt{n}} \quad \text{to} \quad \bar{x} + z^* \frac{\sigma}{\sqrt{n}}$$

For a 95% confidence interval, $z^* = 1.96$.

CONFIDENCE INTERVAL FOR A DIFFERENCE BETWEEN TWO PERCENTAGES

In one sample are n_1 observations, in another n_2 observations. The percentage in the first sample is P_1 and in the second sample it is P_2. The 95% confidence interval for the difference between the two population percentages Π_1 and Π_2 is

$$(P_1 - P_2) - 1.96 \sqrt{\frac{P_1(100 - P_1)}{n_1} + \frac{P_2(100 - P_2)}{n_2}}$$
$$\text{to} \quad (P_1 - P_2) + 1.96 \sqrt{\frac{P_1(100 - P_1)}{n_1} + \frac{P_2(100 - P_2)}{n_2}} \qquad (6.4)$$

CONFIDENCE INTERVAL FOR A DIFFERENCE BETWEEN TWO MEANS

In one sample are n_1 observations with mean $\bar{x}_1$ and standard deviation s_1. In another sample are n_2 observations with mean $\bar{x}_2$ and standard deviation s_2. First an average value of the two standard deviations is found according to the expression

$$s = \sqrt{\frac{(n_1 - 1)s_1^2 + (n_2 - 1)s_2^2}{n_1 + n_2 - 2}}$$

The confidence interval for the difference between the two population means μ_1 and μ_2 is

$$(\bar{x}_1 - \bar{x}_2) - t^*s\sqrt{\frac{1}{n_1} + \frac{1}{n_2}} \quad \text{to} \quad (\bar{x}_1 - \bar{x}_2) + t^*s\sqrt{\frac{1}{n_1} + \frac{1}{n_2}} \quad (6.5)$$

We find t^* from the table of the t-distribution with $n_1 + n_2 - 2$ degrees of freedom such that the probability equals 0.95 of the t-variable falling between $-t^*$ and $-t^*$.

EXERCISES

REVIEW (EXERCISES 6.1–6.21)

6.1 What is the goal of statistical inference?

6.2 If the mean is computed from a sample, what is the number called?

6.3 If the mean is known for a population, what is the number called?

6.4 What kind of data do statisticians use Roman letters for and what do they use Greek letters for?

6.5 How are μ, Π, and σ different from $\bar{x}$, P, and s in terms of defining quantities of various groups?

6.6   a. What is a point estimate of a parameter?

b. What is an interval estimate of a parameter?

c. Name an advantage and a disadvantage of each of these methods of estimating a parameter.

6.7 What does it mean to have an unbiased statistical estimate?

6.8 Explain a confidence interval for a population parameter.

6.9 The mean of a sample is 10. The sampling error is ±2.

a. How is the confidence interval for the population mean calculated?

b. If you calculate the confidence interval in this case, how do you interpret the resulting interval?

6.10 Suppose you have a large number of samples drawn from a population, and you calculate a confidence interval for the population mean from each of the samples. What are the chances that a randomly chosen interval does *not* contain the population parameter?

6.11 Describe the two ways to make a confidence interval shorter.

6.12 a. Give three major reasons for survey designers to use samples instead of populations when they are interested in a national opinion on a new product.

b. Name one major drawback involved in using a sample instead of a population.

6.13 How does using sample data help in learning about a population?

6.14 How does a point estimate differ from an interval estimate of a parameter?

6.15 If you take 20 random samples from the same population, you can compute 20 means and medians. You can also compute the standard deviation of the 20 means as well as the standard deviation of the 20 medians.

a. Which standard deviation will be smaller, the one for the means or the one for the medians?

b. Why is the statistic with the smaller standard deviation a better choice for estimating the population mean?

6.16 Why do you think a *confidence level* is called by this name?

6.17 "The value of a population parameter forever remains unknown." Should statisticians not even try to learn about population parameters? Explain.

6.18 a. Why is it so important that a sample be drawn randomly from a population?

b. What happens if a sample is not drawn randomly?

c. How might a nonrandom sample affect efforts to estimate population parameters?

6.19 a. In the example of Allies trying to estimate how many German tanks had been produced, list at least two assumptions that were necessary regarding the sample of the captured tanks.

b. Can you think of any ways in which the estimates might have been either too small or too large had these assumptions proved false?

6.20 Why do many national surveys contain about 1,200 respondents and not 500 or 2,000 or some other number?

6.21 Before a local election in Dogpatch, the editor of the town newspaper took a survey of the resident voters about their choice for town dogcatcher. A whopping 89% said they planned to vote for Abner Yokum, and 11% said they would write in names of their relatives. What confidence level should the editor use to be precise in reporting the results of the survey?

INTERPRETATION (EXERCISES 6.22–6.35)

6.22 A survey of voters in the 1960 presidential election with John Kennedy running against Richard Nixon showed that 47% of 469 women and 53% of 429 men voted for Kennedy. The difference equals 6%, and a confidence interval for the true difference between the two percentages goes from −3% to 15%. *(Source: The data were made available by the Inter-university Consortium for Political Research. The data for the Survey Research Center 1960 American National Election Study were originally collected by Angus Campbell, Philip Converse, Warren Miller, and Donald Stokes. Neither the original collectors of the data nor the Consortium bear any responsibility for the analysis and interpretations presented here.)*

a. What is the meaning of the statement "We are 95% confident that the true difference lies between −3 and 15%"?

b. What is particularly interesting about this interval?

6.23 A 1990 Roper Organization poll of 3,000 randomly selected women found that 58% of the respondents endorsed the statement "Most men think only their own opinions about the world are important." The margin of error was ±2%. *(Source: Delwin D. Cahoon and Ed M.*

Edmonds, "Comments concerning increased female negativism toward males," Contemporary Social Psychology, vol. 15 (1991), no. 2, p. 53.)

    a. What can you say about the other 42% of the respondents?

    b. What else do you need to know about the way in which these data were collected before you can draw conclusions about these results?

    c. How should you use the margin of error to interpret the result?

6.24 In a study of babies born to mothers with HIV who took a placebo as subjects in a control group in a drug study, 40 of 154 (26%) of the babies were born with HIV. (*Source:* The New York Times, *February 21, 1994, p. A1.*) A 95% confidence interval for the percentage of babies born with HIV goes from 19% to 33%.

    a. How do you interpret this confidence interval?

    b. How large is the sampling error in this study?

6.25 A study that asked Americans how well they liked their jobs indicated that between 70% and 75% (given a confidence interval of plus or minus 2.5) of those who had worked at the same job for more than ten years enjoyed it very much. Given these results, explain whether it is true that the actual population percentage for all such American workers who enjoy their job is located between 70% and 75%.

6.26 In a 1989 Gallup Poll, 13% of the white Americans surveyed reported that there were times during the last year when they did not have enough money to buy food. Among African Americans the same percentage was 33%. The difference between the two percentages is 20%.

    a. Do you think that the confidence interval for the difference between the two percentages in this study would be large enough to be able to argue that there is no difference between the two corresponding population percentages?

    b. How large would the sampling error for the difference between the two percentages have to be to conclude that there may be no difference between the two population percentages?

6.27 In each of the following, determine whether the item is associated with samples or populations. Explain your answers.

a. A random drawing of 1/10th of the student body taken to study the preferences of the students for lunch menus in the dining hall

b. All clarinet players in the band are polled about their preference for section leader.

c. All those who voted in the last election, to find out the President's popularity among voters

d. All recorded civil war deaths among troops, to find out which side lost more soldiers in the war

6.28 In a study of 22 female gymnasts, the mean age was 12.3 years and the standard error of the mean was 0.2 years. This gives a 95% confidence interval for the population mean age of from 11.9 years to 12.7 years. *(Source: G. E. Theintz, H. Howard, U. Weiss, and P. C. Sizonenko, "Evidence for a reduction of growth potential in adolescent female gymnasts," The Journal of Pediatrics, vol. 122 (1993), pp. 306–313.)* How do you interpret this confidence interval?

6.29 In a sample of 205 people from the South, 70 people (34%) classified themselves as professionals, and in a sample of 151 people from the non-South, 62% said they were professionals. The difference between the two percentages of professionals is 62% − 34% = 28%. The computation of a 95% confidence interval for the population difference in the percentage of professionals in the non-South and the South results in the interval from 18 to 38. *(Source: Adapted from J. C. McKinney and L. B. Borque, "Further comments on 'The changing South': A response to Sly and Weller," American Sociological Review, vol. 37 (1972), p. 236.)*

a. How do you interpret this confidence interval?

b. Does it seem possible that the percentages of professionals in the two parts of the country are equal in the two underlying populations?

6.30 Take a look at the winners and losers in the first 24 football Super Bowls. Among the 24 winners, 7 of them (29%) played in the Super Bowl the following year. Among the 24 losers, 4 of them (17%) played in the Super Bowl the following year. The difference between these two percentages is 29% − 17% = 12%. To see if the difference could have occurred by chance alone, a 95% confidence interval is found for the difference of from −12 to 36.

    a. How do you interpret this interval?

    b. Even though more of the winners than losers played next year, could it be that there really is no difference between the winners and losers when it comes to playing next year? Explain.

6.31 In a study of baldness, 55% of a sample of bald men had suffered heart attacks, and 43% of a sample of men who still had hair on their heads had suffered heart attacks. *(Source:* The New York Times, *February 14, 1993, pp. A1 and C12.)* This results in a difference between the two percentages of 55 − 43 = 12. A 95% confidence interval for the difference between the corresponding population percentages goes from 6 to 18.

    a. How do you interpret this interval?

    b. From this interval, does it seem as if the two population percentages could be equal?

6.32 In a sample of college male seniors, the mean height is 71 inches and the standard deviation is 2.1 inches. *(Source:* The New York Times, *July 26, 1992, p. E5.)* A 95% confidence interval for the corresponding mean height in the population from which these data came goes from 70.4 inches to 71.6 inches. The mean height of American men equals 69.1 inches.

    a. How do you interpret this confidence interval?

    b. What does the confidence interval tell you about the height of college seniors as opposed to all men?

6.33 In a study of the speed of flow in artificial heart valves, the mean flow in a sample of valves is 5.96. The corresponding 95% confidence interval for the mean flow in the population of heart valves goes from 5.22 to 6.70.

    a. How do you interpret this confidence interval?

    b. The manufacturer wants to guarantee that the valves have a mean flow of at least 5.00. Does it seem as if the manufacturer can provide this guarantee?

6.34 In a sample of 49 employees, the mean number of days lost to illness per year was 7.0 days. A 95% confidence interval for the population mean goes from 6.3 days to 7.7 days. According to national figures, the mean number of days lost to illness equals 5.1 days.

    a. How do you interpret this confidence interval?

    b. Does it seem as if this sample comes from a population where the mean number of days lost to illness is less than or equal to or more than 5.1 days?

6.35 The sample of swordfish reported in Exercises 3.46 and 4.66 has a mean mercury concentration of 1.09 ppm, and a 95% confidence interval for the population mean concentration goes from 0.90 ppm to 1.28 ppm.

    a. How do you interpret this confidence interval?

    b. Does it seem as if the mercury concentration could satisfy the legal limit of no more than 1.00 ppm?

ANALYSIS (EXERCISES 6.36–6.52)

6.36 Give point estimates of the two corresponding population percentages for the following sample results: In a sample of 682 married couples, 30% of the women and 23% of the men answered "yes" to the following question: "Has the thought of getting a divorce from your husband/wife ever crossed your mind?" *(Source: Joan Huber and Glenna Spitze,* Sex Stratification, Children, Housework, and Jobs, *New York: Academic Press, 1983, p. 98.)*

6.37 In general, how is a confidence interval for a population parameter found?

6.38   a. Use computer software to generate 10 different samples of 50 observations from a population of normally distributed values with mean 0 and standard deviation 1. Record the 10 different sample means.

    b. Find the standard deviation of the 10 means. This can be used as the standard error of the means.

    c. Multiply the standard error by 1.96 and calculate a 95% confidence interval for each of the 10 means.

    d. Display the confidence intervals in a graph similar to Figure 6.3.

    e. How many of the confidence intervals contain the population mean?

6.39 Right after the Republican party convention in 1992, four different polls reported on the support for President George Bush. The *CNN/USA Today* poll found that 42% ± 4% would vote for Bush, the *Newsweek* poll found 39 ± 4, the *Los Angeles Times* reported 41 ± 3, and the *Washington Post* reported 40 ± 4.

    a. Why are you not surprised that the four polls gave four different percentages?

    b. Why was the sampling error only ±3 in one of the polls and ±4 in the other polls?

    c. Construct 95% confidence intervals for the unknown population percentage from each of the polls and display the confidence intervals in a graph to make them easy to compare.

    d. Assume that none of the four intervals are unusual and that they all contain the true population percentage. What is the range of possible values for the population percentage?

    e. Approximately how many people were interviewed in each of the polls?

    f. Combine the results from the four polls and show that the Bush votes make up 40% of those interviewed with a sampling error of ±2.

    g. Construct a 95% confidence interval from the numbers in part f.

    h. How does the confidence interval obtained in part f compare to the interval obtained in part d?

6.40 According to the Census Bureau, the mean number of bedrooms in housing units in this country was $\mu = 2.6$ in 1990. The standard deviation was $\sigma = 0.9$ bedrooms. In a survey of 100 housing units in a Chicago suburb the mean number of bedrooms was 3.1.

    a. Find a 95% confidence interval for the population mean number of bedrooms in this community.

    b. Does it seem as if this suburb has more bedrooms than the country as a whole?

6.41 In a sample of 49 employees of a large business firm, the mean number of days in a year lost to illness by the employees was 7.0 with a standard deviation of 2.5 days.

a. Find a 95% confidence interval for the mean number of days lost to illness for the entire company.

b. How do you interpret this interval?

6.42 In 1989, in answer to the question "Which of two career paths would you choose, (1) one that would enable you to schedule your own full-time work hours and give more attention to your family but with slower career advancement or (2) one with rigid work hours that permit less attention to your family but faster career advancement?" 74% of men and 82% of women chose option 1. *(Source: Juliet B. Schor, The Overworked American: The Unexpected Decline of Leisure, New York: Basic Books, 1991, p. 148.)* Suppose the sampling error for the percentages was plus or minus 3.

a. Find the confidence interval for the population percentage of men who chose option 1.

b. Find the confidence interval for the population percentage of women who chose option 1.

c. From these two confidence intervals, does it seem possible that the two population percentages are equal?

d. Does it look as if either the men or the women who chose option 1 are actually in the minority in the overall population? Explain.

6.43 In a study of 21 swimmers, the mean age was 12.3 years and the standard error of the mean was 0.3 year. *(Source: G. E. Theintz, H. Howard, U. Weiss, and P. C. Sizonenko, "Evidence for a reduction of growth potential in adolescent female gymnasts," Journal of Pediatrics, vol. 122 (1993), no. 2, pp. 306–313.)*

a. Find a 95% confidence interval for the population mean age of the swimmers.

b. What does this interval tell you about the age of the swimmers from which this sample was taken?

6.44 Several older patients were complaining of postherpetic neuralgia. The mean amount of time after diagnosis before the treatment started for 6 men was 30.5 months and for 12 women was 37.9 weeks. The difference in waiting time was therefore 7.4 months. The standard deviation for the men was 17.5 months and for the women was 30.8 months. *(Source: P. R. Layman, E. Agyras, and C. J. Glynn, "Iontophoresis of vincristine versus saline in post-herpetic neuralgia: A controlled trial,"* Pain, *vol. 25 (1986), pp. 165–170. Reported in W. W. Piegorsch, "Complementary log regression for generalized linear models,"* The American Statistician, *vol. 46 (1992), pp. 94–99.)*

　　a. Show that the pooled standard deviation equals 27.4 months.

　　b. Find a 95% confidence interval for the difference between the two underlying population means.

　　c. Is no difference between the two population means possible?

6.45 In 1989, the National Center for Health Statistics asked youths aged from 12 to 18 years who smoked what brand of cigarettes they bought. Of the 41 African Americans in the study, 61% smoked Newport cigarettes; 5% of the 807 whites smoked Newports. This results in a difference between the two percentages of 61 − 5 = 56.

　　a. Show that a 95% confidence interval for the difference between the two corresponding population percentages goes from 41 to 71.

　　b. How do you interpret this confidence interval?

　　c. Does it seem as if the two population percentages are equal to or different from each other?

6.46 Use the same method the Allies used to estimate German tank production during World War II to estimate the production of an imaginary new fighter plane in World War II. Each plane has a number indicating its place in the production line. Suppose the highest number on any plane shot down is 100. The sample of planes is 20. How many planes should you estimate have been produced?

6.47 In a study of 531 large companies by the Wyatt Company, 61% of the business managers surveyed indicated that they expected improved customer service as a result of downsizing, but only 33% of them concluded that better customer service actually occurred as the result of staff reductions. *(Source: R. Reich, "Companies are cutting their hearts out,"* The New York Times Magazine, *December 19, 1993, p. 54.)*

a. Using the quick approximation formula (Formula 6.2), find confidence intervals for the two corresponding population percentages.

b. What conclusion might you draw about downsizing from this information?

6.48 A sample of 117 late adolescents were given a scale on which to rate their attachment to their mothers and fathers. The fathers of 71 of the adolescents were alcoholics. The mean attachment rating of these adolescents to their fathers was 78, with a standard deviation of 25. The mean attachment of the children of nonalcoholic fathers to their fathers was 91, with a standard deviation of 22. *(Source: Timothy Cavell et al., "Perceptions of attachment and the adjustment of adolescents with alcoholic fathers,"* Journal of Family Psychology, *vol. 7 (1993), pp. 204–212.)*

a. Find a confidence interval for the population means for each group.

b. Would you agree that adolescents with alcoholic fathers are less attached to their fathers than adolescent children of nonalcoholic fathers are attached to theirs?

6.49 In a Gallup poll of 502 respondents, 56% said they were "morning" people; 44% of the respondents said they were "night owls." *(Source: USA Today, December 13, 1993, p. 1A.)* Using the quick approximation formula (Formula 6.2), find confidence intervals for the percentages in parts a, b, and c (read cautiously). Then go on to parts d, e, and f.

a. Of the morning people, 53% believed that morning people have more energy than most people, while among the night owls, 39% believed that morning people have more energy than most people.

b. Of the morning people, 45% thought that morning people exercise more than most people, while 37% of the night owls believed that morning people exercise more than most people.

c. Of the morning people, 74% believed that morning people led active lives, while 64% of the night owls thought that morning people led active lives.

Table 6.1 Data for Exercise 6.50

Scale	Men		Women	
	Mean	Standard deviation	Mean	Standard deviation
Marital satisfaction	31.6	8.7	30.0	9.8
Idealistic distortion	16.7	5.1	14.0	5.5

Source: Blaine J. Flowers and David H. Olson, "ENRICH material satisfaction scale: A brief research and clinical tool," Journal of Family Psychology, vol. 7 (1993), pp. 176–185.

  d. When you look at the three sets of confidence intervals, does it seem as if the morning people and the night owls differed in how they perceived the morning people? Describe your findings.

  e. The report indicates that Quaker Oatmeal sponsored the study, and the margin of error was 4.4%. How does this margin of error compare with your results?

  f. Would using the 4.4% margin of error instead of your own calculations change your answers to part e?

6.50 Research on marital satisfaction was carried out with a sample of 2,112 couples using two scales, a marital satisfaction scale to rate how happy each person was with the marriage and an idealistic distortion scale, which rates the tendency of the person to evaluate marriage in an unrealistically positive way. The results are given in Table 6.1. To answer the questions, construct confidence intervals for the differences between the means.

  a. Does it seem as if men are more satisfied in marriage than women?

  b. Would you agree that men have a more distorted view about marriage than women?

6.51 In Chapter 5 we report on the results from one of our introductory statistics classes where 25 students spun pennies 10 times each and counted the number of heads. The mean number of heads was $\bar{y} = 3.96$ and the standard deviation was 1.74, so the standard error of the mean was $1.74/\sqrt{25} = 0.35$.

  a. Use these data to calculate a 95% confidence interval for the population mean number of heads in 10 trials.

b. What does this confidence interval tell you about the possibility that spinning a penny produces heads and tails with equal probabilities, such that we would expect 5 heads and 5 tails in 10 spins?

6.52 The 25 students in Exercise 6.51 spun pennies a total of 250 times. They observed 97 heads and 153 tails, so the proportion of heads equals $97/250 = 0.39$. Find a 95% confidence interval for the probability of getting heads when you spin a penny.

CHAPTER 7

7.1 The hypothesis as a question

7.2 How to answer the question posed by the null hypothesis

7.3 Significance level

7.4 Testing a population proportion

7.5 Difference between two population proportions

7.6 Testing hypotheses versus constructing confidence intervals

7.7 Statistical versus substantive significance

7.8 Applications: When to reject the null hypothesis

7.9 Summary

DRAWING CONCLUSIONS:
HYPOTHESIS TESTING

Do French people know more about geography than other people? Do Americans know less?

In 1988 (July 28) *The New York Times* carried a story on people's knowledge of geography. The article described a study commissioned by The National Geographic Society and carried out by the Gallup Organization in which the researchers asked a large number of people in random samples from different countries to identify 16 locations (13 countries, Central America, the Persian Gulf, and the Pacific Ocean) on a world map. The researchers added up the number of correct identifications (from 0 to 16) each person made.

The mean numbers of correctly identified locations for samples from four countries were

United States 6.9
Mexico 8.2
Great Britain 9.0
France 9.2

On the average, the French respondents were able to identify a larger number of spots on the map than the respondents in the other three countries. The newspaper article goes on to say, "To be considered statistically significant, differences in scores among all adults must be at least six-tenths of a point apart." What does the term "statistically significant" mean?

The difference between the means for France and Great Britain is only 0.2 locations. This small difference hardly seems worth mentioning, and it is in fact not large enough to meet the 0.6 criterion of statistical significance. On the other hand, the difference between the means for Great Britain and the United States is 2.1 locations, large enough to be statistically significant. This is because 2.1 is larger than 0.6. The differences between the means for Mexico and Great Britain and Mexico and France are large enough to be statistically significant as well.

While these differences among samples may be interesting, we want more from this study than the sample means for the different countries. We want to make general conclusions about whether the means for all the adults in these countries are different, not just the sample means; we want to know if the population means in various countries are different. If we can conclude that the population means for two countries are different, then we can say that the difference between the sample means is statistically significant.

One approach to the problem would be to use the ideas in Chapter 6 and estimate the population means to see if they are different. In this chapter we follow another approach called hypothesis testing. In both cases we are interested in population parameters, such as percentages, differences between percentages, means, and differences between means. But in hypothesis testing we focus on a particular value and wonder if the parameter could be equal to this value. In the geography example, the parameters are population means, and to illustrate the topics in this chapter, we pursue the question of whether the population mean in Mexico is equal to the population mean in the United States. We want to find out whether the difference between two population means equals the value zero, even though we observe a difference of 1.3 between the sample means. The difference in the samples could be due to randomness only.

> In **estimation** we try to find what value a parameter might have.
> In **hypothesis testing** we try to find out whether a parameter is equal to a particular value of special interest.

7.1 THE HYPOTHESIS AS A QUESTION

To begin hypothesis testing we ask a question: Is the difference between the population means in Mexico and the United States equal to zero or not? The difference between the two sample means is $8.2 - 6.9 = 1.3$. Thus, on the average, the respondents in Mexico were able to identify 1.3 more spots on the map than were the respondents in the United States. Even if there truly were no difference between the people in the two countries, we would not expect the two sample means to be equal. There would be some sampling variation in the two different random samples, but this variation might not be large enough to explain the observed difference of 1.3 spots.

Null hypothesis

To see if the observed difference of 1.3 is larger than the sampling variation, we examine what would happen to the sample means if the two population means were equal, that is, whether the difference between the two means equals zero. In statistics such a question is known as a *null hypothesis*. The null hypothesis is always expressed in terms of one or more parameters, and it states that the parameter(s) are equal to some specific value. In this example, the null hypothesis asks whether the difference between the two population means equals zero.

In the more formal language of statistics we write the null hypothesis as an equality. Let the population mean in Mexico be denoted μ_M and the population mean in the United States be denoted μ_{US}. (Remember that the Greek μ denotes population mean.) The statistical equality for the null hypothesis in the geography example is

$$H_0: \mu_M - \mu_{US} = 0$$

The letter H stands for hypothesis, and the subscript 0 identifies it as a null hypothesis. The reason for the word *null* is that the hypothesis states that there is *no* difference or *no* change or *no* relationships between the variables.

A statistical **null hypothesis** asks whether a parameter equals a particular value. Formally the null hypothesis is written

$$H_0: \text{parameter} = \text{value}$$

Even though the null hypothesis states that the difference between the two population means equals zero, the statement is not necessarily true. It is simply a way of asking the question whether the difference between the two population means equals zero. Most data are collected to show that groups are different; the two countries would be shown to be different if the null hypothesis were wrong. The question posed by the null hypothesis is answered on the basis of the sample data.

It is important to emphasize that a null hypothesis is always a question about population parameters and therefore contains Greek letters. It would make no sense for a null hypothesis to be about a sample statistic, say the sample mean $\bar{x}$ or the difference between two sample means, because we *know* the sample statistics and can tell whether they are different or not simply by looking at them. In the example, the difference between the two sample means equals 1.3, which obviously is different from zero. But that is not necessarily the difference between the population means, and that is why we ask about the values of the population means.

Alternative hypothesis

The logical alternative to a null hypothesis of no difference is a hypothesis that there is a difference between two parameters—an *alternative hypothesis*. In the example, the alternative hypothesis states that the difference between the two population means is different from zero:

$$H_a: \mu_M - \mu_{US} \neq 0$$

Thus, if the answer to the question posed in the null hypothesis is no, the alternative hypothesis is true. If the two means are not equal, then they must be different; the no in the null hypothesis and the no in the answer to the null hypothesis cancel each other out. If the evidence in the sample data leads to answering no to the question posed in the null hypothesis, then we *reject* the null hypothesis in favor of the alternative hypothesis.

Errors in answering the question

The question expressed in the null hypothesis is a yes/no question. Is the difference between the population means in Mexico and the United States equal to zero? Either it is or it isn't, yes or no. The answer to the question is determined by the information in the sample data. But since the amount of available information in sample data is limited because it comes only from a sample, not the entire population, the answer may not be correct.

Hypothesis testing is like a jury decision on the question of whether the defendant is guilty or not. If the defendant is truly innocent but our verdict is guilty, then we have made an error. On the other hand, if the defendant is truly guilty but our verdict is not guilty, then we have made another type of error.

There are two possibilities for the question posed in the null hypothesis.

1. The difference between the two population means equals zero, and the correct answer is therefore **Yes.**

2. The difference between the two population means is different from zero, and the correct answer is therefore **No.**

If the null hypothesis is true   If the null hypothesis is true and the difference between the two population means really equals zero, then the

> The **type I error (alpha error)** in hypothesis testing is rejection of a null hypothesis that is true.

> The **type II error (beta error)** in hypothesis testing is failure to reject a null hypothesis that is false.

correct answer to the question the null hypothesis poses is yes. If we say yes, we have given the correct answer. But if we answer the question no, then we have made a mistake. We have made an *alpha error* or a *type I error*: we have said no to and rejected a null hypothesis that is true.

If the null hypothesis is false If the null hypothesis is not true and the difference between the two country means is different from zero, then the correct answer to the question posed by the null hypothesis is no. If we answer yes and conclude that the difference between the two means equals zero, we have made another type of mistake, called the *beta error* or a *type II error*.

STOP AND PONDER 7.1

A man is on trial for murdering his wife. He is, in fact, guilty. The jury finds him innocent. The null hypothesis: A person is innocent unless proven guilty beyond a shadow of a doubt. What kind of error has been made in this trial, a type I or a type II? Can you create a court case in which the reverse type of error is committed? Which of the two errors is our legal system more willing to allow?

7.2 HOW TO ANSWER THE QUESTION POSED BY THE NULL HYPOTHESIS

The question posed by the null hypothesis is answered on the basis of the sample data. If the data appear to support the null hypothesis, the hypothesis is not rejected. The null hypothesis is rejected if the data are inconsistent with it. For example, if the null hypothesis states that the difference between two population means is equal to zero and the difference between the two sample means is very different from zero, the null hypothesis is rejected.

The way we decide if the sample is inconsistent with the null hypothesis is to ask if we could expect to get the data we got if the null hypothesis were true. If the population means in Mexico and the United States are equal, could we expect a difference between the two sample means as large as the observed difference of 1.3? Or, put the other way around, because the null hypothesis states that the difference between the two population means equals zero, the difference between the two sample means should be close to zero if the null hypothesis is true.

How often would we get a difference between sample means of 1.3 or more if the samples come from populations where the difference between the population means is zero? If the difference between the population means is zero, does a difference between sample means of 1.3 belong to an unusual set of differences between sample means? In other words, what is the probability of getting a difference between two sample means of 1.3 or more from populations where the difference between the mean equals zero?

Probability: The *p*-value

To determine whether the sample difference of 1.3 locations belongs to an unusual set of data, we compute the probability of getting a sample difference equal to 1.3 or more when the population difference is zero. Such a probability is known as a *p-value*. When the *p*-value is so small that it does not seem possible to get the data we got if the null hypothesis were true, we reject the null hypothesis. The smaller the *p*-value, the more evidence against the null hypothesis. But what do we mean by "small"? Probabilities are numbers on a scale from 0 to 1, so a small probability is some number close to 0. The famous British statistician Ronald Fisher, came up with the standard that 1 in 20 was small. This translates to 0.05, and since his time a probability of .05 or smaller has been considered small. Fisher did not have any profound reasons for deciding that 0.05 is small, but his idea caught on.

> The *p*-value is the probability of getting the observed data or more extreme data when the null hypothesis is true.

How small we want the *p*-value to be before we decide to reject the null hypothesis should in principle be decided on the basis of the consequences of making the mistake of rejecting a true null hypothesis. However, such consequences are often hard to determine.

When Fisher settled on 0.05, he considered the probability of getting a sample statistic either larger or smaller than the parameter value in the null hypothesis. This means that half of the significance level probability (0.025) is on the left side of the parameter value and the other half (0.025) is on the right side of the parameter value.

If the *p*-value is the probability of the observed sample statistic or more extreme ones, then the *p*-value is one-sided, and the criterion for smallness is 0.025. This is the case when the *p*-value is determined by using a value of the *z*-variable or a value of *t*-variable. Sometimes people report two-sided *p*-values and sometimes they report one-sided *p*-values.

WARNING! WATCH THE *p*'s, PLEASE

People sometimes erroneously think that *p*-values have something to do with probabilities that the null hypothesis is right or wrong, but this cannot be. When we toss a coin it lands heads up sometimes and tails up sometimes, and we can therefore talk about the probability of the coin landing either heads or tails. Such a probability tells us how often heads or tails happens when the coin is tossed many times.

But we cannot talk about the probability that a particular null hypothesis is right or wrong. A null hypothesis is not sometimes right and sometimes wrong. A null hypothesis is either right or it is wrong, and it is always one of the two.

Let's say your name is David. We cannot say that the probability is 0.04 that your name is David. That implies that when we ask you 100 times what your name is, you say David 4 times and something else the other 96 times. This is similar to asking if a null hypothesis is true or not.

Instead, *p*-values refer to probabilities about data. The *p*-value tells how often data of a certain kind occur in many different samples from a certain population. Suppose you are in a class of 10 students and 2 of them are named David. Let a sample consist of one student and the data be the name of that student. The *p*-value for the name David is 0.2 because this name will come up 20% of the time if many samples are selected.

> When a null hypothesis is rejected, the sample result is statistically significant.

When data lead to rejection of a null hypothesis, then the empirical results are statistically significant. Put another way, the empirical result is statistically significant if the *p*-value is small.

In the geography example, the null hypothesis states that the mean number of correctly identified geographical locations is the same for people in Mexico and people in the United States. The observed difference between the two sample means equals 1.3 locations. The reason this difference is statistically significant is that the probability of such a sample difference or a larger difference in populations where the two means are equal is less than 0.025. For a difference between sample means of 0.6 the *p*-value is exactly equal to 0.025.

Mechanics of hypothesis testing

To find the *p*-value from the data, statistical theory specifies that we change the observed difference between the two sample means—1.3 locations on the world map—to a standard score. The *standard score* for a difference between two sample means is a value of the statistical *t*-variable. For these data the observed difference between the two sample means $\bar{x}_M - \bar{x}_{US} = 1.3$ becomes $t = 4.25$. Because there were about 1,600 observations from the United States and about 1,200 from Mexico, the degrees of freedom are so large that the value of t can be regarded as equivalent to a value of the standard normal *z*-variable. (The *t*-variable and degrees of freedom are discussed in Chapter 5.)

Changing the scale from one variable $(\bar{x}_M - \bar{x}_{US})$ to another variable (t) is about the same as converting a temperature from Fahrenheit to Celsius degrees. The value of t is found from the sample means, the sample standard deviations, and the number of observations in the samples by using Formula 7.2 at the end of the chapter, either with pencil and paper or using statistical software on a computer.

The procedure for changing the value of the difference between two sample means to a value of the *t*-variable is shown in Figure 7.1. The figure shows the correspondence between the two scales such that a difference of 0 locations corresponds to $t = 0.00$ and the observed difference = 1.3 locations corresponds to $t = 4.25$. The probability that t is equal to or larger than 4.25 is then equal to the probability that observed difference between the sample means is equal to or larger than 1.3.

Any standard score beyond $+2$ or -2 is unusually large, so 4.25 is an unusually large positive value of t. The probability that the difference between the sample means is 1.3 or larger cannot be found di-

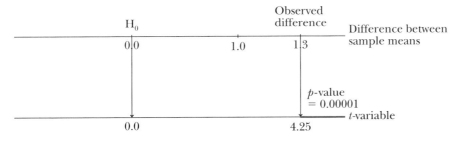

Figure 7.1 Correspondence between a difference between two sample means and the value of the *t*-variable

rectly, but the probability that t equals 4.25 or more can. The probability that the statistical variable t is equal to or larger than any value can always be found. Many statistical computer software packages can find probabilities for t directly, or extensive published statistical tables can be used to find probabilities for t.

For this example, the probability that t is equal to or larger than 4.25 for a sample of more than 2,000 observations is as small as 0.00001. Therefore, the probability that the difference between the sample means is equal to or larger than 1.3 is also equal to 0.00001, if the difference between the two population means is zero. In other words, if 100,000 samples from populations with equal means were drawn, then only 1 of the samples would have a difference between the sample means of 1.3 or more; that is, this large a difference between the sample means would be very unusual if the population means were equal.

As discussed in Chapter 5, the t-variable depends on how large the sample is, and the same value of t has different probabilities depending on the number of observations in the sample. But rather than talking about the number of observations, statisticians talk about the number of degrees of freedom (d.f. or df).

In the case of the difference between two sample means, the number of degrees of freedom for t equals the total number of observations minus 2. In the example, the number of observations is so large (2,800) that we can use the standard normal z-variable instead of the t-value.

To reject or not to reject the null hypothesis

On the basis of the assumption that the two population means are equal, the probability of getting a difference between the two sample means of 1.3 or more equals 0.00001, or 1 in 100,000. Thus, the observed sample difference of 1.3 belongs to a collection of means that are very unlikely and have a very small probability of occurring if the population means are equal.

There are two ways of explaining what has happened. Either the null hypothesis is true and the observed data are unlikely and do not occur very often, or the data are likely and the assumption that the population means are equal is wrong. (This is the same type of discussion as the one on probabilities in Chapter 5, which was on hypothesis testing as well, even though the term was not used there.) Because the probability of the large difference between the sample means is as small as 0.00001 if the population means are equal, we choose the second alternative. We decide that the assumption that led to the small probability—that the two population means are equal—is wrong. We reject the null hypothesis that the difference between the two population means is zero, and we conclude that zero is therefore not the true value of the difference between the two population parameters. Thus, on the basis of the data in the sample we are able to draw a conclusion about particular values of the parameters and thereby what the populations are like.

The analysis in this example can be reported in the following way in the technical jargon: To test $H_0: \mu_M - \mu_{US} = 0$, a sample of $n_1 = 1{,}200$ observations from Mexico and a sample of $n_2 = 1{,}600$ observations from the United States were collected.

$$\bar{y}_M - \bar{y}_{US} = 1.3 \ (t = 4.25 \text{ on } 2{,}800 \text{ d.f.}, p = 0.00001)$$

In plain English: To test the null hypothesis that the difference between the population means in Mexico and the United States equals zero, samples of 1,200 observations from Mexico and 1,600 observations from the United States were collected. The difference between the sample means was 1.3, or a t-value of 4.25 with a very large number of degrees of freedom. The probability of a difference between the sample means of 1.3 or more from populations where the means are equal is 1 in 100,000.

STOP AND PONDER 7.2

Translate this technical statement into English: To test $H_0: \mu = 5.0$, collect a sample of 22 observations and find $\bar{x} = 3.5$ ($t = -2.50$ on 21 d.f., $p = 0.01$).

Causal effect: A step too far

Without additional knowledge, we cannot go the extra step and say that the cultural differences in countries *caused* the difference in ability to recognize locations on a world map. That is a much stronger statement than claiming statistical significance, and we do not have enough evidence to support such a statement. Perhaps a knowledge of other factors such as educational curricula would help us understand the observed difference between Mexico and the United States better.

A little statistical theory and a game on the computer

The *p*-value for the Mexican-American difference in means is 0.00001. In other words, in only 1 of 100,000 samples from the two populations, assuming the same mean, would the difference between the sample means be equal to or larger than the observed difference of 1.3. But the study did not have 100,000 samples; it had only one sample. The idea of 100,000 different samples seems a bit overwhelming, anyway.

The right software can generate many different samples from two populations with the same mean. We actually did that; we selected another 99 pairs of samples, and for each pair we computed the difference between the means. Thus, we have 99 sample differences between means from populations with the same mean and the one observed difference. Figure 7.2 is a histogram of the sample differences. (The samples were chosen from populations with normal distributions, means equal to 8.0, and standard deviations of 6.0.).

The histogram shows the one observed difference between sample means at the far right, and it shows that the other differences range from about -0.5 to 0.5. They cluster around 0, which is not surprising, since they were drawn from populations in which the means were known to be equal. The striking feature of the histogram is that the difference between the means from the actual samples is all on its own at 1.3. None of the other differences from the computer samples are

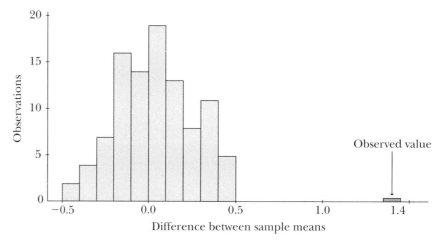

Figure 7.2 Histogram of 99 computer-generated differences between means and the one observed difference

anywhere close to 1.3. If we had used data from 100,000 samples instead of only 100, we would have had a few more extreme sample differences. But even with 100 samples, it is very clear that the observed sample difference between means is very unusual. Another way to say that it is very unusual is to say that it has a very small *p*-value.

With a small *p*-value, it helps to visualize a histogram like the one in Figure 7.2. It is a reminder of the implication of a small *p*-value, and it shows what is meant by an unusual sample. In the histogram, the difference of 1.3 does not belong with the collection of other sample differences. If that is the case, then we must conclude that the observed sample difference did not come from populations where the means are equal the way the other samples did. Thus, the Mexican and the United States populations from which the samples were drawn have different means, and based on the information in the sample data, we reject the null hypothesis.

Figure 7.3 illustrates the statistical theory that underlies the geography example. Figure 7.3a is the case where the null hypothesis is true.

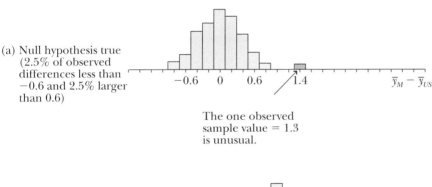

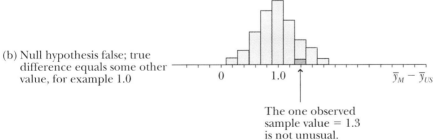

Figure 7.3 Distribution of a sample statistic from many different samples when the null hypothesis is true (a) and when it is false (b)

The parameter (population mean, population percentage difference between population means—here difference between 2 population means) is equal to the value we specified in the null hypothesis (the two population means are equal). If we now draw many different samples, then the corresponding sample statistic (sample mean, percentage, or whatever) clusters around the value of the population parameter. Figure 7.3a shows how the observed sample statistics cluster around the parameter value. Because of the variation from one sample to the next, the results from one sample to the next are not equal. Note how far the observed sample statistic is from the other sample results in Figure 7.3a.

Figure 7.3b shows the other possibility, that the population parameter is equal to some other value instead of the null hypothesis value. Again, the results from many different samples cluster around the true parameter value, and the value of the observed sample statistic falls among the results from the other samples.

In any actual investigation we have only one sample. Also, we do not know whether the null hypothesis is correct or not. The question is whether our one sample result belongs to the collection of results in Figure 7.3a or to the collection of results in Figure 7.3b. If we can tell from which collection it comes, then we can conclude whether the null hypothesis is true or not. What we do is to assume that it comes from the collection in Figure 7.3a and find the p-value for the sample results. Any sample result from the collection of results in Figure 7.3b would have a very small p-value, and based on this very small p-value we would correctly reject the null hypothesis. If the observed value belongs to the histogram in Figure 7.3a, then it is a very unusual value and has a very small p-value. But if it belongs to the histogram in Figure 7.3b, there is nothing unusual about it. Thus, hypothesis testing is answering the question of what group of sample statistics the one observed value belongs to, and the decision is based on the magnitude of the p-value.

STOP AND PONDER 7.3

A newspaper story claims that Australians drink more beer than Americans and supports the claim with one piece of information: in the month of December, Australians drank 2 liters of beer per person, while Americans drank 1.7 liters of beer. What questions might you want answered before you agreed that Aussies are bigger beer drinkers than Yanks?

7.3 SIGNIFICANCE LEVEL

Before statistical software was readily available for the computation of *p*-values, hypothesis testing was often done with a slightly different method. Instead of computing a *p*-value *after* the data have been collected, a small probability is chosen *before* data are collected and used to determine a range of values for the sample data based on which the null hypothesis would be rejected. Such a probability is known as the *significance level* of the test, and 0.05 is the commonly used value. With a significance level of 0.05, null hypotheses that are true are erroneously rejected 5 times out of 100 times, commonly accepted as a reasonable risk.

Figure 7.4 illustrates how the significance level is used in the geography example. The null hypothesis specifies that the difference between the two population means is equal to 0.00. This value is marked off in the middle of the horizontal line in the figure. To test the hypothesis, the difference between two sample means is computed. This computed value of the difference between the two sample means falls somewhere along the horizontal line labeled "Sample statistic."

> The **significance level** α (alpha) of a test is the probability of getting data that will lead to rejection of a null hypothesis that is really true.

If the null hypothesis is true   When the null hypothesis is really true, the sample statistic usually falls close to the value specified in the null hypothesis. If the difference between the two population means is zero, then the difference between the two sample means is close to zero. But occasionally the random drawing of the sample produces a sample statistic that is quite far away from the population value. Using the proper formulas, two values of the sample statistic can be found such that 2.5% (0.025) of the possible sample statistics are larger than one value and

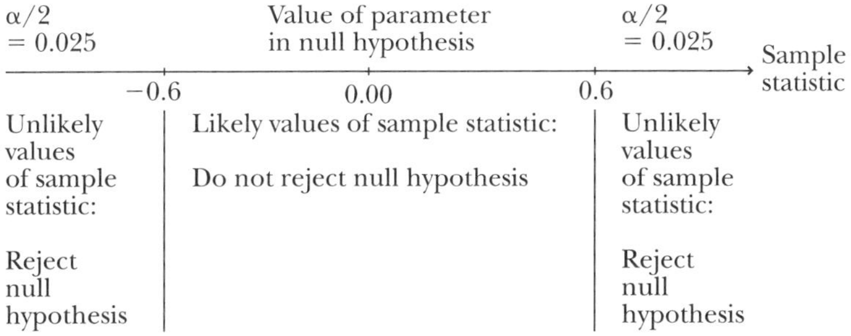

```
α/2                 Value of parameter          α/2
= 0.025             in null hypothesis          = 0.025         Sample
         −0.6              0.00                      0.6        statistic
Unlikely            Likely values of sample statistic:    Unlikely
values                                                    values
of sample           Do not reject null hypothesis         of sample
statistic:                                                statistic:

Reject                                                    Reject
null                                                      null
hypothesis                                                hypothesis
```

Figure 7.4   Hypothesis testing for the difference between two means with 0.05 significance level

another 2.5% are smaller than the other value. These are the extreme and unlikely values of the sample statistic. (Note in the figure that the significance level of 5% (0.05) is split into two parts, one on the left side and one on the right side of the midpoint.) In the geography example, anything more extreme than plus and minus 0.6 leads to a rejection of the null hypothesis, according to computations made for this example. A difference between the sample means of −0.6 or less would occur only 2.5% of the time and a difference of the sample means of 0.6 or more would only occur 2.5% of the time, if the null hypothesis is true. The two scores +0.6 and −0.6 are called the *critical values*.

> For a two-sided test, the **critical values** of a sample statistic are two values chosen such that only 5% of the sample statistics from different samples will exceed these values in either direction, when the null hypothesis is true.

If the null hypothesis is false If the difference between the two population means for Mexico and the United States is not equal to zero, then the null hypothesis is false. In that case the extreme values of the sample statistic are not so unlikely, and they will occur more often. So, if one of these extreme values of the sample statistic is observed, then we conclude that the null hypothesis is false and should be rejected. The observed difference between Mexico and the United States equals a whopping 1.3, greatly surpassing 0.6; the observed difference is statistically significant, and the null hypothesis of equal means is rejected. We conclude that knowledge of geography on this test is greater in Mexico than in the United States.

Smaller significance levels If we use a significance level smaller than 0.05, for example 0.01, then the two critical values move farther away from each other and are larger than ±0.6. With a significance level of 0.01, the critical values are ±0.79. With an observed difference of 1.3, we would still reject the null hypothesis at that level of significance. Indeed, for any critical values that are less than the observed difference 1.3, we would reject the null hypothesis. This means we could reject the null hypothesis for a significance level considerably smaller than 0.05. The smallest value of the significance level for which we would reject the null hypothesis is known as the (two-sided) *p*-value for these data.

Two-sided and one-sided tests of significance The alternative hypothesis states that the difference between the population means in Mexico and the United States is different from zero. Different from zero can mean either smaller than zero or larger than zero. For this reason it is called a two-sided alternative.

Sometimes a one-sided alternative hypothesis is useful. Suppose we have additional knowledge about how people study geography in the two countries and how they score on tests identifying locations on a world map. Suppose we know that Mexicans, on average, do not do worse than people in the United States on these tests. Then Mexicans either do as well or do better. With this additional knowledge, the alternative hypothesis can be restated

$$H_a: \mu_M - \mu_{US} > 0$$

Since we know the difference cannot be negative, the only alternative to it being zero is that it is positive.

This alternative hypothesis determines a new rejection region for the null hypothesis: a large positive difference between the sample means. Any negative difference is due to sampling variations and can be ignored. How large does the difference between the sample means have to be before we reject the null hypothesis?

Because the entire significance level of 0.05 is now on the right side of 0, we reject the null hypothesis if the computed value of t is larger than 1.64. This t-score corresponds to a difference between the sample means of 0.5. Thus, any difference between the sample means of the two countries that is larger than 0.5 is significant. With a two-sided test, a positive or negative difference of more than 0.6 is needed to reject the null hypothesis.

Two-sided and one-sided tests are illustrated in Figure 7.5. The figure shows when the null hypothesis with a two-sided alternative hypothesis is rejected and when the null hypothesis with a one-sided alternative hypothesis is rejected. In both cases the significance level equals 0.05.

7.4 TESTING A POPULATION PROPORTION

Surveys and polls are constantly probing to find out how many of us feel one way or another, do this or that, or know one thing or another. Newspapers reported that 22% of executives in 1995 feared they might be fired, compared to 6% in 1993; that 61% of teenagers who have abortions tell both parents; that 60% of college seniors do not know

282 Chapter 7 • Drawing Conclusions: Hypothesis Testing

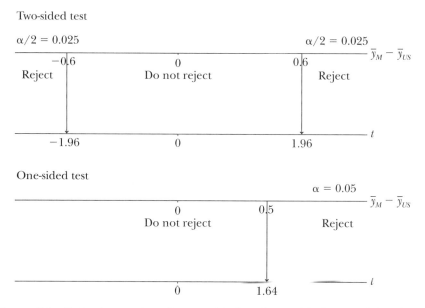

Figure 7.5 Rejection regions for two-sided and one-sided tests of significance

when Columbus landed in the Americas. Just before elections, the polls tell us in detail where the candidates stand with the electorate and how many percentage points one candidate is ahead of the other.

For a political candidate, the 50% mark is especially important in a two-person race. The candidate will win the election with more than 50% of the votes and lose with less than 50% of the votes. A poll before the election can give the candidate a good sense of how the campaign is going. To do this, the statistical expert for the candidate can test the null hypothesis that the percentage of supporters equals 50: More formally this can be written

$$H_0: \pi = 0.50$$

where π is the proportion of the population supporting the candidate. (We are all used to the letter π standing for the number 3.14 . . . , so using it here for a number between 0 and 1 is somewhat confusing. But we want to stay with the convention that parameters are denoted by Greek letters, and no other Greek letter lends itself as well to denote a probability or population proportion.)

The null hypothesis can be written using the corresponding percentage, found by multiplying the proportion by 100. If we denote the percentage of the population supporting the candidate by Π (capital pi), then the null hypothesis is written

$$H_0: \Pi = 50\%$$

The computations are easier with percentages than with proportions, and percentages are almost always used in reporting data on categorical variables.

To test this null hypothesis, the first step is determining how large the sample should be. The larger the sample, the closer the sample percentage will be to the population percentage. A sample of around 1,200 respondents is often used in election and other polls reported in the news media.

Let us see what happens if the null hypothesis of an evenly split electorate is true. With a sample of 1,200 respondents, we can calculate that in 95% of all samples, the sample proportions will fall between 0.47 and 0.53, or the sample percentages between 47% and 53%, when the null hypothesis is true. These are the two critical values for testing the null hypothesis. Once we know these values, they do not have to be calculated again for this sample size. A sampling error of $+3\%$ or -3% is often used in important polls, as you may have noticed from newspapers or TV.

If the observed sample value falls outside the range from 47 to 53, then we reject the null hypothesis with a 0.05 significance level. For example, if in the sample 55% of the voters declared they will vote for the candidate, then we can be quite confident that the voters are not evenly split but favor the candidate: the observed percentage is larger than the critical value of 53%.

We can also take the actual sample percentage of 55% and compute a *p*-value for that percentage. If the *p*-value is small enough, then we reject the null hypothesis. For a sample percentage of 55, 1,200 observations in the sample, and a null hypothesis that the population percentage equals 50, the *p*-value is a small 0.0008. Thus, if the population percentage truly equals 50, then only 8 of 10,000 different samples would give us a sample percentage of 55 or larger. The *p*-value is found by first computing a value of the *z*-variable for the sample percentage (Formula 7.4), then using statistical software or Statistical Table 1 to find the proper *p*-value.

Notice how much more informative it is to say that the null hypothesis was rejected because the *p*-value equals 0.0008 than to say that it was rejected using a 0.05 significance level. The *p*-value approach is much more common now than the significance level approach because modern statistical software routinely gives us exact *p*-values.

> **STOP AND PONDER 7.4**
>
> Look in a newspaper for stories on election or popularity polls for political candidates or office holders. Check to see if there are as many as 1,000 respondents in the sample. If not, how many are there? How does the reporter present the results of the poll? Is the story correct, balanced, and reflective of the proper statistical uncertainty as you interpret it? If not, what are some of the problems?

7.5 DIFFERENCE BETWEEN TWO POPULATION PROPORTIONS

The null hypothesis that the proportions from two different populations are equal is stated

$$H_0: \pi_1 - \pi_2 = 0$$

where π_1 is the proportion in the first population and π_2 is the corresponding proportion in the second population. Again, it is not that the two proportions are necessarily in fact equal. But if we can show with the data that the null hypothesis can be rejected, then we have been able to show that the two proportions are different.

To test this null hypothesis, we collect a sample of observations from each population and compute the two sample proportions p_1 and p_2. If the difference between the two sample proportions is large, then we reject the null hypothesis of equal population proportions. How large the difference between the two sample proportions has to be to reject the null hypothesis depends mainly on how large the two samples are.

As an example, look at the study in Chapter 3 that deals with criminals who did or did not take a literature class. Among those who took the class, 6 of 32 were convicted of new crimes for a proportion of $6/32 = 0.19$. Among those who did not take the class, 18 of 40 were convicted of new crimes for a proportion of $18/40 = 0.45$. The observed difference between the two sample proportions is 0.45 − 0.19 = 0.26. Is this difference large enough to reject the null hypothesis, or is the observed difference simply due to the randomness of the samples?

Testing the null hypothesis

With a proportion from samples this large, we first change the observed difference between the two proportions to a value of the statistical z-variable (Formula 7.5). We then use a table of the z-distribution, also known as the normal distribution (Statistical Table 1), or appropriate statistical software to find the corresponding p-value. If the p-value is small, say, less than 0.025, we reject the null hypothesis of equal population proportions.

The observed difference of 0.26 between the two sample proportions gives $z = 2.35$. The probability of finding z equal to or larger than 2.35 equals 0.009, or 9 in 1,000. This means that if the null hypothesis is true and the two population proportions are equal, then the probability is also 0.009 of finding an observed difference between two sample proportions of 0.26 or more. This probability is so small that it is overwhelming evidence against the null hypothesis. We reject the null hypothesis and conclude that there is a statistically significant difference in the proportion of those convicted of new crimes who took the literature course and those who did not.

Estimating the difference

Since we reject the null hypothesis, the difference between the two proportions is not equal to zero. How large is the difference? The best estimate for the difference between the two true population proportions is the observed difference 0.26.

We could also construct a confidence interval for the true difference. For an observed difference of 0.26, a 95% confidence interval is 0.06 to 0.46. Because the number of observations is small in the two samples, the interval is long. As researchers, we hope this interval is one of the many intervals that contain the true difference between the two population proportions and not one of the flukey few intervals that does not contain it. As we expected, since we have already rejected the null hypothesis that the true difference is equal to 0, the confidence interval does not contain the value 0.

z OR t?

In preceding chapters, the variables z and t have been used at different times as standard scores in place of raw scores. One of the most difficult questions beginning statistics students struggle with is When do we use which one of the theoretical statistical variables? So far, we have primarily used the t-variable for hypothesis tests involving means and the z-variable for tests involving proportions.

> For hypothesis tests involving one or two means, the t-variable is used. For statistical tests involving one or two proportions from large samples, the z-variable is used.

In later chapters we use the chi-square variable and the F-variable. Each of these variables is used to find the p-value for the observed data in certain contexts.

It is possible to figure out mathematically which variables to use, but such derivations go well beyond the scope of this book. We the authors are therefore left in the uncomfortable position of simply stating for each type of problem which variable to use. Many times statistical software is written in such a way that the computer picks the correct variable and the user does not think about it. The important question is whether we correctly interpret the p-values calculated from the analysis.

7.6 TESTING HYPOTHESES VERSUS CONSTRUCTING CONFIDENCE INTERVALS

Testing hypotheses and creating confidence intervals are both methods for drawing conclusions about parameter values, and thereby the real world, from the information in a sample. With hypothesis testing, we focus on a particular value of a parameter and ask if the parameter could possibly be equal to that value. For example, could the population mean for an IQ test be 100? With confidence intervals, we estimate the true value of the parameter. For example, we find a confidence interval of 102 to 107 for the population mean.

As we mentioned in Section 7.5 on proportion differences in a population, the confidence interval ranges from a lower value L to an upper value U, and hopefully this interval contains the true value of

"Shoe" reprinted by permission of Tribune Media Services.

the parameter. If the value of the parameter specified in the null hypothesis lies somewhere between L and U, then we do not reject the null hypothesis. If the value lies somewhere outside the interval, then we reject the null hypothesis.

In many ways confidence intervals are more informative than tests of hypotheses. Confidence intervals give us a range of possible values of the parameter, while a test focuses only on one possible value. For example, if the population parameter is not 100, we are not clear what it might be. Sometimes the one value is a very crucial and interesting value, as when we test to see if a difference between two means equals zero. But even when we reject the null hypothesis and conclude that the difference is not equal to zero, the question that immediately arises is how large the difference then is. Such a question is answered by a confidence interval.

Even though confidence intervals may be more desirable, hypothesis tests are commonly used in most fields. The main reason for this is that statistical software packages do not automatically compute confidence intervals.

7.7 STATISTICAL VERSUS SUBSTANTIVE SIGNIFICANCE

The importance of statistical significance has been stressed in this chapter. But sometimes a statistically significant finding should be regarded as trivial.

It is important to realize that a statistically significant result is not necessarily a substantively significant result. With large samples, for example, most results are statistically significant. The substantive significance of a result can be determined only in the context of what we are studying.

STOP AND PONDER 7.5

You read the following in a research journal of sports psychology:

> In a sample of 75 medalists, a statistically significant difference was found between the scores on the uneven bars of Olympic team members who began their competitive training by the age of 7 and the scores of those who began after 7 years of age ($p = 0.017$).

What does this sentence tell you about the gymnasts' scores—and what does it *not* tell you? How could using the phrase "statistically significant" disguise weak distinctions between groups for the statistically naive reader?

(Source: Gudmund Iversen.)

The null hypothesis specifies that the parameter equals a particular value. In the geography example we ask whether the mean number of correctly identified locations on the map is the same in Mexico and the United States. If the difference between the true population means really equals 0.1 locations, the null hypothesis is false, since the difference is not equal to 0. The correct answer to the question posed in the null hypothesis is no, and the null hypothesis should be rejected. If the sample is very large, the null hypothesis will be rejected.

However, a difference of 0.1 is not very different from 0. A geographer might say that the difference between the two numbers is so small that it is geographically uninteresting and meaningless. Thus, even though a difference of 0.1 is statistically significant, substantively it is not significant. If we focus only on the statistical significance in this case and not on the substantive significance, our analysis of the data is incomplete.

7.8 APPLICATIONS: WHEN TO REJECT THE NULL HYPOTHESIS

To gain some experience in putting all these ideas together, take a look at two simplified research examples using hypothesis-testing principles.

These examples are designed to show how the material from this chapter can be applied to other "real world" situations. The analyses also serve as a warning that any information reported to be found in one sample, or in comparing two or more samples, needs to be carefully scrutinized.

Psychology experiment on cooperation and competition

A psychologist is studying how effective a particular task is in getting small groups to cooperate in their work strategies. The psychologist observes each group at work on the task through a one-way mirror and rates it as either cooperative or competitive at the end of the task. After observing 8 groups, the psychologist classifies 7 of them as cooperative. Could this result occur by chance alone, or did something about the design of the task make them cooperative?

Let π be the probability that a group is cooperative. If chance alone determines the outcome, then this probability equals 0.5. (The results of the study are equivalent to tossing a coin and observing 7 tails in 8 tosses.) The psychologist wonders if cooperation occurred more often than by chance and sets up the null hypothesis

$$H_0: \pi = 0.5$$

To decide whether the null hypothesis should be rejected or not, the psychologist needs to find the p-value for the observed data. The observed data consist of 7 cooperative groups, and to find the p-value, the psychologist needs to find the probability of the observed data and

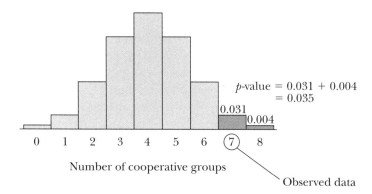

Figure 7.6 Binomial example of hypothesis testing ($n = 8$ observations; probability $\pi = 0.5$)

more extreme data: given that $\pi = 0.5$, what is the probability of getting 7 or 8 cooperative groups?

This is a small sample. Therefore, the probability of any number of cooperative groups can be found from the binomial distribution with $n = 8$ trials and $\pi = 0.5$. This binomial probability can be found from statistical software tables or printed tables of the binomial distribution, or it can be computed from the formula for the binomial distribution (Formula 5.4). The probabilities for 7 and for 8 cooperative groups are shown in Figure 7.6. The p-value for the data equals 0.035. If the groups are equally likely to become cooperative or competitive, then in 35 of 1,000 samples of 8 groups, 7 or 8 cooperative groups would occur by chance alone. This probability is larger than the standard 0.025 for a two-sided test, and it does not give any overwhelming evidence against the null hypothesis. Chance alone may still determine how a group turns out.

If the psychologist doubled the number of groups to 16 and found 14 cooperative groups, the p-value would be a small 0.002. Only in 2 of 1,000 different samples would 14 or more cooperative groups occur just by chance alone—very strong evidence against the null hypothesis that chance rather than the task design determines whether a group becomes cooperative or not.

Community study of blue-collar workers

Suppose that 60% of the work force in a community consisted of blue-collar workers in 1990. With the decline of manufacturing jobs in the

United States, the city labor commissioner wanted to know if there had been a change in the percentage of blue-collar workers in the city.

Let Π be the population percentage of blue-collar workers. The null hypothesis is

$$H_0: \Pi = 60\%$$

This null hypothesis says that there has been no change in the percentage of blue-collar workers.

To test the null hypothesis, the city selected a random sample of 400 people in the work force, which included 215 blue-collar workers. The percentage of the sample that is blue collar is therefore $P = 215/400 = 53.75\%$. This sample percentage is somewhat smaller than the percentage in the null hypothesis, but the city does not know whether 53.75% is much smaller than 60% or not. It may be that the true population value is still 60% and that 53.75 represents a random variation around 60.

To find the p-value for the observed percentage, the value of the percentage is changed to a value of the normal z-variable according to Formula 7.4: $z = -2.55$. The normal distribution table (Statistical Table 1) indicates the probability of z being equal to or less than -2.55 equals 0.005. This tells us that only 5 of 1,000 values of z are smaller than -2.55. Thus, the observed percentage of 53.75 belongs to an unlikely set of percentages, since a percentage of 53.75 or smaller would occur in only 5 of 1,000 different samples when the population percentage equals 60. The city thinks the sample is not particularly unusual and therefore rejects the null hypothesis, concluding that the percentage of blue-collar workers is less than 60%.

7.9 SUMMARY

When we make an estimate of a parameter, we try to find out what its actual value is. When we do hypothesis testing, on the other hand, we ask of the data whether or not the parameter could possibly be equal to a particular value of interest to us because of the problem we are studying.

7.1 The hypothesis as a question

A *null hypothesis* states that the parameter is equal to some specific value. Often the null hypothesis states that there has been no change in the value of a parameter, that there is no difference between two parameters, or that two variables are not related. The logical alternative to a null hypothesis is the *alternative hypothesis*. The alternative hypothesis states that the parameter has changed value, that a difference between two parameters exists, or that two variables are related.

To test the null hypothesis, we first assume that it is true. Then we analyze the data to see if they support the null hypothesis or not. If the evidence in the sample data allows us to question the null hypothesis, then we *reject* the null hypothesis.

Errors can be made by making wrong conclusions about the null hypothesis: The alpha error (type I) is made when the null hypothesis is true and judged false. The beta error (type II) is made when the null hypothesis is false but judged true.

7.2 How to answer the question posed by the null hypothesis

The p-value is the probability of getting the observed data or more extreme data when the null hypothesis is true, that is, when the parameter equals the value specified in the null hypothesis. A p-value refers to how often we can expect to get data of a certain kind from numerous different samples from a population. It does not refer to how likely it is that a null hypothesis is true. The null hypothesis either is or is not true. If the p-value is very small (typically less than 0.05 or 0.025), the null hypothesis is rejected.

An alternative hypothesis can be two-sided or one-sided. For a two-sided alternative hypothesis, we reject the null hypothesis if the sample statistic is much larger or much smaller than the population value specified in the null hypothesis. For a one-sided alternative hypothesis that states that the parameter is larger than the value in the null hypothesis, we reject for large values of the sample statistic. For a one-sided alternative hypothesis that states that the parameter is smaller than the value in the null hypothesis, we reject for small value of the sample statistic.

When a null hypothesis is rejected, the sample result is statistically significant.

Sample scores should be translated to standard scores (such as t- or z-scores) in order to find p-values. Standard scores greater than $+2$ or -2 are unusually large and are associated with very small p-values.

7.3 Significance level

The significance level is often denoted by the Greek alpha (α). The critical values of a sample statistic are values chosen such that only 5%

of the sample statistics from different samples will exceed these values when the null hypothesis is true.

7.4 Testing a population proportion

In testing population proportions, the null hypothesis states that the population proportion (π) is equal to a particular value, such as .50. With a large sample (1,200), 95% of all samples fall within 3% points of the true population percentage. If the observed sample value falls outside the critical value range (+3 to −3), the null hypothesis is rejected.

7.5 Difference between two population proportions

The null hypothesis for the difference between two population proportions is that of no difference. If the difference between the sample statistics is found to be statistically significant, the null hypothesis is rejected.

To estimate the size of the difference between two population proportions, once the null hypothesis is rejected, a 95% confidence interval can be established.

7.6 Testing hypotheses versus constructing confidence intervals

Testing hypotheses and constructing confidence intervals are both useful methods for drawing conclusions about population parameters. Hypothesis testing asks whether a parameter could be equal to a specific value or not. A confidence interval gives an interval within which the parameter is assumed to be. Hypothesis testing is more frequently done because of the design of much statistical software.

7.7 Statistical versus substantive significance

With large samples, a result can be statistically significant but for all intents and purposes not substantively significant.

7.8 Applications: When to reject the null hypothesis

Two applications illustrate the use of hypothesis testing methods: a small-sample psychology experiment, which uses the binomial distribution table (Statistical Table 5), and a larger-sample study of the proportion of blue-collar workers in a city population.

ADDITIONAL READINGS

Henkel, Ramon E. *Tests of Significance* (Sage University Paper Series on Quantitative Applications in the Social Sciences, series no. 07-004). Beverly Hills, CA: Sage, 1976.

Mohr, Lawrence B. *Understanding Significance Testing* (Sage University Paper Series on Quantitative Applications in the Social Sciences, series no. 07-073). Newbury Park, CA: Sage, 1990.

FORMULAS

TESTING A SINGLE MEAN

The null hypothesis asks whether the population mean μ is equal to a particular numerical value μ_0:

$$H_0: \mu = \mu_0$$

A sample consists of n observations with mean $\bar{x}$ and standard deviation s. To test the null hypothesis, the sample mean is changed to a value of the statistical t-value by the formula

$$t = \frac{\bar{x} - \mu_0}{s/\sqrt{n}} \qquad \text{d.f.} = n - 1 \qquad (7.1)$$

Using this observed value of t, the corresponding p-value can be found with statistical software or in statistical tables. If the p-value is small, then the null hypothesis is rejected.

Consider the null hypothesis $H_0: \mu = 4.0$. The data in a sample of $n = 12$ observations with mean $\bar{x} = 2.0$ and standard deviation $s = 1.54$ yield

$$t = \frac{2.0 - 4.0}{1.54/\sqrt{12}} = -4.50 \qquad \text{d.f.} = 12 - 1 = 11$$

Using the *p*-value approach, the probability that *t* is less than the observed -4.50 equals 0.0005. Using a 5% significance level and the *t*-table (Statistical Table 2), for 11 degrees of freedom the probability is 0.05 that $t < -2.20$ or $t > 2.20$. The null hypothesis is rejected because of a more extreme $t = -4.50$.

TESTING THE DIFFERENCE BETWEEN TWO MEANS

Two populations have means μ_1 and μ_2, respectively. The null hypothesis states that the two means are equal:

$$H_0: \mu_1 - \mu_2 = 0$$

To test the null hypothesis sample data are collected from the two populations, one sample with n_1 observations, mean $\bar{y}_1$, and standard deviation s_1, the second sample with n_2 observations, mean $\bar{y}_2$, and standard deviation s_2.

From these quantities the following *t*-statistic is computed:

$$t = \frac{\bar{y}_1 - \bar{y}_2}{s\sqrt{1/n_1 + 1/n_2}} \qquad \text{d.f.} = n_1 + n_2 - 2 \qquad (7.2)$$

If the computed value of *t* is larger than the critical value of *t* found in the *t*-table (Statistical Table 2), the null hypothesis is rejected. (It is more informative to use the *p*-value approach and find the probability of the observed or more extreme values of *t*.)

To compute *t*, we first need to find the standard deviation *s* in the denominator of *t*. This is done by first computing the variance s^2, a weighted average of the variances in the two samples. It is known as the pooled variance for the two samples, and it is found from the expression

$$s^2 = \frac{(n_1 - 1)s_1^2 + (n_2 - 1)s_2^2}{n_1 + n_2 - 2} \qquad (7.3)$$

Note that the computations involve the two sample variances and not the sample standard deviations. This computation is based on the assumption that variances in the two populations from which the samples were drawn are equal. If the population variances are equal, then the two sample variances are estimates of the same quantity, and the two estimates may as well be combined. The pooled variance is such a combined estimate. If the two population variances are not thought to be equal, then the computations need to be modified.

The square root of the pooled variance is the pooled standard deviation s, and it is used together with the two sample means and the two sample sizes to compute the observed value of the t-variable.

In the geography example, we know the two sample means in the numerator for any two countries, but we do not know the pooled s or the sample sizes, so we cannot compute t. But we do know that for France and Great Britain the value of t will be small because the difference of 0.2 locations on the world map is not statistically significant. For Mexico and the United States, t will be large, since the difference of 1.3 locations is statistically significant. Typically, values of t larger than about 2.00 are considered large and statistically significant, results and values less than about 2.00 are small and are not statistically significant.

A shortcut method for finding the t-value for equal variances does not require finding the pooled s. Using subscripts 1 and 2 for the two samples, an approximate value for t can be found from the expression

$$t = \frac{\bar{y}_1 - \bar{y}_2}{\sqrt{s_2^2/n_1 + s_1^2/n_2}} \qquad \text{d.f.} = n_1 + n_2 - 2$$

In the denominator under the square root sign, the variance of the second sample is divided by the number of observations in the first sample, and the variance in the second sample is divided by the number of observations in the second sample. For almost all purposes this second method is as good as the first.

The t-test for comparing two means for paired data is given in Chapter 12.

TESTING A POPULATION PROPORTION

The null hypothesis states that the population proportion π is equal to a particular value π_0:

$$H_0: \pi = \pi_0$$

We change the sample proportion p to a value of the statistical z-variable according to the formula

$$z = \frac{p - \pi_0}{\sqrt{\dfrac{\pi_0(1 - \pi_0)}{n}}} \qquad (7.4)$$

where n is the size of the sample.

As an example, suppose a sample of $n = 1{,}000$ observations has an observed sample proportion of $p = 0.60$. We want to test the null hypothesis

$$H_0: \pi = 0.50$$

This gives the following value of z:

$$z = \frac{0.60 - 0.50}{\sqrt{\dfrac{0.50(1 - 0.50)}{1{,}000}}} = 6.32$$

The probability of getting a z-value of 6.32 or more is less than 0.0001. Thus, the probability of getting a sample proportion of 0.60 or more in a sample of 1,000 from a population where the proportion equals 0.50 is so small that we reject the null hypothesis and conclude that the sample could not have come from such a population. The population proportion must be larger than 0.50.

TESTING THE DIFFERENCE BETWEEN TWO PROPORTIONS

The null hypothesis states that the proportions in two populations are equal:

$$H_0: \pi_1 - \pi_2 = 0$$

To test the null hypothesis, we change the difference between the two observed proportions in the two samples, $p_1 - p_2$, to a z-score according to the formula

$$z = \frac{(p_1 - p_2) - (\pi_1 - \pi_2)}{\sqrt{\dfrac{p_1(1 - p_1)}{n_1} + \dfrac{p_2(1 - p_2)}{n_2}}} \qquad (7.5)$$

where n_1 and n_2 are the numbers of observations in the two samples.

For the example with the criminals, in one group 6 people of a sample of 32 were convicted of new crimes while in another group 18 people of a sample of 40 were convicted of new crimes. This gives $p_1 = 6/32 = 0.19$ and $p_2 = 18/40 = 0.45$. Also,

$$z = \frac{(0.45 - 0.19) - (0)}{\sqrt{\dfrac{0.45(1.00 - 0.45)}{40} + \dfrac{0.19(1.00 - 0.19)}{32}}} = 2.35$$

The p-value for $z = 2.35$ equals 0.0094, so we reject the null hypothesis of equal proportions in the underlying population of criminals.

This approach can also work with a minor variation. If the null hypothesis is true and the two population proportions really are equal, then this common value could be estimated by combining the data from the two samples. The total number of people convicted of new crimes is the overall proportion

$$p = \frac{6 + 18}{32 + 40} = \frac{24}{72} = 0.33$$

We can use this common value of p in the formula for z instead of the two separate values:

$$z = \frac{(0.45 - 0.19) - (0)}{\sqrt{\dfrac{0.33(1.00 - 0.33)}{40} + \dfrac{0.33(1.00 - 0.33)}{32}}}$$

$$= \frac{0.45 - 0.19}{\sqrt{0.33(1.00 - 0.33)}\sqrt{\frac{1}{32} + \frac{1}{40}}}$$

$$= 2.33$$

The p-value is 0.0099, and the same conclusion is reached; the null hypothesis of equal proportions in the populations is rejected.

This expression looks amazingly like the expression for t for the difference between two means; the difference between the two means is in the numerator, the s in the denominator is replaced by the square root of the product of p and $1 - p$, and the square root of the sum of the two inverse frequencies is the same.

Comparing two proportions can also be done as the study of the relationship between two categorical variables in a contingency table with 2 rows and 2 columns. This approach is discussed in Chapter 9.

EXERCISES

REVIEW (EXERCISES 7.1–7.24)

7.1 What does *statistical significance* mean?

7.2 a. When we do hypothesis testing with a set of data, what are we trying to find out about a particular parameter?

b. Give one example of a null hypothesis.

7.3 a. What is a null hypothesis?

b. How is a null hypothesis different from an alternative hypothesis?

c. Write the symbol for each one.

7.4 a. What is the difference between an alpha error and a beta error?

b. Create a mnemonic device for remembering which is which. (A mnemonic device is a memory aid—a clever saying, story, song, visual image, association or whatever that helps you correctly remember something.) For example, the **a**lpha error is made when the null hypothesis is **a**ccurate; the **b**eta error is made when the null hypothesis is **b**ad. You can do better!

7.5 In general, if the sample mean is very different from the population mean as stipulated in the null hypothesis, is the null hypothesis rejected or not rejected?

7.6 a. What is a standard score?

b. What is the typical range of values for t when the null hypothesis is true?

c. How are standard scores used in hypothesis testing?

d. If you were given a *t*-value for a sample mean, how would you find the probability of obtaining that value or a more extreme value, given the null hypothesis? What two ways can be used to answer this question?

7.7 a. What does a *p*-value tell us?

b. Why do we reject a null hypothesis when the corresponding *p*-value is small?

c. What is the difference between a significance level and a *p*-value?

7.8 If the null hypothesis that the population mean is 4 is *falsely* rejected in a study, what type of error has is been made?

7.9 a. What is the most commonly used significance level?

b. What does it mean in everyday language?

7.10 If the null hypothesis is true, then most sample statistics will have values that are [close to; far from] the value of the parameter specified in the null hypothesis.

7.11 a. Significance level is often denoted by what Greek letter?

b. How is that letter spelled in English?

7.12 a. What is the meaning of the statement "The critical values in this study are +2.3 and +2.3"?

b. If in your results you find a standard score of −3.0, what can you conclude?

c. If the significance level were changed so that it was smaller than 0.05—for example, 0.01—what would happen to the critical values?

d. Do you think your conclusion in part b would be altered with the change in significance level?

7.13 a. Write the formal null hypothesis for this statistical test: "The null hypothesis states that the proportion of those in the population supporting the president is equal to 0.50."

b. Write the formal statement for this statistical test: "The null hypothesis states that the percentage of those in the population supporting the president is equal to 50%."

7.14 Produce a formal statement for the null hypothesis that could be used with this statement: "There is no difference between the proportion of the Democrats and the proportion of the Republicans who support the latest tax reform bill."

7.15 What factors are important in determining how large a difference between two sample proportions from two populations must be to reject the null hypothesis that the population proportions are equal?

7.16 You have worked very hard to initiate a new method of training workers in their jobs because you believe it will make them more satisfied with their work. A study is done to compare the old method and your new one. Which result will you hope for: that the data support the null hypothesis or the alternative hypothesis? Explain your feelings.

7.17 We have given half the first-grade classes a 20-minute recess between reading and numbers and the other half a 20-minute workbook exercise at their desks. We are testing the idea that the outdoor play will increase the scores of the children on their numbers worksheets.

    a. What is the null hypothesis in this study?

    b. What is the alternative hypothesis?

7.18 You believe that there is less air pollution in Chicago this year than last because of the various changes in automobile design, gasoline additives, emission control laws, and so on. Your garage mechanic disagrees with you, believing that there is really no change in the air quality despite all these efforts. You agree to a test of your competing ideas. How would you state the null hypothesis?

7.19 The mean salary for the top salary grade in the school system in a suburban township was $43,000 in 1985. In 1995, a sample of teachers' salaries at this level indicated a mean of $53,000.

    a. What is the null hypothesis regarding the top grade of teachers' salaries in this system if you want to see if there really has been an increase in salaries?

    b. Do you think anything has happened to the population mean of this variable in the past decade? Explain your answer.

7.20 According to statistical rules of thumb, which of the following p-values would lead to a rejection of the null hypothesis? Which ones would not? Which might be borderline?

$p = 0.50$ $p = 0.25$ $p = 0.001$ $p = 0.10$ $p = 0.05$ $p = 0.025$

7.21 What does a p-value of 0.50 mean in everyday language?

7.22 Is it easier or harder to obtain statistically significant results with a small sample than it is with a large sample?

7.23 What are the main differences between a one-tailed and a two-tailed test?

7.24 If you do not reject a null hypothesis, have you proven the null hypothesis to be true?

INTERPRETATION (EXERCISES 7.25–7.43)

7.25 On a jar of strawberry jam it says, among other things, "net weight 18 oz."

  a. Explain whether this implies that if you carefully weighed the jam, it would weigh exactly 18 oz.

  b. Explain whether the weight of the jam in another jar would be the same as the weight of the jam in the first jar.

  c. How could you design a study and analyze the resulting data to see if the manufacturer has a right to the claim on the jar about the net weight?

7.26 Data on the height of American adult males came from results of physical examinations reported to the National Center for Health Statistics in the period from 1976 to 1980. From these data the mean height of American males equals 69.1 inches. *(Source:* The New York Times, *July 26, 1992, p. E5.)* In a sample of 50 male college seniors, we find a mean height of 71 inches and a standard deviation of 2.1 inches. These data give $t = 6.40$ on 49 degrees of freedom ($p < 0.0001$).

  a. What is the null hypothesis for the data on the college seniors if we want to find out if college students are different from the general population?

  b. What can we say about the mean height in the population from which this sample came? (Do we reject or not reject the null hypothesis?)

7.27 According to a Gallup poll, the national percentage of people who own a gun is 53%. *(Source:* The New York Times, *July 26, 1992, p. E5.)* To see if gun ownership is equally prevalent in a middle-size town in the Midwest, a survey of 300 respondents finds that 45% of them owned a gun. A test of the null hypothesis that the population percentage in this town equals the national percentage gives $z = 2.78$ with a p-value of 0.003.

    a. Why can we conclude that the gun-owning percentage in this town is not equal to the national percentage?

    b. The difference between the survey percentage and the national percentage is statistically significant, but is the difference large enough to be substantively interesting?

7.28 A telephone survey report testing hypotheses concerning shaving habits is based on interviews with 200 people who were willing to spend 20 minutes on the phone with the interviewer. The fact that 70% of the people called hung up or refused to finish the interview is ignored by the market research group in its report. The conclusions of this study are flawed because of the biases in the sample. Why is it important to have a properly drawn sample in order to test hypotheses?

7.29 Why is the null hypothesis called *null*?

7.30 When statisticians reject a null hypothesis, can they ever be absolutely certain that they made the correct decision? Explain your answer.

7.31   a. Explain what it means when a statistical report indicates that the p-value is 0.025.

    b. What are the possible practical consequences of rejecting the null hypothesis?

7.32 Why was it necessary in the geography study described in the chapter to change the difference between the two sample means to a t-score to find the probability of getting such a difference or larger in a sample, assuming the population means to be equal?

7.33 In a study of taste preferences, a random sample of 200 cola drinkers rated their taste satisfaction with two major colas. The null hypothesis stated that there was no difference between consumer preference for one cola over another. Cola A's mean rating was 5.0 on a 7-point scale for goodness of taste; Cola B's mean rating was 4.6.

  a. The *p*-value equals 0.001; why is the null hypothesis rejected?

  b. What does the rejection of the null hypothesis tell us about the chances that the conclusion drawn from this survey was wrong?

  c. Do you think the cola companies will find the result substantively interesting?

7.34 Government reports indicate that pharmaceutical companies' profits have gone from 62.8% of every dollar spent on prescriptions in 1986 to 69.0% in 1992. If the report indicated that a change of 3% was statistically significant, what could you say about the profit margins of drug companies in this 6-year period? *(Source: Adapted from a report by Stephen Schondelmeyer, economist, Health Care Financing Administration, Office of Technology Assessment, U.S. Government, 1992.)*

7.35 The famous statistician Ronald Fisher established as the most appropriate (maximum) level at which to judge if a *p*-value is small enough to reject the null hypothesis. Recall what this statement means in everyday language, and then give your opinion as to why Fisher might have selected 0.05 and not 0.20 or 0.0001 or some other threshold.

7.36 a. If a sample is relatively small (for example 50 observations) and the null hypothesis is not rejected because the *p*-value is 0.10, what can we change in another similar study to increase the chances that the null hypothesis will be rejected?

  b. Can we argue that wealthier researchers are able to reject their null hypotheses more easily than poor ones?

  c. Is it always desirable to reject the null hypothesis?

7.37 Does rejecting a null hypothesis (having a statistically significant result) mean that you have discovered an exciting new fact about the world, or can results be statistically significant yet substantially trivial or meaningless? Discuss, for example in terms of evaluating advertising and marketing claims.

7.38 In 1987 Diane and George Weiss promised the 112 students graduating from Belmont Elementary School in Philadelphia that they would pay the college expenses for those students who graduated from high school and went on to college. Six years later, 45 percent of the group graduated from high school. As a comparison, in the 1986 Belmont class 28 percent graduated from high school six years later. The newspaper report from which this is taken goes on to say that the difference in the two percentages is statistically significant. *(Source:* The Philadelphia Inquirer, *June 25, 1993, pp. A1, A18.)*

a. What is the null hypothesis used in this story?

b. What statistical test was used to establish the statistical significance?

c. What does the newspaper tell us when it says that the result is statistically significant?

7.39 In a study of the speed of blood flow in artificial heart valves is the following statement:

$$H_0: \mu = 5.0, \bar{x} = 5.96, t = 2.59 \text{ (47 d.f.)}, p = 0.0049$$

a. Translate this statement into English.

b. What does the statement tell us about this type of heart valve?

7.40 The FBI reports that nationally 55% of all homicides were the result of gunshot wounds. In a recent sample taken in one community, 66% of all homicides were the result of gunshot wounds. What three possible conclusions can you draw about the percentage from this community compared to the national percentage? What additional information would you need to begin to choose one conclusion over another?

7.41 In a study of depression among recently married Mexican immigrant women in California, the following statement compares depressed women with nondepressed women in the sample:

Perceived discrimination ($\bar{x} = 17.3$ vs. $\bar{x} = 11.4$), t (df 136) = -3.7, $p < 0.001$, . . . and concern about starting a family in this country ($\bar{x} = 16.3$ vs. $\bar{x} = 12.0$), t (df 117) = -2.5, $p < 0.05$, were identified as important factors in placing this group of immigrant women at risk for the development of depressive disorders.
(Adapted from Salgado de Snyder and V. Nelly, "Factors associated with accul-

turative stress and depressive symptomatology among married Mexican immigrant women," Psychology of Women Quarterly, *vol. 11 (1987), pp. 475–488.*)

a. Translate this statement into English.

b. What does the statement tell you about depression in this group of women?

7.42 A grocery store has recently added automatic bank withdrawal as a form of payment at the check-out counters. A study comparing average amount spent per customer six months prior to the changeover and in December, one month after the changeover, indicated a statistically significant increase in purchases. Can the store manager be told with confidence that the new bank system has caused this increase? Explain your answer.

7.43 A national survey of 900 voters finds that 47% of them support the president in his recent foreign policy decisions. According to the general rule of thumb used for surveys, does this strongly suggest that the president's actions have not been supported by the majority of the voters? Why or why not?

ANALYSIS (EXERCISES 7.44–7.59)

7.44 Standard scores can be useful when we want to determine from which of several possible populations a particular sample comes. In this exercise we work with ten samples from each of two different populations. While we hardly ever have several samples when we work with actual data, the exercise illustrates an important use of standard scores. The means of these samples are as follows:

Population A: 61.2   62.6   40.1   51.7   38.0   59.8   47.6   47.7   56.3   35.0
Population B: 83.0   93.7   82.1   72.4   92.3   68.7   76.5   88.4   79.6   63.3

The means from population A have a mean value of 50.0, and the standard deviation of the means equals 10.0. The means from population B have a mean of 80.0, and the standard deviations of the means also is equal to 10.0.

a. Change all the means of the samples from population A to standard scores.

b. Change all the means of the samples from population B to standard scores.

c. A twenty-first sample has a mean equal to 75.0. What is the value of the standard score for this sample mean if the sample really came from population A? What is the value of the standard score for this sample mean if the sample really came from population B?

d. On the basis of on your answers to parts a–c, do you think the new sample belongs to the set of possible samples from population A or the set of possible samples from population B? Why do you think the new sample comes from population A or B?

7.45 In a sample of 12 families, the mean number of children per family equals 2.0. In a certain country the mean number of children per family equals 1.4 children. You want to know whether the sample could have come from that country.

a. Suppose you took a large number of different samples from this country. Explain why the sample means would not all be equal to 1.4, but the mean of all the sample means would equal the population mean 1.4.

b. The standard deviation of all the sample means (standard error) is found to equal 0.5. Find the standard score for your particular sample.

c. Is the standard score unusually large, or could your sample be one of the possible samples from that population? Explain.

7.46 In a sample survey of 50 respondents, the proportion of voters favoring a political candidate is found to equal 0.38. You want to know whether the corresponding proportion in the population of all voters equals 0.50. If the proportion equals 0.50 in the population, then the mean of the sample proportions from a large number of different samples would equal 0.50. The standard deviation (standard error) of these sample proportions would equal 0.071.

a. Find the standard score for the sample proportion.

b. Is the standard score for the sample unusually large, or could the sample belong to the set of possible samples that could have come from a population in which the proportion favoring this candidate equals 0.50? Explain.

7.47 The mean number of days of work or school lost by a person to illness is 5.1 per year, according to the National Center for Health Statistics. In a small business firm with 49 employees, the mean number of days lost to illness was 7.0 with a standard deviation of 2.5 days. The firm owner wonders if the employees are sick more often than should be expected.

a. What is the null hypothesis the owner examines?

b. Use the reported data as data from a random sample and find the *p*-value for the observed mean number of days.

c. What can the owner conclude from the test of the null hypothesis?

7.48 It is thought that 64% of adult Americans drink alcohol. In a random sample of 25 college students, 19 say they drink alcohol of one kind or another. The dean of students wants to study whether more college students drink alcohol than in the adult population at large.

a. What is the null hypothesis the dean uses?

b. With the null hypothesis, find the *p*-value for the observed number of drinkers in the sample.

c. What does the dean conclude on the basis of the *p*-value?

d. If the *p*-value is large and the dean does not reject the null hypothesis, has the dean proven that the percentage of drinkers among the students is actually equal to the national percentage?

e. Use the sample data to construct a 95% confidence interval for the population percentage of drinkers, and explain whether the interval contains the value 64% or not.

7.49 According to the Census Bureau, 73.2% of workers 16 years or older drive alone in their cars to work. After instituting a car pool program, a city finds in a survey of 300 workers that 67% of them drive alone. The city manager wants to see if the city percentage of workers driving alone is less than the national percentage.

a. What is the city manager's null hypothesis?

b. Use the data from the survey to test the null hypothesis. Is there a statistically significant reduction in the percentage of workers who drive alone?

c. Is the reduction in the percentage who drive alone of any substantive magnitude?

7.50 Refer to the list of national sample means of correctly identified geographical locations at the beginning of this chapter. Compare the differences between the means of each pair of nations, as was done with Mexico and the United States in the chapter. Using the knowledge you gained from the chapter concerning comparisons between two means, decide among which countries significant differences in geographic knowledge exist. The comparisons are as follows: United States and Mexico; United States and Great Britain; United States and France; Mexico and Great Britain; Mexico and France; Great Britain and France.

7.51 A random sample from Detroit includes 103 Baptists and 87 Methodists. *(Source: H. Schuman, "The religious factor in Detroit: Review, replication and reanalysis,"* American Sociological Review, *vol. 36 (1971), pp. 30–48.)* While there clearly are more Baptists than Methodists in the sample, can you conclude from the data that there are more Baptists than Methodists in the population from which the sample was drawn at the time of the study? Set up the proper null hypothesis and use the data to test the null hypothesis.

7.52 The sample of 28 swordfish in Exercise 4.66 has a mean mercury concentration of 1.09 ppm and a standard deviation of 0.48 ppm. Since swordfish is not supposed to be eaten when the concentration of mercury is larger than 1.00 ppm, you want to test to see if the mean concentration in the population of swordfish could be equal to 1.00.

a. Test the null hypothesis.

b. What can you conclude about the population mean of mercury concentration in swordfish?

7.53 Faculty parking lots are usually filled with a motley collection of cars, and it may not be surprising that the make of each car partly is a reflection of the owner's attitudes toward a variety of issues. In one survey, faculty members were asked what kind of cars they drove and how they voted in the previous presidential election. According to the report on the survey, among professors who drove Saabs, 98% of them voted for the Democratic candidate. Similarly, among professors who drove Volvos, 80% of them voted for the Democratic candidate. *(Source: The Ladd-Lipset Survey,* The Chronicle of Higher Education, *April 5, 1976, p. 18.)* The report does not include the number of Saab and Volvo owners in the survey; assume 50 Saab owners and 200 Volvo owners. Test the null hypothesis that the percentages Saab and Volvo owners voting Democratic are equal in the population of all owners of these two Swedish cars.

7.54 NCAA collected data on graduation rates of athletes in Division I in the mid 1980s. Among 2,332 men they found that 1,343 had not graduated from college, and among 959 women they found that 441 had not graduated. Use these data to test the null hypothesis that the proportions of graduation rates for men and women are equal.

7.55 Among the first 200 riders of the Mile High Terrifying Trojans Roller Coaster ride at the local amusement part were 134 men. Do more men than women in general ride roller coasters, or can the observed data be attributed to chance alone?

7.56 To be statistically significant, differences in the number of flaws among sample subgroups of imported silk scarves must be at least 2.1. Each of the groups in Table 7.1 reported the number of flaws detected in groups of imported silk scarves.

a. Assuming that the scarves are randomly flawed, are some groups more accurate than others in detecting flaws?

b. Which of the groups are significantly different from one another?

Table 7.1 Data for Exercise 7.56

Quality control groups	Number of rejected scarves
1	17
2	14
3	12
4	18
5	21
6	20

7.57 In Exercise 4.59 is a list of the number of children of faculty members in a mathematics and statistics department. Of the 10 children, 3 are girls and 7 are boys. Do a statistical test to see if mathematicians and statisticians in general have an equal number of boys and girls.

7.58 Students in one of our introductory statistics class spun pennies a total of 250 times. They observed 97 heads and 153 tails. Test the null hypothesis that the probability equals 0.5 that a penny lands heads up when it is spun.

7.59 In a random sample of college students, 16 of 36 men said they read nutritional labels when they bought food, and 28 of 36 women said they read labels. *(Source: Data used by permission of Jasa Porciello, Swarthmore College.)* Test the null hypothesis that the percentages of female and male college students who read nutritional labels are the same.

CHAPTER 8

8.1 Four questions about two variables and their relationships

8.2 Prediction

8.3 Independent and dependent variables

8.4 Different types of variables: Categorical, rank, metric

8.5 Return to the question of causality

8.6 Summary

RELATIONSHIPS BETWEEN VARIABLES

The ecology club of the university plans to put increased funding into refreshments and entertainment at meetings in hopes of attracting dues-paying members. As a club member, you are in charge of determining whether an increase in entertainment and refreshment expenditures actually does increase income from dues. Is there a relationship between these two variables: expenditures and dues?

A local store is asked to purchase advertising in the school yearbook. They want to know if advertising in the yearbook increases business with students. As business manager, you are asked to respond to this query.

The football team has been asked whether they would like to play on Astroturf or keep the grass. Their response hinges on the injury rates for grass versus Astroturf surfaces. If you were to advise the team, how would you answer this question?

Turning to more global issues, has there been any change in the percentage of women in the labor force over the last decades? Do people of different races smoke different brands of cigarettes? Do fewer or more pregnant women with AIDs who take AZT pass on the virus to their babies? Do places with different mean annual temperatures have different rates of breast cancer? Does the number of executions affect the homicide rate?

In previous chapters, we have looked at ways in which statistical methods are used to collect, summarize, and draw conclusions from data on a single variable at a time. Now we analyze the data on two or more variables at the same time; that is, we study relationships between variables. Most sciences have as a goal establishing relationships between variables, and statistics plays in important role in this task. In this chapter we suggest the scope of concerns for statistical analysis. A good deal of the rest of the book expands on the study of relationships between variables.

When we study two variables, we examined whether certain values on one variable correspond to certain values on the other variable. For example, does increasing the entertainment budget bring in more dues? Does advertising in the yearbook increase business income? Are the number and severity of injuries in football related to the type of playing surface? When we do find patterns in data like these, we say there is a *statistical relationship* between the variables.

Data for the study of two variables come from the two major sources we discussed in Chapter 2: experiments and observations. The conclusions we reach about the nature of the statistical relationships from these two types of data are often very different.

Suppose a biologist is interested in how the amount of light affects the growth of a certain type of plant. To answer this question, she sets up an experiment. She selects plants that are similar in every possible way, and she maintains identical conditions for them all, except for the amount of light they receive. After the plants in the different light

Why would it be difficult to assess the effect of light on these plants? *(Source: Hans Reinhard/Okapia, Photo Researchers.)*

conditions mature, she measures their growth. In this case, the data for each plant consist of observations on two variables: amount of light and growth of plant. The biologist records the data from the variables in two columns (reading down) of numbers in a data file. One column contains the amount of light each plant was exposed to and the other column contains the growth of each plant. Each row (reading across) contains the data on amount of light and growth for a specific plant.

A political scientist is interested in whether a person's age is related to how the person votes. To answer this question, the political scientist uses observational data. He constructs a survey in which the pollsters ask for the respondent's age and who the respondent voted for in the last election. As in the example with the plants, the political scientist records the data in a data file consisting of two columns. One column lists the data on the age variable and the other lists the data on the vote variable. A particular row consists of the age and vote of an individual respondent.

Now that the data in these two examples are collected, they can be analyzed.

Interest in voting preferences has sparked great interest in survey research.
(Source: Rob Crandall, Stock Boston.)

8.1 FOUR QUESTIONS ABOUT TWO VARIABLES

The major task in the analysis of data on two variables is to address four critical questions. These questions provide a framework for how we go about the analysis of statistical relationships. The analysis of data on more than two variables usually involves getting answers to these four questions as well.

Question 1. Is there a relationship between the variables in these data? First we try to establish whether there is a pattern of relationship in the observed data. When we find a relationship between variables, then we go on with the remaining questions.

Question 2. If so, how strong is the relationship between the variables? If a relationship in the data exists, we try to establish how much of a relationship there is. Relationships between variables may be strong or weak.

Question 3. Is there a relationship in the population, not just in the sample? In other words, how well we can generalize from the observed to the real world? The relationship between two variables in sample data from an observational study or an experiment may be interesting, but is usually even more interesting if we can establish the same relationship of the variables in the larger population from which the sample data came. In a political poll, it is one thing to find that two variables are related in a survey of a few hundred respondents and another thing to find that the two variables are related in the entire electorate.

Sometimes the third question is rephrased; we ask instead whether the results could have occurred by chance alone or whether some systematic effect produced the results. If we discover that the results may well have occurred by chance alone, then we usually conclude that the variables are not related in the population from which the sample was taken.

Question 4. Is the relationship a causal relationship? This, of course, is the hardest question to answer. But it is often also the most important question, and, particularly with observational data, it is the question to which statistics contributes least. With observational data, the question often remains unanswered because we cannot know whether an observed relationship between two variables is caused by other variables not included in the analysis, as in the ice cream and children's accident example in the next paragraph. With experimental data, the situation is often different. In an experiment planned according to proper statistical principles, we are often able to eliminate the effects of other variables through the control we have over the variables, thus enabling ourselves to establish causality.

Many variables can have a statistical relationship but no causal relationship. For example, suppose we have data for each month of the year on how much ice cream was consumed and how many children were injured in traffic accidents. In months with a high consumption of ice cream, many children are hurt in car accidents, while in months with lower ice cream consumption, fewer accidents occur. Based on such a pattern, we can conclude that the two variables, ice cream consumption and number of accidents, are statistically related, but we cannot conclude that the relationship is causal. Our finding that in months with a high consumption of ice cream many children get hurt in car accidents does not mean that it is ice cream consumption that causes children to be hurt in accidents, or that being hurt in accidents causes children to eat ice cream. Such a relationship is known as a *spurious relationship,* and it can be explained away if we introduce other variables, for example, temperature. Children eat more ice cream in warmer months, and those are also the months when they are on vacation and

> An observed relationship between two variables that can be explained away by a third variable is known as a **spurious relationship.**

DOES THE STORK BRING BABIES AFTER ALL?

A delightful example of a spurious relationship was the observed relationship between the number of storks and the birthrate in Danish counties. In counties with a large number of storks the birthrate was high, and in counties with fewer storks the birthrate was low. Even though there is a statistical relationship between the two variables, a causal relationship cannot be established—but the statistical relationship may be a reason why the myth that storks bring babies became so commonplace. *(Source: Stock Montage, Inc.)*

running around more and when more people are on the road in cars. While this example is clearly one of a noncausal relationship, in many other situations it is not so obvious whether two variables are causally or only statistically related.

The relationship between children's ice cream consumption and children's accident rates is spurious. *(Source: First Image West, Inc.)*

Table 8.1 Data matrix with observations on two variables for 10 individuals

Person	Gender	Activity preference
1	M	W
2	F	W
3	M	S
4	F	S
5	F	S
6	M	W
7	F	W
8	M	S
9	M	W
10	F	S

Table 8.1 shows a small part of the data in Exercise 9.41 as an example of how we try to answer the four questions. The table is a data file that describes whether men and women select work or social life as their preferred activity. Each row refers to a person, and each column refers to a variable. Each person is given a number instead of a name, as well as a gender designation (F for female and M for male) and an activity preference (W for work and S for social life). Using the data in the table, we shall try to answer each of the four data analysis questions.

Question 1. Relationship between the variables?

Is there a relationship between the two variables gender and activity preference in the data matrix in Table 8.1? Do certain values of the gender variable correspond with certain values of the activity preference variable? Just by looking at the data, we can detect a pattern of sorts. The F's tend to go with the S's, and the M's tend to go with the W's. That is, females tend to prefer social life while males tend to prefer work.

As with most data, we need to simplify and summarize the gender/activity data before the correspondences are clear. As discussed in Chapter 3, we can make a graph or a table or compute a statistic. Here

Table 8.2 Distribution of 10 people on the two variables gender and activity

		Gender		
		Female	Male	Total
Activity	Work	2	3	5
	Social life	3	2	5
	Total	5	5	10

it is convenient to arrange the data in a table with two rows and two columns (Table 8.2). From the table we can see clearly that certain values on one variable correspond with certain values on the other variable. Women tend to prefer social life to work, and men tend to prefer work to social life. From this pattern, we conclude that there is a relationship between the two variables in this data set.

Question 2. Strength of the relationship?

Is the relationship between gender and activity strong or weak? The relationship appears to have some strength: more people fall on the diagonal from the lower left of the table to the upper right than on the other diagonal. Of course, the relationship would have been stronger if all 5 social life preferences had come from the women and all 5 men had preferred work. Then the female column would have been 0 and 5 and the male column 5 and 0 (Table 8.3c). The relationship would not have been as strong if the female column had been 1 and 4 and the male column had been 4 and 1 (Table 8.3b).

Later chapters discuss computing various coefficients that tell us the strength of a relationship between variables. Such a coefficient is a statistic that can have possible values ranging from 0 to 1. When the coefficient is 0, there is no relationship between the variables. When the coefficient is 1, the relationship has maximum strength. For the data in Table 8.2, the coefficient of the strength of the relationship is 0.20. Similarly, Table 8.3b has a coefficient of strength equal to 0.60, and Table 8.3c has a coefficient of strength of 1.00. On a scale from 0 to 1, the observed value of the statistic at 0.20 indicates that the relationship between the two variables is a weak one.

Table 8.3 Data of Table 8.2 with different strengths

(a) Strength = 0.20

2	3	5
3	2	5
5	5	10

(b) Strength = 0.60

1	4	5
4	1	5
5	5	10

(c) Strength = 1.00

0	5	5
5	0	5
5	5	10

Question 3. Relationship in the population?

Could the data have occurred by chance alone? If the relationship is not simply a chance event, then we can generalize from the sample and conclude that there is a relationship in the population from which the sample was drawn.

Imagine Table 8.2, enlarged and drawn on a dart board, with its two rows and two columns. For the sake of simplicity we keep the same number of men and women as well as the same number of work and social life preferences. Then we randomly throw 10 darts at the table. Could we possibly get results similar to the pattern in our data due to some lucky throws, or does the data in Table 8.2 contain something more than randomness? If we find something more than randomness in the data, then we conclude that there is a relationship between the two variables not only in the sample but also in the population from which the sample was drawn.

To answer the question of whether the gender/activity data could have occurred by chance alone, we set up a null hypothesis that there is no relationship between the two variables. Then we see if the data lead us to reject the hypothesis or not. We do not necessarily believe that the two variables are unrelated; indeed, we want to show that two variables *are* related.

We test the null hypothesis of no relationship by computing the p-value for the observed data. If the p-value is small, then we can reject the null hypothesis. But the p-value for the data in Table 8.2 is too large for us to reject the null hypothesis; hence, the data may have occurred by chance. We cannot reject the null hypothesis of no relationship in the population in this example—not very surprising because the sample is so small. If the sample were larger (1,000 or even 100), we would be much more likely to reject the null hypothesis.

> If the null hypothesis of no relationship can be rejected, then we conclude that there really is a relationship between the variables.

Question 4. Causal relationship?

Does a person's gender in any way determine how the person prefers to spend time? From the available data on the two variables alone we cannot answer that question. We would need data on other relevant variables—and even then we might not be able to determine whether the relationship between gender and activity is causal or a byproduct of the effects of other variables.

322 Chapter 8 • Relationships Between Variables

> **STOP AND PONDER 8.1**
>
> You are given the following statistical information.
>
> a. The strength of the relationship between smoking (or not) and getting lung cancer (or not) is 0.53.
>
> b. The strength of the relationship between the amount of alcohol pregnant women drink and the birthweight of babies is 0.34 (women who drink more alcohol have smaller babies).
>
> c. The strength of the relationship between the age of taxpayers and the size of the tax bills is 0.32 (older taxpayers have higher bills).
>
> How would you answer question 4 about causality on the basis of the information in each statement? What else would you like to know to answer the question more confidently?

8.2 PREDICTION: FROM ONE VARIABLE TO ANOTHER

The existence of a relationship between two variables in data is closely connected with *prediction*. Suppose we know that a person in the sample is a woman. On the basis of the information in Table 8.2, can we predict how this woman prefers to spend time? Knowing that the person is a woman means that we use only the information about the 5 people in the female column in the table instead of all ten people in the entire table. Two thirds of the women prefer social life, so we predict that our woman prefers social life. If we made the same prediction for all 5 women, we would not be correct every time, but we would make the correct prediction more often than the wrong prediction. Indeed, we would be right 3 times and wrong 2 times out of 5. The same procedure could be used to predict the preferences of the men in the sample. Predicting that a man prefers work, we would be right 3 times out of 5.

The presence of a relationship between two variables means that we can use information on the value of one variable for an individual to help us predict the value of the other variable. In the gender/activity example, if we know the gender, we can predict how a person likes to spend time. We will not always make the right prediction, but the stronger the relationship between two variables, the better our predic-

> Two variables do not have to be causally related to predict the values of one from the values of the other.

tion will be. Thus, the strength of the relationship indicates the degree to which we can predict from one variable to the other.

> **STOP AND PONDER 8.2**
> Why can you predict the value of a variable if you know the value of a related variable, even if the two variables are not causally related? Why would it be possible to predict childhood car accident rates on the basis of ice cream sales? Do you think it would be possible to predict the rate of city apartment fires from sales of suburban snow blowers?

8.3 INDEPENDENT AND DEPENDENT VARIABLES

Something about the two variables in the gender/activity example is asymmetric. If there is a causal influence of one variable on another, we would say that it is the gender variable that affects the activity variable. Similarly, we would say that light affects growth in the experimental study on plants. We do not think that the activity a person prefers in any way affects the person's gender. Nor do we think that plant growth determines the amount of light the plant gets in the experimental setting.

For both of these examples, one variable comes before the other in time. People are born a certain sex and years later express an activity preference. A plant is first exposed to light, then it grows. When we study the relationship between two variables such as these, we know that one variable usually comes first and may causally affect the other variable. We call the variable that comes first the *independent variable*, and the variable that is influenced the *dependent variable*. The independent variable is also known as the *explanatory variable*, and the dependent variable is known as the *response variable*. We can show this scheme using an arrow pointing from the independent to the dependent variable:

$$\text{independent variable} \rightarrow \text{dependent variable}$$

Sometimes, for the sake of simplicity, the independent, explanatory variable is denoted by the letter X and the dependent, response variable by the letter Y;

$$X \rightarrow Y$$

A generic independent variable can be called the *X*-variable, and a generic dependent variable the *Y*-variable. (As a memory aid, think of the one-legged *Y* as a pushover when influenced by the sturdy *X*!)

> **STOP AND PONDER 8.3**
>
> In the following pairs, which variable is the independent (or *x*) variable and which is the dependent (or *y*) variable?
>
> a. Lightening bolt—thunder clap
>
> b. Amount of sales tax—amount of cost of goods
>
> c. Rate of popcorn sale in movies—rate of trash bag use
>
> d. Rate of electrical output—number of hot days
>
> e. Timing of commercial breaks—water consumption in town

8.4 DIFFERENT TYPES OF VARIABLES: CATEGORICAL, RANK, METRIC

The nature of the two variables determines what type of data analysis we use. Different statistical methods have been developed for different types of variables.

The independent variable and the dependent variable can be any of three types:

1. *Categorical:* The values are nonnumerical categories; example: for the gender variable, the values are female and male.

2. *Rank:* The values are ordered; examples: for an attitude variable, the values are opposed, neutral, in favor; for placement in a race, the values are first, second, and third.

3. *Metric:* Meaningful numerical values can be manipulated mathematically (added, multiplied); examples: income, weight, age.

The independent variable can be of one type and the dependent variable of another. Combining the independent and the dependent variables yields nine possible pairs of variables (3 × 3), which can occur

Table 8.4 Possible pairs of variable types

		Categorical	*Independent variable x* Rank	Metric
	Metric	D (Chapter 12)		B (Chapter 10)
Dependent variable y	Rank		E (Chapter 11)	
	Categorical	A (Chapter 9)		C (Chapter 10)

in a relationship between two variables (Table 8.4). In the table, the independent variable X runs horizontally and the dependent variable Y runs vertically. From left to right, the variables go from the simpler categorical to the more complex metric variables. The dependent variables move up in complexity from categorical to metric.

The example with gender and activity preference belongs in the lower left corner of the table (A), since both variables are categorical; this relationship is examined in Chapter 9. The example with light and growth of plants belongs in the upper right corner of the table (B), since both variables are metric; this relationship is examined in Chapter 10. A study measuring the literacy rate in a country and comparing it to the type of government of the country would belong in the lower right corner of the table (C). This is because the independent variable—literacy rate—is a metric variable, and the dependent variable—type of government—is a categorical variable; the relationship is mentioned in Chapter 10. A study of the effects of three different types of teaching methods (categorical independent variable) on school performance (metric dependent variable) would go in the upper left corner; this kind of relationship, considered in Chapter 12, is quite common.

Relationships involving rank variables do not occur very often, and the only one we consider here is one with two rank variables (E). An example is the ranking of baseball teams in two different years; this relationship is considered in Chapter 11. If a rank variable and a metric variable are combined in a study, then we usually treat them as two metric variables. Similarly, we usually treat a rank variable and a categorical variable as two categorical variables, even though we have to give up some information. (Note that it is always possible to change a metric variable to a rank or categorical variable but not the reverse. For example, an income scale, which is metric, can be converted to poor, average, and rich, a rank variable. But poor, average, and rich cannot be converted to a metric scale unless the exact income information on which to base the conversion is available.)

In considering the relationship between two variables, we need to locate the cell in this table to which the study belongs because different statistical analysis methods have been developed for the different combinations of variables in the table. Data on two metric variables are much more informative than data on two categorical variables, and many statistical procedures can be performed on them that other variables cannot support. When rank and categorical data cannot be avoided, appropriate procedures extract as much information as possible from them.

> **STOP AND PONDER 8.4**
>
> Data have been collected on the following pairs of variables. Where in Table 8.4 would you put them?
>
> **a.** Average temperature in degrees and relative cleanliness in public areas in major European cities
>
> **b.** Ethnic group identification and social class for people in Kuala Lampur
>
> **c.** Average speed of trains in Japan between Kobe and Kyoto and average price of ticket per kilometer

8.5 RETURN TO THE QUESTION OF CAUSALITY

Question 4 asks whether the independent variable causes the dependent variable. Causality is a difficult concept, and philosophers have struggled with its meaning for centuries. We do not solve the problem of causality here, nor does statistics alone establish whether one variable causally determines another. But statistical methods offer a perspective on this complex problem, and in Chapter 14 we discuss in detail some of the paths we can follow to establish whether variables are causally related.

As we have noted, finding a relationship between two variables in a statistical analysis is not the same as proving that the variables are causally related. We may surmise that to the extent to which a person's gender is determined by certain chromosomes in the genes, it is unlikely that the biological components of a person determine how the person prefers to spend time. It is even harder to imagine that the amount of ice cream consumed in a month causally determines the number of children injured in car accidents that month. We could guess that the relationship between literacy and type of government is more than statistical, and we could be pretty sure of a causal relationship between light and the growth of plants.

Role of other variables

Our speculations about the chapter examples are not founded in any statistical methods; they are simply commonsense observations based on what we know about the variables in everyday life. Two variables seem not to be causally related when we presume that *other variables* account for the observed relationship. Thus, in trying to determine if two variables are causally related, we first ask whether other variables could have produced the observed relationship. With gender and activity preference, we ask if something about how we are socialized is at work. For the example of ice cream consumption and car accidents, temperature may explain the observed relationship; when the temperature is higher in the summer, children are not in school and may be more prone to be involved in car accidents, and they also eat more ice cream.

The ice cream and car accident example is illustrated in Figure 8.1. The figure shows the causal effect of temperature as arrows from temperature to the other two variables. The data on ice cream consumption and number of injuries in car accidents show that those two vari-

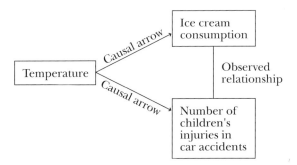

Figure 8.1 Observed spurious relationship between two variables caused by an underlying third variable

ables are statistically related, but the relationship is spurious because we know it is not a causal relationship. If temperature data were included in the example, we could then use multivariate statistical methods to show that the relationship of ice cream consumption and car accidents is spurious. These methods are discussed in Chapter 13.

Role of time

One requirement for a causal relationship between two variables is that the independent variable occurs before the dependent variable in time. Light is directed at the plants, and then the plants grow. Additionally, an independent variable causes a dependent variable when a change in the independent variable produces a change in the dependent variable. If we change the length of time the light is left on and the plant grows accordingly, we reason that the light causes plant growth. If, on the other hand, the relationship is spurious, that is, noncausal, we do not expect a change in one variable to produce a change in the other. We recognize the absurdity of the notion of, for example, a change in ice cream consumption causing a change in car accidents.

This difference between causal and noncausal relationships highlights a contrast between experimental and observational data. With experimental data it is possible to cancel out effects of other variables and to manipulate the independent variable to see if effects are produced on the dependent variable. With observational data we do not have the freedom to make manipulations. It is therefore easier to establish causality between variables based on experimental data than to establish causality based on observational data. With observational data, we are always struggling with the effects of other variables.

A famous claim for the existence of a causal relationship based on observational data is the statement that appears on packs of cigarettes and in cigarette advertising. In various forms, the statement warns that smoking may cause a variety of health problems. Had an experimental study been done, researchers would have randomly assigned newborns to become smokers or nonsmokers. Years later, they would have examined the health patterns in the two groups. Since such a study was not possible, the Surgeon General did the next best thing and asked statisticians to examine observational health data on smokers and nonsmokers. Based on the available evidence, most people now accept that there is a causal link between smoking and health problems.

Despite the number of studies published on the effects of smoking, no exact experimental study can be done to isolate causal factors such as whether or not smoking causes lung cancer. *(Source: 1989, Comstock.)*

Multiple causal factors

In everyday life, understanding causality is complicated by the fact that several variables, not just a single variable, usually determine a dependent variable. A person's salary, for example, is affected by the job description, how much training is needed, how much experience the person has, how talented the person is, and realistically, sometimes the person's gender and race. When we consider the effect of a single variable, we should not expect it to be the only cause of the dependent variable.

Finally, even if two variables are causally related, the relationship may not hold in every case. For example, not all smokers develop health problems, and the health problems of smokers are not all caused by smoking; the relationship holds only for some people. To more fully understand smokers' health problems, other independent variables in addition to smoking must be considered.

8.6 SUMMARY

This chapter discusses relationships between two variables. Data come from two major sources: experiments and observations.

8.1 Four questions about two variables

Once data are gathered on two variables, four questions may be asked of the variables:

Question 1. Is there a relationship between the variables in the data?

Question 2. If so, how strong is the relationship between the variables?

Question 3. Is there a relationship in the population, not just in the sample?

Question 4. Is the observed relationship a causal relationship?

To answer question 1, we look at the patterns in the sample data. If there is a relationship in the sample data, we ask question 2. To answer question 2, we compute a coefficient that measures the strength of the relationship. A strong relationship is indicated by a value nearer 1 and a weak relationship by a value nearer 0. To answer question 3, we set up a null hypothesis that there is no relationship between the two variables and see if we can reject that hypothesis or not. The p-value for the observed data is computed, and if the p-value is small then we reject the null hypothesis.

Answering question 4 on causality is often difficult. A statistical relationship—even a very strong one—can exist between two variables without a causal relationship. It is easier to establish causality for experimental data than for observational data because the effects of other variables can be controlled for in an experiment.

8.2 Prediction: From one variable to another

Even if two variables are not causally related, we can predict the value of one variable for an individual observation if we know the value of the other. The strength of the relationship indicates the degree to which we are able to predict from one variable to the other.

8.3 Independent and dependent variables

In the study of relationships between variables, variables are often called independent variables and dependent variables. In a causal relationship between variables, the causal variable is the independent variable and the variable that is influenced by the independent variable is the dependent variable. Independent variables usually occur in time before dependent variables. (But not all variables that precede in time are independent variables.) A generic independent variable is labeled an *X*-variable and a generic dependent variable is labeled a *Y*-variable.

8.4 Different types of variables: Categorical, rank, or metric

Both the independent and the dependent variable can be one of three types. Categorical variables are those where the values are two or more categories, such as female and male. Rank variables are those where the values are ranked from lower to higher, such as the outcome of a race. Metric variables are those where the values are meaningful numbers that can be manipulated mathematically (added, multiplied). Examples are income, weight, and age. Different statistical methods have been developed to analyze different types of pairs of variables.

8.5 Return to the question of causality

To judge if an independent variable is causally related to a dependent variable (once a relationship established between the two variables is established in the population), we (1) think about the problem in terms of everyday logic to see if the possibility makes sense in the world as we know it; (2) note whether the independent variable has preceded the dependent variable in time; (3) if feasible, alter the independent variable and observe whether the dependent variable is also affected (that is, do an experiment), and (4) accept that even if the independent variable has a causal impact on the dependent variable, other important variables that are not part of the current research program may affect the dependent variable.

ADDITIONAL READINGS

Davis, James A. *The Logic of Causal Order* (Sage University Paper Series on Quantitative Applications in the Social Sciences, series no. 07-055). Beverly Hills, CA: Sage, 1985. A small paperback book on causality.

Liebetrau, Albert M. *Measures of Association* (Sage University Paper Series on Quantitative Applications in the Social Sciences, series no. 07-032). Beverly Hills, CA: Sage, 1983. An introduction to different ways of measuring the strength of a relationship between two variables.

Simon, Herbert A. "Causation." In William H. Kruskal and Judith M. Tanur (eds.), *International Encyclopedia of Statistics*. New York: The Free Press, 1978. A leading expert's view of the connection between causality and statistics.

EXERCISES

REVIEW (EXERCISES 8.1–8.23)

8.1 Describe in other words what is meant by a "statistical relationship" between two variables.

8.2 What are the four questions a researcher asks of the data collected in a study on two variables?

8.3 In answer to question 1, if no relationship exists between two variables in sample data, what can you conclude about the population from which the sample was drawn?

8.4 Why is it important whether or not a relationship in the *sample* is strong or weak?

8.5 Paraphrase question 3, "Is there a relationship in the population?"

8.6 Why is it easier to establish causality with experimental data rather than observational data?

8.7 What is a spurious relationship between two variables?

8.8 Historically, a relationship has been found between the length of women's skirts and the strength of the economy: the shorter the shirts, the stronger the economy. Is this a causal or a spurious relationship?

8.9 Statisticians try to resolve spurious relationships by bringing into the analysis other variables that may have a causal relationship with the variable in question.

   a. What are the major reasons that statistically literate people like you might think a relationship between two variables was spurious?

   b. Why, for example, do we think the relationship between ice cream consumption and childhood car accidents is spurious?

   c. How could a mistake be made in judging whether a relationship is spurious or not?

   d. Can you think of any historical circumstances where a relationship that was causal was presumed to be spurious or vice versa?

8.10 Discuss whether or not you think the following relationships are causal or spurious. If you are undecided, discuss why.

   a. Simmons and Blyth (1987) found early-maturing girls to have the following problems: body image disturbances, lower academic success, and conduct problems in school.

   b. Increasing the number of policemen on the street increases the number of crimes committed.

   c. High school students who smoke marijuana regularly get lower marks in school than those who do not smoke it regularly.

8.11 Researchers compute a coefficient to determine the strength of a relationship between two variables. Give two values of such a coefficient that would indicate a strong or a weak relationship.

8.12 If you were told that the strength of the relationship between two different forms of a test was 0.96, what could you say about these two tests?

8.13 If a researcher is able to reject the null hypothesis, what can the researcher conclude about the relationship between the two variables?

8.14 What size p-value leads to rejection of the null hypothesis?

8.15 True or false: If you cannot prove causal relationships in your data, you cannot make predictions of values on the dependent variable from values of the independent variable. Defend your answer.

8.16 a. What is an independent variable?
b. What is a dependent variable?
c. Which letters are used to designate each?
d. How are you going to remember this?

8.17 In a study of the relationship between two variables, what factors would help you decide which variable is the independent and which is the dependent variable?

8.18 Why is it so difficult to use observational data to establish that a relationship between two variables is causal?

8.19 In studying the relationship between two variables, why does it matter what kind of variables—categorical, rank, or metric—one is working with?

8.20 Name the type of variable—categorical, rank, or metric—for each of the following.
a. Religion
b. NFL standings
c. Height
d. Placement of horses in the Kentucky Derby
e. Winners of the bronze, silver, and gold medals in the luge
f. Olympic team membership
g. Days of the week
h. Age

8.21 a. Give an example of a study in which the independent variable is categorical and the dependent variable is metric.
b. Give an example of a study in which both variables are metric.
c. Give an example of a study in which both variables are rank variables.

8.22 If a study includes one metric variable and one rank variable, how could you analyze the relationship between the two?

8.23 What is the general name of the methods used to analyze three or more variables?

INTERPRETATION (EXERCISES 8.24–8.33)

8.24 According to National Opinion Research Center surveys done in 1972 and 1991, the tendency to be satisfied with one's job is related to the years one has worked at it—the longer the happier. Table 8.5 shows the relationship between work satisfaction and years on the job for two different groups of 270 workers

    a. Which part of the table shows the strongest statistical relationship? Explain your answer.

    b. What is the difference between finding relationships that apply only to the sample and finding relationships that apply to the population from which the sample is drawn?

    c. How might it happen that you would find a relationship between two variables in a sample but not in the corresponding population?

8.25 Explain the following statement: "Finding a relationship between two variables in a statistical analysis is not the same as proving that the variables are causally related."

8.26 Why is one variable usually called "independent" and the other "dependent" in the study of the relationship between two variables?

8.27 On what grounds might statisticians who are employed by tobacco companies argue that it has not been proven that cigarette smoking is hazardous to your health?

Table 8.5 Data for Exercise 8.24

(a)	Years on the Job			(b)	Years on the Job		
	Under 10	Over 10	Total		Under 10	Over 10	Total
Happy	50	100	150	Happy	70	80	150
Unhappy	100	20	120	Unhappy	80	40	120
Total	150	120	270	Total	150	120	270

8.28 How could it be proved that chewing tobacco causes various forms of cancers in people who chew it? Why has it not been done?

8.29 What are major ways to establish causality once a relationship between two variables is known?

8.30 A metric variable can be converted to a rank variable but not vice versa. Why not? Give an example to illustrate your reasoning.

8.31 It has been observed that women who smoke have smaller babies at birth than women who do not smoke. Can you conclude that smoking causes babies to have low birth weights? Discuss.

8.32 In the box on storks and birthrate, it was possible to predict with a fair amount of accuracy the birthrate for a county from the number of storks in the county. Describe why it is not necessary to understand the cause of a variable in order to predict the value of the variable.

8.33 a. Explain the steps by which survey researchers concluded that in Beijing, 74% of women said they were happy with their bodies as opposed to 84% of women in Tokyo, who said they were unhappy with their bodies. *(Source:* Newsweek, *February 12, 1996, p. 41.)*

b. These researchers suggested that the cause of this difference was the introduction in Japan of "racy" western lingerie advertising. Comment on whether you think this is a spurious or a causal relationship.

ANALYSIS (EXERCISES 8.34–8.40)

8.34 In this exercise, one variable is enrollment in a special program for juvenile offenders and the other variable is contact with the police. If we know that a person was enrolled in this program, can we predict whether the person had further police contact? Table 8.6 displays the data for 100 juvenile offenders.

a. Suppose you were introduced to one of these juveniles, and you were not told whether or not this person had been enrolled in the special program. Looking at Table 8.5, why would your best prediction be that this person has not had any further contact with the police?

b. If you were to predict no further police contact for each of the 100 juveniles, for how many of them would you make the wrong prediction?

Table 8.6 Data for Exercise 8.34

		Enrolled in juvenile offender program		
		Yes	No	Total
Further police	No	37	20	57
contact	Yes	13	30	43
	Total	50	50	100

Source: T. Hirschi and M. J. Hindelang, "Intelligence and delinquency: A revisionist review," *American Sociological Review, vol. 42 (1977), p. 575.*

    c. If you were told that a person had been enrolled in the special program, why would your best prediction be that this person has had no further contact with the police?

    d. If you were to predict no further contact for each of the 50 juveniles enrolled in the special program, how many times would you make the wrong prediction?

    e. Now suppose you were told that a juvenile had not been enrolled in the special program. Would your best prediction be that the person did or did not have any further contact with the police?

    f. If you were to predict further contact for each of the 50 juveniles not enrolled, how many times would you make the wrong prediction?

    g. Why is the sum of the numbers in answer to parts d and f the total number of wrong predictions the enrollment status for the people is known?

    h. Explain how ability to predict is improved and 10 fewer wrong predictions made when enrollment status in the special program is known.

    i. Explain why the ratio $10/43 = 0.23$ measures improvement in prediction on a scale from zero to one.

8.35 Of 665 male patients admitted to hospitals in New England for a first heart attack, 214 of them had (vertex) baldness. In another group of 772 patients admitted for noncardiac conditions, 175 had the same kind of baldness. (*Source:* The New York Times. *February 14, 1993, pp. A1, C12.*)

a. Show the data in a table.

b. What percentage of those admitted for a first heart attack were bald?

c. What percentage of those admitted for noncardiac conditions were bald?

d. What does the difference between the two percentages tell you?

e. Why are we more interested in comparing the percentage of bald men who had heart attacks with the percentage of nonbald men who had heart attacks than we are in comparing the percentage who were bald among those who had heart attacks with the percentage who were bald among those who did not have heart attacks? That is, should we compare the percentages in the two columns in the table or should we compare the percentages for the two rows?

f. Even though the percentages in part d can be computed from the given numbers, why could the percentages not be very meaningful?

g. What do the numbers tell you about whether baldness causes heart attacks?

8.36 People in a random sample from the Detroit metropolitan area were asked to choose or not choose the following statement: "Work is important and gives a feeling of achievement." In a study of the relationship between religious affiliation (Baptist or Methodist) and whether or not people chose the statement, Table 8.7 emerged.

a. What is the independent and what is the dependent variable in this study?

Table 8.7 Data for Exercise 8.36

		Religious affiliation		
		Baptist	Methodist	Total
Statement	Chose	36	51	87
	Did not choose	67	36	103
	Total	103	87	190

Source: H. Schuman, "The religious factor in Detroit: Review, replication and reanalysis," *American Sociological Review, vol. 36 (1971), pp. 30–48.*

Table 8.8 Data for Exercise 8.37

		Race		Total
		Black	White	
	Marlboro	4	576	580
Brand	Newport	25	45	70
	Kool	4	5	9
	Other	12	181	193
	Total	45	807	852

Source: Teenage Attitudes and Practices Survey, 1989, by the National Center for Health Statistics, as reported in Chance, *vol. 5 (1992), nos. 1–2, p. 27.*

b. Does there seem to be a relationship between the two variables in these data?

c. Does the relationship between the two variables seem strong or weak?

d. What does the table indicate about the difference between Baptists and Methodists in Detroit at that time?

8.37 In Exercise 6.45, you looked at who among young smokers smoke the Newport brand. Here we include two other brands popular among young smokers. Table 8.8 includes the three brands that were reported by more than 10% of the two groups of smokers.

a. What do you see in this table, without making any further computations?

b. Change the numbers in each column to percentages that add to 100% in each column. What differences do you see between blacks and whites?

c. Does the table show a strong or a weak relationship between the variables?

d. What do you think the three brands are so attractive to young smokers?

8.38 Table 8.9 is a data matrix taken from a study of child caretakers reported in dollars per hour.

a. How could you rearrange the data so that it would be easier to detect group differences in the amount paid for caretakers?

Table 8.9 Data for Exercise 8.38

Baby	Caretaker	Cost per hour (dollars)
1	Relative	$4.90
2	Nanny	$7.00
3	Relative	$5.00
4	Day care center	$6.60
5	Private home	$5.35
6	Nanny	$7.50
7	Private home	$5.50
8	Day care center	$6.75
9	Relative	$5.25
10	Private home	$5.15
11	Nanny	$7.55
12	Day care center	$6.67
13	Relative	$5.10
14	Private home	$5.35
15	Nanny	$7.40
16	Day care center	$6.75

Source: Sandra L. Hofferth, Urban Institute.

b. Does there seem to be a relationship between the two variables caretaker and dollars per hour?

c. If you think there is a relationship, do you think it is strong?

d. Do you think the results are just a fluke of the data, or do you think there might be a relationship in the population as well?

e. How well can you predict from the type of caretaker the amount of pay received?

f. Which is the independent and which the dependent variable?

8.39 In this exercise, one variable is year (1960 to 1995), and the other variable is percent of mothers with children under 6 in the labor force (Table 8.10). From the data for a particular year, can you predict whether or not a mother with a child under 6 is in the labor force?

Table 8.10 Data for Exercise 8.39

	1960	1965	1970	1975	1980	1985	1990	1995
In labor force	20	25	32	38	47	52	58	58
At home	80	75	68	62	53	48	42	42

Source: Bureau of Labor Statistics.

a. Suppose you were introduced to one of these mothers and you were not told whether or not she was in the labor force. Looking at the table, why would your best prediction be that this person had not been in the labor force?

b. If you were to predict that each mother had stayed at home, for what percentage of them would you make the wrong prediction, assuming the same number of women in each five-year period?

c. If you were told that a mother had had a child under 6 in 1960, why would your best prediction be that this woman had stayed at home?

d. If you were to predict that all the mothers in the 1970 sample had stayed at home, how many wrong predictions would you make?

e. Now suppose you were told that a mother had a child under 6 in 1995. With that knowledge, would your best prediction be that she did or did not work outside the home?

f. If you were to predict work outside the home for each of the mothers in 1995, how many times would you make the wrong prediction?

g. Explain how you improve your ability to predict and make fewer wrong predictions when you know the year in which the mothers had young children.

h. Does it appear that a shift in the proportion of mothers of young children in the labor force is occurring? What is your evidence?

i. Is the relationship between year and percentage in labor force a causal relationship?

Table 8.11 Data for Exercise 8.40

		Tone of speaker's comment		Total
		Negative	Nonnegative	
Tone of response	Negative	444	181	625
	Nonnegative	435	679	1114
	Total	879	860	1739

Source: V. L. Walsh et al., "Impact of message valence, focus, expressive style, and gender on communication patterns among maritally distressed couples," Journal of Family Psychology, vol. 7 (1993), pp. 163–175.

8.40 The data in Table 8.11 are taken from a study of communication patterns among unhappily married couples. The question of concern: When one partner says something either negative or nonnegative to the other, what type of response is given, a negative or a nonnegative one? One variable, tone of speaker's comment, has values coded as negative or nonnegative, and the other variable, tone of response, also has values coded as negative or nonnegative. These data summarize 1739 comments from 52 couples.

a. Suppose you selected one of the speaker's comments and did not know whether or not it was negative. What would your best prediction be as to what type of response it was?

b. If you were to predict that every response was nonnegative, for how many responses would you make the wrong prediction?

c. If you were told that a person had made a negative comment, why would your best prediction be that the response was negative?

d. If you were to predict a negative response for all the negative comments, how many times would you make the wrong prediction?

e. Now suppose you were told that a comment was nonnegative. Would your best prediction be that the response was negative or nonnegative?

f. If you were to predict nonnegative for each of response to a nonnegative comment, how many times would you make the wrong prediction?

g. Why would the sum of the numbers in answer to parts d and f tell you the total number of wrong predictions when you know the type of comment made?

h. What do these data suggest about how to have pleasant and unpleasant conversations with a significant other?

i. Can you think of any difficulties in generalizing about the conversations of these maritally distressed couples on the basis of these data?

j. Can you think of any potential difficulty in generalizing these results to conversations among college students taking a statistics course?

CHAPTER 9

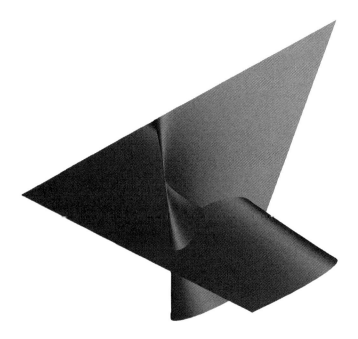

9.1 Analysis of the data: Are there trustworthy differences in attitude?

9.2 Question 1. Relationship between the variables?

9.3 Question 2. Strength of the relationship?

9.4 Question 3. Relationship in the populations?

9.5 Question 4. Causal relationship?

9.6 Larger tables: A banquet of possibilities

9.7 Summary

CHI-SQUARE ANALYSIS FOR TWO CATEGORICAL VARIABLES

Do people in different countries view strangers the same way? In surveys in several European countries, people were asked the following question: "Generally speaking, would you say that most people can be trusted or that you cannot be too careful in dealing with people?"

> A **contingency table** is a table of frequencies showing the distribution of data on two categorical variables.

Let us compare two of the countries, Denmark and France (Table 9.1). Table 9.1 is called a *contingency table.* The table shows how the people in the poll were distributed on the two categorical variables country and attitude toward people. Note that this contingency table has two rows and two columns (as well as a row and a column labeled total), because each of the two variables has two values (categories). The country variable has the two values Denmark and France, and the attitude variable has the two values trust and suspicion. Of course, it is possible for categorical variables to have more than two categories, or values. With four countries the table would have four columns and two rows.

A contingency table shows *frequencies,* that is, the number of elements in the various categories. Analyzing the data in a contingency table takes several steps. First we can consider each variable separately. For the country variable, in Table 9.1, we see that there are 985 Danes and 969 French people. For the attitude variable, 831 people think most people can be trusted, and 1123 say that you cannot be too careful. Then, to figure out whether the countries are different, we consider the two variables together. We find that there are 625 Danes who think most people can be trusted and a little more than half that number who think you cannot be too careful. Among the French, 206 feel people can be trusted and 763 feel you cannot be too careful. Thus, the majority of the Danes are trusting and the majority of the French are suspicious. By skimming the table this way, we have already learned something about how the people in this sample felt. Perhaps there truly are differences between the Danes and the French.

Table 9.1 Country and attitude toward people

		Country		
		Denmark	France	Total
Attitude toward people	Trust	625	206	831
	Suspicion	360	763	1,123
	Total	985	969	1,954

Source: Jacques-René Rabier, Helen Riffault, and Ronald Inglehart, Euro-barometer 25: Holiday Travel and Environmental Problems, *April 1986, Ann Arbor, MI: Inter-University Consortium for Political and Social Research, 1988. Codebook p. 10.*

> **STOP AND PONDER 9.1**
>
> Give an example of two categorical variables that might be related and set up a contingency table for the two variables. Why are your variables categorical variables?

9.1 ANALYSIS OF THE DATA: ARE THERE TRUSTWORTHY DIFFERENCES IN ATTITUDE?

What should be the independent and what should be the dependent variable in this analysis? At least for all the native-born people in the sample, it is clear that country came first in time, and then people developed their attitudes about trust. Thus, we select country as the independent and attitude as the dependent variable.

Note that Table 9.1 is set up with the independent variable country running horizontally across the table and the dependent variable attitude running vertically down the side. Displaying the independent variable horizontally and the dependent variable vertically is a common way of setting up contingency tables. This way is also consistent with the way we display data for other types of variables in later chapters as well as Chapter 3.

> **STOP AND PONDER 9.2**
>
> A local chapter of Phi Beta Kappa is interested in bringing a speaker to campus. The selection committee narrows the choices to two speakers: one is a professor of romance languages whose specialty is the poetry of courtly love, and the other scholar's topic is Lacanian psychoanalysis and the politics of pain. Unable to decide, the committee asks the entire chapter to vote on their preference. Among the more senior members of the chapter, 14 vote for the poetry lecture and 7 for the Lacanian one. Among the more junior members, 10 vote for the Lacanian lecture and 5 for the poetry of courtly love.
>
> Set up a contingency table that represents these results. What does the table tell you about the relationship between the two variables?

This rule is not ironclad. If the independent variable has only a few categories and the dependent variable has many categories, the variables can be reversed, with the independent variable running vertically and the dependent variable running horizontally. The table is set up this way to take up less space on the page, since the table has fewer rows than columns.

Bar graphs

Three-dimensional   The data in Table 9.1 can be displayed in graphs of various kinds, and as you recall from Chapter 3, it is often easier to see patterns in data in a graph than in a table. One possible graph expands on the idea of a bar graph. Since we are dealing with two variables, instead of a bar we use a rectangular prism to show the number of observations for each value of the variables (Figure 9.1).

The graph is harder to draw than a bar graph for one variable, and it is also harder to read. The largest frequency is usually shown at the back of the graph so that it doesn't "block the view" of the other frequencies. Here that prism represents the 763 French who say people cannot be trusted. From the graph we learn that among the Danes there are more trusting people, and among the French there are more suspicious people. We see that the smallest group is the French who

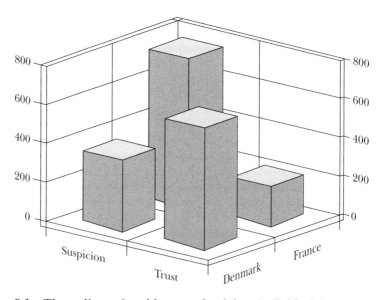

Figure 9.1   Three-dimensional bar graph of data in Table 9.1

say most people can be trusted, the next smallest is the Danes who feel one cannot be too careful, and the third is the Danes who say people can be trusted. In a three-dimensional graph, it is hard to see exactly what the frequencies are equal to, but relative sizes are clear.

Bars same width, different heights Data of this kind can also be displayed in regular bar graphs. We can either stack the bars for each country or place them side by side (Figure 9.2). The bars in both graphs of Figure 9.2 have the same base but are of different heights, showing at a glance that the numbers of Danes and French are not equal. Among the Danes there are more trusting people, and among the French there are more careful people. There is a scale at the left of each graph, but it is hard to read accurately how many people are in each category. For Figure 9.2a it is hard to see how many respondents say that people can be trusted because the bottom bases of the bars do not start at zero. For Figure 9.2b it is hard to see the total number of respondents for each

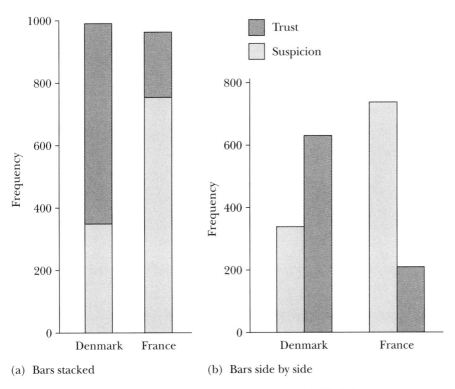

(a) Bars stacked   (b) Bars side by side

Figure 9.2 Two-dimensional bar graphs of data in Table 9.1: Bars same width, different heights

country because both sets of bars start at zero. The advantages and drawbacks of various graph forms, as noted in Chapter 3, are found in these graphs.

Bars different widths, same height The same data can be graphed so that the heights of the bars are the same but the bases are different (Figure 9.3). The area of each bar shows the percentage of observations in each group, but frequencies could be used instead. Again, the graph shows slightly more Danish respondents than French. Among the Danes, about twice as many respondents feel that most people can be trusted. Among the French, a large majority feel that you cannot be too careful in dealing with people. The construction of the bar graphs is discussed in the formula section at the end of the chapter.

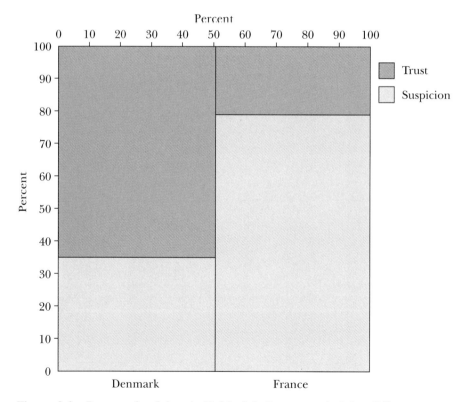

Figure 9.3 Bar graph of data in Table 9.1: Bars same height, different widths

The choice among the three graphs is not obvious. The simplicity of Figure 9.3 is appealing to some experienced viewers, but perhaps less so to those unused to such forms of presentation.

> **STOP AND PONDER 9.3**
>
> Figures 9.1, 9.2, and 9.3 are three different representations of the data on countries and trust. If you were selecting one of them for use in a textbook or business report, which would you most prefer to use and least prefer to use? Why did you make these selections?

Summary computations with categorical variables

Contingency tables are used to study the relationship among categorical variables, and we ask the data in contingency tables the four questions discussed in Chapter 8. (1) We want to know whether there is a relationship in these data between the nationality of those polled and how they expressed their attitudes toward trusting others. (2) We want to know how strong the relationship is between the two variables. Is there just a slight tendency for one nationality to be more trusting or is there a strong tendency? (3) We want to know if the results of the poll apply to the real world of all people in the two countries as well as those in the sample. If we find that there is a relationship between these two variables in the larger population, then we have learned something about the way people feel and about human behavior more generally. (4) Finally, could the relationship be causal?

Refer for a moment to Table 9.1. The three numbers in parentheses below the table are part of the statistical analysis of the data; they describe various aspects of the relationship between the two variables. We discuss how to interpret these numbers in the next several pages. And at the end of the chapter, we explain how the numbers are found and give the formulas for finding them.

9.2 QUESTION 1. RELATIONSHIP BETWEEN THE VARIABLES?

The first question we ask is whether a relationship exists between nationality and attitude—that is, do certain values of one variable tend to occur more often with certain values of the other variable? In other words, are the Danes different from the French in their attitudes?

A quick look at Table 9.1 shows that among the Danes in this sample a majority believe people can be trusted, while among the French a majority feel they cannot be too careful. Note that more observations in the table fall on the diagonal from the upper left corner to the lower right corner than fall on the other diagonal. This pattern of frequencies indicates that the two variables in these data are related. The graphs in Figures 9.1, 9.2, and 9.3 also show that the two variables are related.

A common way to determine whether two variables are related is to change the frequencies to percentages and then compare the percentages. It is easier to compare 64% to 21% than 625 of 985 to 206 of 969.

Percentages are always computed within the values of the independent variable. Thus, we first find the percentages of Danes who feel trusting and who feel the need to be careful, and then we do the same for the French. Table 9.2 shows the results.

A table of percentages should show the total 100% for each group in the independent variable to guide the reader toward the direction in which the percentages were computed. In Table 9.2, 100% is given at the foot to show that the columns add to 100. Also given is the total number of observations in parentheses (n) for each group, making it possible to recover the actual frequencies from the percentages.

The two columns of percentages show that the attitudes for the Danes and the French in this sample are different. From the fact that the two columns are different we conclude that a relationship does indeed exist between country and attitude in these data.

We can express the results of the analysis of this table two ways. We can focus on the values of the independent variable country and say that there is a *difference* between the two countries. Or we can focus on

Table 9.2 Percentage distributions of attitude for two countries

		Country	
		Denmark	France
Attitude toward people	Trust	64	21
	Suspicion	36	79
	Total	100	100
	(n)	(985)	(969)

the variables themselves and say that there is a statistical *relationship* between country and attitude. Sometimes it is tempting to express the results as differences between the groups defined by the independent variable, but it is more consistent with the remaining chapters to speak of relationships between the variables.

9.3 QUESTION 2. STRENGTH OF THE RELATIONSHIP?

The second question we ask is how *strong* the relationship is between country and trust. Strength of a statistical relationship is measured by a coefficient that ranges in possible values from 0 to 1. When the coefficient equals 0, there is no relationship between the variables; the percentages in the two columns are equal, indicating no difference in attitudes. When the coefficient of strength equals 1, the relationship between the variable is of maximum strength: all the Danes would feel people can be trusted, and all the French would feel they cannot be too careful; the attitudes of the two countries could not be any more different. A coefficient in the range from 0.00 to 0.30 or so is thought to indicate a weak relationship, the range from about 0.30 to 0.70 to show moderate strength, and the range from 0.70 or so to 1.00 to show a strong relationship.

STOP AND PONDER 9.4

By just looking at the data in Table 9.1, what value of the coefficient would you guess these data give?

Phi in the sample

For a contingency table such as Table 9.1, composed of two rows and two columns of observational data, the coefficient we compute is called *phi* (Formula 9.1 at the end of the chapter). Phi in Table 9.2 equals 0.43. A relationship definitely exists in these data since the two columns of percentages are different. With phi equal to 0.43, we conclude that the relationship is moderately strong. A moderately strong relationship means that country may have some effect on how people feel about trusting others but that other variables beyond country also determine a person's attitude.

If phi were equal to 0.00, the table would look like Table 9.3a. When phi was equal to 0.43, 625 Danes thought most people could be trusted;

if phi were 0.00, 419 Danes would have that attitude. Thus, 625 − 419 = 206 more Danes actually felt that way compared to the data if phi were equal to 0.00. The same is true with the French sample: 206 more French felt you cannot be too careful in the actual data than in the table where phi equals 0.00. Those 412 people determine an observed phi of 0.43.

Table 9.3b is a table of the data if phi were equal to 1.00. All the Danes would have felt that people can be trusted and all the French would have felt that you cannot be too careful. For phi to equal 1.00, 985 Danes would have to feel that people can be trusted, an additional 360 Danes to the 625 who actually felt that way.

Table 9.3c shows the connection between the number of Danes who felt people can be trusted and the value of phi. As we increase the

Table 9.3 Hypothetical tables with different values of phi

(a) phi = 0.00

		Country		Total
		Denmark	France	
Attitude toward people	Trust	419	412	831
	Suspicion	566	557	1,123
	Total	985	969	1,954

(b) phi = 1.00

		Country		Total
		Denmark	France	
Attitude toward people	Trust	985	0	985
	Suspicion	0	969	969
	Total	985	969	1,954

(c) Correspondence between observations and phi

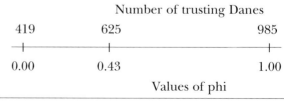

number of Danes feeling that way from 419 to 625 to 985, the value of phi increases from 0.00 to 0.43 to 1.00. The observed value of phi at 0.43 is about four tenths of the way from the minimum of 0 to the maximum of 1, just as the number 625 is about four tenths of the way from the minimum of 419 toward the maximum of 985.

Many other coefficients have been developed for the measurement of the strength of the relationship between two categorical variables when each variable has two categories, but phi is the most commonly used. For a larger table with more than two rows and/or two columns, phi cannot be used, and we choose from among a variety of other coefficients.

Phi in the population

Since we do not have data on the attitudes of all the people in Denmark and France, we do not know how strong the relationship is in the total population in the two countries. But we can use the sample value of phi equal to 0.43 as an estimate of the strength of the relationship between nationality and feeling of trust in the two countries.

9.4 QUESTION 3. RELATIONSHIP IN THE POPULATIONS?

The third question we ask is whether there is a relationship between nationality and attitude toward people in the populations of the two countries. Are the variables related not only in the sample of 1954 people but also among all Danes and French? In round numbers, that's about 60 million people.

Setting up the null hypothesis

If we knew how each citizen in the two countries felt, we could fill in a contingency table with two rows and two columns for all 60 million people. Then, to find out if there were a relationship between the two variables in *that* table, we could simply compare the percentage distributions. If the two columns of percentages were different, then the table would show that the two variables are related in the population of all Danes and French, and we could compute phi to see how strong the relationship was.

But this is a statistical pipe dream. We could never collect all the data needed to construct a population table. The limits are too many, and it would be too costly to ask everyone how they felt. Besides, we would have no reason to believe that everyone would answer the question or, even if they answered, that they told the truth. We can, however, extrapolate the information obtained from the sample to the real world.

The extrapolation from the sample to the population of Danes and French is done through statistical hypothesis testing. As you recall from Chapter 7, we begin with a nonobvious step. We set up the null hypothesis that the two variables are *not* related in the population of all Danes and French, and then we see whether the data provide evidence to reject the null hypothesis. If we can reject the null hypothesis, then we have found evidence that the two variables are related. This is in several ways a backward way of doing things, but it is the best we can do.

Whether we reject the null hypothesis or not depends on two factors: (1) the strength of the relationship in the sample (phi) and (2) the number of observations in the sample (n). In data from any sample, both factors play a role. If we have a very large sample, then even a small value of phi is large enough to reject the null hypothesis. With a small sample, we need a large value of phi to reject the null hypothesis; with a low value of phi and a small sample, we may not be able to reject the null hypothesis. However, that would not necessarily mean that the null hypothesis is true. It might be that the null hypothesis is false, but we do not have enough evidence to support the fact.

The strength of the relationship phi between the two variables country and trust in the sample is equal to 0.43, and there are 1,954 observations in the sample. This combination of strength and sample size contains more than enough evidence to reject the null hypothesis. We therefore conclude that a relationship does exist between country and trust in the real world of all Danes and French.

Testing the null hypothesis

To reach the decision to reject the null hypothesis, we use the method described in previous chapters. We first assume that the null hypothesis is true; we assume that in the hypothetical table of all people in the two countries there is no relationship between the two variables and that phi in the table would equal 0. Then we ask whether a phi of 0.43 or larger from a sample is possible by chance.

Another way to consider the same question is to ask: If we drew many samples of Danes and French, and assumed that the variables are not related, how often would phi be of the magnitude of 0.43 or larger? More formally, we calculate the probability of getting a phi equal to or larger than 0.43 in a sample from a population where there is no relationship between the two variables. This probability indicates whether a value of phi of 0.43 belongs to an unusual set of phis or not, and it is the *p*-value for our data. The *p*-value for the sample is given in parentheses in Table 9.1 as less than 0.0001, or less than 1 in 10,000.

A *p*-value that size is very small, to say the least, and it tells us that it is almost impossible to draw a random sample of 1,954 people from a population where there is no relationship and get a phi of 0.43 or more. Fewer than 1 in 10,000 samples would have a phi of 0.43 or more if we sampled many times from a population in which there is no relationship between the two variables.

We can interpret this *p*-value two ways. Either the null hypothesis is true and a random sample with phi as large as 0.43 is extremely unusual, or there is a relationship in the population and a phi of around 0.43 is not surprising. Because the *p*-value is so small, we have overwhelming evidence against the null hypothesis. We think our sample is not particularly unusual, and therefore we conclude that our sample came from a population where the two variables are related, and we reject the null hypothesis of no relationship.

From chi-square to *p*-value

How do we find the *p*-value? Unfortunately, there is no statistical table in which we could look up the *p*-value for a sample of 1,954 people and phi equal to 0.43. We could construct a population where we know that the variables are unrelated, and then we could actually draw many different samples from this population and see how many of these samples have a phi of 0.43 or more. That would give us the probability of obtaining a phi of 0.43 or more. A computer could be programmed to do this for us, but it would still be cumbersome. Instead, we find the *p*-value by transforming our phi to a value of one of the theoretical statistical variables introduced in Chapter 5.

Look back at the data in Table 9.1. Below the table it says that chi-square equals 355.78 with 1 d.f. The size of the sample and the value of phi have been used together to compute a value of the chi-square variable. As discussed earlier, this computation is a simple translation from one variable to another, not unlike the translation of Fahrenheit to Celsius degrees. In other chapters we have translated data into values of the *z*- or the *t*-variable; here we use the chi-square variable. Such a

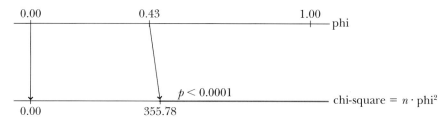

Figure 9.4 Transforming phi to a value of chi-square to find the *p*-value for phi

change of values does not change the value of our results; one variable is simply more convenient to use than another. At the end of the chapter is the formula for computing the chi-square from the number of observations in the sample and the value of phi (Formula 9.3); the translation from phi to chi-square is illustrated in Figure 9.4. The figure shows how the probability of phi equal to or larger than 0.43 in a sample of 1,954 people is equal to the probability of chi-square on 1 d.f. equal to or larger than 355.78.

In describing her views on chi-square, one of our students, Maura McDermott, said this in a paper: "A chi-square is a mysterious thing, a bit like baking soda to an amateur cook; we don't quite know what it does, but we know that we need it!" Chi-square is actually not all that mysterious. It is much like any other variable we measure. Just as we step on the bathroom scale to find how much we weigh, we put our data table on the chi-square scale to find how much our data weigh. Chi-square might be less mysterious if we took the time to go through all the underlying mathematical derivations, but the derivations would not make us better statisticians.

After we have found the value of chi-square we can use statistical software on a computer to find the *p*-value, or we can find the corresponding *p*-value in a statistical table for the chi-square distribution. But a table for the chi-square distribution does not go as far as 355.78, so we found the *p*-value using a statistical software program.

With a *p*-value less than 1 in 10,000, a phi of 0.43 or more is almost impossible in samples that come from a population where there is no relationship between the two variables. About the only possible explanation for a phi this large is that the sample came from a population where there is a relationship between the variables. We reject the null hypothesis because of the small *p*-value and demonstrate that country and attitude toward people are related in the larger population.

Degrees of freedom for chi-square analysis

To find the *p*-value, we need more than the value of chi-square alone. The probability depends not only on how large the value is for chi-square, it also depends on the number of rows and columns in the contingency table. The size of the table is measured using what is called degrees of freedom (d.f., or df). Table 9.1 has two rows and two columns, translating to 1 d.f.

To understand why a table with two rows and two columns has 1 d.f., imagine removing the four cell frequencies from Table 9.1 and keeping the row totals and the column totals. The table then looks like Table 9.4. How many of the four missing frequencies would we have to know to complete the table? Four numbers are missing in the table, but we need to know only one of them. We can find the other three by subtraction from the totals. For example, if we know that there are 625 trusting Danes in the sample, we can fill in the frequency in the upper left corner of the table. To find the frequency of suspicious Danes, we subtract the number of missing Danes from the total: $985 - 625 = 360$ suspicious Danes. By similar subtractions we find the French frequencies to complete the table.

Since we need to know only one of the missing frequencies to find the others, the table and thereby chi-square for the table is said to have 1 degree of freedom. In general, the number of degrees of freedom for a particular contingency table is the product of the number of rows minus 1 and the number of columns minus 1:

$$\text{degrees of freedom} = (\text{number of rows} - 1)(\text{number of columns} - 1)$$

With two rows and two columns,

$$\text{d.f.} = (2 - 1)(2 - 1) = 1$$

Table 9.4 Table 9.1 without cell entries

		Country		
		Denmark	France	Total
Attitude toward people	Trust			831
	Suspicion			1,123
	Total	985	969	1,954

9.5 QUESTION 4. CAUSAL RELATIONSHIP?

The fourth and final question we are interested in is whether the relationship between country and attitude toward people is a causal relationship. Is country a variable that causally affects whether or not people trust others? More specifically, does being Danish cause a person to be trusting and being French cause a person to be suspicious? As we know, this question is much harder in general to answer than the other three questions, and in the country/attitude example, we can do very little statistically to answer it. Even though we have found a statistical relationship, we have no evidence that the relationship is causal. Particularly with observational data from a poll or a survey, other variables may be affecting the outcome. Thus, there may be some other variable that causes or helps cause the Danes to be more trusting than the French.

> **STOP AND PONDER 9.5**
> What other variables do you think might help explain the difference in attitude patterns between Danish and French people?

9.6 LARGER TABLES: A BANQUET OF POSSIBILITIES

When one or both categorical variables have more than two categories, the contingency table has more than two columns or rows. A table showing the attitude toward people of people in four countries would have two rows and four columns (Table 9.5).

With a larger table, we still compare columns of percentages to establish whether a relationship exists in the data between the two variables. But other aspects of the analysis change when the table gets larger. For one thing, phi can no longer be used to check how strong the relationship is; phi is designed for tables with only two rows and two columns. One commonly used coefficient for larger tables is called Cramer's V. V is a generalization of phi, and if we use the formula for V on a table with two rows and two columns, V gives the same result as we get for phi. Formula 9.2 shows how to compute V. Like phi, it ranges in values from 0 to 1.

We still must find chi-square for a larger table to see if we can reject the null hypothesis of no relationship in the larger population. Formula 9.4 shows how to find chi-squares for any contingency table. Finally, the degrees of freedom are larger than 1 because the table has

Table 9.5 Country and attitude toward people

		Country				Total
		Denmark	France	Netherlands	West Germany	
Attitude toward people	Trust	625	206	468	393	1,692
	Suspicion	360	763	463	513	2,099
	Total	985	969	931	906	3,791

($V = 0.32$, chi-square $= 367.94$, 3 d.f., $p < 0.0001$)

Source: Jacques-René Rabier, Helen Riffault, and Ronald Inglehart, Euro-barometer 25: Holiday Travel and Environmental Problems, *April 1986, Ann Arbor, Mich.: Interuniversity Consortium for Political and Social Research, 1988. Codebook p. 10.*

more than two columns. To find the number of degrees of freedom, we adopt the formula we used for two-row, two-column tables: d.f. = (number of rows − 1)(number of columns − 1). In a table with three rows and four columns, the degrees of freedom equal $(3 − 1) \times (4 − 1) = (2)(3) = 6$: we need to know six of the numbers in the cells as well as the totals to be able to fill in the other six numbers.

Chi-square is computed on the basis of the idea that the row and column totals in a contingency table are fixed. With fixed totals, we need to know all the frequencies in the table except the ones in the last row and the last column to be able to finish the table (Figure 9.5). If we know all the frequencies except the ones in the last row and column, we can then find those by subtraction of the other frequencies from the corresponding totals.

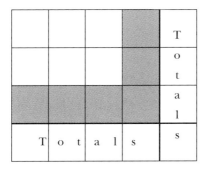

Figure 9.5 Finding degrees of freedom

Table 9.6 Percentage distributions within the four countries

		Country			
		Denmark	France	Netherlands	West Germany
Attitude toward people	Trust	64	21	50	43
	Suspicion	36	79	50	57
	Total	100	100	100	100
	(n)	(985)	(969)	(931)	(906)

After we find V, chi-square, and degrees of freedom, we again find the p-value for the table in order to make a conclusion about the relationship between the two variables in the real world.

Table 9.5 shows the data from The Netherlands and West Germany in addition to the data from Denmark and France. There are four columns for the independent variables and two rows for the dependent variable, making a 2 × 4 contingency table. Let us analyze the data using the four questions about the relationship between variables.

Question 1. Relationship between the variables?

To get a better sense of the existence of a relationship between country and trust and to see the differences between the countries better, we change the frequencies to percentages, as we did with the 2 × 2 contingency table. Since country is the independent variable, we change the frequencies for each country to percentages, as shown in Table 9.6. From the percentages it is clear that the samples in the four countries differ in attitude. Denmark has the highest percentage of people who said that most people can be trusted, while France has the lowest percentage.

Question 2. Strength of the relationship?

To find the strength of the relationship between country and attitude toward people, we compute a Cramer's V coefficient. V is equal to 0.32 for these data, a moderately weak relationship on a scale from 0 to 1.

> ### STOP AND PONDER 9.6
>
> You are interested in the strength of a relationship between the choice of public or private school and the religious affiliation of the parents of college-age students. Parents are categorized as Catholics, Jews, Protestants, or other.
>
> How would you set up a contingency table for this problem? What coefficient would be a good indicator of the strength of the relationship? What coefficient would you use if the religious affiliations were simply Catholic and other?

Question 3. Relationship in the populations?

To see if we can generalize the finding of the relationship in the sample of 3,791 respondents to the population of all adults in the four countries, we again compute chi-square and find the corresponding p-value. If the p-value is small, then we can reject the null hypothesis that these data were generated by chance alone. Here, the value of chi-square equals 367.94 with 3 degrees of freedom, which is a very large chi-square for only 3 degrees of freedom. We now use computer software or a table for the chi-square variable to find a p-value of less than 0.0001: if there is no relationship between the two variables in the four countries, then the probability of getting a V of 0.32 or larger is less than 1 in 10,000. Thus, if we drew many different samples from a population in which there were no relationship between the two variables, for fewer than 1 of 10,000 different samples would we find a V of 0.32 or more.

Since the p-value is so small, we have overwhelming evidence for the argument that these data were not produced by chance alone. Thus, we reject the null hypothesis of no relationship in the population, and we conclude that the relationship does exist in the larger population of the adults in the four countries. We do not know if *all* the countries are different from each other or whether only some of the countries differ from others; finding out would take additional analyses.

Question 4. Causal relationship?

The last question is whether living in a particular country causes people to be more or less trusting. As we repeatedly emphasize, statistical methods cannot be used to answer the question if we have data on only two variables. We can predict, however, that more Danes than French are trusting. We might speculate that trust is more likely to develop where people are fewer and more similar in customs, but we cannot back up the speculation statistically.

9.7 SUMMARY

A contingency table is a table of frequencies that shows how the data are distributed for all combinations of values of two categorical variables.

9.1 Analysis of the data: Are there trustworthy differences in attitude?

In a contingency table, the independent variable usually runs horizontally across the table and the dependent variable vertically down the side. It is often useful to display these data in a graph because patterns in the data are easier to see in a graph than in a contingency table. Contingency tables are useful for studying relationships between categorical variables.

9.2 Question 1. Relationship between the variables?

To find out if a relationship exists between the variables in the observed data, we compare the percentages in the columns of the contingency table. If the percentage distributions are different, we conclude that a relationship exists between the variables in the data.

9.3 Question 2. Strength of the relationship?

The strength of a relationship is measured by a coefficient that is calculated from the data. This coefficient ranges in value between 0 and 1. A coefficient near 0 indicates a weak relationship, and a coefficient near 1 shows a strong relationship.

A commonly used coefficient for the strength of the relationship in a table with two rows and two columns is called phi. A commonly used coefficient for tables with more than two rows and/or two columns is called V.

9.4 Question 3. Relationship in the population?

To find whether there is a relationship between the variables in the population from which the sample came, we extrapolate from what we know about the sample to the population through statistical hypothesis testing. The first step is to set up the null hypothesis that the two variables are not related in the entire population. Whether we reject the null hypothesis or not depends on two factors: (1) how strong the relationship is in the sample (phi or V), and (2) how many observations there are in the sample (n). If the sample is very large, a small value of phi or V is enough to reject the null hypothesis. If the sample has only a few observations, a large value of phi or V is needed to reject the null hypothesis.

To find the p-value for a particular sample, it is necessary to transform the phi or V coefficient to a value of the chi-square variable. The size of the sample and the value of phi or V are used to compute a value of the chi-square variable.

To find the p-value associated with a chi-square value, it is necessary to know the degrees of freedom of the contingency table. The degrees of freedom equal (number of rows − 1)(number of columns − 1).

9.5 Question 4. Causal relationship?

Whether or not there is a causal relationship between two variables is impossible to answer given data on only two variables. A statistical relationship does not provide evidence that the relationship is causal.

ADDITIONAL READING

Reynolds, H. T. *Analysis of Nominal Data*, 2nd ed. (Sage University Paper Series on Quantitative Applications in the Social Sciences, series no. 07-007). Beverly Hills, CA: Sage, 1984. Chi-square tests and measures of association for categorical variables.

FORMULAS

BAR GRAPH WITH BARS OF DIFFERENT WIDTHS

Figure 9.3 is a bar graph of a 2 × 2 contingency table where the bars are of different widths and the area of each rectangle is proportional

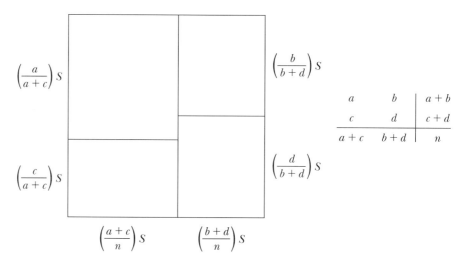

Figure 9.6 Construction of a bar graph where the bars are the same height and different widths in a square with sides S

to the frequency for that category. Such a graph can be drawn for any frequency table for two variables. Figure 9.6 illustrates how such a graph is drawn for a 2×2 table.

The graph is based on the frequencies in the contingency table to its right. The letters a, b, c, and d represent the number of observations in each cell. The total for the entire table is denoted by n. From this table we want to make a graph consisting of a square whose area is proportional to the total frequency n. Each side of the square has length S. The sides of the square can be measured in inches, centimeters, or any other unit.

Next we divide the square into two vertical bars, each representing a column of the table. There are $a + c$ observations in the first column, and the proportion of observations in that column equals $(a + c)/n$. We divide the base of the square using the same proportion. The left bar has width $[(a + c)/n]S$ and the right bar has width $[(b + d)/n]S$. The sum of these two widths is S, as it should be.

Now we divide each of the columns in the graph by the data in the corresponding columns in the table. The top category in the first column of the table contains the proportion $a/(a + c)$ of the total number of observations in the column, $a + c$, so we make the height of the top rectangle equal to $[a/(a + c)]S$ and the height of the bottom rectangle equal to $[c/(a + c)]S$. Similarly, we divide the second column into the proper proportions, as shown in the figure.

Table 9.7 Contingency table with letters for the frequencies

			Total
	a	b	a + b
	c	d	c + d
Total	a + c	b + d	n

The area of the entire square with sides S is S^2. The area of each of the four rectangles is a fraction of this total, and the fraction corresponds to the proportion of observations in the corresponding cell of the table. For example, the area of the lower right rectangle is $(d/n)S^2$, and the proportion of observations in the corresponding cell of the table equals d/n.

The total area does not have to be a square. It can be a rectangle with base B and height H. The base and height are divided using the same proportions as described for S.

Phi is used to measure the strength of the relationship between two categorical variables on a scale between 0 and 1 when each of the variables has two values (categories). When we replace the observed frequencies by the letters a, b, c, d, and n, the contingency table looks like Table 9.7.

To see if there is a relationship between the two variables, we convert the frequencies to percentages and compare the two columns of percentages. In the country/attitude example, no relationship exists between the two variables if the proportions of trusting people in Denmark and France are equal. This can be written

$$\frac{a}{a+c} = \frac{b}{b+d}$$

$$a(b+d) = b(a+c)$$

$$ad - bc = 0$$

The product ad is the product of the two frequencies on one diagonal in the table, and bc is the product of the frequencies on the other diagonal. When there is no relationship between the two variables, the two products are equal. The more different the two products are, the stronger is the relationship.

Phi can then be computed from the difference between the two products on the diagonals according to the following formula:

$$\text{phi} = \frac{ad - bc}{\sqrt{(a + b)(c + d)(a + c)(b + d)}} \qquad (9.1)$$

The denominator is included to make certain phi never becomes larger than its maximum value of 1.00. The numbers in Table 9.1 give a phi of

$$\frac{625 * 763 - 206 * 360}{\sqrt{831 * 1{,}123 * 985 * 969}} = 0.43$$

If phi is negative, we usually disregard the negative sign and report the positive value. This is because the sign of phi changes if we interchange the two columns (or the two rows) in the table. Since the variables are categorical, we can make this change without changing the meaning of the table.

CRAMER'S V

V is used to measure the strength of the relationship on a scale between 0 and 1 between two or more categorical variables when at least one of them has more than two values. *V* is found from the formula

$$V = \sqrt{\frac{X^2}{n(L - 1)}} \qquad (9.2)$$

where *n* is the number of observations in the table and *L* is the smaller of the number of rows and columns in the table. For the data in Table 9.5, with two rows and four columns, *L* is the smaller of 2 and 4, so *L* = 2:

$$V = \sqrt{\frac{367.94}{3{,}791(2 - 1)}} = 0.31$$

V is a generalization of phi, and it becomes phi if it is used for a table with 2 rows and 2 columns.

CHI-SQUARE

2 × 2 table For a table with two rows and two columns, chi-square (X^2) can be found from the following formula:

Table 9.8 Expected frequencies

		Country		Total
		Denmark	France	
Attitude toward people	Trust	$\dfrac{(831)(985)}{1{,}954} = 418.90$	$\dfrac{(831)(969)}{1{,}954} = 412.10$	831
	Suspicion	$\dfrac{(1{,}123)(985)}{1{,}954} = 566.10$	$\dfrac{(1{,}123)(969)}{1{,}954} = 556.90$	1,123
	Total	985	969	1,954

$$X^2 = n(\text{phi})^2$$
$$= \frac{n(ad-bc)^2}{(a+b)(c+d)(a+c)(b+d)}$$
$$= \frac{1{,}954(625*763 - 206*360)^2}{831*1{,}123*985*969} = 355.78 \qquad 1 \text{ d.f.} \qquad (9.3)$$

Larger tables For larger tables we use another way to find chi-square based on the so-called *expected* frequencies. This method can be illustrated using a 2 × 2 table. A table with expected frequencies has the same row and column totals as the original table, but phi for the new table equals 0.00. Chi-square measures how much the observed frequencies differ from the expected frequencies.

The expected frequency for a particular cell in the table is found from the expression

$$\text{expected frequency} = \frac{\text{row total} * \text{column total}}{\text{table total}}$$

For the data in Table 9.1, we get the expected frequencies shown in Table 9.8. The expected frequencies are the same frequencies as in Table 9.3a.

Note that because the row and column totals are the same as in the original table, it is necessary to compute only one of the expected frequencies by multiplying row and column totals and dividing by the table total. The other three expected frequencies can be found by subtraction using the totals. This is the reason the corresponding chi-square has 1 degree of freedom.

Chi-square is then found by comparing the observed with the expected frequencies to see how far away the observed table is from a table without any relationship. We compute chi-square according to the following expression:

$$X^2 = \sum \frac{(\text{obs} - \text{exp})^2}{\text{exp}}$$

$$= \frac{(625 - 418.9)^2}{418.9} + \frac{(206 - 412.1)^2}{412.1}$$

$$+ \frac{(360 - 566.1)^2}{566.1} + \frac{(763 - 556.9)^2}{556.9}$$

$$= 355.78 \quad 1 \text{ d.f.} \tag{9.4}$$

For tables with more rows and columns the procedure is the same. First we find all the expected frequencies, and then we compute chi-square using Formula 9.4. There are, of course, more than four terms in the sum.

Chi-square as an approximation We use the chi-square value to find the p-value for our data. But the chi-square method only provides us with an approximately true p-value. Particularly if the p-value only borders on being significant, an approximation is worrisome. The more observations in the data table, the better the approximation becomes. One way to see if we can use chi-square to find the p-value is to look at the magnitudes of the expected frequencies. In a 2×2 table all the expected frequencies should be larger than 5. For tables with more rows and columns, this requirement is not as important.

EXERCISES

REVIEW (EXERCISES 9.1–9.8)

9.1 Find a report of a survey in a newspaper, news magazine, or other source, and construct a contingency table showing the relationship between two categorical variables used in the survey. Use the table to discuss the relationship between the two variables.

9.2 a. When can you use phi to measure the strength of a relationship between two variables? When can you not use phi?

b. What is an example of a small value of phi? What does a small value of phi tell you about the relationship between two categorical variables?

c. What does a small *p*-value tell you about the relationship between two variables?

d. Discuss in some detail how it is possible to have a low value of phi and still get a small *p*-value.

9.3 Describe a contingency table.

9.4 a. Which axis (horizontal or vertical) is used for the independent variable?

b. Which axis is used for the dependent variable?

c. Is this arrangement of independent and dependent variables a statistical "law" that can never be broken? Explain your answer.

9.5 What is the name of the statistic we compute to measure the strength of the relationship between the variables in a contingency table with two values for each of the two variables?

9.6 a. When we compute chi-square, what is the name of the statistical term used to measure the size (number of rows and columns) of the table? What is a common abbreviation of this term?

b. How do you find the magnitude of this quantity?

9.7 If you find a statistical relationship between two variables from a large chi-square, can you be almost certain that the relationship is causal? Explain.

9.8 A contingency table includes row and column totals for each variable.

a. When you add the rows or the columns, what can you say about the sums?

b. Why is this a useful check on the construction of the contingency table?

INTERPRETATION (EXERCISES 9.9–9.27)

9.9 In the fall of 1986, the House of Representatives voted on whether to give aid to the Nicaraguan contras (a proposal favored by President Reagan). They also voted on a general spending bill. A no vote was seen as support for President Reagan's policies, and a yes vote was seen as opposition to the President's policies. Leaving out those members who did not vote gives Table 9.9 (page 372).

a. Is there a relationship between the two votes in these data?

b. How strong is the relationship in these data?

Table 9.9 Data for Exercise 9.9

		Spending bill		
		Yes	No	Total
Aid to contras	Yes	42	167	209
	No	156	33	189
	Total	198	200	398

(phi = 0.62, chi-square = 156.75, 1 d.f., p = 0.000)

Source: Kenneth Janda and Philip A. Schrodt, Crosstabs: Student Workbook for American Government, Boston: Houghton Mifflin, 1987. Data disk.

c. Could the relationship in the sample have occurred by chance alone?

d. Is the relationship causal?

9.10 In a 1950 study of the relationship between location of residence in the South and non-South parts of the country and occupation as professionals or farmers, frequencies were found in a sample of survey respondents as shown in Table 9.10.

a. Is there a relationship between location and occupation in these data?

b. How strong is the relationship?

Table 9.10 Data for Exercise 9.10

		Location		
		South	Non-South	Total
Occupation	Professional	70	93	163
	Farmers	135	58	193
	Total	205	151	356

(phi = 0.27, chi-square = 26.38, 1 d.f., p = 0.0000)

Source: Adapted from J. C. McKinney and L. B. Borque, "Further comments on 'The changing South': A response to Sly and Weller," American Sociological Review, vol. 37 (1972), p. 236.

Table 9.11 Data for Exercise 9.11

		Outcome		Total
		Winner	Loser	
Next year	Play	8	7	15
	Not play	22	23	45
	Total	30	30	60

(phi = 0.04, chi-square = 0.089, 1 d.f., p-value = 0.76)

c. Is there a relationship between the two variables in the larger population from which these data came?

d. Is the relationship causal?

9.11 In the first 30 Super Bowls, 60 teams played, half of them as winners and the other half as losers. Of the winning teams, 8 played in the Super Bowl the following year, and of the losers 7 played in the Super Bowl the following year. These data can be arranged as in Table 9.11. Report on the relationship between the two variables.

Table 9.12 Data for Exercise 9.12 (page 374)

		Region of birth						
		New England	Middle Atlantic	North Central	South Atlantic	South Central	Mountain and Pacific	Total
Region of residence	New England	306	11	2	2	0	0	321
	Middle Atlantic	18	806	8	14	1	0	847
	North Central	30	127	1,180	39	55	2	1,433
	South Atlantic	2	11	5	717	8	0	743
	South Central	1	5	23	76	790	0	895
	Mountain and Pacific	8	13	24	3	2	72	122
	Total	365	973	1,242	851	856	74	4,361

(V = 0.85, chi-square = 15,733, 25 d.f., $p < 0.0001$)

Source: U.S. Bureau of the Census, Historical Statistics of the United States: Colonial Times to 1957, *Washington, DC, 1960, Series C 15-24, pp. 42, 44.*

9.12 Table 9.12 (page 373) shows data from 1880. The United States is divided into six regions. The columns show regions of birth, while the rows show regions of current residence. The data come from the 1880 census, but we have divided the census figures by 10,000 and treat the data as if they were from a sample of 4,361 people in a population of about 43 million people. The resulting 6 × 6 contingency table is also what sociologists call a *mobility table;* it shows how many people moved from their region of birth to their current region. The numbers are read, "306 people who were born in New England still live in New England; 18 people who were born in New England now live in a middle Atlantic state"; and so on.

a. Report on the relationship between the two variables.

b. What were the major patterns of geographic mobility in the late 1800s (compare the frequencies above the main diagonal of the table with the frequencies below the diagonal)?

9.13 Does the amount of education vary from one group to another? In a random sample of 988 people we find the educational attainments for Asians, Hispanics, and whites shown in Table 9.13.

a. Is there a relationship in this table between the two variables?

b. How strong is the relationship?

c. Is there a statistically significant difference in educational attainment between the three groups?

d. Is this a causal relationship?

Table 9.13 Data for Exercise 9.13

		Group			
		Asian	Hispanic	White	Total
	High school or less	24	98	419	541
Education	Some or complete college	27	34	310	371
	Professional or graduate	9	6	61	76
	Total	60	138	790	988

($V = 0.11$, chi-square = 23.26, 4 d.f., $p = 0.0001$)

Source: Column percentages equal those found by the U.S. Bureau of the Census, as reported in The Chronicle of Higher Education, *vol. XXXIX, no. 1, August 26, 1992, p. 12.*

Table 9.14 Data for Exercise 9.14

		Group			Total
		White	Black	Hispanics	
	Intact family	2,583	526	292	3,401
Situation	Mother only	297	239	75	611
	Step family	317	107	25	449
	Other	175	106	34	315
	Total	3,372	978	426	4,776

Source: L. L. Wu and B. C. Martinson, "Family structure and the risk of a premarital birth," *American Sociological Review,* vol. 58 (1993), p. 217.

9.14 One of the questions in the National Survey of Families and Households asks women what their family situation was at the time they were 14 years old. Part of the data for three groups of women are displayed in Table 9.14. For these data, V equal to 0.16 and chi-square on 6 degrees of freedom is equal to 255.29 with a *p*-value less than 0.0001.

a. What do these numbers tell you about the relationship between the two variables?

b. Typical values of the chi-square variable with 6 degrees of freedom range from 0 to as much as 15. How do you reconcile the enormous value of chi-square here with the low value of V?

9.15 The reported frequencies in the study linking baldness and heart attacks in Exercise 8.35 are arranged in Table 9.15.

Table 9.15 Data for Exercise 9.15

		Baldness		Total
		Yes	No	
	Yes	214	451	665
Heart attack	No	175	597	772
	Total	389	1,048	1,437

(phi = 0.11, chi-square = 16.37, 1 d.f., p = 0.0001)

Source: The New York Times, *February 14, 1993, pp. A1, C12.*

Table 9.16 Data for Exercise 9.17

		Religious affiliation	
		Protestant	Catholic
Statement	Ranked first	62%	50%
	Not ranked first	38	50
	Total	100%	100%
	(n)	(165)	(145)

a. What is the independent variable? Explain your choice.

b. What do these results tell you about the relationship between the two variables?

9.16 For Table 8.6 in Exercise 8.36, which shows data on religious affiliation and choice of a statement about the importance of work, phi equals 0.24, chi-square equals 10.64 on (1 d.f.), and p equals 0.0011. What can you conclude about the relationship between the two variables based on these computed numbers?

9.17 In the same study that provided the data for Exercise 9.16, the sociologist also looked at a difference between Catholics and Protestants. The respondents in the study were asked to rank a set of statements, one of which was "Work is important." Table 9.16 shows the percentages in the two religious groups who ranked this statement first.

a. Is there a relationship between religious affiliation and ranking in these data?

b. Does the relationship seem strong? (It may help to convert the percentages to frequencies and complete the usual 2 × 2 table.)

c. These data can be analyzed either by using a test for the difference between two percentages or by using chi-square on a table with two rows and two columns. Either way, the p-value is equal to 0.013. What can you say about the difference between Protestants and Catholics in Detroit at the time of this study?

9.18 The United States has long been a country of volunteer organizations, and two sociologists were interested in whether there was any change in the number of volunteer organizations to which people be-

Table 9.17 Data for Exercise 9.18

		Year		Total
		1955	1962	
	0	1,523	1,012	2,535
Number of	1	476	390	866
organizations	2	214	195	419
	3	95	106	201
	4+	71	71	142
	Total	2,379	1,774	4,163

Source: H. H. Hyman and C. R. Wright, "Trends in voluntary association memberships of American adults: Replication based on secondary analysis of national sample surveys," American Sociological Review, vol. 36 (1971), pp. 191–206.

longed from the middle 1950s into the early 1960s. Table 9.17 shows the number of volunteer organizations people reported they belonged to in surveys at two different times. The mean number of organizations people belonged to in 1955 was 0.64 and in 1962 it was 0.80, so there had been an increase in the number of organizations people belonged to. Strictly speaking, the number of organizations is a metric variable, and the study of these two variables belongs in Chapter 12. Here, as a starter, we analyze these data as a contingency table with five rows and two columns. The table yields the following results: chi-square = 25.44 on 4 d.f., $p = 0.00004$, and $V = 0.08$. What can you conclude about the relationship between the two variables?

9.19 In a study in Baltimore around 1970, children at different levels of their education were asked about their views of social stratification in the United States. In answer to the question "Do all kids in America have the same chance to grow up and get the good things in life?" the sociologists reported the results in Table 9.18 (page 378).

  a. What are a couple of striking things about how the children answered the question? (You may want to change the frequencies to percentages before answering this question.)

  b. What do the statistical computations tell you about the relationship between level of education and the children's views of social stratification?

9.20 Exercise 8.36 is based on a table showing data on the two variables race and brand of cigarettes smoked. For those data, chi-square =

Table 9.18 Data for Exercise 9.19 (page 377)

		Education			
		Elementary school	Junior high school	Senior high school	Total
Answer	Yes	207	110	67	384
	No	496	327	234	1,057
	Do not know	330	79	34	443
	Total	1,033	516	335	1,884

(chi-square = 99.94, 4 d.f., $p < 0.0001$, $V = 0.16$)

Source: R. G. Simmons and M. Rosenberg, "Functions of childrens' perceptions of the stratification system," American Sociological Review, vol. 36 (1971), pp. 235–249.

181.93 on 3 d.f., $p < 0.0001$, and $V = 0.46$. What do these numbers tell you about the relationship between the two variables?

9.21 Judge Robert Kane of New Bedford District Court in Massachusetts, with the encouragement of Professor Robert P. Waxler at the Dartmouth campus of the University of Massachusetts, gave some people found guilty in his court the choice of going to jail or taking a literature course taught by Professor Waxler. Professor G. Roger Jarjoura, Indiana University, followed 32 men who took the class and found that 6 were convicted of new crimes, while among 40 other men with similar backgrounds 18 were convicted of new crimes. *(Source:* The New York Times, *October 6, 1993, p. B10.)*

a. Display these data in a contingency table with two rows and two columns.

b. For these data, phi = 0.28, chi-square = 5.51 on 1 degree of freedom, and $p = 0.019$. What do these numbers tell you about the relationship between participation in the literature course and recidivism?

c. Does it seem appropriate to apply the statistical methods in part b to data of this kind?

9.22 About 100,000 patients each year in this country are candidates for either heart bypass surgery or a procedure called angioplasty, where balloons are inserted into the arteries to clear blockages. A comparison

Table 9.19 Data for Exercise 9.22

(a)

		Treatment		
		Angioplasty	Bypass	Total
Further surgery	Yes	122	24	146
	No	74	172	246
	Total	196	196	392

(phi = 0.52, chi-square = 104.82, p-value < 0.0001)

(b)

		Treatment		
		Angioplasty	Bypass	Total
Additional attack	Yes	29	39	68
	No	167	157	324
	Total	196	196	392

(phi = 0.07, chi-square = 1.78, p-value = 0.18)

(c)

		Treatment		
		Angioplasty	Bypass	Total
Flow restored	Yes	110	67	177
	No	86	129	215
	Total	196	196	392

(phi = 0.22, chi-square = 19.05, p-value = 0.00001)

of the two procedures was done in a study on 392 patients followed for 3 years. *(Source: "Study finds angioplasty as good as heart bypass," The New York Times, November 11, 1993. p. A19.)* Assuming that half the patients were treated by angioplasty and half by bypass operations, we can construct Tables 9.19a, b, and c from the percentages given in the newspaper article. Table a shows how many patients in the two groups needed further surgery, Table b shows whether the patients had additional heart attacks, and Table c shows whether the blood flow had been completely restored.

Table 9.20 Data for Exercise 9.23

	Delinquent		
	Yes	No	Total
Yes	1	5	6
No	8	2	10
Total	9	7	16

Source: A. M. Weindling, F. N. Bamford, and R. A. Whittall, "Health of juvenile delinquents," British Medical Journal, *vol. 292 (1986), p. 447.*

a. Why does it make sense to keep the sign of phi and report that the first and the third table have positive phis while the second table has a negative phi?

b. What can you conclude about the relationships between the variables in each of the three tables?

c. What do you conclude about the two procedures based on these data?

9.23 Among a group of boys with bad vision, some used glasses and some did not, and some were delinquents while others were not. Table 9.20 shows the number of boys in each category. For these data, phi equals 0.62, chi-square on 1 d.f. equals to 6.11, and p equals 0.013. (These frequencies are so small that the appropriateness of chi-square is questionable, but the so-called Fisher's exact test gives approximately the same result, so we stay with chi-square for these data.) What do the data tell you about juvenile delinquency and the use of glasses?

9.24 Does cranberry juice cut down on urinary infections in older women? In a study at the Boston Women's Hospital, half of a group of 153 women drank a glass of cranberry juice each day for six months, while the other half drank a placebo drink with the same color and flavor. At the end of the six months, the bacteria that cause infections was present in the urine of 15% of the cranberry juice drinkers and in the urine of 28% of the placebo drinkers. *(Source:* Discover Magazine, August 1994, p. 13.) For these data, phi equals 0.16, chi-square on 1 d.f. equals 3.96, and p equals 0.047.

a. Display these data in 2 × 2 table.

b. Discuss the results.

Table 9.19 Data for Exercise 9.22

(a)

		Treatment		
		Angioplasty	Bypass	Total
Further surgery	Yes	122	24	146
	No	74	172	246
	Total	196	196	392

(phi = 0.52, chi-square = 104.82, p-value < 0.0001)

(b)

		Treatment		
		Angioplasty	Bypass	Total
Additional attack	Yes	29	39	68
	No	167	157	324
	Total	196	196	392

(phi = 0.07, chi-square = 1.78, p-value = 0.18)

(c)

		Treatment		
		Angioplasty	Bypass	Total
Flow restored	Yes	110	67	177
	No	86	129	215
	Total	196	196	392

(phi = 0.22, chi-square = 19.05, p-value = 0.00001)

of the two procedures was done in a study on 392 patients followed for 3 years. *(Source: "Study finds angioplasty as good as heart bypass," The New York Times, November 11, 1993. p. A19.)* Assuming that half the patients were treated by angioplasty and half by bypass operations, we can construct Tables 9.19a, b, and c from the percentages given in the newspaper article. Table a shows how many patients in the two groups needed further surgery, Table b shows whether the patients had additional heart attacks, and Table c shows whether the blood flow had been completely restored.

Table 9.20 Data for Exercise 9.23

	Delinquent		
	Yes	No	Total
Yes	1	5	6
No	8	2	10
Total	9	7	16

Source: A. M. Weindling, F. N. Bamford, and R. A. Whittall, "Health of juvenile delinquents," British Medical Journal, *vol. 292 (1986), p. 447.*

a. Why does it make sense to keep the sign of phi and report that the first and the third table have positive phis while the second table has a negative phi?

b. What can you conclude about the relationships between the variables in each of the three tables?

c. What do you conclude about the two procedures based on these data?

9.23 Among a group of boys with bad vision, some used glasses and some did not, and some were delinquents while others were not. Table 9.20 shows the number of boys in each category. For these data, phi equals 0.62, chi-square on 1 d.f. equals to 6.11, and *p* equals 0.013. (These frequencies are so small that the appropriateness of chi-square is questionable, but the so-called Fisher's exact test gives approximately the same result, so we stay with chi-square for these data.) What do the data tell you about juvenile delinquency and the use of glasses?

9.24 Does cranberry juice cut down on urinary infections in older women? In a study at the Boston Women's Hospital, half of a group of 153 women drank a glass of cranberry juice each day for six months, while the other half drank a placebo drink with the same color and flavor. At the end of the six months, the bacteria that cause infections was present in the urine of 15% of the cranberry juice drinkers and in the urine of 28% of the placebo drinkers. *(Source:* Discover Magazine, August 1994, p. 13.) For these data, phi equals 0.16, chi-square on 1 d.f. equals 3.96, and *p* equals 0.047.

a. Display these data in 2 × 2 table.

b. Discuss the results.

Table 9.21 Data for Exercise 9.26

		Differences between parties		
		Very important	None	Total
Who will win	Stevenson	86	234	320
	Eisenhower	79	340	419
	Total	165	574	739

Source: Data utilized in this exercise made available by the Inter-University Consortium for Political and Social Research, data originally collected by Angus Campbell, Gerald Gurin, and Warren Miller. Neither the original collectors of the data nor the Consortium bear any responsibility for the analyses or interpretations presented here.

9.25 Each year the American Statistical Association honors some of its members by electing them Fellows of the organization. In 1994, of the 77 nominated men 36 were elected Fellows, while of the 20 nominated women 13 were elected Fellows. *(Source: Daniel L. Solomon, "Turning women into Fellows—Continued," Newsletter,* Caucus for Women in Statistics, *vol. 4 (1977), no. 4, p. 11.)*

a. Display these data in a 2 × 2 contingency table.

b. For these data, phi equals 0.17 and chi-square equals 2.12 on 1 degree of freedom for a *p*-value of 0.15. What do these results tell you about the relationship between gender and election?

9.26 It is 1952. Adlai Stevenson is the Presidential nominee of the Democratic party and Dwight Eisenhower is the nominee of the Republican party. In a survey before the election is held, the respondents are asked, among other things, whether they think there are differences between the two parties and who they think will win the election. Responses of people who say there are very importance differences or no differences between the two parties are shown in Table 9.21. For these data phi equals 0.10, chi-square equals 6.73 on 1 d.f., and *p* equals 0.01. What do you conclude about how people saw the two major parties and people's expectations of how the election would turn out?

9.27 A random sample of 72 college students were asked whether they read nutritional labels when they buy food. The interviewer also recorded the gender of each student (Table 9.22, page 382). For these data, phi equals 0.48 and chi-square on 1 degree of freedom equals 8.48. What can you conclude about the relationship between gender and nutritional label reading?

Table 9.22 Data for Exercise 9.27 (page 381)

		Gender		
		Female	Male	Total
Read labels	Yes	16	28	44
	No	20	8	28
	Total	36	36	72

Source: Data used by permission of Jasa Porciello, Swarthmore College.

ANALYSIS (EXERCISES 9.28–9.59)

9.28 It was thought that taking the drug AZT would reduce the chances that HIV-positive pregnant women would pass on the virus to their children. In a study performed at several medical centers in the United States and France, some HIV-positive mothers-to-be were given AZT and others were given a placebo. The results for 364 newborn babies are shown in Table 9.23.

a. Is there a relationship between the treatment and the result in these data?

b. How strong is the relationship?

c. Show that chi-square (1 degree of freedom) for these data equals 27.95 ($p < 0.0001$).

d. What can you conclude about the relationship between the treatment and the outcome?

Table 9.23 Data for Exercise 9.28

		Treatment of mothers		
		AZT	Placebo	Total
Condition of newborns	HIV positive	13	40	53
	HIV negative	197	114	311
	Total	210	154	364

Source: The New York Times, February 21, 1994, p. A1.

e. These are actually preliminary results. Would you stop the experiment at the time these data became available, or would you let the study run until more babies were born?

9.29 In 1969 the United States reinstituted the draft lottery. The purpose of the lottery was to determine a young man's eligibility for the military draft by a random mechanism. Each day of the year was assigned a supposedly random integer between 1 and 366 as the draft number for that day. For example, September 14 was assigned draft number 1 in the lottery that year. Each man was then assigned the draft number that corresponded to his birthday. Induction into the armed forces was done by calling men of draft age in order of their lottery number, starting with the number 1. The probability of receiving a low draft number should not have depended on what time of year a person is born but in the 1969 lottery it did. Serious questions were raised about the randomness of the first draft lottery because among the low draft numbers (1–183), 73 were assigned to birthdays in the first half of the year and 110 to birthdays in the second half of the year. For the high draft numbers the pattern was reversed. Table 9.24 shows the relationship between birthdays and draft numbers.

a. What should the contingency table have looked like for a completely random lottery?

b. From the table, does the 1969 lottery seem random?

c. Set up the expression for chi-square for these data.

d. We find that chi-square (1 d.f.) equals 14.16 ($p < 0.0001$). What do you conclude about the lottery? Could the lottery have been random?

9.30 In the 1976 Republican convention in Kansas City, then-President Gerald R. Ford received a plurality of delegate votes from the

Table 9.24 Data for Exercise 9.29

		Birthday by months		
		January–June	July–December	Total
Draft	1–183	73	110	183
number	186–366	109	74	183
	Total	182	184	366

following states: AL, CO, DE, FL, HI, IL, IO, KS, KY, MD, MA, MI, MN, MS, NH, NJ, NY, ND, OH, OR, PA, RI, VT, WV and WI. Then-Governor Ronald Reagan won a plurality from the remaining states.

a. Set up a contingency table showing how the states east and west of the Mississippi River voted for Ford and Reagan.

b. What is the independent variable in the table? The dependent variable?

c. Determine whether a relationship exists between the two variables in these data.

d. How strong is the relationship between the two variables?

e. Find the expected frequencies.

f. How strong is the relationship between the two variables in the table containing the expected frequencies?

g. Find chi-square. Is it large enough to prompt you to think that the difference between the voting patterns east and west of the Mississippi did not occur by chance alone?

h. Comment on the implications of your answers to parts 0-9 for the Republican party in the late 1970s.

9.31 A continuing debate involves whether relationship exists between IQ and crime. One study reports that 24.3% of 486 white males with low IQs committed two or more delinquent acts. Similarly, 9.4% of 1,053 white males with high IQs, 37.6% of 702 black males with low IQs, and 23.3% of 266 black males with high IQs committed two or more delinquent acts. *(Source: T. Hirschi and M. J. Hindelang, "Intelligence and delinquency: A Revisionist review,"* American Sociological Review, *vol. 42 (1977), p. 575.*

a. Arrange these data in a 2 × 2 table.

b. Is there relationship between the two variables?

c. How strong is the relationship?

d. Assuming that the data come from a proper random sample, could there be a relationship in the populations from which these data came?

e. Comment on whether intelligence affects crime or not.

9.32 It is often thought that juvenile offenders should receive special, personalized treatment to stop them from committing criminal acts. Of a group of 100 young offenders, half were randomly assigned to a

Table 9.25 Data for Exercise 9.32

		Special treatment		Total
		Yes	No	
Further criminal acts	No	37	20	56
	Yes	13	30	43
	Total	50	50	100

Source: T. Hirschi and M. J. Hindelang, "Intelligence and delinquency: A revisionist review," *American Sociological Review,* vol. 42 (1977), p. 575.

special treatment while the other half were not and were used as a control group. After four years, the results in Table 9.25 were obtained.

  a. How strong is the relationship between the two variables?

  b. Is there any reason to believe that the program would have any effect in the population from which these data came?

9.33 Is there a difference in religious affiliation between North America and South America? A sample of 500 respondents from North America consisted of 190 Catholics, 10 Jews, 120 Protestants, and 180 other. A sample of 450 respondents from South America consisted of 310 Catholics, 10 Jews, 30 Protestants, and 100 other.

  a. Set up a contingency table for these data.

  b. Change the distribution of religious affiliation within each region to percentages.

  c. Does a relationship exist between the two variables in these data?

  d. How strong is the relationship between the two variables?

  e. Is the relationship stronger, the same, or weaker than you thought it would be?

  f. On the basis of the sample data, can you conclude that a relationship exists in the population from which the data came?

  g. The number of Jews in the data is very small. Leave out the Jewish category and redo the chi-square analysis.

  h. Compare the two chi-squares and describe how they differ.

9.34 Two groups of people are to be compared on scores of a socioeconomic variable. In the first group, 80 people have values smaller than the overall median value and 40 people have values larger than the median. In the second group, 10 observations are smaller and 50 are larger than the overall median.

a. Set up a 2 × 2 contingency table with groups 1 and 2 as one variable and smaller and larger than the median as the other variable.

b. Why is it not surprising that the two rows have the same number of observations?

c. Is there a difference between the groups on socioeconomic values?

d. How strong is the relationship between the two variables?

e. Can you conclude that there is a difference between the two groups in the populations from which these data came?

9.35 The data on religious affiliation and party vote from a random sample of registered voters are set up in Table 9.26.

a. From looking at the table and comparing the columns, is there a relationship between the two variables in these data?

b. Does the relationship seem to be strong or weak?

c. Find the value of chi-square and discuss whether there is a relationship between the two variables in the population.

d. Find V.

9.36 A study was done some years ago on the relationship between taking aspirin and having heart attacks. 22,071 doctors participated in this study. Among the 10,037 doctors who took an aspirin a day, 104 had a heart attack during the time period of the study. Among the 10,034 doctors who took a placebo, 189 had a heart attack.

Table 9.26 Data for Exercise 9.35

		Religious affiliation				
		Catholic	Jewish	Protestant	Other	Total
	Democrat	575	75	275	90	1,015
Party vote	Republican	325	25	325	110	785
	None	150	10	120	50	330
	Total	1,150	110	720	250	2,130

Table 9.27 Data for Exercise 9.37

		Gender		Total
		Female	Male	
Rank	Professor	9	63	72
	Associate professor	27	20	47
	Assistant professor	32	30	62
	Total	68	113	181

Source: Swarthmore College Bulletin, 1995–1996.

    a. Arrange these data effectively in a table.

    b. How strong is the relationship between the two variables?

    c. Based on what you know about issues that arise in data collection, do these results mean that everyone should take an aspirin a day?

9.37 At a small liberal arts college in eastern Pennsylvania, the faculty is distributed for gender and rank as shown in Table 9.27.

    a. Is there a relationship between gender and rank in the data?

    b. How strong is the relationship?

    c. Could the data have occurred by chance alone?

    d. How do you explain the pattern in the data?

9.38 Medical malpractice claims have become increasingly common. Table 9.28 shows how 567 obstetrical–gynecological claims are distrib-

Table 9.28 Data for Exercise 9.38

		Medical school		Total
		English-speaking	Non-English-speaking	
Board certification	Yes	443	42	485
	No	50	32	82
	Total	493	74	567

Source: Bruce Cooil, "Using medical malpractice data to predict the frequency of claims: A study of Poisson process models with random effects," *Journal of the American Statistical Association, vol. 86 (1991), p. 286.*

Table 9.29 Data for Exercise 9.39

		Country			
		Soviet Union	East Germany	United States	Total
Gender	Men	41	15	23	79
	Women	14	22	13	49
	Total	55	37	36	128

Source: Information Please Almanac, *1992.*

uted on two variables: whether the doctor was board certified and whether the doctor was trained in a foreign, non-English-speaking medical school. What does the table tell you about medical claims?

9.39 The last Olympic Games before the fall of the communist regimes in Europe was held in 1988. Much had been made of how the communist countries stressed sports and the role of women in sports. Table 9.29 shows the number of gold medals by gender for the three countries that received the largest number of gold medals that year.

 a. Is there a relationship between country and gender in these data?

 b. How strong is the relationship?

 c. Could the relationship have occurred by chance alone?

 d. What does the table tell us about sports and gender in these countries at that time?

9.40 In a study of how Americans spend their time, people were asked what activity they liked most. Table 9.30 shows the number of employed

Table 9.30 Data for Exercise 9.40

		Gender		
		Men	Women	Total
Preferred activity	Work	64	37	101
	Social life	27	33	60
	Total	91	70	161

(phi = 0.18)

Source: John P. Robinson, How Americans Use Time, *New York: Praeger, 1972, p. 122.*

men and women with high school or less education who mentioned work or social life as their preferred activity. Is there a significant gender difference in these data?

9.41 From the study of how Americans spend their time described in Exercise 9.40, Table 9.31 shows the number of employed men and women with high school or less education who mentioned watching television or participating in social life as their preferred activity. What do these data tell you about the relationship between gender and preferred activity?

9.42 NCAA collected data on graduation rates of athletes in Division I in the mid-1980s. Among 2,332 men, 1,343 had not graduated from college, and among 959 women, 441 had not graduated. *(Source: The Chronicle of Higher Education, July 10, 1991, p. A30.)*

   a. Set up a 2 × 2 contingency table with gender as the independent variable and graduation as the dependent variable.

   b. Analyze the relationship between the two variables.

9.43 How many degrees of freedom are there for a contingency table with three rows and four columns?

9.44 A phi value of 0.69 has a *p*-value for a sample of less than 0.0001. How would you explain the meaning of this statement to someone else in your class? (Try to think of two very different ways of interpreting the finding.)

9.45 Set up the data in Exercise 7.53 in a 2 × 2 table and analyze the relationship between car ownership and presidential vote. (When you use the pooled percentage of Democrats, 209/250 = 83.6, the square root of the chi-square in this exercise equals the *z* in Exercise 7.53, and the two *p*-values are the same if you make the *p*-value for *z* into a two-sided *p*-value.)

Table 9.31 Data for Exercise 9.41

		Gender		
		Men	Women	Total
Preferred activity	Watching TV	43	21	64
	Social life	27	33	60
	Total	70	54	124

Table 9.32 Data for Exercise 9.46

		States		
		Northern	Southern	Total
Executions	0	14	2	16
	1 or more	7	11	18
	Total	21	13	34

Source: NAACP Legal Defense and Educational Fund, as reprinted in The New York Times, *April 21, 1992, p. A14.*

9.46 In the spring of 1992, 34 states had executed prisoners in the last 15 years. A comparison of northern and southern states and whether or not the states had performed any executions in the last 15 years is shown in Table 9.32.

  a. Analyze the relationship between the two variables.

  b. What are some other variables you might want to consider before you draw any conclusions about causality?

9.47 Analyze the data in Exercise 8.37.

9.48 Analyze the data in Exercise 8.40.

9.49 During pregnancy, women may show toxemic signs through hypertension or proteinuria or both. In a sample of English women, the data give the number of women in each of four categories.

Hypertension and proteinuria	Proteinuria only	Hypertension only	Neither sign	Total
28	82	21	286	417

Source: P. J. Brown, J. Stone, and C. Ord-Smith. "Toxaemic signs during pregnancy," Applied Statistics, *vol. 32 (1983), pp. 69–72.*

  a. Display these data in a contingency table with two rows and two columns. Make the first row women who showed hypertension and the second row women who did not show hypertension. Similarly, make the first column women who showed proteinuria and the second column women who did not show this sign.

  b. Analyze the relationship between hypertension and proteinuria.

Table 9.33 Data for Exercise 9.50

	Never snores	Snores every night	Total
Has heart disease	24	30	54
Does not have heart disease	1,355	224	1,579
Total	1,379	254	1,633

Source: P. G. Norton and E. V. Dunn, "Snoring as a risk factor for disease: An epidemiological survey," British Medical Journal, *vol. 291 (1985), pp. 630–632.*

9.50 Snoring is not only unpleasant to listen to, it may also not be good for the people who snore. Table 9.33 shows part of the data obtained in a survey. Analyze these data.

9.51 Hodgkin's disease is a cancer of the lymph nodes. Table 9.34 shows how patients were classified by histological type and response to treatment. Analyze these data.

9.52 In 1994, the Church of England installed its first women priests after long disputes. Table 9.35 (page 392) shows the results of a vote on the issue in 1967, 27 years earlier. What do the data tell you about the vote in 1967?

9.53 Are nonrespondents different from those who do respond in a survey? In a follow-up study of 293 young African-American women in Philadelphia, the researcher was able to contact 95 of them. Table 9.36 (page 392) shows how much education these women said they expected to get when they were first interviewed.

Table 9.34 Data for Exercise 9.51

		Histological type				
		Lymphocyte predominance	Nodular sclerosis	Mixed cellularity	Lymphocyte depletion	Total
Response to treatment	Positive	74	68	154	18	314
	Partial	18	16	54	10	98
	None	12	12	58	44	126
	Total	104	96	266	72	538

Source: B. W. Hancock et al., Clinical Oncology, *vol. 5 (1979), pp. 283–297.*

Table 9.35 Data for Exercise 9.52

		Voting body			
		House of Bishops	House of Clergy	House of Laity	Total
Vote	Aye	1	14	45	60
	No	8	96	207	311
	Abstain	8	20	52	80
	Total	17	130	304	451

Source: The Daily Telegraph, *July 4, 1967.*

Table 9.36 Data for Exercise 9.53

		Contacted		
		Yes	No	Total
	Partial high school	2	4	6
	High school	17	32	49
Expected	College	32	60	92
education	Professional degree	8	15	23
	Technical/Trade school	34	87	121
	Do not know	2	0	2
	Total	95	198	293

Source: Roberta R. Iversen, Income and Employment Consequences for African-American Participants of a Family Planning Clinic: A Seven-Year Follow-Up, *Doctoral dissertation, Bryn Mawr College, 1991. Dissertation Abstract International 52:1522A.*

Table 9.37 Data for Exercise 9.54

		Performance		
		Top half	Bottom half	Total
	Letter	37	29	66
Grade	Pass/Fail	5	13	18
	Total	42	42	84

Table 9.38 Data for Exercise 9.55

		Name		
		A–K	L–Z	Total
Grade	Letter	35	31	66
	Pass/Fail	7	11	18
	Total	42	42	84

a. Compare the percentages for the two groups to see if they are different.

b. Consider the two groups as random samples and test the null hypothesis that they are not different.

9.54 Some students elect at the beginning of a course to take a Pass/Fail grade instead of a letter grade. We divided one of our introductory statistics classes into a top and a bottom half based on the performance of the students, and we counted in each group the students who had elected the Pass/Fail option (Table 9.37). Analyze the relationship between standing and type of grading.

9.55 For the same class as in Exercise 9.54, we also divided the class into two halves based on the last names of the students (Table 9.38).

a. Analyze the relationship between last name and type of grading.

b. Comment on differences between the data and results here and in Exercise 9.54.

Table 9.39 Data for Exercise 9.56

		Gender		
		Male	Female	Total
Division	Natural science/Engineering	13	4	17
	Humanities	8	7	15
	Social sciences	7	8	15
	Total	22	28	50

Source: Data used by permission of Heather Repenning, Swarthmore College.

9.56 In a random sample of students at Swarthmore College in the fall of 1995, the students were asked which division of the college their major was in. The interviewer also noted the gender of the students. The data are shown in Table 9.39. Analyze the relationship between the two variables.

9.57 In a study of Alzheimer's disease among nuns of the Milwaukee convent of the School Sisters of Notre Dame, the researchers also studied the writing styles of the nuns as expressed in statements the nuns made when they entered the convent as young women. Among nuns whose brains were checked for Alzheimer's disease after their deaths, the results in Table 9.40 were found.

   a. The paper reports that of the 10 nuns in whom Alzheimer's was confirmed by autopsy, 90% had shown low linguistic ability; among the nuns who died without developing the disease, 13% had shown low linguistic ability. What does this imply about the way the paper chose the independent and the dependent variable? What would you choose for independent and dependent variable in this study?

   b. Analyze the relationship between the two variables.

9.58 Ideally, statistical analyses are performed on properly collected random samples. However, to keep the data collection simple, observe any 20 or so students on two categorical variables (female/male, tall/short, thin/not thin, etc.).

   a. Set up the contingency table for the data.

   b. Create a bar graph similar to Figure 9.1 for the data.

   c. Create a new table by changing the frequencies to percentages. Note: It is possible to compute percentages in different ways, so take care that you follow the guidelines in the chapter

Table 9.40 Data for Exercise 9.57

		Writing style		
		Low linguistic ability	High linguistic ability	Total
Alzheimer's disease	Yes	9	1	10
	No	2	13	15
	Total	11	14	25

Source: The Philadelphia Inquirer, *February 21, 1995, p. A1.*

(and refer to Table 9.2). Would you say that the magnitude of differences in the percentages from your data is very large or not?

d. How strong is the relationship between the two variables?

e. Is there a statistically significant relationship between the two variables?

9.59 a. As a class project, find the median height of the students in the class.

b. Construct a contingency table that shows the gender of each student and whether the student's height is above or below the median value.

c. How strong is the relationship between gender and height?

d. Is the relationship statistically significant?

CHAPTER 10

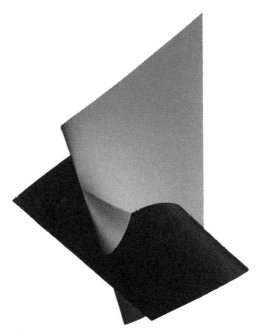

10.1 Question 1. Relationship between the variables?

10.2 Question 2a. Strength of the relationship?

10.3 Question 2b. Form of the relationship?

10.4 Question 3. Relationship in the population?

10.5 Warning: What you measure is what you get

10.6 How to be smart using dummy variables

10.7 Summary

REGRESSION AND CORRELATION FOR TWO METRIC VARIABLES

Do foods with higher fat content contain more calories than foods with lower fat content? What is the relationship between the weight of a car and the mileage it gets? How do robbery rates relate to larceny rates in the various states? Is there a relationship between cigarette consumption and cancer rates in different countries? How does the percentage of people with low education relate to the percentage of people with low income in different states? Is there a relationship between how tall parents are and how tall their children grow? Has there been any change in malignant melanoma cases in the last decades? How has the world record in the men's mile race changed since Roger Bannister broke the four-minute barrier in 1954?

Regression analysis describes the way in which a dependent variable is affected by a change in the values of one or more independent variables. **Correlation analysis** describes how strong the relationship is between metric variables.

Questions about relationships between metric variables with well-defined units of measurement, such as food calories and fat content, gas mileage and vehicle weight, are answered using the statistical methods know as *regression analysis* and *correlation analysis*. Regression and correlation analyses represent two major and complementary aspects of the analysis of the relationship between metric variables.

In statistics, regression is a more specialized term than it is in ordinary language. Normally we think of regression as going backward in ability or performance. Statistical regression received its name from an early study in which the method was used on the heights of parents and children (see the box). The study found a tendency for children to be more average in height than very short or very tall parents. This tendency toward the middle was labeled a regression effect.

REGRESSION TOWARD THE MIDDLE HEIGHT

The term *regression* was coined by the famous British statistician Francis Galton in his late nineteenth-century study of the heights of children and their parents. Galton found that tall parents tended to have tall children, as we would expect. But the children were not, on the average, quite as tall as their parents. The same thing was found for short parents: they tended to have short children, as we would expect, but on the average the children were taller than the parents. (This is just as well, because if children of tall parents were even taller than their parents and if children of short parents were even shorter than their parents, we would all continue to grow more and more apart over the generations.) It was the tendency of the children's heights to move toward the middle that made Galton call it a regression effect, and the methods he developed for the study of data on two metric variables became known as regression analysis.

As instructors, we see the same regression phenomenon in our classes. Good students who get top scores on the midterm do well on the final examination but on the average not quite as well. Similarly, students who do badly on the midterm on the average do better on the final. In sports the regression effect is a well-known phenomenon: the rookie who has an exceptionally good first season does not perform as well the second year. The "regressive" feature of regression analysis is not equally obvious, however, in every case where regression analysis is used.

Correlation analysis measures the strength of the relationship between metric variables. Two variables may have a high correlation or they may have a low correlation, depending on how strongly they are related, and the word correlation in statistics corresponds well with how the word is used in daily speech.

STOP AND PONDER 10.1

Give an example of two metric variables that may be related, and list some of the values of the variables. Why are your variables metric variables?

In this chapter, we study regression analysis for two variables; this type of analysis is known as *simple regression analysis*. You may find that name ironic because at first glance there is nothing simple about it. But "simple" here is a way of denoting two variables instead of more than two. Simple regression analysis is as simple as regression analysis can get! Since Galton's days the methods have been extended to more than two variables, and we examine some of them in Chapter 13.

Let's start with a dieter's dilemma. You are trying to diet but have a bad case of the "munchies." Standing in front of a vending machine,

Food for thought for regression analysis. *(Source: 1992, Comstock.)*

Table 10.1 Calories and fat in snack foods

Food	Calories (kcal)	Fat (g)
Tortilla chips (15)	110	4
Light potato chips (18)	120	6
Cheese-flavored snacks (34)	120	6
Doughnut (1)	164	8
Apple pie (1/6 of 8-in. pie)	430	19
Popcorn (3 cups)	192	11
Ice cream (1/2 cup)	175	12
Chocolate chip cookie (1 large)	236	12
Cheese and crackers (2 oz. and 10 thin)	429	26
Chicken wings (2)	318	21
Bagel with cream cheese	249	11
Peanut butter cups (2)	281	16
Dry roasted peanuts (1 oz.)	160	14
Chocolate bar (1 oz.)	147	9
Cheese or peanut butter crackers (6)	210	9
Granola bar (1)	120	5

Source: ASDA data and manufacturer's data shown as an advertisement in The New York Times Magazine, *April 20, 1990, p. 20.*

you stare at the delectable choices: potato chips, pretzels, popcorn, candy bars. Fat and carbohydrates seem to beckon you as you ponder. Which snack will do your diet the least damage? Which have the least calories, those with high fat content or those with low? Does it make a difference? What you need to know is how calories increase or decrease with increasing fat content in the foods. Regression analysis gives you a way to find out. We tackle the problem of fat content and calories by first looking at the data we have (Table 10.1). To do a simple regression analysis (and correlation analysis), the data file must consist of two columns of numbers, one column for each variable. In addition, the data file should contain a column which makes it possible to identify each row. Each row in Table 10.1 contains data on one particular snack. Column 1 gives the name of the snack, column 2 gives the calories, and column 3 gives the fat content.

> **STOP AND PONDER 10.2**
>
> Look at the calorie values in Table 10.1. Why is it that all the foods do not have the same number of calories? Could it be because they have different fat content? If it is so, then we are interested in the *variation* in the calorie values. What can you compute from the data that will tell you how different the observations are from each other?

From the data in Table 10.1, how well are we able to answer the question of how fat and calories are related? Just scanning the data, we see that foods with high fat content also seem to have a high number of calories, and foods with low fat content seem to have fewer calories; the two variables seem to be related. But to access the detailed information the data contains—for example, whether one food that has twice as much fat as another also has twice as many calories as the other—we turn to regression and correlation analysis.

10.1 QUESTION 1. RELATIONSHIP BETWEEN THE VARIABLES?

We just answered this question when we observed that smaller values of the fat variable tend to correspond to smaller values of the calorie variable, and larger values of the fat variable tend to correspond to larger values of the calorie variable, indicating a relationship between the two variables. To get the details of the relationship, we need to analyze the data. As before, we can make a graph or a table and in addition compute a number or two from the data.

Graphing the data in a scatterplot

With two metric variables, we always start the analysis of the data with a graph. The purpose of the graph is to get a visual impression of the relationship between the variables. The graph is also used to see if the data are of a kind that permits us to use statistical correlation and regression methods. Not all data from two metric variables lend themselves to this kind of analysis, and a graph usually indicates whether we can proceed or not.

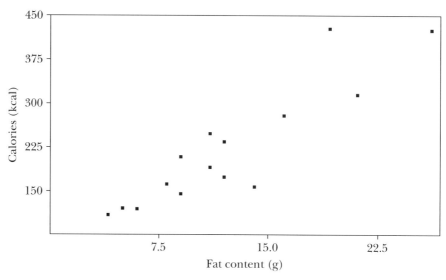

Figure 10.1 Scatterplot of data in Table 10.1 (page 400)

> **STOP AND PONDER 10.3**
>
> On a scatterplot, which variable would be on the *y*-axis and which on the *x*-axis, and when would it be unclear?
>
> a. Number of curve balls thrown and number of home runs hit off a pitcher
>
> b. Size of helmet and size of baseball glove
>
> c. Number of strikeouts at bat and yearly salary
>
> d. Number of stolen bases and number of games played

> A **scatterplot** is a graph with a horizontal axis for the independent variable and a vertical axis for the dependent variable. Each pair of observations is represented by a point in the graph.

The graph we make is a *scatterplot*. The horizontal *x*-axis is used for the independent variable, and the vertical *y*-axis is used for the dependent variable. For the data in Table 10.1, the *x*-axis is used for the fat content and the *y*-axis is used for the number of calories because we hypothesize that fat content (independent variable) affects calories (dependent variable). Looking at the data, tortilla chips have an *x*-value of 4 and a *y*-value of 110. We plot these two numbers (4, 110) in the graph as a point at the 4 and 110 coordinates. We also plot the other snacks as points in the graph, ending up with the scatterplot shown in Figure 10.1.

STOP AND PONDER 10.4

Following are data on funding, in millions of dollars, for public broadcasting from subscribers and businesses (not federal government and other sources) between 1983 and 1993.

Year	From subscribers	From businesses
1983	180	110
1984	195	135
1985	225	175
1986	240	175
1987	275	195
1988	300	210
1989	320	240
1990	340	260
1991	370	290
1992	390	300
1993	395	300

Source: Foundation for Public Broadcasting.

a. Create a scatterplot from the data.

b. Does there seem to be a relationship between the two variables in the data?

c. Is it a negative or a positive relationship?

d. What would you expect the scatterplot to look like if the variables have a strong relationship? Does this relationship seem to be strong?

e. What seems to have happened to the relationship between the two variables at the end of this period? Does it suggest that a ground-swell of support was already developing among big business for the 1995 Republican Congress's initiative to cut off funds for public broadcasting?

Scatterplots are always made with the independent variable running horizontally and the dependent variable vertically—the same way we set up contingency tables for two categorical variables. Of course, there are times when it is not clear which is the independent and which is the dependent variable. With height and weight on students in the class, for example, it is not obvious which affects which. The choice matters for regression analysis, but it does not matter for correlation analysis; correlation analysis yields the same result either way.

Learning from the scatterplot

The scatterplot in Figure 10.1 shows that the greater the fat content in a snack, the greater the number of calories. The pattern in the points in the graph is evidence for a solid yes answer to the question of whether there is a relationship between the two variables in the data. The graph lends support to the conclusion we drew on the basis of the data table alone. From the graph we are confident that the two variables are related.

In addition, because the points scatter from the lower left to the upper right in the graph, the relationship between these two variables is *positive:* the *more* fat in a snack, the *more* calories. Some variables, such as the weight of a car and the gas mileage, have a *negative* relationship: the *more* the car weighs, the *fewer* miles per gallon. In a scatterplot with a negative relationship between the two variables, the points scatter from the upper left corner to the lower right corner.

Most statistical software is designed to make scatterplots, and Figure 10.1 was made by a computer after the data had been entered into a computer file. If the data do not contain too many observations, we can make scatterplots by hand.

Linear relationships

We continue the analysis with another question: In what way do the *y*-values (calories) differ as the *x*-values (fat) increase or decrease? To find out whether there was a relationship between two categorical variables, we took each value of the independent variable and found the corresponding percentage distribution of the dependent variable. Here, the dependent variable is a metric variable, so we find means instead of percentages.

What we do for two metric variables is not very different from what we did for categorical variables. We choose values of the independent variable, fat content, and find the corresponding values of the dependent calorie variable. For example, for fat values around 7.5 grams, the calories have a mean of about 150; for fat content around 15.0 grams, the corresponding mean of the calories is about 250. Regression analysis is based on the fact that different values of the independent variable correspond to different values of the corresponding mean of the dependent variable. With more data—enough to show several values of the calorie variable for each value of the fat variable—we could find the actual mean number of calories for each fat value.

STOP AND PONDER 10.5

Which of the scatterplots seem to lend themselves to regression and correlation analysis? Describe in a sentence what kind of relationship each scatterplot shows.

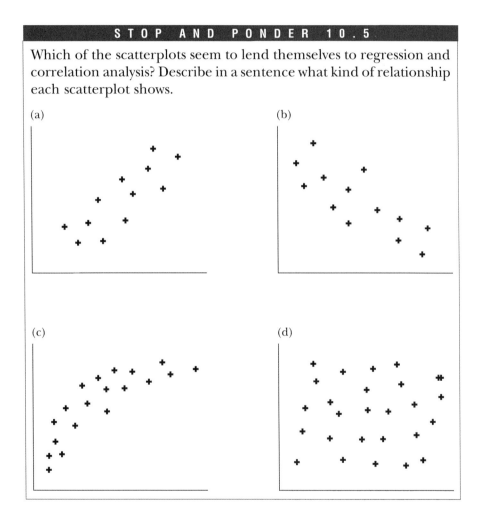

If the mean calorie values were different from one another for different values of the fat content, then we could establish that the two variables are related. In addition, if the points representing means in a scatterplot grouped around a straight line through the middle of the scatter, then we could do regression and correlation analysis on the data. We do not have enough data here to find such means, but the points for the data we have more or less group along a straight line, permitting us to continue.

If the points in the scatterplot had grouped along what looked like a curve, then we could not continue with the analysis. If the points in the scatterplot had formed a cloud with no pattern, then the data might have been random with no relationship between the variables.

10.2 QUESTION 2A. STRENGTH OF THE RELATIONSHIP?

When the data points are strung out along a line, we can compute a number that measures how strong the relationship is between the two variables. For two metric variables this computation creates a coefficient that is denoted r, and we call it simply the *correlation coefficient*, although it has many names. It can be called the *linear correlation coefficient*, *Pearson's correlation coefficient*, in honor of the English statistician Karl Pearson who did important work with it, or the *product-moment correlation coefficient*, reflecting the way in which the numerical value of the coefficient is computed.

> The **correlation coefficient r** measures the strength of the relationship between two metric variables on a scale of -1 to 0 to 1.

Is r positive or negative? Large or small?

The correlation coefficient r for the fat/calorie data equals 0.91 (produced by Formula 10.1). What is much more important than the mechanics of the computation is the meaning of this number. The first thing we notice is that r is positive for these data. This means that small values of one variable correspond with small values of the other, and large values of one variable correspond with large values of the other. Thus, tortilla chips, which have a low fat value, also have a low calorie count; cheese and crackers have a high fat value and a high calorie count. The positive value for r corroborates the pattern in the scatterplot.

The second thing we notice about r is its size. Obviously, 0.91 is almost equal to the maximum positive value of 1. This means a very strong relationship between the two variables. By most criteria, any value of r between -0.75 and -1.00 represents a strong, negative relationship, and any value between 0.75 and 1.00 represents a strong positive relationship. Similarly, r values between -0.70 and -0.30 as well as between 0.30 and 0.70 are considered moderate, and anything in the range from -0.25 to 0.25 is considered weak.

10.2 Question 2a. Strength of the Relationship? 407

These rules are merely "rules of thumb." People working in different disciplines tend to find r's in different ranges, and high and low must be seen as relative to common values of r found in a field. A sociologist often considers an r of 0.50 quite high, while an economist may consider an r of 0.50 low. However, 0.91 is high and represents a very strong relationship by almost any account.

Four different scatterplots: From strong to weak relationships

Let us look at a few scatterplots to see how different scatter patterns lead to different values of r. Figure 10.2 shows four different scatterplots, each with 100 observations. The data were generated by a com-

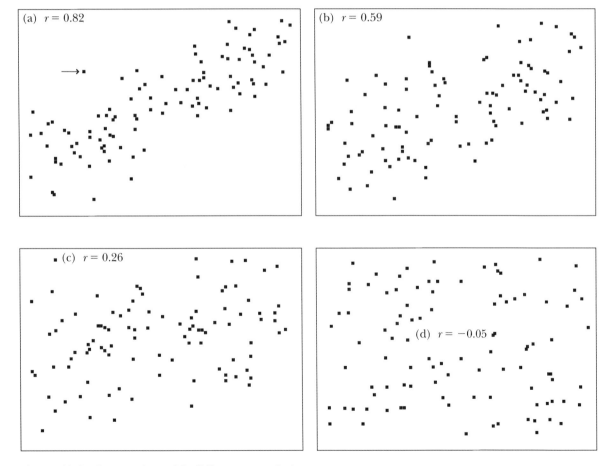

Figure 10.2 Scatterplots with different correlations

AN UNUSUAL OBSERVATION

In Figure 10.2a, the point marked with an arrow stands out from the others. It is isolated from the other points by a good bit of space. Points that do not lie near other points in a scatterplot tend to have a disproportionately large impact on the correlation coefficient. When the isolated point in Figure 10.2a is included in the analysis, $r = 0.818$, to three decimals. When the point is not included in the computation, $r = 0.835$ (these would usually be reported as 0.82 and 0.84). If even 5 of the 100 data points were outliers like this one, the difference in r computed with them and without them would be significant. And points that are even further away from the other points can be even more influential in determining the value of the correlation coefficient.

These differences between correlation coefficients illustrate how sensitive the correlation coefficient is to observations that lie off the main trend in the data. This is a reason we should always look at the scatterplot before computing r. By looking at the scatterplot first, we can see whether there are questionable data points to worry about.

puter and do not refer to anything in particular, so there are no labels or scales on the two axes.

In Figure 10.2a, the points are quite clustered and close to each other, and we can see a definite straight-line trend in the pattern from the lower left corner to the upper right corner in the graph. The points are located in a regular fashion along the diagonal. The strength of the relationship between the two variables ought to be quite strong, and the correlation coefficient bears this out, with $r = 0.82$.

In Figure 10.2b, the points are not as clustered as in Figure 10.2a. But we can still see a definite positive relationship in these data, and the scatterplot translates to a correlation coefficient r equal to 0.59. In Figure 10.2c, the correlation coefficient decreases to 0.26, indicating a weaker relationship. With an r of this size, it is almost impossible to see a pattern in the points and whether there is any relationship between the two variables. In Figure 10.2d, the points are randomly scattered and there is almost no relationship between the two variables.

Figure 10.2 illustrates positive relationships between two variables. It is also possible to generate data with negative values of r. In such a scatterplot, the points would distribute themselves along the other diagonal, the one from the upper left corner to the lower right corner. In that case, large values of the x-variable would have small values of the y-variable, and small values of the x-variable would have large values of the y-variable. For example, if x represented the cost of a car and y the number of cars sold, the value of r would be negative: costlier cars tend to be sold in fewer numbers than cheaper cars.

Here are some other values of r to demonstrate magnitudes of correlation. In a sample of cars, $r = -0.90$ for weight and mileage; for horsepower and mileage, $r = -0.87$. The correlation between the cost of several nuclear power stations and their electrical output is a surprisingly low $r = 0.47$. The r's for the variables in Table 10.1 plus two more variables are shown in Table 10.2.

In Table 10.2 there is a diagonal of 1.00 correlation coefficients. Each 1.00 represents the correlation of a variable with itself. The correlation of a variable with itself is always 1.00. Imagine a scatterplot with the same variable on both the horizontal and vertical axes. All the observations would lie along a 45-degree line, and all the points would be located directly on that line. The relationship between observations on a straight line has maximum strength and $r = 1.00$.

We already have the scatterplot for fat and calories (Figure 10.1), and for the other five correlation coefficients we can picture approximate scatterplots from knowing the r's. The smallest r—for sodium content and cholesterol—is equal to 0.41. In the scatterplot for these data, we would expect to see a general upward trend of the points but a fairly large scatter because the relationship is not very strong.

Table 10.2 Correlations between the variables in Table 10.1 plus cholesterol and sodium

Variable	Calories	Fat	Cholesterol	Sodium
Calories	1.00			
Fat	0.91	1.00		
Cholesterol	0.62	0.69	1.00	
Sodium	0.73	0.59	0.41	1.00

The correlation between the cost of nuclear power stations and their electrical output is a surprisingly low $r = 0.47$. *(Source: 1992, Comstock.)*

STOP AND PONDER 10.6

For the following relationships between two variables, would the correlations be positive or negative?

a. Popularity of a compact disc CD and the price of the CD

b. Size of office and salary of occupant

c. Price of a hamburger and number of hamburgers sold

d. Outdoor temperature (between 40 and 80 degrees Fahrenheit) and number of tickets sold at a swimming pool club

Interpretation of r: An issue of inexactness

We have described the correlation coefficient r as a number computed to measure the strength of the relationship between two metric variables; the values of r range from -1 to $+1$. However, it is hard to come up with an exact interpretation of r. We know that $r = 0.91$ indicates a strong relationship between two variables, and an r of 0.41 represents a moderately strong relationship. But beyond the words strong or moderately strong, what does the value of r mean?

For an exact interpretation of the strength of the relationship between two metric variables, we look at the square of the correlation coefficient instead of the coefficient itself. For fat and calories, $r^2 = 0.91^2 = 0.83$; for sodium and cholesterol, $r^2 = 0.41^2 = 0.17$; for horsepower and mileage in a sample of cars, $r^2 = (-0.87)^2 = 0.76$. These numbers, 0.83, 0.17, and 0.76, have a very specific interpretation, which we discuss later.

10.3 QUESTION 2B. FORM OF THE RELATIONSHIP?

Regression analysis is the other part of the analysis of two metric variables. Correlation and regression are equally important, and a complete analysis of the relationship between two variables includes both.

One of the basic ideas of statistics is that to understand the data better we replace them by one or more numbers computed from them. Let us illustrate this with a single variable first. If we scan the column of calorie values in Table 10.1, we see that they range from a low of about 100 calories for tortilla chips to a high of about 400 for apple pie. Because it is hard to comprehend all the calorie values at the same time, we replace the data by the mean for the variable. The mean equals 216.3 calories, and for many purposes the mean can be used in place of the original data.

What number can we compute to replace the observed data in *two* variables that will capture the relationship between the variables? The discussion of the scatterplot in Figure 10.1 implied that a line running through the middle of the points could represent all the points. We could use the line instead of all the data in discussing the relationship between the two variables.

STOP AND PONDER 10.7

In preparation for the discussion that follows, What is the mathematical equation for a line? What are the two numbers called that we need for such an equation? What does it mean to say that the steepness of the line is measured by the "rise" over the "run"?

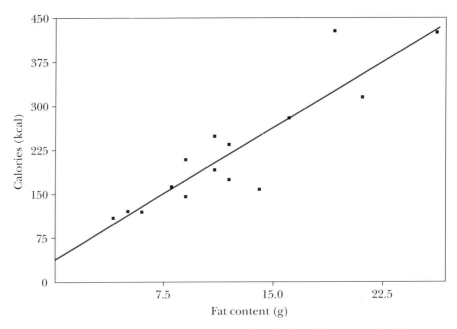

Figure 10.3 Scatterplot of fat content and calories with regression line

A line through the middle of the points

Figure 10.3 repeats the scatterplot of Figure 10.1, with the addition of a line running through the middle of the points. Such a line is known as the *regression line*. If we erased the data points and kept just the line, we would still have a good idea of how fat content is related to calories. This line represents the data on two variables quite well, just as a mean represents the data on one variable quite well.

Just as the points show a positive correlation, the line, running from the lower left to the upper right corner of the graph, has a positive slope: foods with low fat content have a low number of calories, and foods with high fat content have a high number of calories. The steeper the line, the more difference in the calories for a one-unit difference in the fat content. Steepness of a regression line is measured by its slope; if we knew the slope of the line in Figure 10.3, then we would know exactly how many calories difference there are for a one-unit difference in fat content.

To find an approximate value of the slope of the line we could use a ruler and measure the "rise" over the "run" of the regression line.

We can also compute the slope from the observed data (Formulas 10.2 and 10.3). The slope of the line equals 15.3 calories per gram. Thus, two snacks that differ in fat content by one gram will differ by 15.3 calories, on the average.

This point is illustrated in Figure 10.4. The figure shows two foods, A and B, and B has one more gram of fat than A has. Because the slope of the line equals 15.3, we see that on the average B has 15.3 calories more than A. The reason we say "on the average" is that the observed data points for the two foods do not lie exactly on the line. For two particular foods the difference in calories would be either more or less than 15.3 calories. But the mean difference in calories between many pairs of foods that differ by one gram would be 15.3.

The line in Figure 10.3 has a y-intercept, that is, the point where it cuts through the y-axis when x equals 0. Figure 10.3 shows the line extended to the left beyond the observed points, and the line cuts through the vertical axis where the fat content equals 0. The intercept

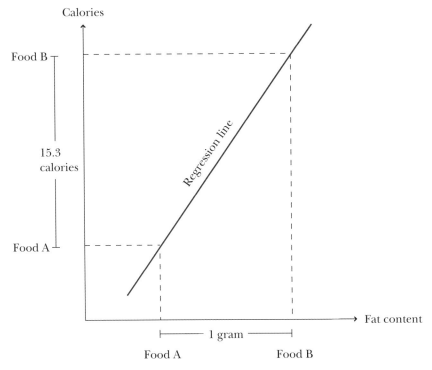

Figure 10.4 Interpretation of the regression coefficient 15.3 calories per gram

for this line equals 36.1 calories. The full equation for this regression line is therefore

$$\text{calories} = 36.1 + 15.3 \text{ fat content}$$

The equation for a regression line is called a *regression equation*. The equation for a regression line is written

$$\text{dependent variable} = \text{intercept} + \text{slope} \times \text{independent variable}$$

—in symbols

$$y = a + bx$$

where a is the intercept of the line, b is the slope of the line. In Figure 10.4, the slope 15.3 is the *regression coefficient* for the independent variable, here fat content. The regression equation tells us much more than the two columns of numbers in Table 10.1. It summarizes the relationship between the two variables in a very compact form.

> **STOP AND PONDER 10.8**
>
> Frequently students are urged to continue their education so that they can earn more money when they have completed their schooling. One way to analyze how much more one could make if one stayed in school extra years is to create a regression line that measures the relationship between years of school and yearly salary.
>
> Let years of school be the x- or independent variable on the horizontal axis. Let salary be the y- or dependent variable on the vertical axis. Could you draw a regression line to display the relationship if you had data on education and income? The x-variable could start at 8 years of education and end at 20 years of education: annual salaries could range from what you think of as a low figure to what you think of as a high figure.

How to find the regression line: The least squares principle

The regression line is determined by its slope and intercept. Those two numbers are computed according to formulas given at the end of the chapter. Formula 10.2 shows how to compute the slope b, and Formula 10.4 shows how to compute the intercept a. Deriving these formulas requires more mathematics than we want to go into, but we can explain the principle that leads to the formulas.

If each of us took the scatterplot in Figure 10.1 and with a ruler drew a line running through the middle of the points, we would each draw a slightly different line. But all the lines would coincide more or less with the line drawn in Figure 10.3, which was drawn by the computer software that did the analysis. The computer-drawn line has a feature that makes it special. The line is based on how far each data point is from the line. The vertical distance from each point in the scatterplot to the regression line can be found using a ruler to measure the distances or by numerically computing the distance for a more accurate result. After we find all the differences, we square each distance and add all the squares. For the snack food data, the sum of squared distances equals 27,182. This number gives us an overall measure of how far the points are from the line.

If we try the same procedure for any other line, the number is always larger than 27,182. No other line gives a smaller sum of squared distances. Thus, the line the computer software drew is the line that has the smallest possible sum of squared vertical distances between the points and the line. In the sense of *least squares,* this regression line is therefore the one that is closest to all the points and in that sense represents the points better than any other line.

As an illustration, let us find a few of the distances and their squares. A granola bar, for example, has 5 grams of fat. When we substitute this value into the equation for the regression line, we get the number of calories predicted by the fat content of the granola bar:

$$\text{predicted calories} = 36.1 + 15.3(5) = 112.4$$

This is the value of calories we find on the regression line, meaning that the point with coordinates (5, 112.4) lies on the line. (You can

"Cathy" copyright 1995 Cathy Guisewite. Reprinted with permission of Universal Press Syndicate. All rights reserved.

check this out for yourself by looking at the scatterplot and the line in Figure 10.3.) For more accuracy in the actual computation, we used more decimals for both the slope and the intercept. Since the granola bar actually has 120 calories, the vertical distance from the observed point to the point on the regression line is 120.0 − 112.4 = 7.6. The actual data point for a granola bar lies 7.6 calories above the estimated point on the regression line. The other vertical distances can be found in the same way. The points above the line will have positive distances and the points below the line will have negative distances. When we square these distances and add the squares, we get the sum 27,182.

The granola bar contributes $7.6^2 = 57.7$ to this sum. That is not a large contribution, because the point for the granola bar lies close to the line. Some of the other points lie farther away, and the squares of these distances are larger numbers.

If all of us agree to use the least squares method to find the line, then all of us will come up with the same line. If we use some other principle, we get different lines. For example, the line where the sum of the absolute values of the distances from the points to the line is the smallest is a different line. This is another example where statistical method, not just data, determines the outcome of an analysis.

Predicting with regression analysis: From fat to calories

> Substituting the value of the independent variable into the equation for the regression line yields the **predicted value** of the dependent variable.

You've just seen that the regression line can be used for prediction purposes. When we know how much fat there is in a food, we can use the regression line to predict the number of calories in the food. (We always predict from the independent to the dependent variable.)

Because of this predictive feature of regression analysis, the regression equation is sometimes expressed this way:

$$\text{predicted calories} = 36.1 + 15.3 \text{ fat content}$$

The word "predicted" on the left side stresses that only the predicted values, not the actual, observed values, of the dependent variable are on the left side. Another way this is sometimes done is to place a "hat" over the term on the left side of the equation:

$$\widehat{\text{calories}} = 36.1 + 15.3 \text{ fat content}$$

The hat symbol means "predicted." Also, sometimes *y* replaces the words for the dependent variable and *x* the words for the independent variable. Then the equation for the predicted value of *y* as a function of *x* can be written

$$\hat{y} = 36.1 + 15.3x$$

In the snack example, the predicted value of the calorie variable for a granola bar with 5 grams of fat is 112.1 calories. The predicted values for some of the other foods are

Tortilla chips	36.1 + 15.3(4) = 97.1
Light potato chips	36.1 + 15.3(6) = 127.6
Cheese-flavored snacks	36.1 + 15.3(6) = 127.6
Doughnut	36.1 + 15.3(8) = 158.1
⋮	⋮
Granola bar	36.1 + 15.3(5) = 112.4

In addition, we could use the regression equation to predict the number of calories in a new food with a known fat content. We would substitute the value of the fat variable into the regression equation and compute the predicted value.

STOP AND PONDER 10.9

You pick up a candy bar at the market. The label indicates that the bar has 3 grams of fat. Estimate the number of calories of the candy bar using the results from this example.

Magnitudes of effects: Interpretation of *r*-square

The number of calories in different snack foods is affected by many other variables in addition to fat content. The other variables, called the *residual variable,* and fat content together determine the calories in these foods (Figure 10.5). The two arrows indicate how we think the influence flows from the fat variable and the residual variable to the calorie variable, and the question marks next to the arrows show that the amount of influence of the fat variable and the residual variable is not known. Can we measure the effects on calories of the fat variable and the residual variable?

> The **residual variable** is the combined net effect on the dependent variables of all variables other than the independent variable.

418 Chapter 10 • Regression and Correlation for Two Metric Variables

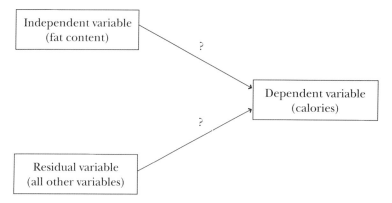

Figure 10.5 Effect of fat content and residual variable on calories

First, imagine that no variables affect the calories in snack foods; then all snack foods would have exactly the same number of calories. The fat content would not make any difference, and no other variable would make any difference. For this thought experiment, assume that

Table 10.3 Observed values of calories and fat if no variables affect the calorie variable

Food	Fat (g)	Common calorie value
Tortilla chips (15)	4	216.3
Light potato chips (18)	6	216.3
Cheese flavored snacks (34)	6	216.3
Doughnut (1)	8	216.3
Apple pie (1/6 of 8-in. pie)	19	216.3
Popcorn (3 cups)	11	216.3
Ice cream (1/2 cup)	12	216.3
Chocolate chip cookie (1 large)	12	216.3
Cheese and crackers (2 oz. and 10 thin)	26	216.3
Chicken wings (2)	21	216.3
Bagel with cream cheese	11	216.3
Peanut butter cups (2)	16	216.3
Dry roasted peanuts (1 oz.)	14	216.3
Chocolate bar (1 oz.)	9	216.3
Cheese, peanut butter and crackers (6)	9	216.3
Granola bar (1)	5	216.3

this common number of calories would be the overall mean of the observed calorie values, or 216.3 calories. All the calorie values would be 216.3, as shown in Table 10.3.

In a scatterplot of the fat content and the values in the last column of Table 10.3, all the points lie on a horizontal line (Figure 10.6). But of course the observed data do not lie on a horizontal line. They are scattered, as Figure 10.6 also shows. This means that the *variation* in the values of the dependent variable calories shows that the number of calories is affected by other variables. How do we measure how much variation there is? How much of the variation is associated with the independent variable fat content and how much is associated with the residual variable?

The observed number of calories of a peanut butter cup, for example, is not on the horizontal line but at 281. The independent variable and the residual variable together pushed the calories of a peanut butter cup up from 216.3 to where we find them at 281, a total of 64.7 calories. So 64.7 is the combined effect of the fat content and the residual variable. Similarly, the combined effect on dry-roasted peanuts

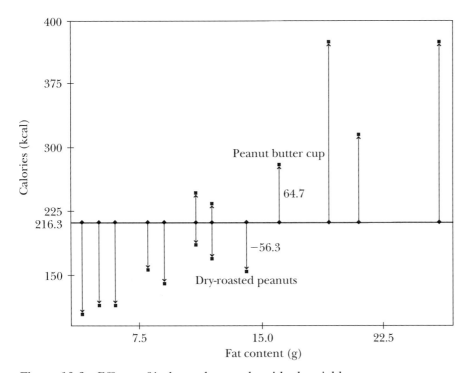

Figure 10.6 Effects of independent and residual variables

is −56.3, because the observed calorie point is below the horizontal line. We can find the combined effect for all the different foods the same way.

Next, we want to summarize all these effects in a single number. For various historical and mathematical reasons, we square each effect (observation minus overall mean) and add the squares. For the foods this sum is 159,060; it is known as the *total sum of squares*.

> The **total sum of squares** measures the effect of both the independent variable and the residual variable on the values of the dependent variable. It is found as
>
> $$\text{sum (observation minus overall mean)}^2$$

Now we want to find out how much of the total sum of squares is due to the effect of the independent variable (fat) and how much is due to the effect of the residual variable (everything else). Suppose that the residual variable has no effect on the calories and that the calories are affected only by the fat content. Then all the points in the scatterplot would lie directly on the regression line. But according to Figure 10.3, that is not where the data lie. The points are scattered around the regression line, so the residual variable must be pushing the points off the regression line.

The effect of the residual variable is the vertical distance from an observed point to the regression line.

The difference between the observed data point and its corresponding position on the regression line is the effect of the residual variable. We can find each of those differences. To summarize their magnitudes, we square each of them and add the squares. This sum represents the effect of the residual variable, and it is known as the *residual sum of squares*, sometimes called the error sum of squares. For the snack foods example, the residual sum of squares equals 27,182. A small residual sum of squares in comparison to the total sum of squares results when the points lie close to the line. Similarly, a large residual sum of squares results when at least some of the points lie quite far away from the regression line.

Since the combined effect of the independent variable and residual variable equals 159,060 and the effect of the residual variable alone is 27,182, the effect of the independent variable is the difference 159,060 − 27,183 = 131,878. This sum of squares is also known as the *regression sum of squares*. Formula 10.5 at the end of the chapter shows how to compute the different sums of squares.

Table 10.4 Sums of squares and proportions for the snack data

Source	Sum of squares	Proportion
Fat content	131,878	0.83
Residual	27,182	0.17
Total	159,060	1.00

Sums of squares are often displayed in a table like Table 10.4. In the table is a row for each variable and a row for column totals. The first column shows the name of the variable, and the second column shows the magnitude of the effect of each variable as measured by the appropriate sum of squares. To make it easier to see how large the effects are compared to each other, the third column shows each sum of squares as a proportion of the total sum of squares. To find the proportions, we divide each individual sum of squares by the total sum of squares. Here, the independent variable contributes 0.829, or 83%, of the effect and the residual variable contributes the remaining 0.171, or 17%. Thus, the effect of the fat content on the dependent variable calories is much larger than the effect of the residual variable.

The proportion of the effect due to the independent variable, here 0.83, is always equal to the square of the correlation coefficient between the independent variable and the dependent variable (shown in Formula 10.6). In Section 10.2, we found the correlation coefficient r between fat content and calories to be 0.91, so r squared is $r^2 = 0.91^2 = 0.83$, the proportion of the effect attributed to the fat variable in Table 10.4! Using regression analysis, we have been able to show how important the fat content of foods is compared to other influences in determining calories and that the closer the points are to the regression line, the smaller the residual sum of squares and the larger the correlation coefficient.

All this means that when we come across a correlation coefficient, we should immediately square it. The square tells us what proportion of the total effect on the dependent variable comes from the independent variable. One minus r-square is the proportion of the total effect that comes from the residual variable. Sometimes the results are disappointing. Suppose we get a middle-range value of r equal to 0.50. The square of 0.50 is 0.25, meaning that the independent variable is associated with only a quarter, not half, of the effect on the dependent variable.

In a report of the snack foods analysis, we might write a sentence such as "The fat content variable explains 83% of the variation in calories," or "83% of the variation in the calorie variable is due to the fat variable." Instead of the terms *explains* or *is due to* we could also use a term like *accounts for*. All these terms are widely used to mean the same thing.

Correlation and/or regression? The more the merrier

What can we conclude from a very high correlation between two variables when we know nothing about the regression line? We tend to be impressed by large correlation coefficients and conclude that we have learned something important about the two variables. Similarly, what are we to think if we are told the regression line, but we are not told what the correlation coefficient equals? The larger the slope and the steeper the line, the more important the independent variable may seem. But knowing only the correlation or the regression is not enough to analyze two variables properly. We should know both.

In the example with fat and calories, the correlation coefficient is equal to a high 0.91. The difficulty in interpreting this correlation coefficient is that this value of the correlation coefficient can be the same for many different data sets that have very different regression lines. Suppose the regression line had the equation

$$\text{calories} = 248 + 1.1 \text{ fat content}$$

instead of the equation we found (calories = 36 + 15.3 fat content). With the new regression line, we are no longer so impressed by the large correlation coefficient of 0.91. The high value of r tells us that the points are closely clustered around the line, but because of the small slope of 1.1, the line is almost horizontal.

Let us pick two foods, one with a low 7.5 grams of fat and another with a high 22.5 grams of fat. By substituting these numbers into the equation for the line, we find that the predicted number of calories for the first food is 256 and for the second food is 273. The difference between 256 and 273 calories is only 17 calories, a very small and uninteresting difference in calories. Even though one food has three times the amount of fat as the other, they hardly differ in calories. From a nutritional point of view, the result is inconsequential in spite of the fact that there is the very strong relationship between the two variables, as shown by the correlation coefficient (Figure 10.7). This example

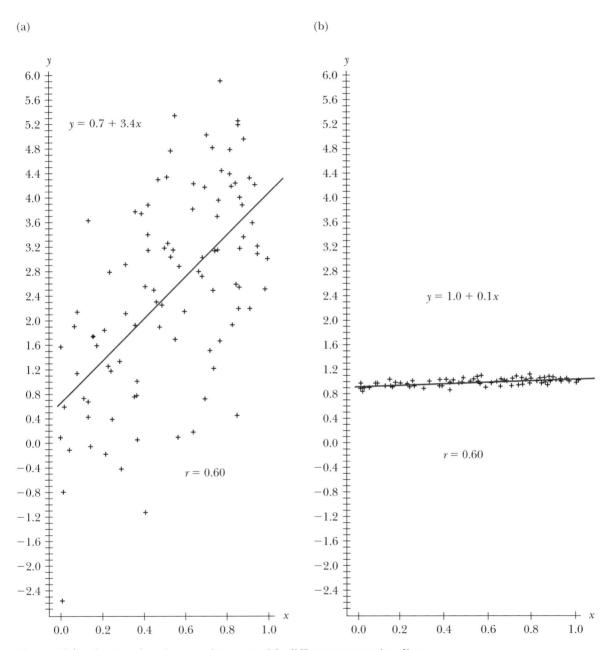

Figure 10.7 Scatterplots for two data sets with different regression lines and the same correlation coefficient

helps illustrate that it is not enough to know the correlation coefficient; we need to know the regression line also.

Figure 10.7 shows two scatterplots in which the two data sets have the same fairly high correlation coefficient. In graph a, the line is quite steep, and in graph b the line is almost horizontal. If we knew only the correlation coefficient, we could not distinguish between the two data sets, in spite of their obvious differences.

The reverse occurs when we know the regression line but not the correlation coefficient: we know how steep the line is, but we do not know how close the data points are to the line. If the points scatter widely around the line, the correlation coefficient is small and the line conveys less than it does when the points are close to the line. Figure 10.8 shows scatterplots of two different data sets with the same regression line and different correlation coefficients.

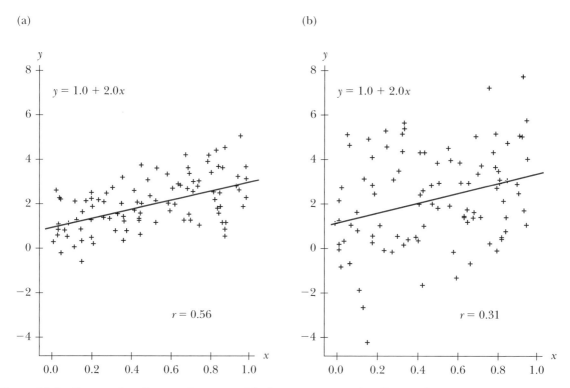

Figure 10.8 Scatterplots for two data sets with the same regression line and different correlation coefficients

Most statistical computer programs give the correlation coefficient or its square when we do regression analysis, but they do not give the regression line when we do correlation analysis. In one particular situation, research reports give correlation coefficient but not the regression line. This occurs when it is not clear which variable should be the independent variable and which should be the dependent variable. For example, in studying verbal and mathematics test scores, the choice of independent and dependent variable is not obvious. Therefore, only the correlation coefficient would be reported.

Regression analysis for data on change

So far we have been careful to interpret the regression coefficient (slope) b by saying that if two observations differ by one unit on the independent variable, then they will differ on the average by b units on the dependent variable. In the snack food example, b equals 15.3; if two snack foods differ by 1 gram of fat, they differ by an average of 15.3 calories. We have data on *different* foods, so the interpretation of b has to be in terms of *different* foods. The data do not permit us to conclude that if the fat content of a snack food *increases* by 1 gram, then the calories would increase on an average by 15.3. For this interpretation we need data on change of fat and calories for a particular food.

When we do have data on change, we can interpret the regression coefficient b in terms of change. For example, when the pediatrician measures the weight of a baby at each office visit, then we can do a regression analysis of the data and conclude that the baby's weight increased by so many ounces per month.

In another example, *The Philadelphia Inquirer* reported (April 7, 1993, p. A2) on a story in the *Journal of the American Medical Association* about the association of reduction of lead in the blood of children with an increase in scores on a "cognitive index" scale. The lead level was measured in milligrams per deciliter of blood, and the cognitive index was derived from standardized intelligence tests. From the data in the article it is possible to conclude that the regression equation for the two variables can be expressed as

$$\text{cognitive index} = 90 - 0.33 \text{ lead content}$$

The negative value of the regression coefficient -0.33 indicates that if lead content goes up, then the cognitive index goes down; if the lead content goes down, then the cognitive index goes up. Specifically, the coefficient shows that if the lead content in the blood is reduced by 1 milligram per deciliter of blood, then the cognitive index goes up on the average 0.33 points—or, as the newspaper report says, if the lead content goes down 3 milligrams, then the cognitive index goes up by 1 point. When public policy is directed at creating social change, a regression analysis of this sort is often appropriate.

10.4 QUESTION 3. RELATIONSHIP IN THE POPULATION?

It is one thing to find that there is a relationship between fat content and calories in a sample of snack foods, and it is another thing to find that there is a relationship between the two variables in the population of all snack foods. Because we do not have population data, we use sample data to make generalizations about the population. This is done in two ways: by constructing a confidence interval for the population regression coefficient β and by setting up and testing a null hypothesis of no relationship. For purposes of the argument, let us treat our sample of snack foods as a randomly selected sample of all snack foods.

Confidence interval approach

In Chapter 6 we use confidence intervals to estimate the value of an unknown population parameter. Here, we first find the observed regression coefficient and then add and subtract a sampling error term to the regression coefficient. In the snack food example, the sample regression coefficient is 15.3 calories per gram, and we compute that the sampling error equals 4.0. The 95% confidence interval for the population slope β therefore is $15.3 - 4.0 = 11.3$ calories per gram to $15.3 + 4.0 = 19.3$ calories per gram. Hopefully, the interval from 11.3 to 19.3 is one of the 95% of all intervals that contain the population regression coefficient β and not one of the few intervals that do not contain β. Formula 10.7 shows how to compute the confidence interval.

The most noticeable feature of this interval is that it does not contain the value 0. We take that to mean that 0 is not a possible value of the population regression coefficient. Since 0 is not a possible value of the slope β, we conclude that β must be different from 0. If the slope is not 0 for the line in the population, there must be a relationship between the two variables—fat content and calories—in the population of all snack foods, not just in the sample.

Hypothesis testing using t

The hypothesis testing approach in Chapter 7 is based on the null hypothesis that there is no relationship between the two variables. To test the null hypothesis we use either the observed sample regression coefficient b or the observed sample correlation coefficient r. They both transform to the same value of the statistical t-variable. From the resulting value of t, we can find the p-value for our data and then make a decision about the null hypothesis. The p-value is the probability of getting the sample data or more extreme data from a population where there is no relationship between the variables.

Figure 10.9 illustrates the process. Here, $b = 15.3$ calories per gram, and this value of b corresponds to $t = 8.24$ with $n - 2 = 14$ degrees of freedom (Formula 10.8). Similarly, $r = 0.91$ also corresponds to $t = 8.24$ (Formula 10.9). From the computer output or from a table of the t-distribution, we find that the probability of getting a value of t equal to or larger than 8.24 is less than 0.0001. Thus, if we have a population where there is no relationship between the two variables, fewer than 1 in 10,000 different samples will have a value of t of 8.24 or larger. This means that the observed sample relationship or stronger is almost impossible by chance alone. Because the p-value is so small, we reject the null hypothesis of no relationship. The small p-value conveys that if there is no relationship in the population of all the data, a sample can almost never have a slope of 15.3 or larger or a correlation coefficient of 0.91 or larger.

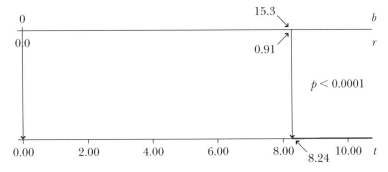

Figure 10.9 Changing the regression coefficient b and the correlation coefficient r to their values of t

Hypothesis testing using F

In addition to finding a value of the t-variable from either the slope b or the correlation coefficient r, there is a third way of testing a hypothesis. Two different ways is already one more than we need, but the third way can be used to generalize, and it is included here in anticipation of Chapters 12 and 13.

The third approach is based on Table 10.4. Adding three more columns to that table (Table 10.5) enables us to compute a value of the statistical F-variable, introduced in Chapter 5 with the z-, t-, and chi-square variables. The value of F can be used to judge the null hypothesis.

The column headed "Mean square" is obtained by dividing each sum of squares by its corresponding degrees of freedom. Since the regression sum of squares has only 1 degree of freedom, the regression mean square is also equal to 131,878. The residual mean square (RMS) is 27,182/14, which equals 1941.6. Then, the regression mean square is divided by the residual mean square (131,878/1941.6), giving an F of 67.90 with 1 and 14 degrees of freedom. Finally, the probability of getting an F of 67.90 or more by chance is less than 0.0001, and the null hypothesis is rejected.

The p-value is the probability of getting a value of the F-variable larger or equal to 67.90 in sample data from a population where there is no relationship between the variables. We know that $t = 8.24$, and we now find that the p-value for F equals the probability that t is less than -8.24 plus the probability that t is larger than 8.24. Thus, the p-value for F is the same as a two-sided p-value for the t-variable.

Table 10.5 Testing the null hypothesis using F

Source	Sum of squares	Proportion	Degrees of freedom	Mean square	F-ratio	p-value
Fat content	131,878	0.829	1	131,878.0	67.90	0.0000
Residual	27,182	0.171	14	1,941.6		
Total	159,060	1.000	15			

10.5 WARNING: WHAT YOU MEASURE IS WHAT YOU GET

With observational data, we usually cannot choose the range of values we observe of any particular variable. We ask people how old they are and we record their answers. With experiments, however, we often have a choice of values to use for an independent variable. Sometimes this choice affects the results of the analysis.

Let us illustrate the point with a small example using the following two sets of data. We want to study the relationship of $Y1$ and $X1$ and the relationship of $Y2$ and $X2$.

Data set 1		Data set 2	
$X1$	$Y1$	$X2$	$Y2$
1	3	1	3
1	7	1	7
3	7	15	31
3	11	15	35

First we make a scatterplot and draw the regression lines for each data set (Figure 10.10). The two lines have the same slopes and the same intercepts, but the line in the graph for data set 2 is longer. The points in both graphs are all the same distance away from the line—one unit either above or below the line.

When we look at the two lines, in graph b the points seem closer to the long line than the points in graph a to the short line. As in a perception experiment, things look different even though they are the same. This apparent difference is reflected in the two correlation coefficients: in graph b with the longer line, $r = 0.99$, while in graph a with the shorter line, $r = 0.71$. This means that the correlation coefficient measures not only how close the points are to the regression line but how spread out the x and y values are.

There is also a difference in the p-values in the two data sets. The short line is not significantly different from 0 since t is equal to 1.41 and the p-value is a large 0.15. The longer line, however, is significantly different from 0 since t is a large 9.90 and the p-value is a small 0.005.

This means that we have to be skeptical about the meaning of statistical significance and the amount of correlation between two variables. Significance and a large value of r can always occur if we have control over the values of the independent variable x and choose values

(a) Data set 1

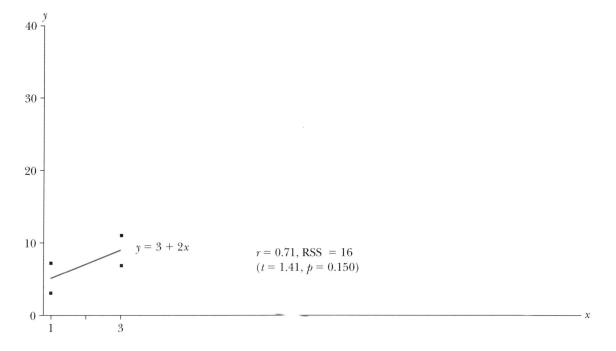

(b) Data set 2

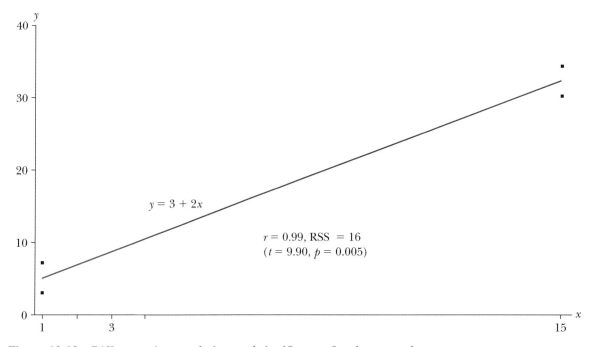

Figure 10.10 Difference in correlation and significance for the same slope

that are spread out. But this usually happens only with experiments where we are free to choose the values of the independent variable. In an observational study, we have to use the values of the independent variable that the sample provides.

10.6 HOW TO BE SMART USING DUMMY VARIABLES

So far we have used regression and correlation with metric variables. However, there are situations where it would be handy to use correlation and regression with other types of variables. In this section, we use them in problems involving one categorical variable and one metric variable.

Categorical independent variable with two values and metric dependent variable

If you are looking for a new place to live, climate is important in your choice. Choosing a location can be difficult, however, if you rely only on yearly mean temperatures, because different regions have different ranges of temperatures in summers and winters. This problem can be studied with the use of correlation and regression.

Suppose you want to be near an ocean, on either the east or west coast of the United States, and you would like a warm climate. To begin the analysis, let the range of temperature be equal to the difference between the mean temperatures in July and January. For example, the mean temperature in Philadelphia in July is 76 degrees and in January it is 32 degrees. You really feel the change in the seasons, and the range in temperature is $76 - 32 = 44$ degrees. In San Diego the same numbers are 70 and 55, so there the range is only 15 degrees. You want to find out whether the range is generally different for coastal cities on the East Coast and the West Coast.

The independent variable region has two values, East Coast and West Coast. This is a categorical variable. The dependent variable is range, and it is a metric variable. Since the region variable has only two values, the analysis can be done by defining a *dummy variable* for region. The two numerical values for a dummy variable can be anything, but we commonly use 0 and 1. This scheme works only when the original categorical variable has two categories. If the categorical variable has more than two categories, we have to turn to other methods.

432 Chapter 10 • Regression and Correlation for Two Metric Variables

> A **dummy variable** is a variable with only two numerical values, and it is used to represent a categorical variable with two categories. One value of the dummy variable is assigned to all the observations in the first category of the categorical variable, and the other value of the dummy variable is assigned to all the observations in the second category.

All the dummy variable really does is to identify the two categories. For computing purposes, the use of the dummy variable makes the categorical variable region into a variable with the two numerical values 0 and 1. Here we assign cities on the West Coast a value of 0 on the dummy variable and cities on the East Coast a value of 1. The scatterplot of the dummy variable for region and the range variable are shown in Figure 10.11. If we now ask the computer software to do a regression analysis for these variables, it produces the regression line shown in the figure. The computer software does not know that the dummy

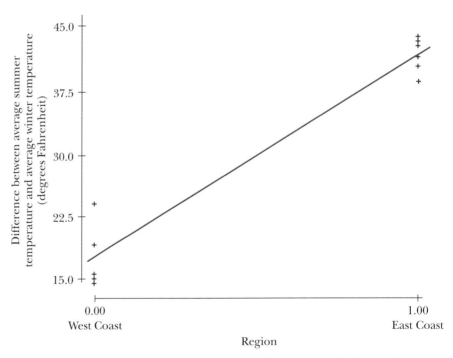

Figure 10.11 Categorical independent variable and metric dependent variable *(Source: Data Desk SMSA file.)*

variable is a categorical variable. The software simply computes all the quantities we need for a regression analysis.

In the scatterplot, the temperature ranges in six cities on the East Coast vary from about 39 to 44 degrees, while on the West Coast temperature ranges in five cities vary from about 15 to 24 degrees. Since the ranges on the West Coast are smaller than those on the East Coast, the regression line has a positive slope. If we had assigned 0 to the East Coast cities and 1 to the West Coast cities, the points in the scatterplot would have been reversed and the regression line would have had a negative slope.

The regression equation for these data is

$$\text{range} = 17.6 + 24.7 \text{ region}$$

The intercept is simply the mean of the ranges of temperature for the West Coast cities, since these cities have the value 0 on the dummy variable region; that mean equals 17.6. When region equals 1 and we substitute that value into the equation for the regression line, we get a predicted range of 42.3, the mean of the temperature ranges for the East Coast cities. The slope of 24.7 is the difference between the means of the East Coast and the West Coast cities.

The fact that the slope of the regression line is different from zero indicates a relationship between region and range of temperature for these data. The strength of the relationship is measured by $r = 0.98$. To test the null hypothesis that there is no relationship between the two variables in the population of all cities, we change the correlation coefficient r or the regression coefficient b to a t-value. For these data, $t = 13.30$ on 9 degrees of freedom. Using statistical software, we find a significant t-value and a p-value less than 0.0001. Statistical Table 2 for the t-distribution goes only to $t = 4.30$ on 9 degrees of freedom, and that corresponds to a p-value of 0.001. Since the observed t-value is much larger than 4.30, the p-value for the data must be much less than 0.001.

The small p-value is overwhelming evidence for the fact that the range in temperature from summer to winter is larger for East Coast cities than for West Coast cities. Less than one in 10,000 samples would produce this or more extreme data if there were no difference between the cities on the two coasts. Thus, there is evidence for the fact that the seasons vary more on the East Coast than on the West Coast. If we had studied this problem using the methods developed in Chapters 7 and 11 for the difference between two means, the results would have been identical, with the same value of t and the same p-value.

Categorical dependent variable with two values and metric independent variable

This problem is the reverse of the temperature range/region problem. In this example, the dependent variable (national origin of cars) is a categorical variable with two categories, and the independent variable (drive ratio) is a metric variable. When we use a dummy variable for the dependent variable, a scatterplot of the data looks something like the plot in Figure 10.12. The dependent variable, national origin, has the two values foreign and domestic.

In this case the scatterplot of the data is so nonlinear that we cannot fit a straight line through the points. The reason is that all the points in the scatterplot are located along two horizontal lines, one line for

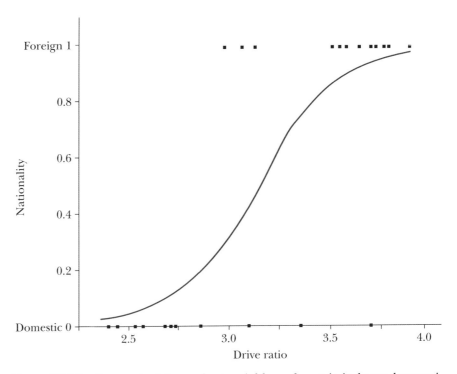

Figure 10.12 Categorical dependent variable and metric independent variable

$y = 0$ and the other for $y = 1$. Instead of a line, we fit an S-shaped curve to the data, as shown in the graph. This type of analysis is called *logistic regression*. The curve starts at the points in the upper right part of the graph and follows these points for a while, then rapidly scoots down to pick up the points in the lower left part of the graph. The curve has much smaller residuals than would any straight line through the points.

The dummy variable represents the national origin of a car, with 1 for foreign and 0 for domestic. Any value between 0 and 1 is interpreted as the probability that a car is foreign. Thus, if we know that a car has a drive ratio of 3.5, we locate that value on the horizontal axis and go up to the curve and over to the vertical axis to find the value 0.87. This number is the estimated probability that a car with a drive ratio of 3.5 is a foreign car.

10.7 QUESTION 4. CAUSAL RELATIONSHIP?

Statistical methods to help us answer this question exist, and we consider those methods in Chapter 13. The methods are based on bringing in other variables and finding out if they can explain the observed relationship between the original two variables. At this point, intuitively it may make sense that fat influences calories, but we cannot tell statistically. Even though the relationship may not be causal, we can still use the results of the analysis and predict calories if we knew the fat contents in a new variable.

10.8 SUMMARY

Correlation and regression are the two complementary methods for analyzing the relationship between two metric variables. Correlation describes the extent to which the two variables are related. Regression analysis describes the way in which a dependent variable is affected by one or more independent variables. Simple regression refers to regression analysis with one independent variable.

10.1 Question 1. Relationship between the variables?

For a visual impression of whether there is a relationship between the variables, the data can be displayed in a scatterplot. The scatterplot is used to see if the data are suitable for correlation and regression analyses. Scatterplots are made with the independent variable running horizontally and the dependent variable running vertically. If the data points seem to scatter along a line from the lower left to the upper right in the graph, there is a positive relationship between these two variables. In a scatterplot with a negative relationship between the two variables, the points start in the upper left corner and fall along a line to the lower right corner. If the points in the scatterplot seem to be randomly distributed, there is very little or no relationship between the variables.

10.2 Question 2a. Strength of the relationship?

If there is a relationship between the x- and the y-variable, we can find out in what way the y values differ as the x values increase or decrease. The statistic that measures how strong the relationship is between two metric variables is the correlation coefficient r. The value of r always lies somewhere between -1 and $+1$. When the value of r is close to $+1$ or -1, there is a very strong relationship between the two variables. When r equals 0, there is no relationship between the two variables. When r is positive, the values of the variables increase and decrease together; when r is negative, the values of one variable increase as the other values decrease.

10.3 Question 2b. Form of the relationship?

In regression analysis, a regression line represents the relationship between the two variables. This line is drawn through the middle of the data points in a scatterplot. The slope of the line measures how steep the line is. The larger the slope of the line, the more difference there is in the dependent variable for each unit difference in the independent variable. The intercept of the regression line is the point on the vertical axis, when the independent variable equals 0, where the regression line cuts through that axis. The regression line is determined by the slope and intercept of the line: an estimated value of the dependent variable *equals* the intercept value *plus* the value of the slope *times* the value of the independent variable. The regression line is found by the least squares principle.

A regression equation can be used to predict the value of the dependent variable from a value of the independent variable. The predicted values of the dependent variable are estimates of the actual values.

The combined effect of all independent variables other than the selected independent variable is known as the residual variable. The total sum of squares measures the effect of all variables on the dependent variable. It is found as the sum of (observation − mean)² for all the observations of the dependent variable. The residual sum of squares measures the effect of all variables other than the independent variable on the dependent variable. It is found as the sum of (observation − estimated value)² for all the observations of the dependent variable. The regression sum of squares measures the effect of the independent variable on the dependent variable. It is found as the sum of (estimated value − mean)² for all values of the dependent variable. The total sum of squares equals the regression sum of squares plus the residual sum of squares. The proportion of the effect on the variation in the values of the dependent variable that is due to the independent variable is always equal to the square of the correlation coefficient between the independent variable and the dependent variable.

To learn about the relationship between two variables, it is important to know the results of both the correlation and the regression analysis. A correlation coefficient can be high, but the regression line almost flat and therefore usually uninteresting. A regression line may be steep, but the points scattered far from the line making the correlation low.

To predicting changes in a variable, the regression coefficient b must be computed on data measured over time. It is harder to predict change if data observed over time are not available.

10.4 Question 3. Relationship in the population?

To find if there is a relationship between the two variables in the population from which the sample data came, we generalize from the sample data by constructing a confidence interval for the population regression coefficient β or by testing a null hypothesis of no relationship. To find the p-value for the data, we change the regression coefficient b or the correlation coefficient r to a value of the t variable. It is also possible to use the data to compute a value of the F-variable and thereby find the p-value.

10.5 A word of warning: What you measure is what you get

We must be cautious in interpreting statistical significance and the size of correlation coefficients between two variables. Statistical significance can more easily be obtained from large samples, and a large value of r can more easily be obtained if the values of the independent variable are spread out.

10.6 How to be smart using dummy variables

Simple correlation and regression analyses can be used with a metric variable and a categorical variable that has two categories. A dummy variable represents the values of the categorical variable, and most often the values 0 and 1 are used for the dummy variable. When the dependent variable is a categorical variable, we can fit an S-shaped curve to the data. This method of analysis is called logistic regression.

10.7 Question 4. Causal relationship?

From the data on only two variables we cannot tell if the relationship is causal, but we can still predict values of the dependent variable if we know the values of the independent variable.

ADDITIONAL READINGS

Draper, N. R., and H. Smith. *Applied Regression Analysis,* 2nd ed. New York: John Wiley & Sons, 1981. This well-known book has an introductory chapter on simple regression.

Kleinbaum, David G., Lawrence L. Kupper, and Keith E. Muller. *Applied Regression Analysis and Other Multivariable Methods,* 2nd ed. Boston: PWS-KENT, 1988. More applied than Draper and Smith with a longer introduction to simple regression.

Lewis-Beck, Michael S. *Applied Regression* (Sage University Paper Series on Quantitative Applications in the Social Sciences, series no. 07-022). Beverly Hills, CA: Sage, 1980. Short introduction to regression analysis.

Tufte, Edward R. *Data Analysis for Politics and Policy.* Englewood Cliffs, NJ: Prentice-Hall, 1974. Chapter 3 includes a good discussion of issues that arise in simple regression analysis.

The data for n observations on two variables x and y can be expressed using these symbols:

x	y
x_1	y_1
x_2	y_2
x_3	y_3
.	.
.	.
.	.
x_i	y_i
.	.
.	.
.	.
x_n	y_n

CORRELATION COEFFICIENT AND REGRESSION COEFFICIENT (SLOPE)

The correlation coefficient r is found by the expression

$$r = \frac{\Sigma(x_i - \bar{x})(y_i - \bar{y})}{\sqrt{\Sigma(x_i - \bar{x})^2 \Sigma(y_i - \bar{y})^2}} = \frac{n\Sigma x_i y_i - \Sigma x_i \Sigma y_i}{\sqrt{[n\Sigma x_i^2 - (\Sigma x_i)^2][n\Sigma x_i^2 - (\Sigma x_i)^2]}} \quad (10.1)$$

The slope b of the regression line is found by the expression

$$b = \frac{\Sigma(x_i - \bar{x})(y_i - \bar{y})}{\Sigma(x_i - \bar{x})^2} = \frac{n\Sigma x_i y_i - \Sigma x_i \Sigma y_i}{n\Sigma x_i^2 - (\Sigma x_i)^2} \quad (10.2)$$

The left-hand formulas for r and b are sometimes used to define r and b. They are not easy to use for calculations because the mean must be subtracted from each of the observations. The right-hand formulas are quicker to use when doing the calculations without a computer.

Notice that the numerators for r and b are the same, and the denominators are almost the same. It follows from the formulas for the correlation coefficient r and the regression coefficient b that the two are related according to the expression

$$b = r\frac{s_y}{s_x} \qquad (10.3)$$

where the s's are the standard deviations of the two variables x and y.

INTERCEPT

The intercept a for the regression line is found by the expression

$$a = \bar{y} - b\bar{x} \qquad (10.4)$$

As an example of a regression analysis, consider the following data on x and y. According to Formulas 10.1 and 10.2 for the correlation coefficient and the regression coefficient, we need to multiply x and y and add the products, to square the x-values and add the squares, to square the y-values and add the squares, as well as to add x and add y. These computations are best set up in a table:

	x	y	x^2	xy	y^2
	1	3	1	3	9
	2	2	4	4	4
	3	5	9	15	25
	4	6	16	24	36
Total	10	16	30	46	74

With these numbers,

$$b = \frac{(4)(46) - (10)(16)}{(4)(30) - 10^2} = \frac{184 - 160}{120 - 100} = \frac{24}{20} = 1.20$$

$$a = \frac{16}{4} - 1.20\left(\frac{10}{4}\right) = 4.0 - 1.20(2.5) = 4.0 - 3.0 = 1.0$$

$$r = \frac{(4)(46) - (10)(16)}{\sqrt{[(4)(30) - 10^2][(4)(74) - 16^2]}} = \frac{184 - 160}{\sqrt{[120 - 100][296 - 256]}}$$

$$= \frac{24}{\sqrt{[20][40]}} = \frac{24}{28.28} = 0.85$$

SUMS OF SQUARES

The various sums of squares are found as follows:

$$\text{total sum of squares (TSS)} = \Sigma(y_i - \bar{y})^2$$

$$\text{regression sum of squares (RegrSS)} = \Sigma(a + bx_i - \bar{y})^2 \qquad (10.5)$$

$$\text{residual sum of squares (RSS)} = \Sigma(y_i - a - bx_i)^2$$

$$r^2 = \frac{\text{RegrSS}}{\text{TSS}} \qquad (10.6)$$

CONFIDENCE INTERVAL FOR THE POPULATION REGRESSION COEFFICIENT β

The interval is

$$b - t*s_b \quad \text{to} \quad b + t*s_b \qquad (10.7)$$

Here, b is the observed regression coefficient, $t*$ is the $(1 - \alpha/2)$ value of t with $n - 2$ degrees of freedom from the t-table, and s_b is the standard error of b. The standard error of b is most often found by having statistical software, but it can be computed from the expression

$$s_b = \sqrt{\frac{\text{RSS}/(n-2)}{\Sigma(x_i - \bar{x})^2}}$$

HYPOTHESIS TESTING

From the value of the regression coefficient b,

$$t = \frac{b}{s_b} = \frac{15.3}{1.85} = 8.24 \qquad (10.8)$$

for our example with fat content and calories. The same value of t is found from the correlation coefficient r according to the formula

$$t = \frac{r}{\sqrt{\frac{1-r^2}{n-2}}} = \frac{0.910}{\sqrt{\frac{1-0.829}{16-2}}} = 8.24 \qquad (10.9)$$

EXERCISES

REVIEW (EXERCISES 10.1–10.31)

10.1 How do you define the correlation between two variables?

10.2 The word regression usually refers to a movement backward. What movement does regression refer to in statistical analysis?

10.3 What do we mean by the term "simple" regression analysis?

10.4 a. What did Francis Galton find in his study of short and tall parents and children?

b. What effect does this study illustrate?

10.5 a. Give an example of two variables for which you would use correlation and regression analysis.

b. What two statistics are first computed in regression?

10.6 What is the purpose of creating a scatterplot?

10.7 a. In a scatterplot, which variable is measured on the x-axis and which on the y-axis?

b. You want to plot data on the number of TV antismoking commercials watched and rate of quitting smoking among high school TV viewers. Which variable would you put on the x-axis?

c. What is x-axis variable called?

d. What does a scatterplot look like that shows a positive relationship between two variables?

10.8 Suppose you made a scatterplot for the rates of illiteracy within city blocks and the rates of drug-related crime.

a. How would you expect the scatterplot to look?

b. In which direction would the points seem to scatter, upward or downward from the x-axis?

c. What would this directional flow of data points mean to the statistically acute observer?

10.9 A look at a scatterplot indicates that there are 3 points (out of 100) that are far away from the main clusters of points. What impact, if any, would there be on the correlation coefficient if these three points were removed?

10.10 a. What are some names for the correlation coefficient r?

b. What are the largest and smallest possible values of r?

c. Which is stronger, a correlation of $+1.00$ or -1.00? Explain your answer.

d. How is the correlation coefficient r different from phi or V?

10.11 In Table 10.2, the correlation coefficients on the diagonal of the table are all equal to 1.00. Why is this?

10.12 a. How strong is the relationship between two variables when the correlation coefficient r falls between 0.75 and 1.00?

b. How strong is the relationship between two variables when the correlation coefficient r falls between -0.70 to -0.30?

c. How strong is the relationship between two variables when the correlation coefficient r falls between 0 and 0.25?

10.13 If you could design a better society than the present one, what would be the desired correlation coefficient between the following variables? Give a numerical value as well as a verbal description.

a. Level of income and level of taxes

b. Number of years of education and amount of illiteracy

c. Height of a person and size of pay check

10.14 a. Name the line that runs through the middle of the points of a scatterplot.

b. What does the line convey about the two variables?

c. If the line has a positive slope, from where to where on the scatterplot does the line run?

d. If the line has a negative slope, from where to where on the scatterplot does the line run?

e. What does a line with a negative slope convey about the correlation between the two variables?

f. What does the steepness of the line indicate?

10.15 The y-intercept of the regression line is the spot where the line cuts through the y-axis when x equals 0.

a. What is the y-intercept for the fat/calorie examples?

b. Explain this value so that a dieter might understand it.

10.16 The regression line found by the least squares method is the line that is closest to all the points in the scatterplot. What does the term *least squares* refer to?

10.17 The regression line and the regression equation can be used to predict a value of the dependent *y*-variable from the independent *x*-variable. What symbol do we use to designate a predicted value of the dependent variable?

10.18 What are some of the variables that might go into the residual variable in the study of the effect of fat content on calories?

10.19 The total sum of squares can be seen as composed of two parts.
   a. What are the two parts?
   b. How are the two parts calculated?

10.20 Write down the regression equation for the fat/calorie problem. Describe each part of the equation and the use of the equation for understanding the relationship between fat content and calories.

10.21 If all the points in a scatterplot were found directly on the regression line, what conclusions could you draw about
   a. the effect of the independent variable on the dependent variable?
   b. the effect of the residual variable on the dependent variable?
   c. the correlation coefficient between the *x*- and the *y*-variable?

10.22 Give an example of two variables where you would want to find out whether or not the intercept of the regression line equals zero.

10.23 a. Compare the merits of regression and correlation analyses.
   b. What does each one do that is special in an analysis of two metric variables?

10.24 What is the connection between r^2 and the proportion of the variation of the dependent variable due to the independent variable?

10.25 "The more the merrier" can mean that the more people there are at a party, the merrier the party is. Is the relationship between the size of the party and the merriment positive or negative?

10.26 a. What does the Greek letter β stand for in regression?
   b. When is it used?

10.27 Name two methods that can be used in regression analysis to decide whether the relationship between two variables in a sample is statistically significant.

10.28 What two theoretical statistical variables are useful in finding the *p*-value that indicates whether an observed sample relationship is statistically significant?

10.29 a. What effect does extending the values of the *x*-variable in an experiment have on the size of the correlation coefficient?

b. Does this have any effect on the *p*-value for the correlation coefficient?

10.30 a. What is a dummy variable?

b. When is a dummy variable used?

c. Give an example of how you might use a dummy variable in studying the differences between golf scores on windy and calm days.

10.31 a. When do we use the method known as logistic regression?

b. Create a problem using a dummy variable for which this type of approach would be helpful.

INTERPRETATION (EXERCISES 10.32–10.48)

10.32 You have data on crime rates for each of the 48 contiguous states, and you would like to know whether different types of crimes are related. When you do a regression analysis of larceny rates as dependent variable on robbery rates as independent variable, you find

larceny = 2,682 + 1.49 robbery   ($t = 2.05$ with 46 d.f., $p = 0.023$)

(*Source: Bureau of the Census*, Statistical Abstracts of the United States: 1995, *115th ed., Washington, D.C., 1995.*)

a. What conclusion can you draw about the relationship between larceny and robbery from this analysis?

b. What are some of the conclusions you cannot draw about the relationship from this analysis?

10.33 Another way to consider the draft data in Exercise 9.29 is to let the independent variable *x* be equal to 1 for January, 2 for February, up to 12 for December. Let the dependent variable y be the mean of the draft numbers for each month.

a. If you make a scatterplot of the data points (x, y) for the 12 months and perform a regression analysis, what values would you expect to get for the intercept, the regression coefficient for x, and the correlation coefficient for a lottery that was truly random?

b. What do you conclude about the draft lottery when the analysis gives the following results?

$$\text{mean draft number} = 230 - 7.1 \text{ month} \qquad (r = -0.87)$$

c. The value of t for the correlation coefficient equals -5.50. Could this value have occurred by chance alone?

10.34 Sometimes it seems that the better things taste, the worse they are for us. Table 10.6 shows some data on different types of chocolate frozen yogurts. The first column of numbers shows the percentage of the calories in the yogurts that come from the fat in the yogurts, and the second shows the rating of the flavor of the yogurts determined by a panel of trained tasters on a scale from 0 to 100. A regression analysis

Table 10.6 Data for Exercise 10.34

Brand	Percent calories from fat	Flavor rating
Breyers	24	85
Honey Hill Farms	33	85
Elan	21	80
Crowley Silver Premium	20	78
Edy's/Dreyer Inspirations	25	74
Häagen-Dazs	21	71
Kemps	20	65
Lucerne	23	63
Yoplait Soft	20	61
Albertsons	12	51

Source: "Low-fat frozen desserts: Better for you than ice cream?" Consumer Reports, vol. 57, no. 8 (August 1992), pp. 483–487.

of flavor as the dependent variables and the percentage of calories from fat as the independent variable gives the following result:

$$\text{flavor} = 37 + 1.6 \text{ (percentage of calories from fat)}$$

$$(r = 0.74, t = 3.11, 8 \text{ d.f.}, p = 0.0073)$$

a. Make a scatterplot of these data.

b. Draw in the regression line.

c. What do you conclude about the relationship between the two variables?

d. Why might we prefer to eat the desserts that lie above the regression line rather than the ones that lie below the line?

10.35 Is there any difference in cost between chocolate and vanilla frozen desserts? In data collected by *Consumer Reports,* the mean cost per serving of the chocolate desserts studied equals 29.4 cents, and for the vanilla desserts the mean cost equals 30.4 cents. *(Source: "Low-fat frozen desserts: Better for you than ice cream?" Consumer Reports, vol. 57, no. 8 (August 1992), pp. 483–487.)* To see if the difference between the two means is statistically significant, you introduce a dummy variable for the type of dessert and do a regression analysis. You give all the chocolate desserts the value of 1 on the dummy variable type and all the vanilla desserts the value of 0 on the same dummy variable. A regression analysis with cost as the dependent variable and type as the independent variable gives the result

$$\text{cost} = 29.4 + 1.0 \text{ type}$$

a. How could you have found the intercept and slope of this analysis directly from the two means?

b. The slope $b = 1.0$, and this value of b translates $t = 0.18$ on 42 d.f. Is there a statistically significant effect of type on cost?

10.36 Many factors affect poverty, and education may be one of them. One way to study the relationship between those two variables would be to collect data from individuals on the amount of education they have and the size of their income. In this exercise, the data are not on individuals but on states. From the Census Bureau we find the percentage of the adult population in each state that has a ninth-grade education or less and the percentage of the population with income below the official poverty level.

Regressing the percentage below the poverty level on the percentage with ninth-grade education or less for the 50 states and the District of Columbia gives

$$\text{percent poor} = 4.6 + 0.8 \text{ percent low education}$$

$(r = 0.70, t = 6.72 \text{ on } 49 \text{ d.f.}, p < 0.0001)$

a. What does the value 0.8 of the regression coefficient tell us?

b. What is the probability of a correlation coefficient of 0.70 or more by chance alone in a sample of 51 observations from a population where the correlation coefficient equals 0.00?

c. The scatterplot of these data is shown in Figure 10.13. Why are some of the points above the regression line and some below the line?

d. The states with the largest negative residuals are New Jersey, Connecticut, Hawaii, Rhode Island, and Virginia. What might

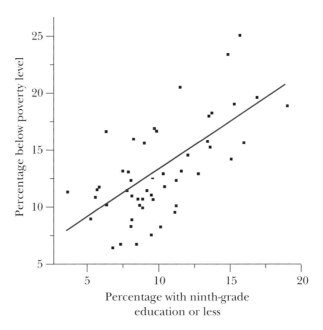

Figure 10.13 Scatterplot of people below poverty level and people with ninth-grade education or less (Exercise 10.36) *(Source: 1990 U.S. Census data reported in* The Chronicle of Higher Education, *vol. 34, no. 1 (August 26, 1992), p. 4.)*

some of the reasons be that these states lie below the regression line?

e. The states with the largest positive residuals are Mississippi, Louisiana, New Mexico, Montana, and Utah. What might some of the reasons be that these states lie above the regression line?

10.37 When we regress the percentage of the population below the poverty level on the percentage of the population with at least a college degree, using the state as the unit,

$$\text{percent poor} = 26.9 - 0.7 \text{ percent college or more}$$
$$(r = -0.62, t = -5.54 \text{ on 48 d.f.}, p < 0.0001)$$

a. Why is it not surprising that the regression coefficient for the college variable is negative?

b. Could this relationship have occurred by chance alone?

c. Is the relationship between the two variables causal?

d. Why does it make sense to leave out Washington, D.C., in analyzing these two variables?

10.38 Many people who finish high school do not go on to college. Here we look at the percentage of people 18 to 24 years old who were enrolled in college each year from 1980 to 1990. *(Source: Data from annual Census Bureau surveys of 60,000 households as reported in* The Chronicle of Higher Education, *vol. XXXIX, no. 1 (August 26, 1992), p. 12.)* We look separately at blacks, Hispanics, and whites. To simplify the analysis we code 1980 as the value 0, 1981 as 1, and so on, up to 1990 as 10. When we do a regression analysis with the yearly percentages enrolled in college as the dependent variable and years from 0 to 10 as the independent variable, we get these three regression lines:

Blacks: percent = 26.5 + 0.42 year $r = 0.71$

Hispanics: percent = 29.9 − 0.08 year $r = -0.22$

Whites: percent = 31.1 + 0.75 year $r = 0.97$

a. Draw the three regression lines on one graph.

b. What is the meaning of the coefficient 0.42 for blacks?

c. The college enrollment percentages of which of the three groups increased fastest from year to year in this period?

d. Describe the three lines and how they differ from each other.

e. Why do you think the line for Hispanics has a negative slope and shows a declining percentage of 18- to 24-year-old Hispanics attending college across the decade of the 1980s?

10.39 In a regression problem, Sam discovers that for every unit of change in the independent variable X, on the average there is a change of 10.2 for the dependent variable $Y1$. Anne discovers that for every unit change in the independent variable, on the average there is a change of 4.2 for another dependent variable $Y2$.

a. In a figure showing these data, which regression line slope is steeper?

b. Do all of the units of the dependent variable increase by 10.2 with a unit change in the independent variable?

c. Do you think there is a relationship between the two dependent variables $Y1$ and $Y2$?

10.40 A well-known statistician collected and analyzed data on the height (in inches) of fathers and their children. *(Source: Class data, introductory statistics course, Swarthmore College, 1992.)* She found the following results:

$$\text{offspring height} = 1.52 + 0.75 \text{ father height}$$
$$(r = 0.59, r^2 = 0.34, t = 2.90, p = 0.0051, n = 18)$$

a. What do the results of the statistical analysis tell you about the relationship between the two variables?

b. The 95% confidence interval for the population regression coefficient is 0.20 to 1.30. What is the interpretation of this interval?

10.41 The effects of chemicals on animals is often studied by giving increasingly larger doses to groups of animals and seeing how many animals respond in each group. The data in Table 10.7 show how many test animals responded to different dose rates in a study of the effect of dieldrin (a white crystalline insecticide $C_{12}H_8C_{16}O$).

a. How does increasing the dose seem to affect the proportion of mice responding to dieldrin?

Table 10.7 Data for Exercise 10.41

Dose rate (ppm)	Proportion responding	Number of animals
0.00	0.11	156
1.25	0.18	60
2.50	0.43	58
5.00	0.73	60

Source: A. I. T. Walker, E. Thorpe, and D. E. Stevenson, "The toxicology of dieldrin (HEOD): I. Long-term oral toxicity studies in mice," *Food and Cosmetics Toxicology*, vol. 11 (1972), pp. 415–432.

b. In a regression analysis of the proportion responding to the dose rate (without taking into account the number of animals in each group),

$$\text{proportion responding} = 0.08 + 0.13 \text{ dose rate}$$

$$(t = 9.24, p = 0.006)$$

What do these numbers tell you about the relationship between the two variables?

c. What do these numbers *not* tell you about the relationship between the two variables?

10.42 Figure 10.14 (page 452) shows a scatterplot of the number of incidences of malignant melanoma cases reported in Connecticut per 10,000 inhabitants from 1936 to 1972. Melanomas are skin tumors containing dark pigment and may be cancerous. A regression analysis of the data results in the regression line shown on the graph. The equation for this line is

$$\text{incidences} = -2{,}127 + 1.1 \text{ year} \qquad (r = 0.963)$$

(The reason for the large negative value of the intercept is that the independent variable year has such large values, ranging from 1936 to 1972, and the line has to be extended a long way down to the left before we find the actual *y*-intercept.) The regression analysis also results in Table 10.8 (page 452).

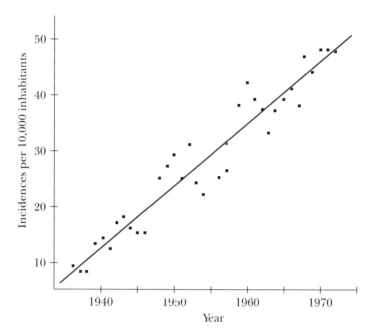

Figure 10.14 Number of incidences of melanoma in Connecticut 1936–1972 (Exercise 10.42, page 451) *Source: A. Houghton. E. W. Muenster, and M. V. Viola, "Increased incidence of malignant melanoma after peaks of sunspot activity," The Lancet, April 8, 1978, pp. 759–760, as reported in D. F. Andrews and A. M. Herzberg, Data: A Collection of Problems from Many Fields for the Student and Research Worker, New York: Springer-Verlag, 1985, p. 201.*

Table 10.8 Data for Exercise 10.42 (page 451)

Source	Sum of squares	Degrees of freedom	Mean square	F-ratio	p-value
Year	5131	1	5131.0	453	0.0000
Residual	396	35	11.3		
Total	5527	36			

Source: A. Houghton, E. W. Meunster, and M. V. Viola, "Increased incidence of malignant melanoma after peaks of sunspot activity," The Lancet, April 8, 1978. pp. 759–760, as reported in D. F. Andrews and A. M. Herzberg, Data: A Collection of Problems from Many Fields for the Student and Research Worker, New York: Springer-Verlag, 1985, p. 201.

a. Describe the regularities in the scatterplot and what they suggest about possible further analyses of these data.

b. What does the equation for the regression line tell you about the relationship between the two variables?

c. What do the numbers in the table tell you about the relationship between the two variables?

10.43 The British runner Roger Bannister made history in 1954 when he became the first person to run 1 mile in less than 4 minutes at a track meet. Since that time, new world records in the mile race for men have been set more than a dozen times up through 1993, when Noureddine Morceli from Algeria ran the mile about 15 seconds faster than Bannister, at a time of 3:44.39. A scatterplot of the record time as the dependent variable and the year the record was set as the independent variable shows a remarkably straight-lined relationship. The correlation between the two variables $r = -0.968$. To make the results of the regression analysis easier to work with, we first subtract 3 minutes from all the records and 1900 from all the years. Morceli's numbers become 44.39 for the time and 93 for the year. With these changes, the equation for the regression line is

$$\text{record} = 70.07 - 0.3468 \text{ year}$$

a. Based on this regression analysis, how much would you expect the record to change in a 10-year period?

b. How much would you expect the record to change in a 39-year period compared to the actual change of 15 seconds?

c. How much faster or slower did Morceli run compared to what you predict his time to be?

d. What do you predict the world record in the mile will be in the year 2000?

10.44 The annual college guide from *U.S. News and World Report* has a wealth of data on a large number of colleges and universities. The meaning of the rankings is debatable, but schools do like to be ranked high. Look at the dozen top-ranked national universities (excluding California Institute of Technology) and how much they spend on each student in a year in Table 10.9 (page 454). The spending figure comes from dividing the annual budget by the number of students, and it is higher than tuition and fees because the schools subsidize students from other incomes.

Table 10.9 Data for Exercise 10.44 (page 455)

University	Rank	Spending per student per year (thousands of dollars)
Harvard	1	36
Princeton	2	28
Yale	3	39
MIT	4	33
Stanford	5	36
Duke	6	26
Dartmouth	7	30
Chicago	8	37
Cornell	9	21
Columbia	10	31
Brown	11	20
Northwestern	12	25

Source: America's Best Colleges 1994 College Guide, U.S. News and World Report, *pp.* 20–21.

In a scatterplot of these data, some schools lie above the regression line and some lie below it. A regression analysis of rank on spending gives a line with the equation

$$\text{rank} = 17 - 0.34 \text{ spending} \qquad r = -0.60$$

The relationship is negative because highest-ranked Harvard has the lowest numerical rank and is therefore located at the bottom of the scatterplot, and Northwestern with rank 12 is located at the top.

    a. What is the implication for a school if its data point is located below the regression line?

    b. Do the schools located below the regression line have anything in common?

    c. What is the average difference in rank between two schools that differ by $3,000 in spending per student per year?

10.45 In a 1907 study of 16 steamships ranging in tonnage from 192 to 3,246 tons and in crew size from 5 to 32 men, the following results occur when crew size is regressed on tonnage:

$$\text{crew size} = 9.5 + 0.00062 \text{ tonnage}$$

$$(r = 0.87, t = 6.79 \text{ on } 14 \text{ d.f.}, p < 0.0001)$$

(Source: R. Floud, An Introduction to Quantitative Methods for Historians, *London: Methuen, 1973, Table 4.1.)*

   a. What do the numbers tell you about the relationship between the two variables?

   b. On the average, how large is the difference in crew size for two ships that differ in tonnage by 1,000 tons?

   c. What is the estimated number of crew members for the smallest ship, and what is the estimated number of crew members for the largest ship?

10.46 *The New York Times* once commissioned a laboratory to analyze a dozen slices of pizza from different stores in New York City for calories and fat content. The slices ranged in weight from 5.25 to 10.5 ounces, in calories from 366 to 613, and in fat from 11 to 25 grams. *(Source:* The New York Times, *September 14, 1995; p. C1.)* A regression analysis of calories on grams of fat yields the following results:

$$\text{calories} = 280 + 13.4 \text{ grams of fat} \qquad r = 0.78$$

$$(t = 4.01 \text{ on } 10 \text{ d.f.}, p = 0.0012)$$

   a. What do these numbers tell you about the relationship between the two variables?

   b. Suppose you had divided the calories and fat for each slice by the weight of the slice. How do you think that would effect the analysis?

10.47 Mean academic salaries vary across disciplines from agriculture at $36,900 to library science at $23,600 (1984 figures from a study of 25 fields). *(Source: M. Bellas, and B. F. Reskin, "On comparable worth,"* Academe, *vol. 80 (1989), no. 5, pp. 83–85.)* Women faculty members are unevenly represented in the fields, from 94% in nursing to about 5% in engineering. In a regression analysis of mean salary as the dependent variable and percentage of women as the independent variable,

$$\text{mean salary} = 34{,}300 - 120 \text{ percent women} \qquad r = -0.829$$

$$(t = -7.10 \text{ on } 23 \text{ d.f.}, p < 0.0001)$$

a. Make a graph of the regression line.

b. What do the results of the analysis tell you about the relationship between these two variables?

c. What other variables might explain this relationship?

d. The scatterplot of the data shows that mathematics, sociology and anthropology, music, journalism, English, foreign languages, and drama lie noticeably below the regression line. Social work, nursing, life sciences, agriculture, and engineering lie noticeably above the regression line. What might explain these patterns?

10.48 At the end of Chapter 4 are some data on death rates in selected countries for the liver ailment known as cirrhosis. For the same countries we also have yearly consumption of pure alcohol in quarts per capita, since excessive alcohol is supposed to be harmful to the liver. The yearly alcohol consumption in the countries ranges from a high of 13.3 quarts per capita in Luxembourg to a low of 1.0 quart in Israel. The United States is in the middle at 7.2 quarts. A regression analysis of these data gives the equation for the regression line

$$\text{cirrhosis deaths} = 2.1 + 2.09 \text{ quarts of alcohol} \qquad r = 0.45$$

$$(t = 2.67 \text{ on } 28 \text{ d.f.}, p = 0.006)$$

What do the results of the regression analysis tell you about alcohol consumption and cirrhosis deaths?

ANALYSIS (EXERCISES 10.49–10.74)

10.49 Go to the Springer Web site (http://www.springer-ny.com/supplements/iversen/) to find files relating to this book. Open the data file called Baseball Team Scores. Column 1 shows the number of games each team won during the season and column 3 the mean number of runs scored per game.

a. How strong is the relationship between the two variables?

b. Is the relationship statistically significant or could it have occurred by chance alone?

c. Do a regression analysis to see how the number of runs scored (independent variable) affects the number of wins (dependent variable).

d. If a team could score one more run per game, how many more games would the team expect to win?

10.50 The road distance from one city to another is usually longer than the direct distance. In a sample of English cities the regression equation for these distances was

$$\text{road distance} = 7 + 1.17 \text{ direct distance}$$

(Source: Neville Hunt, "A tale of six cities," *Teaching Statistics*, vol. 16, (1994) no. 1, pp. 5–8.) What does the equation tell you about the direct distance versus the road distance in England?

10.51 For a random sample of American cities, road distance/direct distance data are shown in Table 10.10.

a. Find the regression equation for the data.

b. Why might you expect the intercept of the regression line to be equal to 0?

c. Can you reject the null hypothesis that the population intercept equals 0?

d. Can you explain why the patterns of roads and distances in the United States make the two distances almost equal for some pairs of cities and very different for other pairs of cities?

Table 10.10 Data for Exercise 10.51

	Cheyenne	*Fargo*	*Los Angeles*	*Oklahoma City*	*St. Louis*
Atlanta	1,482, 1,235	1,394, 1,105	2,121, 1,940	833, 765	565, 476
Cheyenne	—	780, 553	1,124, 894	694, 560	942, 790
Fargo		—	1,808, 1,430	870, 782	850, 647
Los Angeles			—	1,339, 1,205	1,836, 1,590
Oklahoma City				—	500, 465

Source: Road Atlas, Boston: Rand McNally, 1991.

10.52 We hear much about the increasing number of divorces in the United States. One way to examine this phenomenon is to compare the number of divorces with the number of marriages, since people must marry before they can divorce. The following data show the number of marriages and divorces in thousands for 1890 and every fifth year up to 1980. The year variable has been recoded to 1 for 1890, 2 for 1895, and up to 19 for 1980 to make it easier to enter the data in a computer.

Year	1	2	3	4	5	6	7	8	9	10
Marriages	570	620	709	842	948	1,008	1,274	1,188	1,127	1,327
Divorces	33	40	56	68	83	104	170	175	196	218

Year	11	12	13	14	15	16	17	18	19
Marriages	1,596	1,613	1,667	1,531	1,523	1,800	2,159	2,153	2,413
Divorces	264	485	385	377	393	479	708	1,036	1,182

Source: National Center for Health Statistics, Public Health Service, in The World Almanac *1986, p. 779.*

a. Show the data on marriages and divorces in a scatterplot.

b. Comment on the shape of the scatterplot. Would it make sense to do correlation and regression analyses on these data?

For the more mathematically inclined, take the logarithm of each observation. Construct a new column in the computer file consisting of the logarithms of the marriage data and another new column consisting of the logarithms of the divorce data.

c. Show the two logarithmic variables in a scatterplot.

d. Comment on the shape of this scatterplot.

e. Divide the divorces by the marriages for each year and plot this ratio against the time variable.

f. What does this scatterplot tell us?

10.53 On the Springer Web site (http://www.springer-ny.com/supplements/iversen/), the data file called Baseball Team Scores contains data on all 28 baseball teams for the 1996 season. The columns contain data on the following variables.

1. Number of games the team won
2. Team earned-run average (measure of pitching)
3. Mean number of runs scored per game
4. Total number of stolen bases
5. Total number of home runs
6. Team batting average

What matters in the end is how many games a team wins during a season, so make the variable in column 1 the dependent variable.

    a. Correlate variable 1 with each of the other variables and find the importance of the other variables in determining the number of wins.

    b. Regress variable 1 as dependent variable with each of the other variables as independent variables and find how much an increase in one unit affects the number of winning games.

10.54 Following are data on percentage of people literate and per capita income for a sample of 20 countries. The countries are Afghanistan, Boliva, Cambodia, Chile, Cuba, Ecuador, Ghana, Guyana, Ivory Coast, North Korea, Mali, Malawi, Nepal, Pakistan, Philippines, Senegal, South Africa, Tanzania, Uganda, and Yemen.

Country	1	2	3	4	5	6	7	8	9	10
Percent literate	6	43	50	87	80	71	30	77	9	77
Per capita income	61	165	125	645	398	208	289	311	246	86

Country	11	12	13	14	15	16	17	18	19	20
Percent literate	10	6	6	22	80	6	46	11	30	6
Per capita income	46	72	73	107	246	158	600	174	92	66

Source: Arthur S. Banks, *Cross-Polity Time Series Date,* Cambridge, MA: MIT Press, 1971, pp. 237–255, 269–282.

a. How are the two variables related?

b. How much do two countries differ in literacy percentage (dependent variable) if per capita income (independent variable) differs by $100?

c. How much do two countries differ in per capita income (dependent variable) if their literacy rates (independent variable) differ by 10%?

10.55 On the Springer Web site (http://www.springer-ny.com/supplements/iversen/), the data file called Baseball Individual Scores contains data on 480 baseball players from both leagues for the 1996 season. The columns contain data on the following variables:

1. Number of times at bat
2. Number of runs scored
3. Number of hits
4. Number of home runs
5. Number of runs batted in
6. Batting average

    a. For which of these variables should you use the mean and for which variables should you use the median as the measure of central tendency?

    b. Study the relationships between some of these variables.

10.56 In a discussion of the relationship between population growth and the acceleration of cereal production in a textbook written in 1858 is the following passage:

> The change in the quantity of the several kinds of foods is given in the following passage from a recent work of much ability, by which it is shown, that the supply has grown twice more rapidly than population; and that, therefore, the Malthusian theory finds small support in the course of events in France: . . . For the cereals, our agricultural statistics give the . . . figures [in Table 10.11]. (Source: H. C. Carey, *Principles of Social Science*, vol. 2, Philadelphia: Lippincott, 1858, p. 54.)

    a. Regress quantity on population.

    b. Use the results from the analysis and comment on the quotation.

Table 10.11 Data for Exercise 10.56

Year	Population (millions)	Quantity (millions of hectoliters)
1760	21	94.5
1784	24	115.8
1813	30	132.4
1840	34	182.5

Source: H. C. Carey, Principles of Social Science, *vol. 2, Philadelphia: Lippincott, 1858, p. 54.*

c. Regress population on year, and regress quantity on year.

d. Construct a new variable

$$\text{ratio} = \text{quantity}/\text{population}$$

and regress ratio on year.

e. What do the analyses in parts c and d tell you?

10.57 From data on the Calabrian Mafia in Table 10.12, it is possible to study whether Mafia groups (coscas) choose their leaders such that

Table 10.12 Data for Exercise 10.57

Cosca name	Mean age of members	Age of chief
Cataldi-Marafioti	37	42
Nirta-Romeo	40	67
Ursino-Jerino	34	53
Ruga	31	29
D'Agostino	39	54
Mazzaferro	32	38
Aquino-Scali	33	36
Cordi	35	29
Macri	39	43

Source: Pino Arlacchi, Mafia Business, *London: Verso, 1986, p. 132. Brought to our attention by Matthew Werner.*

the age of the leader is related to the mean age of the members of the cosca. Analyze the relationship between the two variables.

10.58 Draw a scatterplot, using for data the amount of money you spent on 10 gifts you recently gave to people from whom you had received a gift. Follow the example for Fred, who gave a gift that cost $10 and received a gift that cost $5.

Name	Amount spent on person's gift	Amount person spent on me
Fred	$10	$5

You may use any exchanges you wish, factual or fictitious. Analyze the relationship between the two variables.

10.59 From a regression analysis, the predicted success in college at a state university as measured by grade point average is found from the equation

$$\text{college GPA} = 0.6 + 0.74 \text{ (high school GPA)}$$

a. Using your own high school GPA, calculate what grade point average you might expect if you were attending this college.

b. Now estimate what GPA you might have in your current circumstances. Change the regression equation so that it better fits your particular case.

c. What does the new equation look like?

d. Does the regression equation fit every person? Why not?

e. An admissions officer is looking at the high school record of Elmer Ebert, who has a high school GPA of 1.9. People cannot graduate from the university with a GPA lower than 2.0. What do you predict will happen to Elmer if he is allowed to enroll?

f. If Elmer and all other applicants with low GPA's are not admitted to the college, how might the regression equation change?

10.60 To develop a table of sums of squares, the following data about the effect of attending training camp on scoring percentages of basketball players were written down (Table 10.13). Training camp is a dummy variable with 0 for not attending and 1 for attending.

Table 10.13 Data for Exercise 10.60

Source	Sums of ?????	Proportion
Training camp	90,999	????
Residual variable	???????	0.21
Total	115,189	????

a. Some items were inadvertently smudged by a careless coffee drinker. Repair the table for your untidy friend.

b. Find the correlation coefficient r between the training camp variable and the performance variable for the basketball players.

c. Did the training camp seem to have any effect on performance?

10.61 Livers from 4 female and 4 male rats were given oleic acid. The data in Table 10.14 show the uptake and the amount incorporated into keotone bodies.

a. Make a scatterplot of the data with uptake as the independent variable and amount incorporated as the dependent variable; use a different symbol for female and male rats.

Table 10.14 Data for Exercise 10.61

Uptake	Incorporated	Gender
29.3	1.82	Female
25.5	0.84	Female
26.3	1.09	Female
31.0	1.45	Female
20.6	1.56	Male
17.9	0.93	Male
23.6	1.54	Male
25.4	1.76	Male

Source: C. Soler-Agilaga and M. Heimberg, "Comparison of metabolism of free fatty acid by isolated perfused livers from male and female rats," Journal of Lipid Research, vol. 17 (1976), pp. 605–615.

b. Describe in words the relationship between uptake and amount incorporated for each gender.

c. Find the regression line for each gender.

d. How does the relationship differ in the two sets of observations?

e. Suppose the uptake or both a female and a male rat is the value 25. How different are the predicted values of the dependent variable for the two rats?

f. Does there seem to be a significant difference in amount incorporated between female and male rats?

10.62 If you had measured your height at each birthday since babyhood and made a scatterplot of the data, the points would not lie along a straight line. But growth data over a shorter period of time sometimes can be analyzed using linear regression. The table shows age and height data for the son of the Count de Montebeillard from 1762 to 1789.

Age (years)	3	4	5	6	7	8	9
Height (centimeters)	98.8	105.2	111.7	117.8	124.3	130.8	137.0

Source: R. E. Scammon, "The first seriatim study of human growth," American Journal of Physical Anthropology, vol. 10 (1927), pp. 329–336, as reported in R. L. Sandland and C. A. McGilchrist, "Stochastic growth curve analysis," Biometrics, vol. 35 (1979), pp. 255–271.

a. Make a scatterplot of the data.

b. Do the data display a linear pattern?

c. Find the equation for the regression line.

d. How do you interpret the value of the regression coefficient for this example?

e. Find how much the count's son grew each year by subtracting the height each year from the height the next year, and find the mean growth per year.

f. Explain the connection between the regression coefficient and the mean growth per year.

10.63 In a sample of ten states, the values for the percentage of the state population that receive Medicaid benefits as well as the number of hospital beds per 100,000 population are shown in Table 10.15.

a. Why do the numbers of hospital beds per 100,000 population vary from state to state, from a high in North Dakota of 507 to a low in Utah of only 255)?

Table 10.15 Data for Exercise 10.63

State	Percentage receiving Medicaid benefits	Number of hospital beds per 100,000 population
Arkansas	11.3	430
Florida	8.0	392
Indiana	6.3	382
Maine	10.8	335
Mississippi	16.8	457
New Hampshire	4.0	290
North Dakota	7.7	507
Rhode Island	11.7	319
Utah	6.3	255
Wisconsin	8.0	342

Source: Medicaid data: U.S. Department of Commerce, Bureau of the Census and the Health Care Financing Administration, Form-2082. Hospital bed data: American Association of Retired Persons, Reforming the Health Care System: State Profiles 1990, *Washington D.C.: AARP, 1991. These data are reprinted in the report* Medicaid Hospital Payment Congressional Report, *The Prospective Payment Assessment Commission, C-91-02, October 1, 1991, pp. 27 and 39.*

b. Make a scatterplot of the data with percentage receiving Medicaid benefits as the independent variable and number of beds per 100,000 population as the dependent variable. Label the points with the names of the states.

c. Comment on patterns you see in the data.

d. Analyze the relationship between the two variables.

10.64 The deterrent effect of the death penalty is a widely discussed question. Table 10.16 (page 465) shows the number of people executed for homicides and the homicide rate in this country for each year in the decade starting in 1950. What do these numbers add to the discussion of the deterrent effect of capital punishment?

10.65 In a study conducted through the National Toxicology Program, about 100 female mice were given ethylene glycol, and then their litters were observed. Four different dosages were used, and the data in Table 10.17 show the mean number of offspring for each group of

Table 10.16 Data for Exercise 10.64 (page 465)

Year	Number of executions	Homicide rate
1950	68	5.3
1951	87	4.9
1952	71	5.2
1953	51	4.8
1954	71	4.8
1955	65	4.5
1956	52	4.6
1957	54	4.5
1958	41	4.5
1959	41	4.6

Source: W. C. Bailey and R. D. Peterson, "Murder and capital punishment: A monthly time-series analysis of execution publicity," American Sociological Review, *vol. 54 (1989), p. 740.*

mice receiving a particular dosage together with the percentage of the offspring with malformations and the mean fetal weights.

  a. Make three scatterplots of dose as the independent variable and each of the other three variables as dependent variables.

Table 10.17 Data for Exercise 10.65

Dose (g/kg)	Mean litter size	Percentage of animals with malformations	Mean fetal weight (g)
0.00	11.90	0.3	0.972
0.75	11.50	9.3	0.877
1.50	10.40	39.0	0.764
3.00	9.83	57.0	0.704

Source: C. J. Price, C. A. Kimmel, R. W. Tyl, and M. C. Marr, "The developmental toxicity of ethylene glycol in rats and mice," Toxicological Applications in Pharmacology, *vol. 81 (1985), pp. 113–127,* in P. J. Catalonao and L. M. Ryan, "Bivariate latent variable models for clustered discrete and continuous outcomes," Journal of the American Statistical Association, *vol. 87 (1992), pp. 651–668.*

b. What do the scatterplots show?

c. Analyze the relationship between dosage and each of the other three variables using regression and correlation analyses.

d. Can you compare the three regression coefficients to see for which of the three dependent variables dosage is more important?

e. Both litter size and weight are means of about 25 observations each. What effect do you think it would have had on the analyses if you had used the original individual data instead of the means for each dose?

10.66 For their Medicaid inpatient hospital payments, states can either use a retrospective cost-based payment method or they can use a prospective payment system. Table 10.18 shows the percentage of the states that use a form of prospective payment methodology at four different times. Analyze the relationship between the two variables.

10.67 In a study of body fat percentage and age the following data were found. Analyze the data.

Age	23	23	27	27	39	41	45	49	50
Percent fat	9.5	27.9	7.8	17.8	31.4	25.9	27.4	25.2	31.1

Age	53	53	54	56	57	58	58	60	61
Percent fat	34.7	42.0	29.1	32.5	30.3	33.0	33.8	41.1	34.5

Source: R. B. Mazeness, W. W. Peppler, and M. Gibbons, "Total body composition by dual-photon (^{153}Gd) absorptiometry," American Journal of Clinical Nutrition, vol. 40 (1984), pp. 834–839.

Table 10.18 Data for Exercise 10.66

Year	Percent
1977	14
1981	32
1985	84
1991	92

Source: Medicaid Hospital Payment Congressional Report, *The Prospective Payment Assessment Commission*, C-91-02. October 1, 1991, p. 44.

10.68 The data in Table 10.19 (page 468) show the number of bird species in isolated areas of paramo vegetation in the northern Andes and the altitude of the areas in thousands of feet. The data here include only areas with an altitude less than 5,000 feet. Analyze the data to see if the number of species is related to the altitude of the regions.

10.69 Below Table 10.19 on page 468 are data on mean annual temperature and mortality rate from breast cancer in some regions of Great Britain, Norway, and Sweden.

Table 10.19 Data for Exercise 10.68 (page 467)

Area	Number of species	Altitude (thousands of feet)
Chiles	36	4.1
Las Papas-Cocunuco	30	3.8
Sumapaz	37	3.5
Parmillo	11	1.5
Pamplona	11	2.3
Cachira	13	2.4
Tama	17	2.0
Batallon	13	2.2
Merida	29	4.9
Perija	4	2.5
Cende	15	1.8

Source: F. Vuilleumier, "Insular biogeography in continental regions: I. The northern Andes of South America," American Naturalist, *vol. 104 (1970), pp. 373–388.*

Region	1	2	3	4	5	6	7	8
Temperature	51.3	49.9	50.0	49.2	48.5	47.8	47.3	45.1
Mortality	102.5	104.5	100.4	95.9	87.0	95.0	88.6	89.2

Region	9	10	11	12	13	14	15	16
Temperature	46.3	42.1	44.2	43.5	42.3	40.2	31.8	34.0
Mortality	78.9	84.6	81.7	72.2	65.1	68.1	67.3	52.5

Source: A. J. Lea, "New observations on distribution of neoplasms of female breast in certain European countries," British Medical Journal, *vol. 1 (1965), pp. 448–490.*

a. Analyze the data.

b. Could this be a causal relationship, or can you think of other variables that might explain why the two variables are related?

10.70 Table 10.20 (page 469) shows a decade of crime rates per 100,000 population in California and the yearly population in the state.

a. Why does using the Year column or the Population column, as the independent variable make little difference?

Table 10.20 Data for Exercise 10.70

Year	Homicide	Rape	Robbery	Assault	Total	Population (millions)
1983	10.5	48.2	342.3	374.6	775.6	25.1
1984	10.5	45.7	328.3	379.9	764.4	25.6
1985	10.7	43.0	331.1	388.2	773.0	26.1
1986	11.4	45.3	346.0	526.1	928.7	26.7
1987	10.7	44.2	304.4	568.5	927.8	27.4
1988	10.5	41.9	307.2	574.0	933.6	28.1
1989	11.0	41.6	335.1	599.5	987.2	28.8
1990	12.1	43.0	380.5	619.8	1,055.4	29.6
1991	12.6	42.2	408.2	616.7	1,079.7	30.6
1992	12.5	40.7	418.1	632.5	1,103.8	31.3

Source: California Department of Justice, as reported in The Economist, *March 19, 1994, p. 31.*

b. Plot the assault rate versus year and describe the pattern you see.

c. How do you explain the pattern in the scatterplot?

d. Do a regression analysis of rape on year and report on the results.

e. Do a regression analysis of homicide on year and report on the results.

f. Do a regression analysis of robbery on year and report on the results.

10.71 Collect data on two metric variables, analyze the data, and write a report on your findings.

10.72 The table shows two measurements, in millimeters, on a sample of 16 littleneck clams from Garrison Bay, Washington.

Clam	1	2	3	4	5	6	7	8
Length	530	517	505	512	487	481	485	479
Width	494	477	471	413	407	427	408	430

Clam	9	10	11	12	13	14	15	16
Length	452	468	459	449	472	471	455	394
Width	395	417	394	397	402	401	385	338

Source: D. F. Andrews and A. M. Herzberg, Data: A Collection of Problems from Many Fields for the Student and Research Worker, *New York: Springer-Verlag, 1985, p. 336.*

a. In a study of the relationship between the two variables, which, if any, of the two variables should be used as the independent variable and which as the dependent variable?

b. Make a scatterplot of the data.

c. How strong is the relationship between the two variables?

d. One point in the scatterplot is isolated from the others. Compute the strength of the relationship without that point.

e. Does the point in the lower left corner have much of an impact on the strength of the relationship?

10.73 Each week the local newspaper publishes the names and ages of people who have applied for marriage licenses at the county court house. Here are the ages for grooms and brides one week, written as (groom age, bride age):

(37, 30) (30, 27) (65, 56) (45, 40) (32, 30) (28, 26) (45, 31) (29, 24)
(26, 23) (28, 25) (42, 29) (36, 33) (32, 29) (24, 22) (32, 33) (21, 29)
(37, 46) (28, 25) (33, 34) (17, 19) (21, 23) (24, 23) (49, 44) (28, 29)
(30, 30) (24, 25) (22, 23) (68, 60) (25, 25) (32, 27) (42, 37) (24, 24)
(24, 22) (28, 27) (36, 31) (23, 24) (30, 26)

Source: The Philadelphia Inquirer, *September 10, 1995, p. MD12-d.*

a. If each groom and bride were the same age, what would be the slope and intercept for the regression line through the points?

b. If each groom were, say, 5 years older than his bride, what would be the slope and intercept for the regression line through the points?

c. If each groom were, say, 10% older than his bride, what would be the slope and intercept for the regression line through the points?

d. Find the regression line for the actual ages.

e. From the regression line, what can you conclude about the age patterns of these brides and grooms?

f. What does the scatterplot tell you about the data that you do not learn from the way the data are presented in Exercise 3.20?

10.74 This exercise explores the idea that an independent variable x is associated with a certain part of the variation in a dependent variable y. Suppose you have the following data:

$$x: 1\ 2\ 3\ 4$$
$$y: 3\ 2\ 5\ 6$$

a. Show that the total variation in the y-values equals 10.0.

b. The regression line has the equation $y = 1.0 + 1.2x$. Find the predicted values of y, and show that the variation in those values equals 7.2.

c. Find the four residual values and show that the variation in the residual values equals 2.8.

d. What proportion of the total variation in y is associated with x and what proportion is associated with the residual variable?

e. How large is the correlation between x and y?

CHAPTER 11

11.1 Analysis of variance: Comparing the mean-ings of things

11.2 Question 1. Relationship between violent crime rate and region?

11.3 Question 2. Strength of the relationship?

11.4 Question 3. Could the relationship have occurred by chance alone?

11.5 Question 4. Causal relationship?

11.6 Analysis of a variance: A bird's-eye review

11.7 Matched pair analysis: Two observations per unit

11.8 Summary

11
ANOVA: ANALYSIS OF VARIANCE FOR A CATEGORICAL AND A METRIC VARIABLE

According to many surveys, one of the major social concerns of people today is crime. Consequently, much discussion is focused on crime in political debates, in daily news reports, and among neighbors and friends, and it is hard to get a balanced sense of the importance of this issue. Are crime rates alarmingly high, or do we have less to worry about than we think?

We might ask many questions about crime. One is whether the crime rates are different in different regions of the country. Is it "safer" to live in one region versus another? Another question is whether the number of crimes has been increasing.

In this chapter we focus on one particular kind of crime—violent crime: murder, forcible rape, robbery, and aggravated violent crime. We address the question of whether the chances of being a victim of violent crime are the same from one part of the country to another. Then, if we do find that the number of violent crimes is not the same in different parts of the country, where are the high and low incidences of violent crime? These questions are more than theoretical. Insult and injury are at stake.

To answer these questions, we need data on the number of violent crimes in each geographical area for a specific time period. But it is not easy to find out how many violent crimes occurred in Pennsylvania last year. The police count the violent crimes that are reported, but there is every reason to believe that many violent crimes go unreported for one reason or the other. Another approach to getting data on violent crimes is to ask people in sample surveys whether they have been victims of violent crimes, but such surveys suffer from the same errors all surveys suffer from. (You may recall that we discussed some of these problems in Chapter 2 on the collection of data.) We probably get more accurate data by asking people in a proper statistical sample about violent crimes than by relying on data reported to the police. (Sexual crimes, for example, tend to be underreported to the police.)

Nonetheless, the analysis in this chapter is based on the number of violent crimes reported by the FBI in their Unified Crime Reports for each of the contiguous 48 states for the years 1986 and 1992. Since the states vary a great deal in population size and therefore in the number of violent crimes, we use the violent crime *rate*, how many violent crimes there were in the state for each 100,000 population. For example, in Pennsylvania the violent crime rate was 359 violent crimes per 100,000 inhabitants in 1986. We group the states in seven regions—New England, Mid-Atlantic, Midwest, South, Southwest, Rocky Mountains, and Pacific Coast—to see how different parts of the country compare in their rates of crime. Table 11.1 (pages 476–477) shows the violent crime rate for each state in these regions.

We're going to study whether the seven regions differ in their violent crime rates based on the data from the states in each region. Note that although we have chosen to examine regional differences using state data, we could have chosen other geographical areas for the units in the analysis. For example, we could have used violent crime rates in each of the 3,000 or so counties instead. While it is not clear whether it is better to use state instead of county data, one reason for using state data is that it keeps the number of observations manageable.

The dependent variable in the analysis is the violent crime rate, which is a metric variable (its numbers increase in meaningful intervals). Region is the independent variable, and it is a categorical variable. Each region can be named by a number, but if South is equal to 6, we do not assume it is twice as great as New England at 3, even if we are southerners. Since we are studying the effect of region on violent crime rate, we are studying the effect of a categorical independent variable on a metric dependent variable.

The special case when the categorical independent variable has only two categories is studied in Chapter 7 as a *t*-test for the difference between two means. It is also studied in Chapter 10 when the independent variable is a dummy variable. In this case the independent variable has seven categories, so the *t*-test, for two variables, is not appropriate.

> **STOP AND PONDER 11.1**
>
> We want to know whether unemployment rates differ by region of the country. How is this problem similar to the one we are discussing about violent crime rates? Name the variables and their types (categorical, metric) for this problem.
>
> Can you invent another, similar problem using different variables?

11.1 ANALYSIS OF VARIANCE: COMPARING THE MEAN-INGS OF THINGS

When we study the effect of one (or more) categorical independent variables on a metric dependent variable, we use a statistical method known as *analysis of variance,* often abbreviated *anova* (pronounced ə-nō′-və). Analysis of variance is closely related to regression analysis (Chapter 10), although this may not seem obvious at first glance. Both can be considered as special cases of a more general statistical model.

Analysis of variance was originally developed in the 1920s for the analysis of agricultural data, and it is a very commonly used statistical tool in many different disciplines. In particular, it is often used for the

Analysis of variance is a statistical method used for the comparison of the means of a dependent variable across different groups.

Table 11.1 Violent crimes in the contiguous 48 states in 1986

State	Crime rate per 100,000 population	Region
Maine	147	
New Hampshire	140	
Vermont	149	New England
Massachusetts	557	
Rhode Island	336	
Connecticut	426	
New York	986	
New Jersey	572	Mid-Atlantic
Pennsylvania	359	
Ohio	423	
Indiana	308	
Illinois	800	
Michigan	804	
Wisconsin	258	
Minnesota	285	
Iowa	235	Midwest
Nebraska	263	
Missouri	578	
North Dakota	51	
South Dakota	125	
Kansas	369	
Delaware	427	
Maryland	833	South
Virginia	306	

Source: F.B.I. Uniform Crime Report for the United States.

analysis of experimental data in fields such as psychology, biology, engineering, and medicine. With experiments we often think of the independent variable as the treatment variable and the dependent variable as the response variable. In an agricultural experiment the

Table 11.1 Violent crimes in the contiguous 48 states in 1986 *(continued)*

State	Crime rate per 100,000 population	Region
West Virginia	164	
North Carolina	476	
South Carolina	675	
Georgia	588	
Florida	1,036	
Kentucky	334	South
Tennessee	540	
Alabama	558	
Mississippi	274	
Arkansas	395	
Louisiana	758	
Oklahoma	436	
Texas	659	Southwest
Arizona	658	
New Mexico	726	
Wyoming	293	
Colorado	524	
Montana	157	Rocky Mountains
Idaho	222	
Utah	267	
Nevada	719	
Washington	437	
Oregon	550	Pacific Coast
California	920	

treatment variable might consist of different types of fertilizers used on a field of corn and the response variable the yield from the different fertilizers. As we see in our example, analysis of variance can also be used on observational data.

11.2 QUESTION 1. RELATIONSHIP BETWEEN VIOLENT CRIME RATE AND REGION?

Scatterplot

The first question we ask is if there is a relationship between the two variables in these data; that is, are there differences between the regions? We answer the first question the same way we did for correlation and regression analyses. To see if the regions differ in their violent crime rates, we first display the data in a scatterplot, with region as the independent variable on the horizontal axis and violent crime rate as the dependent variable on the vertical axis. The scatterplot is shown in Figure 11.1; each dot in the figure represents one of the 48 continental states, and the states are grouped by regions.

The main difference between analysis of variance and regression analysis is that in analysis of variance the independent variable along the horizontal axis is a categorical variable, while in regression the

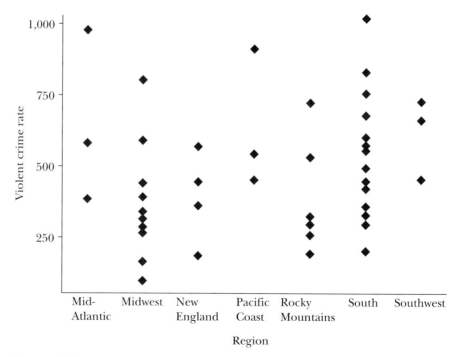

Figure 11.1 Scatterplot of crime rates

independent variable on the horizontal axis is metric. With a categorical variable, we can place the categories (or values) of the variable on the axis anywhere we want and in any order we want. Here the values (regions) are simply placed alphabetically along the horizontal axis. If we had used another system of arranging the regions, for example from northeast to southwest, the data points would have made a different pattern. Thus, since the pattern of points is arbitrary, it makes no sense to draw a line through these points as we do in regression analysis. With a metric variable, the placement of the values on the axis is determined from low to high values, and there is only one way to display the points and draw a line through the points. In the end, this difference between the arbitrary placement of the values of the categorical independent variable and the fixed placement of the values of a metric independent variable on the horizontal axis of a scatterplot is the crucial difference between analysis of variance and regression analysis.

Even a quick look at the scatterplot tells us that there are considerable differences in the violent crime rates from one region to the next. Also, within each region the states differ considerably from one another. The figure shows that the New England states overall have a lower level of violent crimes, while in the Mid-Atlantic and the Pacific Coast states the violent crime rates generally seem higher. These differences indicate a relationship between region and violent crime rate, at least in these data. If all seven regions had violent crime rates of the same magnitudes, then there would not be a relationship between region and violent crime rate in these data.

Boxplot: A simpler view of the data

To get a better sense of the differences in rates from one region to the next, we need to simplify the data displayed in the scatterplot. The 48 dots in the scatterplot tell the whole story about these data, but there are too many to show the relationship between the two variables clearly. Simplification can be done in various ways, and one is to make a boxplot of the data in each region, in which we replace the data for the individual states in a particular region by five numbers: the median, the 25th and 75th percentiles, and the minimum and maximum violent crime rates. These boxplots are shown in Figure 11.2.

The use of boxplots reduces the data in each region to five values. Since there are seven regions, the original data of 48 observations is reduced to 35 numbers. That is not much of a reduction, but at least the data are displayed the same way for each region. If there had been

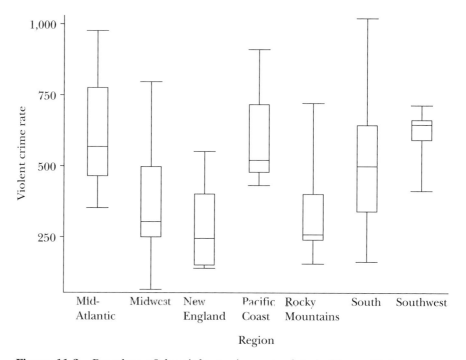

Figure 11.2 Boxplots of the violent crime rate data in Figure 11.1

more observations within the groups, the reduction would have been larger. Boxplots make it much easier to compare the regions. Because the boxplots for the regions are next to each other, we can easily compare the differences in the regions.

What does a comparison of the boxplots reveal about violent crime rates based on the data? First we compare the medians in the different regions, since they represent the central values. When we scan across the lines in the middle of the boxes representing the medians, we see that the states in the Southwest, Mid-Atlantic, and Pacific Coast regions of the country have the highest medians, so they have the highest average violent crime rates. The New England states have the lowest average violent crime rates, followed closely by the Rocky Mountain states.

Another feature of these boxplots is that the boxes have different heights. For example, the boxes for the Southwest and the Rocky Mountain states are shorter than the boxes for the other regions. This shows that the violent crime rates in the states in those regions are more alike than the rates in other regions.

11.3 QUESTION 2. STRENGTH OF THE RELATIONSHIP?

The boxplots show more graphically than the scatterplot that the regions differ in their violent crime rates and that there is a relationship between the two variables. But we also want to find how strong the relationship is between the two variables, and we want to know whether the relationship could have occurred by chance or not. To answer these questions we need to dig deeper—using an analysis of variance.

Formal analysis of variance is based not on medians but on the *means* within each group of observations. This is mainly because the means lend themselves better to a mathematical analysis of the differences between the groups.

The name analysis of variance in some ways is a misleading one for what we are doing. A more appropriate name would be analysis of means: we are concerned with whether the means of the dependent variable (violent crime rate) differ across the groups defined by the independent variable (region). Thus, we are interested in the means, but we will be using variances to find out if the means differ in any interesting way.

The first step in answering the second question is to calculate the mean violent crime rate for each region and an overall mean for all the states combined. These means are shown in Figure 11.3, the same scatterplot as in Figure 11.1 with the addition of a horizontal line across the points showing the overall mean violent crime rate for all the states ($\bar{y} = 460$) and a zigzag line connecting the means of the violent crime rates of the seven regions. (Since the crime rate is the dependent, or y-variable, we use the letter y to designate this mean.) Note that the means range from a low of about 292 for New England to a high of about 639 for the Mid-Atlantic states.

Region variable

Analysis of variance of the data is based on the notion that the violent crime rate in a particular state is determined by two factors, the region the state is located in and the combined effect of everything else. These two factors completely determine the violent crime rate in a particular state. (Logically, it cannot be otherwise.)

One way to understand what this means is to think about how each state acquires its violent crime rate. First, imagine that all the states started with a crime rate of the same value. The best estimate of a common crime rate would be the overall mean derived from the ob-

482 Chapter 11 • Anova: Analysis of Variance for a Categorical and a Metric Variable

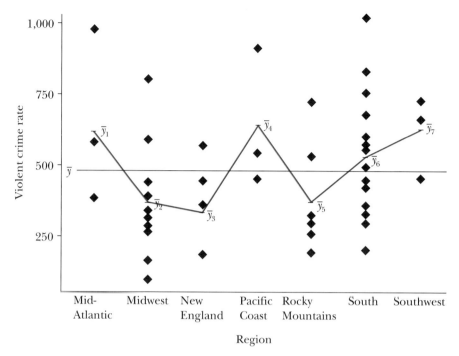

Figure 11.3 Scatterplot of crime rates with overall and region means

served data from each of the 48 states, the value denoted $\bar{y}$ in Figure 11.3. Numerically, $\bar{y}$ equals 460 for these data. Thus, all the states would have a violent crime rate of 460 if no variables had any effect.

Next, imagine that each state is influenced by the effect of being in a particular region. This influence changes each state's crime rate from the common overall mean to a common value for each region. It has to be a common value, because all the states in a given region would be affected the same way. The best estimate of that common value is the observed mean in that region. For example, the three Mid-Atlantic states of New York, New Jersey, and Pennsylvania would all have crime rates of 639, the mean of the three observed state rates 986, 572, and 359. The similar values for the other regions would be the means connected by the zigzag line in Figure 11.3.

Thus, the effect of the region variable for each state can be found to equal the difference between the region mean and the overall mean. For example, the region variable moves Pennsylvania from the overall mean of 460 to the regional mean of 639, for a difference of 639 − 460 = 179. Region has the same effect on the other two Mid-Atlantic states. When we find this difference for every state and add the squares of these distances, we get a sum equal to 662,641. This sum is the effect of the region variable, and we call it the region sum of squares.

> The **independent variable** (here region) **sum of squares** is found as
>
> $$\text{sum(group mean} - \text{overall mean})^2$$
>
> across all the observations.

Residual variable

But the violent crime rates in all the states in a region do not have the same value. Other variables act together to shift the crime rate for a given state away from the region mean to the observed value for that state. The net effect of the other variables is called the residual variable, just as in regression analysis. For example, Pennsylvania has a rate of 359, while the mean for the region it is in equals 639. Thus, the residual variable moved the rate from 639, where it would have been without the residual variable, down to 359, the observed state value. The effect of the residual variable is the difference between the observed value for a state and mean for the region the state is in.

Next, we calculate a single, overall number summarizing how large these differences are. Because some of them are negative and some are positive and the mean difference is zero, the mean of the differences is not of much help. But if we square each difference and add the squares for the states in a particular region, then we have a measure of the effect of the residual variable in the region. We do the same in the other regions and add the squares for all the states to get the effect of the residual variable for all the states. This sum of squares is known as the residual sum of squares, and in this example it equals 2,145,613. Thus, for these data, the magnitude of the effect on the violent crime rates by the residual variable—the combined effect of all variables except for region—is equal to 2,145,613.

The residual variable is sometimes known as the *error variable*. Error here does not mean there is something wrong with the variable. Much of the original work in analysis of variance was done with data resulting from several attempts to measure the same quantity. It was assumed that a true value existed and that the extent to which a particular observed value was not equal to the true value was due to errors in the measurement. It is the error term that now is often called the effect of the residual variable.

> The residual variable, also known as the error variable and sometimes called the "everything else" variable in this text, is the name given to the combined effect on the dependent variable of all variables other than the independent variable.
> The **residual sum of squares** is found as
>
> $$\text{sum(observation} - \text{group mean})^2$$

across all the observations.

Effect of both region and residual variables: Total sum of squares

The states do not all have the same violent crime rate because each state has been affected by both the region and the residual variables. Thus, the difference between the violent crime rate for a particular state and the overall mean value of 460 measures the magnitude of both the effect of the region variable and the effect of the residual variable. Pennsylvania, for example, has a state violent crime rate of 359. The difference between Pennsylvania's rate and the overall or grand mean is $359 - 460$, or -101. Thus, the combined effect of region and residual variables on the violent crime rate in Pennsylvania equals -101. Using this procedure for each state, we end up with 48 differences.

To summarize in one number the magnitude of the differences between the individual observation and the overall mean, we square each difference and add up all the squares. The sum of the squared differences is known as the total sum of squares. The total sum of squares for the violent crime rates equals 2,804,254. Thus, the magnitude of the combined effect of all variables that affect the violent crime rates is equal to 2,804,254.

> The effect of both the independent variable and the residual variable is the **total sum of squares** and is found as
>
> $$\text{sum(observation} - \text{overall mean})^2$$
>
> across all the observations.

Remarkably, the effect of both variables equals the sum of the effects we found earlier for each of the two variables. The effect of the region variable was 662,641, and the effect of the residual variable was 2,145,613: 662,641 + 2,145,613 = 2,804,254.

Measuring the strength of the relationship

Table 11.2 shows the effects of the variables as measured by the sums of squares. Large numbers like these are hard to compare, so the third column of the table shows the proportion of the total that is contributed by each of the two separate effects. The proportions are found by dividing each of the two sums of squares by the total. The effect of the region variable is 662,641/2,804,254 = 0.24, or 24%, of the total effect. This proportion, 0.24, is known as R^2. (This number is directly comparable to the squared correlation coefficient in regression analysis.)

The region variable accounts for only about one quarter of the total effect on crime rates. The rest of the effect is that of the residual variable. Looking at the scatterplot in Figure 11.1 or the boxplots in Figure 11.2, it is not surprising that the residual variable has such a large effect. In each of the seven regions, the range of values is quite large and the

Table 11.2 Effects of region and residual variables

Source	Sum of squares	Proportion
Region	662,641	0.24
Residual	2,141,613	0.76
Total	2,804,254	1.00
	$R = 0.49$	

states vary a good deal from each other. Most of the variation comes from the residual variable. This means that for a fuller understanding of the crime rates, we would have to identify some of the other variables that make the residuals so large and bring them into the analysis as additional independent variables. We touch briefly on this kind of analysis in Chapter 13. The computation of R^2 is shown in Formula 11.2 at the end of the chapter.

The square root of R^2 is, of course, R itself. Here, by taking the square root of 0.24, we find R equal to 0.49. This number measures the strength of the relationship between the independent and the dependent variable. (This number is directly comparable to the correlation coefficient r in regression analysis; the main difference is that in analysis of variance R cannot be negative.) An R ranges in possible values from 0 to 1. With a value of 0.48 we have a moderately strong relationship between the two variables region and violent crime rates.

Explained amounts of variation

It is often stated that the independent variable *explains (produces, accounts for,* even *causes)* a certain percentage of the variation in the dependent variable. In the example, we can say that region explains 24% of the variation in the violent crime rates. What does this mean?

If region and the residual variable had no effects on the crime rates, then all the states would have the same rate and there would be no variation in the rates from state to state. The best estimate of this common value would be the overall $\bar{y}$ of 460, the number listed in the third column in Table 11.3. The last line of the table notes that when the values are equal, then there is no variation in them.

When the region variable is permitted to affect the rates, then the rates in each region move from the overall mean of 460 to the mean of the region. Thus, if region were the only variable affecting the rates, then the observed data would look like the fourth column of the table. There we see that within each region the state rates are the same. The variation in these rates is produced by the region variable. As before, we measure the amount of variation of the numbers in a column by subtracting the overall mean from each term, squaring the differences, and adding the squares: the variation in the 48 rates produced by the region variable is equal to 662,641.

In the final column are the data as they were observed, after the residual variable also produced its effect. The rates are even more different from each other than they were when affected by region only.

PYTHAGORAS'S TRIANGLE

For the more mathematically inclined, let us take a closer look at the numbers in Table 11.2. How can we display the three sums of squares in a graph? We could show the figures in a pie graph divided into two slices, a larger one for the residual variable and a smaller one for the region variable. We could also use bar graphs of different kinds.

There is another way to display these numbers. All three are sums of squares, but let us think of them simply as squares. Then we can draw on the thinking of the wise Greek mathematician who said that when two squared numbers add up to a third squared number, the three numbers can be displayed in a triangle with one 90-degree angle. The lengths of the three sides in the triangle are the three numbers.

The sum of two squares equaling the third square

$$662{,}641 + 2{,}141{,}613 = 2{,}804{,}254$$

can be written

$$814^2 + 1{,}463^2 = 1{,}674^2$$

We draw a right triangle with the length of the hypotenuse equal to 1,674 and the lengths of the two sides equal to 814 and 1,463.

This triangle has some nice features. For one thing, we have modified the effects of all the squaring we did. By using squares we emphasized large differences; the squares of large numbers get very large. Perhaps we really should not give that much weight to observations that are far away from the means. Square roots change the perspective on the effects. Now region, at 814, is clearly seen as about half as important as the residual, at 1,463. When we measure variation by sums of squares, we see that about one quarter of the crime rate variation is associated with region.

Another feature of the triangle can be seen in the lower left corner with the angle marked θ (theta). This is the angle formed by the hypotenuse and the side representing the independent variable region. The trigonometric function cosine of an angle is defined as the length of the adjacent side divided by the length of the hypote-

(Box continued on following page)

PYTHAGORAS'S TRIANGLE *(continued)*

nuse. That means here that $\cos(\theta) = 814/1{,}674 = 0.49$. With a cosine of 0.49, we can find from mathematical tables that the angle itself is 61 degrees.

Remarkably, the cosine of the angle formed by the hypotenuse and the side for the independent variable equals the correlation coefficient R between the independent and the dependent variable. As the size of the angle gets smaller, the cosine, and thereby R, becomes larger. When the angle is down to zero degrees and there is no residual effect, then the correlation coefficient becomes 1. Similarly, when the angle approaches 90 degrees, then the correlation coefficient approaches zero. The cosine connection clarifies the relationship of the correlation and the effects of the independent and the residual variables.

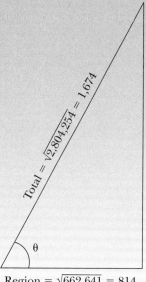

$\cos(\theta) = R = 0.49$

$(\theta = 61 \text{ degrees})$

Triangle representing sums of squares

11.3 Question 2. Strength of the Relationship?

Table 11.3 Variations in violent crime rates under different assumptions

State	Region	Crime rate if no variable had an effect (overall mean)	Crime rate if only region had an effect (region mean)	Crime rate when both region and residual have effects (observed rate)
Maine		460	292	147
New Hampshire		460	292	140
Vermont	New England	460	292	140
Massachusetts		460	292	557
Rhode Island		460	292	336
Connecticut		460	292	426
New York		460	639	986
New Jersey	Mid-Atlantic	460	639	572
Pennsylvania		460	639	359
Ohio	Midwest	460	385	423
⋮		⋮	⋮	⋮
⋮	⋮	⋮	⋮	⋮
⋮	Pacific Coast	⋮	⋮	⋮
California		460	636	920
Variation in the rates		0	662,641	2,804,254

$$R^2 = 662{,}641/2{,}804{,}254 = 0.24 \qquad R = 0.49$$

To find how much additional variation the residual variable produces, we subtract the overall mean from the rate for each state, square the differences, and add the squares, for a total variation of 2,804,254. The region variable accounts for 650,000 of the total. From the computation at the bottom of the table, region accounts for 0.24, or 24%, of the total variation in the crime rates for these 48 states. This is the quantity we call R^2. The residual variable therefore explains the remaining 76% of the total variation in the rates. But because we don't know what it is about different regions that makes violent crime rates lower in some regions than in others—climate, poverty, social factors—perhaps "explaining" is too strong a word. We could say region is *associated* with 24% of the variation in the rates and the residual variable is *associated* with the remaining 76% of the variation.

11.4 QUESTION 3. COULD THE RELATIONSHIP HAVE OCCURRED BY CHANCE ALONE?

The third question we ask in a statistical analysis is whether a relationship exists between the two variables not just in the sample but in the entire population. But this question can be asked only if the data are from a sample of a larger population. In this example, the data are on all the 48 continental states; they are population data, not sample data. So for this example we phrase the question of statistical inference slightly differently: we ask whether the relationship between the region and the violent crime rate variables could have occurred by chance alone or whether the relationship really does exist.

Events took place that produced the observed crime rates. Were they simply chance events or not? If only chance events led to all these crimes, then there would be only random variation in the rates from one region to the next.

> **STOP AND PONDER 11.2**
>
> Research often makes distinctions among various categories of independent variables, such as educational level, gender, age, race, residence, income, national background, region, and so on. Given the difficulty in determining why region makes a difference in determining rates of violent crimes, why should we always be careful to explain differences based on demographic variables such as gender. Why is it difficult to formulate a simple and straightforward independent variable? Why do people do it anyway?

The null hypothesis

The typical null hypothesis for the study of the relationship between two variables asks whether there is no relationship in the population. In analysis of variance, the null hypothesis is usually stated in terms of the means of the dependent variable for the various categories of the independent variable. In the example, that would be a statement that the crime rates in the seven regions are equal. Formally, this null hypothesis is written

$$H_0: \mu_1 = \mu_2 = \mu_3 = \mu_4 = \mu_5 = \mu_6 = \mu_7$$

The logical alternative to seven things being equal is that at least some of them are different. Thus, if we reject the null hypothesis, we have shown that at least some of the means are different, even though we have not proven that all the means are necessarily different.

The null hypothesis states that the independent and dependent variables are not related, which means that the relationship we found occurred by chance. To reject this hypothesis (or not), we must go back to the data and compare the magnitudes of the effects we found. If the effect of the independent variable (region, in this case) is large in comparison with the effect of the residual variable, then we reject the null hypothesis. If the effect of the independent variable is small in comparison with the effect of the residual variable, then we do not reject the null hypothesis. Another way of saying the same thing is that we reject the null hypothesis if R^2 is large and do not reject it if R^2 is small. What do we mean by large and small?

Things get a bit more complicated, however, because the meaning of large and small depends not only on the numbers we compute but also on how many groups and how many observations we are working with. In general, the more groups and the more observations, the smaller R^2 can be and still result in rejecting the null hypothesis.

In the example, we are working with 7 groups of data and 48 observations all together. According to the rules of statistical reasoning (and a statistical table), with this many groups and this many observations we need an R^2 at least as large as 0.28 to reject the null hypothesis.

p-value from *F*

To make a decision about the null hypothesis in our problem, we find the *p*-value for the observed data. We convert R^2 into a value of one of the four standard statistical variables. In previous chapters, we have made conversions to *z*-scores, *t*-scores, and chi-square scores. In analysis of variance, we use the *F*-variable (Formula 11.5). After converting the value of R^2 to the corresponding value of *F*, we use statistical software to find the *p*-value, or we look up the value in the table of the *F*-distribution (Statistical Table 5 in the appendix). If our *F* is larger than the critical value of *F* given in the table, then we reject the null hypothesis.

We also can observe the p-value for this F and thereby for the observed R^2. If this p-value is small, say less than 0.05, then we reject the null hypothesis. Either way, the different sample means are extremely unlikely to be as different as they are or more different simply because of chance alone.

The computer output for the analysis of the violent crime data is shown in Table 11.4. This type of table is known as an analysis of variance table (and it is the same kind of table we get with a regression analysis; refer to Table 10.5). As we look closely at the construction of this table, we see how the F-statistic is produced from this array of large numbers.

> An **analysis of variance table** shows the numbers involved in finding R and the F-value. The F-value comes from comparing the sizes or the effects of the independent and the residual variable while compensating for the different degrees of freedom. An analysis of variance table usually also contains the p-value for F.

Some of the numbers in Table 11.4 come from Table 11.2, and some of them are new. The fourth column contains the degrees of freedom for the region and residual variables. In this example there are 7 regions, so the region variable has 6 degrees of freedom: the degrees of freedom for the independent variable is always one less than the number of categories of the variable. The total degrees of freedom is always one less than the number of observations, so with 48 observations there are 47 degrees of freedom (Formula 11.3). The number of degrees of freedom of the residual variable always is the number of observations minus the number of groups—in the example, 48 − 7 = 41.

Table 11.4 Analysis of variance table

Source	Sum of squares	Proportion	Degrees of freedom	Mean square	F-ratio	p-value
Region	662,641	0.24	6	110,440	2.11	0.072
Residual	2,141,613	0.76	41	52,234		
Total	2,804,254	1.00	47			

> **STOP AND PONDER 11.3**
>
> An analysis of variance is done to determine if the quantity of antidepressants taken by a group of elderly people depends on in which of 5 nursing homes they reside. Each of the homes has 20 residents. What are the degrees of freedom for the nursing home independent variable and the residual variable?

The value of F (Formula 11.5) amounts to a comparison of the effects of the independent and the residual variables. Instead of comparing the sums of squares, however, we divide each of them by their degrees of freedom to get a sum of squares per degree of freedom (Formula 11.4). This quantity is called a *mean square*, not because it is nasty, of course, but because it is a way of averaging the effects by the degrees of freedom. The table indicates that the mean square for the region variable equals $662{,}641/6 = 110{,}440$. Similarly, we find that the residual mean square equals $2{,}145{,}613/41 = 52{,}234$. The two mean squares still are two large and seemingly meaningless numbers, despite our efforts, but they are a step toward simplification.

If the data represent only chance variations and there is no difference between the regions, then the underlying formal theory states that the two mean squares are about equal. We therefore compare the mean squares to see if they are about the same or not. We can subtract one from the other and see if the difference is approximately equal to zero, or we can divide one by the other and see if the ratio is approximately equal to one.

We do the comparison by dividing the mean square for the independent variable (region) by the mean square for the residual variable. In this case, the region's mean square is about twice as large as the residual's mean square. More exactly, the ratio is equal to $110{,}440/52{,}234 = 2.11$. This ratio is our observed value of the F-variable with 6 and 41 degrees of freedom.

The probability of getting a value of this F-variable equal to or larger than 2.11 by chance alone is only 0.072; that is, we expect an F this large or larger in 72 of 1,000 different samples if there truly is no relationship between the two variables and the data were produced by chance alone. Thus, the probability of getting the R^2 of 0.24 or more for 7 groups and 48 observations by chance alone is marginally statistically significant. The probability does not quite reach the magic significance level of 0.05, but it comes close.

Now we begin to see why this kind of analysis is known as analysis of variance. The effects are computed as sums of squares. The numerator in the computation of the variance is also a sum of squares. To find variance in Chapter 4, we divided the sum of squares by $n - 1$, the proper degrees of freedom. Similarly here, when we divide the sums of squares by their degrees of freedom to get the mean squares, we find variances. Thus, we actually compare the means by comparing variances when we find F.

> **STOP AND PONDER 11.4**
>
> The F-ratio in the nursing home example in Stop and Ponder 11.3 equals 3.50, which is statistically significant with $p = 0.01$. What does this suggest about the use of antidepressants in the nursing homes? What can we not say, given only these statistical results? What seems to be missing from the analysis that we would like to know?

Going beyond the F-test: Making mean comparisons

The result we have found so far, that the 7 regions differ in their mean violent crime rates with $p = 0.072$, is not very interesting. The null hypothesis states that the means are equal, and the alternative to everything being equal is that not everything is equal. But this can mean that one or some or all of the means are different from one another. We know from the value of F in our analysis of variance that the population means are not equal, but which of the population means are not equal? Is it safer on the Pacific Coast than in the Mid-Atlantic region? Is only the violent crime rate in the Great Lakes region different and that in the other 6 regions about the same? We should always ask these kinds of questions about an analysis of variance if we want to know which means are different from each other.

If there were only two means in the analysis and a significant F (or t from the t-test for the difference between two means), we would know right away that all the means are different. We could do a similar comparison of all pairs of means to find out which of them are different. With 7 means there are 21 possible pairs to compare. If we did 21 different, independent statistical tests, even if the means in all of them were equal, statistical theory says that with a 5% significance level, we would make a mistake 5% of the time and find 5% of the pairs statistically significant even though they truly are not different. Since 5% of

Table 11.5 Mean violent crime rates and number of states in seven regions

Region	Mean number of violent crimes	Number of states
Mid-Atlantic	639	3
Midwest	375	12
New England	292	6
Pacific Coast	636	3
Rocky Mountains	364	6
South	526	14
Southwest	620	4
Overall	460	48

21 is 1.0, we could expect about one of the differences between means to be different just by chance alone. Here, not all the tests would be independent of each other, but we would still have the problem of what happens when we do many statistical tests.

To show which of the means are different, we list the means and number of states in each region in Table 11.5. (The means are also shown in Figure 11.3.) It is still hard to tell which of the means are statistically different and which are not. Figure 11.4 gives a better sense of the regional differences in means. The figure makes it clear that the Mid-Atlantic, Pacific Coast, and Southwest are the three regions with the highest means, while the Midwest and the Rocky Mountains cluster lower, and New England has the lowest mean.

Looking at a few comparisons among the regions, we note that the difference between the means for the Mid-Atlantic and New England regions equals $639 - 292 = 347$. This difference converts to a *t*-value of 2.14 on 41 degrees of freedom, and such a value is statistically significant with $p = 0.02$. Similarly, if we continue to use the *t*-test for the comparison of two means, we find that the differences between the Pacific Coast region and New England, the difference between the Southwest and New England, and the difference between the South and New England all are statistically significant. These differences are a reflection of the fact that the overall *F*-value is at least marginally significant, since the *p*-value at 0.072 is close to the common cutoff of 0.05.

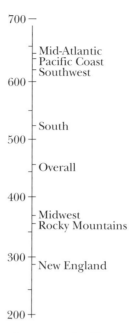

Figure 11.4 Mean violent crime rates for 7 regions and overall mean for 48 states

11.5 QUESTION 4. CAUSAL RELATIONSHIP?

There are statistical methods available to help us answer this question. They are based on bringing in other variables and finding out if they can explain the presence of the relationship. But at this point the answer to the fourth question can only be speculative. It is certainly hard to imagine that a region per se can determine its violent crime rate. The differences between regions in violent crime rates can probably be explained by other variables, such as percentage of population living in urban areas, poverty level, and population density. Despite the fact that we cannot enumerate why the crime rates are different, we can still use this analysis for the prediction of which states will have higher and lower crime rates.

11.6 ANALYSIS OF VARIANCE: A BIRD'S-EYE REVIEW

Because it takes many steps to perform an analysis of variance, we briefly review them here before we introduce a new method of analysis.

Analysis of variance is part of the process of studying the association between a categorical independent variable and a metric dependent variable. In the example, we study the two variables violent crime rate and region. Other examples are the study of the relationship between religious affiliation and income and the study of the relationship between different types of teaching methods and student learning.

The analysis is based on the computations of how different the means of the dependent variable are for different categories of the independent variable, as well as how different the observations in each category are from each other. The conclusions we draw from the analysis are based on the magnitudes of various sums of squares. The typical output from a computer analysis is shown in Table 11.4 (naturally, the looks of tables vary from one computer program to another).

We can draw several conclusions from an analysis of variance table. As long as the sum of squares for the categorical variable is different from zero, we know that there is a relationship between the two variables in the observed data. The larger that sum of squares is relative to the other sums of squares, the stronger is the relationship between the two variables. This is also reflected in the proportion obtained when we divide the categorical variable sum of squares by the total sum of squares. This proportion tells us how much of the variation in the values of the dependent variable is explained by the categorical variable, and it can be as small as 0 or as large as 1.

The F-ratio and the p-value that goes with it tell us whether the group means of the dependent variable are significantly different from each other. If F is large and therefore p is small, then we reject the null hypothesis of no differences and conclude that there is a relationship between the two variables in the population from which the data were sampled. By a small p we usually mean anything less than 0.05. Newer statistics software programs for the computer and some hand calculators give us the exact value of p.

If we see $F*$ or $F**$, with a footnote explaining that one asterisk means that p is less than 0.05 and two asterisks mean that p is less than 0.01, a statistical table was used to find an approximate p-value. The shortcoming of statistical tables is that they are never detailed enough to provide the exact p-value; they state only that p is less than some value. Exact p-values are much more informative because then we know how much less than 0.05 or 0.01 the p-value is and how much evidence there is against the null hypothesis.

11.7 MATCHED PAIR ANALYSIS: TWO OBSERVATIONS PER UNIT

The second question we raise in the introduction to this chapter is whether there has been any change in the crime rates over time. Perhaps the most common example of data on change is before-and-after data. This name implies an initial measurement—say, of subjects on an attitude scale—then exposure of the subjects to some kind of stimulus—say, a movie—then a measurement of the subjects again on the same attitude scale to see if there has been a change in their attitudes brought on by the stimulus. Because two paired observations belong to a particular individual, the data are called matched pair data.

A *t*-test

Matched pair analysis is used when we have repeated a measurement with two observations on each element in a study.

Table 11.6 shows matched pair data on violent crime rates of the 48 continental states. The 1986 data from Table 11.1 are given, plus the violent crime rates six years later, in 1992. To see if there has been any change in the crime rates between 1986 and 1992, we can compare the mean scores for the two points in time. If we simply compare the two means without taking the pairing into account, the mean in 1992 minus the mean in 1986 equals $579 - 460 = 119$, which translates to a *t*-value of 2.13 on 94 degrees of freedom and corresponds to a *p*-value of 0.04. This is not quite significant at a two-sided 5% level, since significance requires a one-sided *p*-value of 0.025 or smaller. In this case the independent variable is time, and the residual variable is the effects of all other variables.

One of these other variables is the state variable. For example, New York is a state with two high values, 986 and 1130, while Mississippi is a state with two low values, 274 and 418. They both have changed by 114 crimes per 100,000 inhabitants; the change has been the same for both states. The difference in the values between the two states, one high and the other low, is included in the effect of the residual variable. The effect of the residual variable occurs in the denominator for the *t*-value, and the differences from one state to the next therefore deflate the value of *t*.

To get around this problem, we look at the *differences* in rates at the two points in time. After all, we want to study change. Both New York and Mississippi end up with a difference score of 114, since they both have had the same change in crime rates. Now the data consist of one column of 48 difference scores.

Table 11.6 Violent crimes in the contiguous 48 states per 100,000 population in 1986 and 1992

State	Crime rate 1986	Crime rate 1992	Difference	Region
Maine	147	132	−15	
New Hampshire	140	126	−14	
Vermont	149	111	−38	New England
Massachusetts	557	777	220	
Rhode Island	336	395	59	
Connecticut	426	494	68	
New York	986	1130	144	
New Jersey	572	630	58	Mid-Atlantic
Pennsylvania	359	432	75	
Ohio	423	534	111	
Indiana	308	519	211	
Illinois	800	519	194	
Michigan	804	825	21	
Wisconsin	258	282	24	
Minnesota	285	346	61	
Iowa	235	281	46	Midwest
Nebraska	263	355	92	
Missouri	578	757	179	
North Dakota	51	89	227	
South Dakota	125	199	74	
Kansas	369	520	151	
Delaware	427	643	216	
Maryland	833	816	−17	
Virginia	306	386	80	
West Virginia	164	214	50	South
North Carolina	476	703	227	
South Carolina	675	976	301	
Georgia	588	764	176	
Florida	1,036	1,258	222	

Source: F.B.I. Uniform Crime Reports for the United States

(Table continued on following page)

Table 11.6 Violent crimes in the contiguous 48 states per 100,000 population in 1986 and 1992 *(continued)*

State	Crime rate 1986	Crime rate 1992	Difference	Region
Kentucky	334	546	212	
Tennessee	540	769	229	
Alabama	558	892	334	South
Mississippi	274	418	144	
Arkansas	395	588	193	
Louisiana	758	1,000	242	
Oklahoma	436	636	200	
Texas	659	838	179	
Arizona	658	701	43	Southwest
New Mexico	726	976	241	
Wyoming	293	329	36	
Colorado	524	610	86	
Montana	157	175	18	Rocky
Idaho	222	298	76	Mountains
Utah	267	306	39	
Nevada	719	770	51	
Washington	437	564	127	
Oregon	550	534	262	Pacific Coast
California	920	1,161	241	

To see if there is any difference in these data, we find the mean of the differences. The mean of the differences equals 119, so there has been some change in the crime rates. To see if these differences could have occurred by chance alone, we set up a null hypothesis that the overall mean equals 0. We can test this null hypothesis the same way we did a *t*-test for a single mean in Chapter 7. For these data, $t = 8.74$ on 47 degrees of freedom. This corresponds to a *p*-value less than 0.0001, and now we have overwhelming evidence against the null hypothesis. The computation of t is shown in Formula 11.6.

The sign test: A simple yes or no

Another and simpler approach to the question of whether there was a change over time is to apply a sign test—that is, to consider only whether the scores were more likely to increase or decrease, instead of asking how much they changed. The logic of such a situation is simple as well: if there were no true change, then there would be only random variation in the differences. In our example, each difference would be equally likely to be positive or negative. Thus, we would expect 24 negative and 24 positive differences. Instead, there are 4 negative differences and 44 positive differences.

Now we call on the binomial distribution from Chapter 4 to study the sign of the differences and find the number of positive and negative differences. This problem is like tossing a coin 48 times and getting 4 heads and 44 tails. With only random variation from 1986 to 1992, the probability equals 0.5 that a difference will be positive and 0.5 that a difference will be negative. Thus, we can use the binomial distribution to test the null hypothesis that the probability of a positive difference equals 0.5. To test this hypothesis, we find the probability of getting 44 or more positive differences.

Tables for the binomial distribution do not go as far as 48 observations, but we can find the p-value we need by computing a z-score. We find that $z = 6.21$, and that gives p less than 0.0001, overwhelming evidence also against the null hypothesis that there has been no change.

The t-test for the paired data give a smaller p-value than the sign test using the binomial distribution and the normal approximation. But that is not surprising. The t-test uses the actual numerical values of the observed differences, which is much more information than whether a difference is positive or negative. The t-test also is based on the additional information that the original scores follow a normal distribution. The more of the information in the data we can use, the more significant are the results.

One advantage of the sign test is that it is easier to use. It requires considerably fewer computations than the t-test, and for small sample sizes we can use tables for the binomial distribution directly. If we have any doubts about whether the data follow a normal distribution and whether we should use the paired t-test, then it is safer to use the sign test.

11.8 SUMMARY

11.1 Analysis of variance: Comparing the mean-ings of things

When we study the effect of one (or more) categorical, independent variables on a metric, dependent variable, we can use a statistical method known as analysis of variance or anova, as it is frequently called. This statistical procedure compares the means of the dependent variable for each value of the independent variable(s).

11.2 Question 1. Relationship between violent crime rate and region?

We first make a scatterplot, with the independent variable along the horizontal axis and the dependent variable along the vertical axis. We can also create boxplots to further compare differences in the dependent variable for groups of observations defined by the independent variable.

11.3 Question 2. Strength of the relationship?

Next, we find the mean of the dependent variable for the entire data set and separately for each value of the independent variable. As in regression analysis, it is assumed that the difference between an observation and the overall mean is composed of the effects of the independent plus the residual variable.

To find the overall effect of the independent variable, we subtract the overall mean from each group mean. Then we square all these differences, multiply the squares by the number of observations in the group, and add all the products. To find the overall effect of the residual variable, we find the difference between each observation and the mean of the group it belongs to. Next we square all these differences and add all the squares. This sum of squares is known as the residual sum of squares. To summarize in one number how large the differences are between the observed values and the overall mean, we square the differences between the means of the region and residual variables and add them up. The sum of squared differences is known as the total sum of squares.

For an easier understanding of the sums of squares, we compute the proportion of the total effect that is contributed by the two separate effects. The proportion for the independent variable is found by dividing the group sum of squares by the total sum of squares. This proportion is known as R^2, and it measures the proportion of the variation in the dependent variable that is associated with the independent variable. This proportion is directly comparable to the squared correlation coefficient in regression analysis.

The square root of R^2 is R. This number measures the strength of the relationship between the independent variable and the dependent variable. R ranges in possible values between 0 and 1, and it is directly comparable to the magnitude of the correlation coefficient r for the strength of the relationship between two metric variables in regression and correlation analysis.

An analysis of variance tells us what percentage of the variation in the dependent variable is associated with the independent variable and what percentage with the residual variable. Sometimes words like *explain, produce, account for,* or *cause* are used instead of *associated with*.

11.4 Question 3. Could the relationship have occurred by chance alone?

We test the null hypothesis that the population means of the dependent variable within each category of the independent variable are equal. If the effect of the independent variable is large compared to the residual variable, then we reject the null hypothesis. If the effect is small, then we do not reject the null hypothesis.

To find the *p*-value for the test, R^2 is converted into a value of the *F*-variable. Statistical software or tables can be used to find the *p*-value from *F*. The degrees of freedom for the independent variable equals the number of groups minus one, and the degrees of freedom for the residual variable equals the number of observations minus the number of groups. The independent variable and residual variable are both divided by their degrees of freedom, and the results are called mean squares. The mean square for the independent variable is divided by the mean square for the residual variable to find the *F*-ratio. Once we know there is a statistically significant difference between the means of the dependent variable in the various groups, then we would like to know *which* means are different from each other.

11.5 Question 4. Causal relationship?

Although we cannot ascertain from these data what causes crime rates to differ from one region to another, we can predict which states will have higher and lower crime rates.

11.6 Analysis of variance: A bird's-eye review

This section reviews the process of doing an analysis of variance.

11.7 Matched pair analysis: Two observations per unit

With matched pair data, we subtract one observation from the other to see if there has been any change. For statistical significance, we test the null hypothesis that the mean change in the population equals zero using the t-test for a single mean. For a quick test of the change, we can count the positive and negative differences and use the binomial distribution to see if the probability of a positive difference equals the probability of a negative difference. This sign test is also appropriate if our data do not follow a normal distribution.

ADDITIONAL READINGS

Iversen, Gudmund R., and Helmut Norpoth. *Analysis of Variance*, 2nd ed. (Sage University Paper Series on Quantitative Applications in the Social Sciences, series no. 07-001). Newbury Park, CA: Sage, 1987. Short introduction to analysis of variance.

Toothaker, Larry E. *Multiple Comparison Procedures* (Sage University Paper Series on Quantitative Applications in the Social Sciences, series no. 07-089). Newbury Park, CA: Sage, 1993. How to compare means of some of the groups after the test of overall significance.

FORMULAS

ANALYSIS OF VARIANCE

In the days when the computations for analysis of variance were done by hand or desk calculator, the data were laid out in a table with one column of observations for each group (Table 11.7). A particular observation is labeled with the two subscripts, one identifying the row and one identifying the column the observation is in. The computing formulas for analysis of variance are typically written out for this kind of arrangement of the data. On the other hand, when the data are laid out in a computer file as in Table 11.1, with one column for each variable and one row for each unit, then one column lists the values of

Table 11.7 Layout of data for analysis of variance

	Group			
	1	2	3	4
	y_{11}	y_{21}	...	y_{k1}
	y_{12}	y_{22}	...	y_{k2}
	y_{13}	y_{23}	...	y_{k3}
	.	.		.
	.	.		.
	.	.		.
	y_{1n_1}	y_{2n_2}	...	y_{kn_k}
Mean	$\bar{y}_1$	$\bar{y}_2$	...	$\bar{y}_k$

Overall mean: $\bar{y}$

the dependent variable and another column identifies the group to which a particular observation belongs. In the following formulas, we use the first layout of the data, the layout shown in Table 11.8. For very small data sets, the formulas can be used directly; for larger data sets, analysis of variance is best done on a computer.

The sums of squares are found from these expressions:

$$\text{categorical variable sum of squares} = \Sigma n_i (\bar{y}_i - \bar{y})^2$$

$$\text{residual sum of squares} = \Sigma\Sigma (y_{ij} - \bar{y}_i)^2 \qquad (11.1)$$

$$\text{total sum of squares} = \Sigma\Sigma (y_{ij} - \bar{y})^2$$

$$R^2 = \frac{\text{categorical variable sum of squares}}{\text{total sum of squares}} \qquad (11.2)$$

Table 11.8 Analysis of variance table for one categorical variable

Source	Sum of squares	Degrees of freedom	Mean square	F-ratio	p-value
Categorical variable	CSS	$k - 1$	CMS	F	p
Residual variable	RSS	$n - k$	RMS		
Total	TSS	$n - 1$			

With k groups and n observations in the total sample, the various degrees of freedom are found from these expressions:

$$\text{categorical variable degrees of freedom} = k - 1$$

$$\text{residual degrees of freedom} = n - k \qquad (11.3)$$

$$\text{total degrees of freedom} = n - 1$$

The mean squares are found by dividing the sums of squares by their degrees of freedom.

$$\text{categorical variable mean square} = \frac{\text{categorical variable sum of squares}}{k - 1}$$

$$\text{residual mean square} = \frac{\text{residual sum of squares}}{n - k} \qquad (11.4)$$

Finally, the F-ratio is found by

$$F = \frac{\text{categorical variable mean square}}{\text{residual mean square}} \qquad (11.5)$$

with $k - 1$ and $n - k$ degrees of freedom.

The computations are often summarized in an analysis of variance table (Table 11.8; C denotes categorical variable and R denotes residual variable).

PAIRED DATA

For the ith element, we first find the difference

$$d_i = y_{Bi} - y_{Ai}$$

where y_{Bi} is the first observation (before) and y_{Ai} is the second observation (after). Then we find the mean $\bar{d}$ and the standard deviation s of the d's. The corresponding value of the t-variable is found from the expression

$$t = \frac{\bar{d}}{s/\sqrt{n}} \qquad n - 1 \text{ d.f.} \qquad (11.6)$$

For the example,

$$t = \frac{118.47}{93.93/\sqrt{48}} = 8.74 \qquad 47 \text{ d.f.}$$

EXERCISES

REVIEW (EXERCISES 11.1–11.17)

11.1 Think about the main example in the text on violent crime rates.

a. How do we calculate the violent crime rate for a given state?

b. Why do we prefer to work with violent crime *rates* rather than *number* of violent crimes?

c. Why do we divide the states into regions for the study of violent crimes rather than considering each state individually?

11.2 a. Judging from Table 11.1, which five states have the highest rates of violent crimes?

b. Which five states have the lowest violent crime rates?

c. From Table 11.1 and other displays in the text, are some regions more crime ridden or crime free than other regions?

11.3 Imagine that for each state we know the rate of violent crimes committed by women and the rate of violent crimes committed by men. We want to study the difference between the rates across all the states.

a. Which is the independent and which is the dependent variable?

b. What types (categorical, rank, metric) of variables are we comparing?

11.4 Analysis of variance, abbreviated anova, is a statistical method used to study what kind of data?

11.5 Give an example of an interesting study in which the relationship between two variables could be analyzed using analysis of variance.

11.6 In the analysis of variance for the violent crime rate problem, what are the two factors that completely determine the violent crime rate in a given state?

11.7 In the study of violent crime rates, if both the region and the residual variable had no effect on the rates, how would the rates compare across the states?

11.8 Suppose we find the difference between the violent crime rate in each state and the overall mean violent crime rate for all the states, square all these differences, and add up all the squares.

    a. What is the name of the sum?

    b. How do we find the value of the residual variable for a particular state?

    c. What do we call the sum of the squares of all the residual terms?

11.9 What do we call the question we ask to determine if there is a relationship between the categorical and the metric variable in the larger population from which a sample was drawn?

11.10 a. Do we reject the null hypothesis in analysis of variance for a large or a small R^2?

    b. In addition to the actual value of R^2, what determines whether the null hypothesis is rejected or not?

    c. In analysis of variance, to which statistical variable do we convert R^2 in order to find the p-value for the data?

11.11 a. If the independent variable has 6 values, which means that we are working with 6 groups, how many degrees of freedom does the independent variable have?

    b. How many degrees of freedom does the residual variable have if there are 50 observations in the sample, divided into 6 groups?

11.12 a. What is a mean square for a variable?

    b. How do we calculate a mean square?

    c. How do we compare the mean square of the independent variable with the mean square of the residual variable?

    d. What do we call the result of comparing the mean squares?

11.13 a. Give an example of matched pairs data.

    b. Why might a researcher want to collect such data?

11.14 What does a sign test do?

11.15 a. Why is the residual variable jokingly called the uninteresting variable?

b. Why is the residual variable somewhat misnamed when it is called the error variable?

11.16 a. Why is a *t*-test generally preferred over a sign test?

b. What is the advantage of a sign test?

11.17 How can the analysis of paired data also be seen as a regression analysis where we study whether the regression line is a 45-degree line with intercept 0 and slope 1?

INTERPRETATION (EXERCISES 11.18–11.28)

11.18 An analysis of variance of data on the per capita income in each of the 48 contiguous states for 1994, with the states grouped into 8 regions, gives the results in Table 11.9.

a. What is the independent variable and what is the dependent variable in this analysis?

b. What conclusions can you draw from the results?

c. What are some things you might want to know about regions and incomes that these results do not convey?

11.19 In the Oakland Growth Study, the staff rated a group of high school girls on a good physique variable. For a sample of 35 middle-class girls, the mean and standard deviation were 56.6 and 13.5, and for a sample of 43 working-class girls, the corresponding values were 48.6 and 14.2. *(Source: G. H. Elder, Jr., "Appearance and education in marriage mobility," American Sociological Review, vol. 34 (1969), p. 524.)*

Table 11.9 Data for Exercise 11.18

Source	Sum of squares	Degrees of freedom	Mean square	F-ratio	p-value
Region	195.0	7	27.86	5.33	0.0002
Residual	209.1	40	5.23		
Total	404.1	47			

Source: Bureau of Census, Statistical Abstract of the United States: 1995 *(115th edition), Washington, DC: U.S. Government Printing Office, 1995.*

a. How large is the difference in good physique between the two groups of girls?

b. The difference translates to $F = 6.40$ on 1 and 76 degrees of freedom ($p = 0.013$). Is there a significant difference between the two means?

11.20 Is there any difference in the cost between chocolate and vanilla frozen desserts? In data collected by *Consumer Reports*, the mean cost per serving of chocolate desserts equals 29.4 cents, and the mean cost per serving of vanilla desserts equals 30.4 cents. To see if the difference between the two means is statistically significant, the *t*-test for the difference between two means gives $t = 0.18$ on 42 d.f. An analysis of variance with type of dessert as the independent variable and cost as the dependent variable gives $F = 0.033$ on 1 and 42 d.f. (*Source:* "Low-fat frozen desserts: Better for you than ice cream?" Consumer Reports, vol. 57, no. 8 (August 1992), pp 483–487.)

a. Show numerically that in this case with a nominal variable with only two categories (chocolate and vanilla) $t^2 = F$.

b. Is the difference in cost between the two types of dessert statistically significant?

11.21 Answer the following questions about Table 11.10 on the effects of region and residual on violent crime rates.

a. Which is the independent variable?

b. What is the residual variable composed of?

c. Why does subtracting R^2 from 1.00 give the proportion of the variation in the dependent variable explained by the residual variable?

d. If there were no effect of the region variable, what would the sum of squares for the independent variable be equal to?

Table 11.10 Data for Exercise 11.21

Variable	Effect	Proportion
Independent (region)	Independent sum of squares	R^2
Residual (other)	Residual sum of squares	$1.00 - R^2$
Total	Total sum of squares	1.00

Table 11.11 Data for Exercise 11.23

Source	Sum of squares	Degrees of freedom	Mean square	F-ratio	p-value
High school	1,450	9			
Residual	9,000	91			
Total	10,450	100			

e. If there were no effect of the residual variable, what would the value be of R^2?

f. What would be the values of all the violent crime rates if there were no effect of the region and the residual variables?

11.22 If the value of the F-variable exceeds the critical value of F given in the statistical table, or if the p-value for this F is smaller than 0.05, what can you conclude about the relationship between a metric dependent and a categorical independent variable?

11.23 An analysis of variance of data on the mean GPA scores from a sample of seniors from 10 different high schools in Minneapolis gives the results in Table 11.11. Without making any further computations, what do the numbers tell you about the differences between the high schools in this study?

11.24 The numbers of female and male piglets in litters concerns people who raise pigs. The data in Table 11.12 show the number of female and male pigs in 6 different litters. The difference between the mean number of females and the mean number of males results in $t = -0.29$, 10 d.f., and $p = 0.39$.

    a. What is the null hypothesis being studied here?

    b. Can you conclude that there is a difference in the mean numbers of female and male piglets in the larger population from which these data came?

    c. What aspect of the data is not taken into account in this analysis?

11.25 In their annual review of colleges and universities, *U.S. News and World Report* ranks graduate departments in a dozen fields. People are asked to assign a rank of from 1 to 5 in evaluating a particular department, and the ranks are then averaged to find a score for the

Table 11.12 Data for Exercise 11.24

Litter	Females	Males
1	4	5
2	6	4
3	5	5
4	3	6
5	4	4
6	6	5

Source: S. M. Free, Jr., "Response: The consultant's forum," Biometrics, vol. 33 (1977), no. 3, p. 561.

Table 11.13 Data for Exercise 11.25

Source	Degrees of freedom	Sum of squares	Proportion	Mean square	F-ratio	p-value
University	5	4.40	0.40	0.88	8.85	0.000002
Residual	66	6.57	0.60	0.10		
Total	71	10.97	1.00			

department. For example, Stanford gets a score of 4.9 in biology according to this procedure. Six universities appear on the list of all 12 departments evaluated. The mean scores for the six universities across all 12 departments are as follows: Columbia 4.13, Harvard 4.55, Princeton 4.44, Stanford 4.78, University of California at Berkeley 4.79, and University of Wisconsin at Madison 4.24. To see if there are significant differences between the schools, we perform a one-way analysis of variance on the scores. The results are in Table 11.13.

   a. What does the table tell you about the differences in scores between the universities?

   b. What other analyses might you want to perform to assess how different the mean scores are from one school to the next?

11.26 The Highway Loss Data Institute collects data on the number of insurance claims per insured vehicle for different car models. The data used here are from 1991–1993, and they are scored in such a way that 100 represents the mean number of claims for all models. With this scoring, the numbers range from a low of 44 insurance claims for the Chevrolet Suburban to 201 for the Hyundai Elantra; the number of insurance claims for Suburban is less than half the average for all cars, while the number of claims for Elantra is about double the average for all cars. Is there a difference in the number of claims between small, mid-size, and large cars? In a random sample of cars there is a mean of 155 claims for 5 small cars, 95 claims for 12 mid-size cars, and 60 claims for 5 large cars. An analysis of variance for these cars using number of claims as the dependent variable and size as the independent variable gives the results in Table 11.14.

   a. Use size on the horizontal axis and number of claims on the vertical axis and graph the three means.

Table 11.14 Data for Exercise 11.26

Source	Degrees of freedom	Sum of squares	Proportion	Mean square	F-ratio	p-value
Size	2	23,110	0.75	11,555	28.20	<0.0001
Residual	19	7,786	0.25	410		
Total	21	30,897	1.00			

Source: Highway Loss Data Institute, as reported in Motor Trend, *vol. 47, no. 1 (January 1995), p. 77.*

      b. What does the analysis of variance tell you about the relationship between size of car and number of insurance claims?

      c. What might possibly explain these results?

11.27 In a comparative study, people were sampled in several different European countries on a wide variety of issues. One of the questions asked was "Altogether, you made how many holiday trips, each lasting four days or more, in 1985?" *(Source: Jacques-René Rabier, Helen Riffault, and Ronald Inglehart,* Euro-barometer 25: Holiday Travel and Environmental Problems, *April 1986. ICPSR ed. Ann Arbor, MI.: Inter-University Consortium for Political and Social Research, 1988. Codebook p. 20.)* The mean number of trips for Denmark was 1.06, for France 1.11, for Ireland 0.81, and for Portugal 0.41. An analysis of variance to see if these means are significantly different gives $F = 85.77$ on 3 and 4,019 degrees of freedom and $p < 0.0001$.

      a. What do the results tell you about the differences in number of vacation trips people in different countries took?

      b. What else might you want to know about these data to better understand the differences in numbers of vacation trips?

11.28 In the study of Europeans cited in Exercise 11.27, people were also asked about the satisfaction they felt with the lives they were leading. Ranking very satisfied as 1, fairly satisfied as 2, not very satisfied as 3, and not at all satisfied as 4, we find a mean satisfaction of 1.41 in Denmark, 2.16 in France, 1.65 in The Netherlands, and 1.88 in West Germany. The overall mean equals 1.77. An analysis of variance to study these country differences gives $R^2 = 0.16$ and $F = 250$ on 3 and 3,995 degrees of freedom. What do the results tell you about the differences between the four countries?

ANALYSIS (EXERCISES 11.29–11.46)

11.29 Is there difference in the mandible (lower jaw) lengths of prehistoric female and male golden jackals (Canis aureus)? The following lengths, in millimeters, are from the jackal collection of the British Museum.

Female	110	111	107	106	110	105	107	106	111	111
Male	120	107	110	116	114	111	113	117	114	112

Source: C. F. Higham, A. Kijngam, and B. F. J. Manly, "An analysis of prehistoric canid remains from Thailand," Journal of Archaeological Science, *vol. 7 (1980), pp. 149–165.*

    a. Find the mean lengths for the two groups.

    b. Do a *t*-test for the difference between the two means or an analysis of variance to see if the difference in the means is statistically significant.

11.30 Go to the Springer Web site (http://www.springer-ny.com/supplements/iversen) to find files relating to this book. The data file "Singers" contains data on heights of sopranos, altos, tenors, and basses in the New York Choral Society. *(Source: J. M. Chambers et al.,* Graphical Methods for Data Analysis, Boston: *Duxbury, 1983. p. 350.)*

    a. Are there differences in heights among the four groups of singers?

    b. Could the differences have occurred by chance alone?

    c. Does type of voice affect height, or could there be other variables involved?

11.31 Table 11.15 shows the flavor quality of different vanilla frozen desserts measured on a scale from 0 to 100 by a panel of trained tasters.

    a. Make a scatterplot of the data. What does the scatterplot show you?

    b. Find the mean flavor quality for each group to see if there is any difference in flavor quality among the three types of desserts.

    c. Do an analysis of variance to find the strength of the relationship is between type of dessert and flavor quality and to see if the differences between the means are statistically significant.

11.32 During the Cold War, the countries involved spent considerable amounts of money on national defense. Different countries make

Table 11.15 Data for Exercise 11.31

Frozen yogurts	Ice milks	Frozen desserts
87	83	33
74	76	31
70	76	31
68	70	31
68	58	27
67	52	10
64	50	
64	47	
63		
57		
54		
50		
48		

Source: "Low-fat frozen desserts: Better for you than ice cream?" *Consumer Reports,* vol. 57, no. 8 (August 1992), pp. 483–487.

different decisions about how to allocate their resources, and the final allocation each year is a reflection of a complex set of values. Here we want to compare resources allocated to education and defense by a few countries during the Cold War using a paired data approach (see Table 11.16).

    a. How large is the difference between the two mean percentages?

    b. Change the mean difference into a *t*-value. (As it happens, it makes very little difference whether the data in this exercise are analyzed using the paired method or not.)

    c. If these countries were a random sample of countries, what could you conclude from this value of the *t*-variable?

11.33 Does more racial discrimination occur in northern or southern cities? To answer this question, each of ten northern cities are matched on similarities of racial composition, median income, and amounts and types of industries with a similar southern city. By matching cities on these variables, the possible effects of these variables on the results of the analysis are eliminated. Each city is then scored on a scale from 0

Table 11.16 Data for Exercise 11.32

Country	Education expenditures as per cent of national income 1969	Defense expenditures as per cent of GNP 1969
Australia	4.0	3.6
China (Taiwan)	3.8	8.8
Hungary	4.4	3.5
Korea, Republic of	3.8	4.0
Norway	6.3	2.9
Sudan	4.9	6.0
United States	6.3	7.8
Yugoslavia	5.1	5.4

Source: U.S. Bureau of the Census, Statistical Abstract of the United States: 1972, *93rd ed. Washington, DC: U.S. Government Printing Office, 1972, pp. 809, 831.*

to 100 for racial discrimination. Discrimination is based on indices like amount of open housing legislation, integration of the schools, differences in incomes for racial groups, and differences in unemployment of racial groups. A high score indicates more discrimination. The discrimination scores are shown in Table 11.17. (In this exercise it makes a large difference for the analysis whether the data are correctly analyzed using the paired method or incorrectly analyzed by just comparing the means of the southern and the northern cities.)

    a. How do the different pairs of cities compare on this variable? (One way to compare them is to graph each pair of cities in a scatterplot with North/South as the independent variable and score as the dependent variable.)

Table 11.17 Data for Exercise 11.33

	Pair of cities									
	1	2	3	4	5	6	7	8	9	10
Southern	72	52	59	36	67	25	80	41	62	55
Northern	79	68	45	45	59	38	75	56	72	60

b. What does the mean of the differences tell you about the difference between cities in the two parts of the country?

c. Is the mean difference statistically different from zero?

11.34 Much discussion has taken place about the extent to which the population in this country shifted from the northern and eastern parts to the southern and western parts in the 1960s. The data in Table 11.18 give the percentages of change in population in that period for a random sample of standard metropolitan statistical areas (SMSAs) with populations of 200,000 or more. The SMSAs are classified by their census regions. For these data the total sum of squares equals 12,738, and the residual sum of squares within the four groups of observations equals 8,822.

a. Make a scatterplot of the data and include the four group means on the plot.

b. Judging from the scatterplot, did the four groups of cities have the same growth?

c. Find the F-value for these data.

d. On the basis of F, is there a significant difference in the mean percentages of growth in the regions?

Table 11.18 Data for Exercise 11.34

	Region		
West	North Central	Northeast	South
20.4	9.2	6.1	36.7
32.1	21.4	5.2	24.0
89.0	13.6	10.8	19.4
41.2	22.4	−0.2	85.7
39.0	15.2	7.6	40.0
3.3	12.8	12.6	16.1
32.4		4.7	8.8
			18.9
			14.4
			30.1

Source: U.S. Bureau of the Census, Statistical Abstracts of the United States: 1972, *93rd. ed. Washington, DC: U.S. Government Printing Office, 1972, pp. 838–878.*

11.35 The researchers cited studied a sample of 22 gymnasts and a sample of 21 swimmers. For each athlete they computed adult height predictions based on several observed variables. For the gymnasts, the mean height prediction was 5.48 cm/year, and for the swimmers, it was 8.00 cm/year. The standard error of the mean for the gymnasts was 0.32 cm/year and for the swimmers 0.50 cm/year. The researchers reported that the mean for the gymnasts was significantly lower than the mean for the swimmers, with $p < 0.05$. (Source: G. E. Theintz, H. Howard, U. Weiss, and P. C. Sizonenko, "Evidence for a reduction of growth potential in adolescent female gymnasts," The Journal of Pediatrics, vol. 122 (1993), no. 2, pp. 306–313.)

    a. What is the independent variable in this study?

    b. What is the dependent variable in this study?

    c. Find the exact p-value for the difference between the two means.

    d. Why is the exact p-value more informative than the authors' statement that $p < 0.05$?

11.36 A set of data shows the number of paid vacation days per year in several European countries, according to agreements between workers and employers: Austria 25, Belgium 25, Denmark 25, Spain 32, Finland 30, France 30, Great Britain 20, Iceland 24, Ireland 18, Italy 25, Norway 21, Netherlands 25, Portugal 30, Sweden 40, Switzerland 23. (Source: Juliet B. Scor, The Overworked American: The Unexpected Decline of Leisure, New York: Basic Books, 1991, p. 82.)

    a. Make a boxplot of the data.

    b. Guess, estimate, or find the number of paid vacation workdays for workers in the following countries in North America: United States, Canada, Mexico, Cuba, Haiti. Create a boxplot for North America similar to the one for Europe.

    c. What are some of the comparisons you can make from the two boxplots of the two groups of countries?

11.37 Table 11.19 is an analysis of variance table from a study on climactic conditions and homicide rates.

    a. Fill in the blanks in the table.

    b. Does climate zone seem to have an effect on the dependent variable? Explain your answer.

Table 11.19 Data for Exercise 11.37

Source	Degrees of freedom	Effect	Proportion	Mean square	F-ratio	p-value
Climate zone	7	88,866				0.002
Residual	40					
Total		218,031				

c. How might you go further in determining which climate zones are different from each other and which are not different from each other?

11.38 Table 11.20 shows data on times in minutes of relief of headaches for a standard and a new treatment. We want to see if there is a difference between the two treatments. An ordinary t-test for the difference between the two means results in a nonsignificant $t = 1.15$ with

Table 11.20 Data for Exercise 11.38

Person	Standard treatment	New treatment
1	8.4	6.9
2	7.7	6.8
3	10.1	10.3
4	9.6	9.4
5	9.3	8.0
6	9.1	8.8
7	9.0	6.1
8	7.7	7.4
9	8.1	8.0
10	5.3	5.1
Mean	8.43	7.68

Source: A. J. Gross and V. A. Clark, Survival Distributions: Reliability Applications in the Biomedical Sciences, *New York: John Wiley & Sons, 1975, p. 232.*

Table 11.21 Data for Exercise 11.41

Baby	Caretaker	Cost per hour
1	Relative	4.90
2	Nanny	7.00
3	Relative	5.00
4	Day care center	6.60
5	Private home	5.35
6	Nanny	7.50
7	Private home	5.50
8	Day care center	6.75
9	Relative	5.25
10	Private home	5.15
11	Nanny	7.55
12	Day care center	6.67
13	Relative	5.10
14	Private home	5.35
15	Nanny	7.40
15	Day care center	6.75

Source: Sandra L. Hofferth, Urban Institute.

18 d.f., $p = 0.13$. This value of t is based on the average of the standard deviations within the two groups. These standard deviations are heavily influenced by the extreme reactions of some of the people. In particular, the scores of persons 3 and 10 contribute heavily to the standard deviations because the scores are so different. One way around this problem is to use the paired aspect of the data and look at the differences in minutes between the standard and the new treatment.

a. Find the difference in minutes between the standard and the new treatment for each person and the mean of the ten differences.

b. How does the mean of the differences compare to the difference between the two means in the table?

c. Find the value of t for the test of the null hypothesis that the population mean difference equals zero.

d. Why is the value of t in this case based on only 9 degrees of freedom?

e. What is the *p*-value for the new value of *t* and what do you conclude about the difference between the standard and the new treatment?

11.39 a. Analyze the data in Exercise 11.38 using the sign test.

b. How do the *t*-test and the sign test compare?

11.40 a. Complete the table in Exercise 11.23.

b. What can you now conclude about the relationship between the two variables?

11.41 Table 11.21 summarizes data on the cost of child care. Use analysis of variance to see if the different types of caretakers charge different rates for child care.

11.42 This is a famous data set collected by Charles Darwin. The data show in inches the heights of plants grown in 15 pairs. One member of each pair was cross-fertilized and the other was self-fertilized. Darwin was interested in whether there was any difference in heights between the two groups.

Pair	1	2	3	4	5	6	7	8
Cross-fertilized	23.5	12.0	21.0	22.0	19.1	21.5	22.1	20.4
Self-fertilized	17.4	20.4	20.0	20.0	18.4	18.6	18.6	15.3

Pair	9	10	11	12	13	14	15
Cross-fertilized	18.3	21.6	23.3	21.0	22.1	23.0	12.0
Self-fertilized	16.5	18.0	16.3	18.0	12.8	15.5	18.0

Source: Charles Darwin, The Effect of Cross- and Self-fertilization in the Vegetable Kingdom, *2d ed., London: John Murray, 1876, p. 451.*

a. Find the difference in height for each pair and test the null hypothesis that the mean difference equals zero.

b. Consider these data as two independent groups of observations and do an ordinary *t*-test for the difference between the two groups.

c. How do the two analyses compare?

11.43 Several societies were rated on two psychological variables, "degree of oral socialization anxiety" and whether "oral explanations of illness" were present or absent. The societies without oral explanations had the following anxiety scores:

$$6\ 7\ 7\ 7\ 7\ 7\ 8\ 8\ 9\ 10\ 10\ 10\ 10\ 12\ 12\ 13 \quad (\bar{y} = 8.9, s = 2.14)$$

The societies with oral explanations had the following anxiety scores:

$$6\ 8\ 8\ 10\ 10\ 10\ 11\ 11\ 12\ 12\ 12\ 12\ 13\ 13\ 13\ 14\ 14\ 14\ 15\ 15\ 15\ 16\ 17$$
$$(\bar{y} = 12.2, s = 2.73)$$

Source: J. W. M. Whiting and I. L. Child, Child Training and Personality, *New Haven: Yale University Press, 1953, p. 156.)*

Analyze the data, using the observations as if they were metric data.

11.44 Cork comes from the bark of the cork tree. Is there any difference in the weight of cork deposits on the north and the south sides of trees? Following are data in grams of cork deposits for 28 trees. Is there any difference in deposits between the two sides of the trees?

Tree	1	2	3	4	5	6	7	8	9	10	11	12	13	14
North	72	60	56	41	32	30	39	42	37	33	32	63	54	47
South	76	66	64	36	35	34	31	31	31	27	34	74	60	52

Tree	15	16	17	18	19	20	21	22	23	24	25	26	27	28
North	91	56	79	81	78	46	39	32	60	35	39	50	43	48
South	99	47	70	68	67	37	34	30	67	48	39	37	39	57

Source: C. R. Rao, "Tests of significance in multivariate analysis," Biometrica, *vol. 35 (1948), pp. 58–79.*

11.45 Is there a difference in how much time teachers spend in the classroom in lower and upper secondary schools? Table 11.22 shows number of hours per year spent teaching in a sample of countries.

a. Analyze these data.

b. Do we gain much by taking into account that these are paired data?

11.46 Exercise 10.73 shows these data on ages for grooms and brides applying for marriage licenses in the form (groom age, bride age):

Table 11.22 Data for Exercise 11.45

Nation	Lower secondary	Upper secondary
Germany	761	673
Ireland	792	792
Italy	612	612
Norway	666	627
Spain	900	630
Sweden	576	528
United States	1,042	1,019
Mean	764	697

Source: OECD, from The New York Times, *May 28, 1995, p. E7.*

(37,30) (30,27) (65,56) (45,40) (32,30) (28,26) (45,31) (29,24) (26,23)
(28,25) (42,29) (36,33) (32,29) (24,22) (32,33) (21,29) (37,46) (28,25)
(33,34) (17,19) (21,23) (24,23) (49,44) (28,29) (30,30) (24,25) (22,23)
(68,60) (25,25) (32,27) (42,37) (24,24) (24,22) (28,27) (36,31) (23,24)
(30,26)

Source: The Philadelphia Inquirer, *September 10, 1995, p. MD-12d.*

a. Treat the data as paired data and determine if there is a significant difference in the ages of the brides and the grooms using the period *t*-test.

b. Use the sign test on these data to see if there is a difference in the ages.

CHAPTER 12

12.1 Two rank variables with words as the values

12.2 Ranking numbers as values: How are the Phillies doing?

12.3 Summary

RANK METHODS FOR TWO RANK VARIABLES

12

What makes some people more interested in political elections than others? Maybe it has something to do with how close they feel to a political party.

How do the standings of baseball teams change over time? Do the good teams stay good season after season? These and other questions can often be answered using rank variables.

Most of the variables mentioned so far have been categorical or metric variables, and by now you are used to distinguishing between them. Have you ever wondered how you might measure variables such as class rank, runners in a race, or an attitude? These variables are neither categorical nor metric but *rank variables.*

As the name implies, rank variables are variables that compare individual elements on some feature in terms of quantity (more and less). For example, voters can be very much interested, somewhat interested, or not very interested in the outcome of an election. Social class can be treated as a rank variable; people can be ordered as to whether they are upper, middle, or lower class. Social psychologists who study attitudes use rank variables to assess the strength as well as the direction of people's opinions. Attitude variables might be ranked on a value scale consisting of strongly opposed, opposed, neutral, in favor, or strongly in favor. Not only can people be shown to have different attitudes, but individuals or groups can be shown to have more strongly held attitudes than others, according to the value of the rank variable. For these rank variables we use *words* to describe the values of the variables.

During the season we look in the sports pages for the standings of our favorite baseball team. The team may be in first place, second place, or some place farther down. In the annual Kentucky Derby race, horses are ranked as they cross the finish line. We know which horse won, which came in second, third, and so on down to the horse that came in last. Similarly, in a draft of athletes, professional sports teams rank the players, and the best players are chosen first. Another, more unusual example of a rank variable comes from China where local officials as early as the 1700s ranked the harvest each year on a scale from 1 to 10, with 10 being the best. For these rank variables we use *numbers* as values.

Sometimes we deliberately create rank data from metric data. Say the original scatterplot for two metric variables shows a nonlinear pattern so that it does not make sense to try to fit a line to the data. A linear pattern might emerge when we change the data on the two variables to rank variables. Then we can analyze the data using the rank method described in this chapter.

Data on one rank variable and one metric variable can also be analyzed this way. If we want to compare the finishing positions of horses in the Kentucky Derby and their financial values, we could find how strongly the two variables are correlated by changing the financial values to rankings.

Rank variables are more complex than categorical variables. With rank variables, we order the values of the variable. Not only is one observation different from another observation on a rank variable, one observation is *more* or *less* than another observation. But notice that with a rank variable we do not know how much more or less one observation is from another. In a horse race we do not know whether the second horse was right behind the first horse or several lengths behind. For the attitude variable we do not know if the amount of difference between one person who agrees and another who is neutral is the same as the amount of difference between one person who is neutral and another who disagrees.

Rank variables are sometimes known as *ordinal variables* because of the ordering of the values. Rank variables are not used as often as categorical variables or metric variables.

> For two observations on a rank variable we can determine whether they are the same or different. In addition, we can determine if one observation is more (or less) than another observation.

STOP AND PONDER 12.1

Give an example of a rank variable and list the values of this variable. Why is your variable a rank variable?

12.1 TWO RANK VARIABLES WITH WORDS AS THE VALUES

Why are some people interested in political elections while others are not? In a famous report on American politics, the authors wondered if interest had to do with how closely people identified with either of the major political parties. Table 12.1 shows the distribution

Table 12.1 Party identification and interest in the 1956 presidential election

		Party identification			
		Independent	Weak	Strong	Total
	Very much	104	150	262	516
Interest	Somewhat	178	273	237	688
	Not much	133	228	125	486
	Total	415	651	624	1690

Source: Angus Campbell, Philip E. Converse, Warren E. Miller, and Donald E. Stokes, The American Voter: An Abridgement, *New York: John Wiley & Sons, 1964, p. 84.*

of a sample of American people on these two variables at the time of the presidential election in 1956 when Adlai Stevenson ran as a Democrat against the incumbent Republican Dwight Eisenhower. Stevenson had been governor of Illinois and had lost the presidential election four years earlier. Eisenhower had been president for four years and was a popular military figure from World War II. The country was at peace, after the Korean war, and the Beatles were about to hit the scene. What did the people think about the coming election?

In the survey, people were asked if they identified themselves as a strong Democrat or Republican, weak Democrat or Republican, or independent and how interested (very much, somewhat, or not much interested) they were in the 1956 presidential election. In the table, party identification is the x- or independent variable, placed horizontally, and interest is the y- or dependent variable and placed, vertically. Strong Democrats and strong Republicans are combined as strong identifiers, and weak Democrats and weak Republicans are combined as weak identifiers.

Question 1. Relationship between identification and interest?

By skimming the table from right to left, we see that from the strong identifiers to the weak identifiers to the independents there seems to be a decrease in the interest in the election. The decrease indicates a relationship between the two variables party identification and interest in these data.

The data can, of course, be displayed in various graphs. Figure 12.1 shows four ways of graphing the data. Figure 12.1a, a bar graph in which the bars have the same widths and different heights, shows that fewer people are independents than people who identify with a political party. It also shows that more people with a weak identification are not much interested in the election compared to the other two identification groups. Figure 12.1b, a bar graph in which the bars have different widths and the same heights, shows that there are fewer independents than people with either weak or strong interest in the group and that of strong identifiers is a smaller percentage of people with not much interest in the election. Figure 12.1c is a series of circles; the area of each circle corresponds to the number of observations in the same cell of the table. In this graph it is harder to see the totals, but the sizes of the circles clearly show where there are many people and where there are only a few. Finally, Figure 12.1d shows the nine frequencies as prisms of varying heights corresponding to the number of observa-

12.1 Two Rank Variables with Words as the Values   529

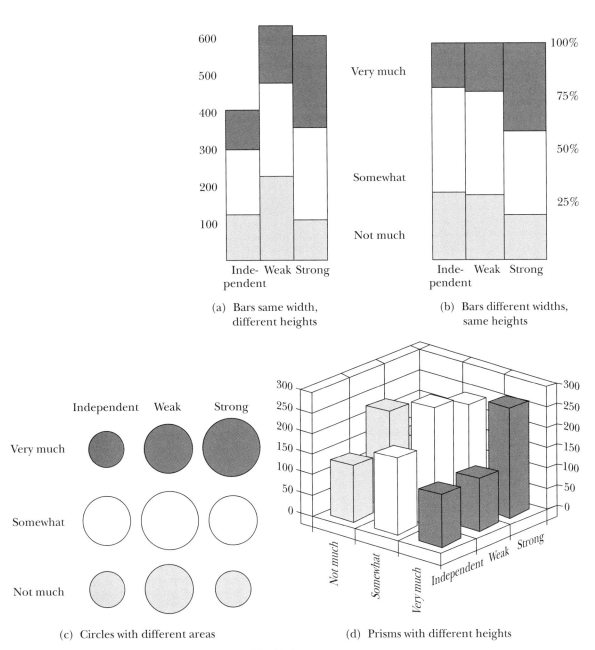

Figure 12.1 Four graphs of the data in Table 12.1

tions. In this graph, too, it is hard to see the totals, but it is simple to compare the frequencies within rows and within columns.

Table 12.1 can be modified to display the data as percentages, as shown in Table 12.2. The three columns of percentages in Table 12.2 and the graphs in Figure 12.1 all show a relationship between the two variables in the data.

STOP AND PONDER 12.2

In a study of sex-role identity and occupational attainment, 161 employed Asian-American women were given a psychological sex role inventory test. High, medium, and low masculinity scores were compared with high or low scores on an occupational attainment variable. Among the high masculine women (45), 81% scored high in occupational attainment and 19% scored low in attainment. Among the medium masculine women (45), the corresponding percentages were 71 and 29. Among the low masculine respondents, the corresponding percentages were 51 and 49. *(Source: Esther Ngan-Ling Chow, "The influence of sex-role identity and occupational attainment on the psychological well-being of Asian American women,"* Psychology of Women Quarterly, *vol. 11 (1987), pp. 69–81.)*

How might you illustrate the relationship between the two rank variables? How would you summarize the relationship between the two variables in everyday language?

Question 2. Strength of the relationship?

A commonly used measurement of strength for two rank variables is a coefficient called *gamma*. For the election interest data, gamma is equal

Table 12.2 Percentage distributions of interest for each identification group

		Party identification		
		Independent	Weak	Strong
	Very much	25%	23%	42%
Interest	Somewhat	43	42	38
	Not much	32	35	20
	Total	100%	100%	100%

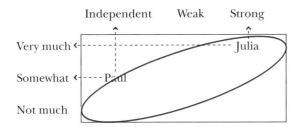

Figure 12.2 Julia ranks higher than Paul on both variables.

to 0.21. Like other coefficients for the strength of a relationship, the value of gamma lies between −1 and 1. A gamma of 0.21 indicates a weak and positive relationship between the two variables in the data.

As with other coefficients of strength, the meaning of gamma has to do with prediction. With rank variables, we try to predict how two individuals rank relative to each other on the dependent variable based on how they rank relative to each other on the independent variable.

Suppose Julia is a strong identifier and is very much interested in the election. Suppose Paul is an independent and somewhat interested in the election. If the identification variable runs horizontally from left to right and the interest variable runs vertically from bottom to top, Julia ranks higher than Paul on the party identification variable because she is located to the right of Paul (Figure 12.2). At the same time, she also ranks higher than Paul on the interest variable because she is located higher than Paul.

Suppose we know only how Julia and Paul compare on the identifier variable. Can we predict how they rank on the interest variable? Assuming that all the people are located in the ellipse from lower right to upper left in Figure 12.2, then Julia will rank higher on both variables than any other person except people who have the same value on one or both variables. Thus, if we know that Julia ranks higher on the independent variable, then we can predict without error that she will rank higher on the dependent variable as well. In this case of perfect prediction, gamma equals 1.00.

The opposite extreme occurs when the observations are all located within an ellipse from the upper left corner the table to the lower right corner, illustrated in Figure 12.3. Here, Julia is a strong identifier and not very interested. Paul is an independent who is somewhat interested. Julia now ranks higher than Paul on the party identification variable since she is located farther to the right than Paul. But Paul ranks higher on the interest variable since he is above Julia. The same pattern will

> The coefficient **gamma** measures the strength of the relationship between two rank variables that have words as values of the variables.

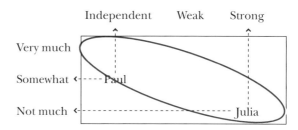

Figure 12.3 Julia ranks higher on party identification, Paul ranks higher on interest.

occur for any two people we match up for observations located in the ellipse from the upper left to the lower right unless they have the same value on one or both variables. If we know the ranking on the independent variable, then we can predict perfectly the ranking on the dependent variable. In this case, gamma equals −1.00.

Gamma is found by counting how many pairs of observations there are of each of the two kinds shown in Figures 12.2 and 12.3. Gamma tells us how much better we can predict the rankings for two people on the dependent variable when we know how they compare on the independent variable than when we do not have this knowledge. With gamma equal to 0.21, knowing how they compare on the independent variable improves our prediction 21% over not knowing how they compare. A more detailed interpretation of gamma and directions for computing it is shown in Formula 12.1 at the end of the chapter.

Question 3. Relationship in the population?

As usual, the null hypothesis states that there is no relationship between the two variables in the real world, and it is rejected if the *p*-value for the data is small. The *p*-value is the probability of getting a gamma of 0.21 or larger in a sample that comes from a population where there is no relationship between the two variables.

The *p*-value is found by first changing the observed value of gamma to a value of the standard normal variable *z*. For the example, *z* equals 6.47. Using statistical software or tables, we find that the *p*-value for a *z* of this magnitude is less than 0.0001. The probability of data leading to a gamma of 0.21 or larger by chance alone is incredibly small; in fewer than 1 of 10,000 different samples from a population where there is no relationship between the two variables would gamma be 0.21 or larger. Such a small *p*-value is very strong evidence against the null hypothesis, and we conclude that the two variables are related in the population of all adults. Formulas 12.2 and 12.3 show how to change gamma to a value of the z-variable.

Question 4. Causal relationship?

From the data alone we cannot determine whether the relationship is causal or not. Here, even which variable comes first is questionable. Do people who are strong party identifiers become interested in an election because of their strong identification, or do people who are interested in elections come to identify strongly with one of the parties?

"Cathy" © 1995 Cathy Guisewite. Reprinted with permission of Universal Press Syndicate. All rights reserved.

> **STOP AND PONDER 12.3**
>
> The gamma for the relationship between two rank variables equals 0.47. The corresponding *p*-value is 0.15. What can you say about the relationship between the two variables in terms of prediction and statistical significance?

12.2 RANKING NUMBERS AS VALUES:
HOW ARE THE PHILLIES DOING?

How do baseball teams compare with each other over time? Do the good teams remain good, or is there change over time? Here we study the change in standings of the six baseball teams in the Eastern Division of the National League, which includes our home team, after the 1987 and 1992 seasons. Table 12.3 shows the standings of the teams at the beginning and end of the 5-year period. The two columns of rankings show many changes in this period. To get a better understanding of the data, we do some statistical analyses.

Question 1. Relationship in the data?

Even though the variables are only rank variables, we can display the data in a scatterplot, with 1987 along one axis and 1992 along the other, because we have numbers for the values of the variables. In Figure 12.4, each team appears as a point in the scatterplot. With only

Table 12.3 Standings of the teams in the National League East after the 1987 and 1992 seasons

	Standing	
Team	1987	1992
Chicago Cubs	6	4
Montreal Expos	3	2
New York Mets	2	5
Philadelphia Phillies	4	6
Pittsburgh Pirates	5	1
St. Louis Cardinals	1	3

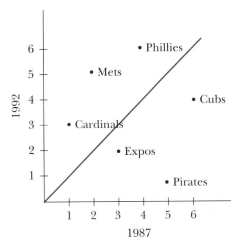

Figure 12.4 Scatterplot of team standings after the 1987 and 1992 seasons

a few observations, it is often helpful to label them, in this case with the names of the teams. If a team had a high standing both years, the team would appear in the lower left corner of the graph; no team appears there. A team with a low standing both years would appear in the top right corner of the graph; our Phillies and the Cubs are near that corner.

What would the scatterplot have looked like if there had been no change in standings? Each team would have had the same standing at the two points in time, so the 6 points would have been on the 45-degree line drawn across the plot, in Figure 12.4 showing a very strong pattern in the data. But the observed points in the graph are scattered and show very little pattern. There doesn't seem to be much of any relationship between the two years.

Question 2. Strength of the relationship?

The strength of the relationship between two rank variables with numbers as values is measured by a coefficient called the *rank order correlation coefficient*, often denoted r_s. The subscript "s" is in honor of the British psychologist and statistician Charles Spearman, who did pioneer work in this area in the early 1900s. For this reason the coefficient is sometimes called the Spearman rank correlation coefficient.

The **rank order correlation coefficient,** often denoted r_s for its originator, Charles Spearman, measures the strength of the relationship between two rank variables with numerical values.	The rank correlation r_s is computed the same way the correlation coefficient r is computed for metric data. We use the numerical ranks as if they were metric data, so r_s is only a special case of r. To find r_s with statistical software, we simply select the two rank variables and ask for the ordinary correlation coefficient r. There is one small difference between the two coefficients. You can see from the formula at the end of the chapter that r_s is much easier to compute using paper and pencil than is r. This is because the values of both variables consist of the integers from 1 to whatever many observations we have, which simplifies the formula for r and gives the formula for r_s.

For the baseball data, r_s is equal to -0.09, which on a scale from -1 to 1 indicates a very weak relationship. If there had been no change and the points had been on the 45-degree line, then the r_s would have been equal to 1.00. If there had been a complete reversal in the standings, then the points would have been on a 45-degree line from the upper left part of the graph to the lower right part. In that case, r_s would have been equal to -1.00. Formula 12.4 shows how to compute the Spearman rank order correlation coefficient.

From the observed coefficient, we conclude that there has been change in the rankings because the value of the coefficient is so far from 1. We also conclude that there is hardly any pattern in the change because the value is so close to 0. If anything, since the coefficient is negative, the good teams in 1987 showed a small tendency to become the bad teams 5 years later, and the bad teams in 1987 improved during the 5 years.

STOP AND PONDER 12.4

How might being able to use Spearman's correlation coefficient be helpful if one were an avid racetrack gambler?

Question 3. Did the relationship occur by chance?

To see whether the changes in standings could have occurred by chance or not, we formulate a null hypothesis and then compute the p-value for the data. In this case, the null hypothesis states that the relationship between the two variables is due to chance alone.

The p-value is found by first changing the value of r_s to the corresponding value of one of the standard statistical variables. In this case we change to a t-variable with $n - 2 = 6 - 2 = 4$ degrees of freedom.

We find that t is equal to -0.17, and the probability of getting a value of t and thereby a value of r_s equal to or larger than -0.09 by chance alone equals 0.44: in 44 of 100 samples from a population where there is no relationship between the two variables would we get an r_s of -0.09 or a more negative value. This p-value is nowhere near significant. Apparently whichever teams were good or bad in 1987 had very little to do with which teams were good or bad 5 years later.

Question 4. Causal relationship?

As usual, we cannot go directly from statistical correlation to causation, but here it certainly looks as if the low correlation was produced by chance and there was no causation involved.

12.3 SUMMARY

A rank variable is created when a group of elements is placed in order according to some comparison, such as size, quality, age, or speed. Rank variables can have words or numbers as values. A rank variable with words as values is ordered on a verbal sliding scale, such as very much, moderately, slightly, not at all. There are usually many observations with the same value. A rank variable with numbers as values is ordered on a numerical sliding scale, such as 1, 2, and so on up to however many observations are ranked. Rank variables are more informative than categorical variables and less informative than metric variables, and they can therefore not be analyzed with methods designed for the other two types of variables.

12.1 Two rank variables with words as values

A commonly used measure of the strength of the relationship between two rank variables with words as values is called gamma. Gamma measures the similarity of paired observations in terms of their relative ranks on the two variables. The null hypothesis of no relationship in the larger population is evaluated by converting gamma to a z-score and then finding the p-value. Finding correlations between rank variables allows us to predict the rank value of one variable from the rank value of another. A rank correlation does not imply that there is a causal relationship between the two variables.

12.2 Ranking numbers as values: How are the Phillies doing?

When the observations are ranked by numbers, with one number for each observation (unless there are tied observations), the strength of the relationship is measured by the Spearman rank order correlation coefficient, often denoted r_s. The null hypothesis of no relationship in the larger population is evaluated by converting the r_s to a *t*-score and then finding the *p*-value.

ADDITIONAL READINGS

Clogg, Clifford C., and Edward S. Shihadeh. *Statistical Models for Ordinal Variables* (Sage Advanced Quantitative Techniques in the Social Sciences, Volume 4). Thousand Oaks, CA: Sage, 1994. An advanced book on rank variables.

Hildebrand, David K., James D. Laing, and Howard Rosenthal. *Analysis of Ordinal Data* (Sage University Paper Series on Quantitative Applications in the Social Sciences, series no. 07-008). Beverly Hills, CA: Sage, 1977. A brief introduction to the analysis of rank variables.

Kendall, Maurice, and Jean Dickinson Gibbons. *Rank Correlation Methods,* 5th ed. New York: Oxford University Press, 1990. A classic book on the topic.

FORMULAS

GAMMA

Gamma is based on pairing and comparing the ranks of the two elements on the variables. Using the political party identification/election interest example, if we pair a strong identifier who is very much interested with a weak identifier who is somewhat interested, we find that the first person ranks higher than the second person on both variables; in a table the second person lies below and to the left of the first person. If all possible pairs are of this kind, then gamma equals 1.00. If we pair an independent who is very much interested with a weak identifier who is not much interested, the first person ranks lower on the party variable but higher on the interest variable; in a table, the second person lies below and to the right of the first person. If all possible pairs are of this kind, then gamma equals −1.00. Gamma is based on the number of pairs of these two kinds.

To find how many pairs there are of the first kind, where one person ranks higher on both variables than another, we pair the 262 people in Table 12.1 who are very interested with all the people who lie below and to the left in the table, or 262 times 178 + 273 + 133 + 228 = 212,744 pairs. The people in each cell of the table can be paired with everyone below and to the left. Thus, the number of pairs where the ranking is the same on the two variables is

Same rankings
   = sum of cell frequency times sum of frequencies below and to the left
   = 262(273 + 178 + 228 + 133) + 150(178 + 133)
     + 237(228 + 133) + 273(133) = 394,256

To pair people so that one person in the pair ranks higher on one variable and the other ranks higher on the second variable, we pick a cell in the table and look below and to the right. The number of pairs where the rankings are different on the two variables is found from the expression.

Different rankings
   = sum of cell frequency times sum of frequencies below and to the right
   = 104(273 + 237 + 228 + 125) + 150(237 + 125)
     + 178(228 + 125) + 273(125) = 258,268

Gamma is then defined as

$$\text{gamma} = \frac{\text{same rankings} - \text{different rankings}}{\text{same rankings} + \text{different rankings}} \qquad (12.1)$$

For the example,

$$G = \frac{394{,}256 - 258{,}268}{394{,}256 + 258{,}268} = \frac{135{,}988}{652{,}524} = 0.21$$

We can think about gamma this way. There is a total of 652,524 possible pairs of people. For each pair, let us predict who ranks higher on the dependent variable. With no knowledge, this is like tossing a coin: we will be right half the time and wrong half the time. Half of 652,524 is 326,262, and so we expect to make the wrong prediction

326,262 times. Now suppose we know which of the two people in a pair ranks higher on the independent variable. Let us predict that the same person also ranks higher on the dependent variable. But we know that there are 258,268 different rankings, so for 258,268 of the pairs this prediction is not true; we will make the wrong prediction 258,268 times. By knowing the ranking on the independent variable we have therefore improved our prediction for $326{,}262 - 258{,}268 = 67{,}994$ pairs. Our improvement is $67{,}994/326{,}262 = 0.21$, the same gamma computed by Formula 12.1.

To test the null hypothesis that gamma equals zero in the population, we compute the test statistic

$$z = \frac{G}{\sqrt{\dfrac{4(\text{number of rows} + 1)(\text{columns} + 1)}{9n(\text{number of rows} - 1)(\text{columns} - 1)}}} \quad (12.2)$$

where n is the number of observations in the table. For the example,

$$z = \frac{0.21}{\sqrt{\dfrac{4(3+1)(3+1)}{9(1690)(3-1)(3-1)}}} = \frac{0.21}{0.0324} = 6.47$$

This expression for z is only an approximation, and it works best for large tables with many observations. A more exact value of z can be found from the expression

$$z = \frac{G}{\sqrt{1 - G^2}} \sqrt{\frac{\text{same rankings} + \text{different rankings}}{n}}$$

$$= \frac{0.21}{\sqrt{1 - 0.21^2}} \sqrt{\frac{394{,}256 + 258{,}268}{1690}} = 4.22 \quad (12.3)$$

The advantage of the first expression is that we can find z if we know gamma, the number of observations, and the size of the table. To compute z from the second expression, we need to know gamma as well as the numbers of same and different rankings. The two values of z are somewhat different in this example, but both values of z have a p-value of less than 0.0001, and we draw the same conclusion from both values of z.

(You may wonder what would have happened if we had treated Table 12.1 as a contingency table with categorical variables and used a chi-square analysis to check for significance. We would have found a chi-square of 71.69 on 4 degrees of freedom, and that value of chi-square would not have produced as small a *p*-value as that produced by the *z*-values computed by Formulas 12.2 and 12.3. The reason is that chi-square is not sensitive to the ordering of the rows and columns of the original table and therefore does not use all the available information in the data. But even this value of chi-square is highly significant, and we could have used chi-square if we had wanted only to establish significance.)

SPEARMAN'S r_s

The rank order correlation coefficient can be computed using the two sets of ranks as *x* and *y* in the formula for the ordinary correlation coefficient *r* (Formula 10.1). Since the two sets of values for the variables are simply the integers from 1 to *n*, the formula for *r* simplifies to Formula 12.4.

We have the two sets of ranks. Now we find the difference between the two ranks for each element, and then we square each of the differences. The computations are shown in Table 12.4. From this sum of squared differences, we find r_s by the formula

$$r_s = 1 - \frac{6\Sigma d_i^2}{n(n^2 - 1)} \qquad (12.4)$$

where the number 6 in the numerator is a constant always present in the formula and *n* is the number of pairs of ranks.

For the example with the baseball teams we get the following result, based on the squared differences shown in Table 12.5. To see if we can

Table 12.4 Computation of r_s

Rank on x	Rank on y	Difference	Squared difference
x_1	y_1	$d_1 = x_1 - y_1$	d_1^2
x_2	y_2	$d_2 = x_2 - y_2$	d_2^2
.	.		.
.	.		.
.	.		.
x_n	y_n	$d_n = x_n - y_n$	d_n^2
		Sum	Σd_i^2

Table 12.5 Example computation of r_s

Team	1987	1992	Difference	Squared difference
Chicago	6	4	2	4
Montreal	3	2	1	1
New York	2	5	−3	9
Philadelphia	4	6	−2	4
Pittsburgh	5	1	4	16
St. Louis	1	3	−2	4
			Sum	38

$$r_s = 1 - \frac{6 \cdot 38}{6(6^2 - 1)} = 1 - 1.09 = -0.09$$

reject the null hypothesis that the rank order correlation coefficient equals 0 in the population from which the sample was drawn, we compute the following *t*-test statistic:

$$t = \frac{r_s}{\sqrt{1 - r_s^2}} \sqrt{n - 2} = \frac{-0.09}{\sqrt{1 - (-0.09)^2}} \sqrt{6 - 2} = -0.17 \quad (12.5)$$

on $n - 2 = 4$ d.f.

Then we use this value of t to find the *p*-value for the data. Here, $p = 0.44$, and we obviously do not reject the null hypothesis for a value of t this small.

EXERCISES

REVIEW (EXERCISES 12.1–12.10)

12.1 a. What is a rank variable?

b. Give two examples of rank variables from your own life experience.

12.2 Rank variables can have values that are in words or in numbers.

a. Give an example of a rank variable using words, and give an example of a rank variable using numbers.

b. What is usually the major difference between these two types of rank variables when it comes to the number of observations for each value of the variable?

c. Does this difference show up in your examples?

12.3 a. What is a gamma?

b. How large and how small can the value of a gamma be?

c. Name another statistic that has the same range of possible values.

d. Looking at a table such as Tables 12.1 and 12.2, how can you tell whether the gamma will have a positive or a negative value?

12.4 a. Give an example of two variables for which you would use gamma to measure the strength of the relationship between the variables.

b. In general, how does one calculate gamma?

c. What would a gamma of 0.75 say about your ability to predict the dependent variable from a knowledge of the independent variable?

d. If the p-value for this gamma were very small, what could you say about the relationship of the two variables in the population from which the sample was drawn?

e. What would you conclude about the null hypothesis of no relationship?

12.5 In general, what steps are taken to find the p-value for a particular gamma?

12.6 a. What is a rank order correlation coefficient?

b. Why is the rank order correlation coefficient denoted r_s?

c. When should you use r_s?

d. What other statistic is r_s most similar to?

12.7 Give an example of two variables for which you would use r_s to measure the strength of their relationship.

12.8 a. What is the major difference between rank variables and categorical variables?

b. Why is it said that rank variables carry more information than categorical variables?

c. How are rank variables different from metric variables?

d. Why is it said that rank variables carry less information in them than metric variables?

12.9 a. What kind of variable is class rank?

b. Why is class rank a less sensitive gauge of academic performance for a group of students than grade point average?

12.10 With data on two metric variable, such as annual sales and profits for the top 100 companies in the country, why do we sometimes change the figures to ranks and analyze the relationship between the two variables using the ranks instead of the original values?

INTERPRETATION (EXERCISES 12.11–12.20)

12.11 The Abbreviated Injury Scale attempts to measure the severity of motorcycle accidents. The values of the variable are (1) minor, (2) moderate, (3) severe, not life-threatening, (4) severe, life-threatening, survival probable, (5) critical, survival uncertain, (6) fatal, currently untreatable. *(Source: Andrew A. Weiss, "The effects of helmet use on the severity of head injuries in motorcycle accidents,"* Journal of the American Statistical Association, *vol. 417 (1992), p. 496)* Explain whether this injury scale is a nominal, ordinal, or metric variable.

12.12 The 1984 summer Olympic Games were unusual because they were boycotted by several eastern European bloc countries. In this problem we study the number of medals won by countries that participated in both the winter and the summer games that year to see if there was any relationship between the number of medals won in the summer and winter games. Since large countries often win many medals, we rank the countries on the number of medals they won. There were 12 countries that won medals in both the winter and summer games, so the data consist of two columns of ranks from 1 to 12. For these data, $r_s = 0.78$ ($t = 3.81$ on 10 d.f., $p = 0.002$). (*Source:* The World Almanac and Book of Facts, 1988, pp. 834, 837.)

a. What does the positive rank correlation mean?

b. Would a negative relationship between the two variables have been surprising?

c. What do the numbers tell us about the relationship between the number of medals won in the winter and summer Olympic Games that year?

Table 12.6 Data for Exercise 12.13

		\multicolumn{4}{c}{Social class}		
		Lower lower	Upper lower	Lower middle	Upper middle	Total
	3+	4	7	12	9	32
Number	2	2	13	5	5	25
of children	1	10	9	7	6	32
	0	4	6	7	4	21
	Total	20	35	31	24	110

Source: Daniel R. Miller and Guy E. Swanson, *The Changing American Parent: A Study in the Detroit Area,* New York: John Wiley & Sons, 1958, p. 205.

12.13 In a study done in Detroit in the 1950s, the researchers reported on bureaucratic (versus entrepreneurial) wives 39 years old or younger with family incomes larger than $6,000, and they found the following distribution of social class and number of children, displayed in Table 12.6. For this table gamma = 0.24 ($z = 1.34$, $p = 0.09$). What can you say about the relationship between social class and number of children for these types of mothers from this analysis?

12.14 In Exercise 10.34, we study the relationship between flavor and percentage of calories from fat for different frozen chocolate yogurts. In a scatterplot for the two variables, the values for the Albertson yogurt are a good deal smaller than the values for the other yogurts. This data point could have a large effect on the analysis, and one way to lessen the impact of such a data point is to change the values of the two variables to ranks. In Table 12.7 are the rank values for the two variables. (Tied values share ranks, and that is the reason the ranks do not run consecutively from 1 to 10.) For the two variables, $r_s = 0.68$ ($t = 2.62$ on 10 d.f., $p = 0.015$).

    a. What can you conclude about the relationship between the two rank variables?

    b. How do these conclusions compare with the conclusions you drew from the analysis of the same data in Exercise 10.34?

    c. What happens to the extreme data from the Albertsons yogurt when you use rank variables?

Table 12.7 Data for Exercise 12.14

Yogurt brand	Rank	
	Percent calories from fat	Flavor
Breyers	9.5	8
Honey Hill Farms	9.5	10
Elan	8	5.5
Crowley Silver Premium	7	3
Edy's/Dreyer Inspirations	6	9
Häagen-Dazs	5	5.5
Kemps	4	3
Lucerne	3	7
Yoplait Soft	2	3
Albertsons	1	1

12.15 Researchers have found that the gamma value for the relationship between amount of sunny weather and the sweetness of grapes is 0.81. The sunnier the weather, the sweeter the grapes in a sample of grapes from Europe.

a. What needs to be done if the researchers wish to argue that this finding applies to the population of grapes from which their sample was drawn?

b. What do you think the outcome of this analysis would be?

c. What is the null hypothesis being considered here?

12.16 The probability that the value of a particular r_s or more extreme value occurs by chance alone is 0.021. How would you explain what this means to a nonstatistical friend?

12.17 When we rank in order of wins the 25 best college basketball teams in the nation in one season and then rank the same teams the following season, we find that the r_s is 0.79.

a. How would you interpret this result for your nonstatistical friend? Does it seem as if the winners stay on top from one season to another?

b. If r_s were the same 0.79 with twice as many teams (50) being ranked over two seasons, would the *p*-value from hypothesis testing be greater, the same, or less?

12.18 The newsmagazine *The Economist* (June 12, 1993, p. 62) ranked 12 British regions on several variables. One variable was death rate. East Anglia had the lowest death rate and therefore ranked number 1, and Scotland had the highest death rate and therefore ranked number 12. Another variable was housing costs. Northern Ireland had the least expensive housing costs and therefore ranked number 1, and Greater London, not surprisingly, had the highest housing costs and therefore ranked number 12. For these two rank variables, death rate and housing costs, $r_s = -0.76$.

a. What does the negative value of the correlation coefficient tell you about the two variables?

b. Why do you think the relationship between the two variables is as strong as it is?

c. To see if the pattern in the two sets of rankings could have occurred by chance alone, we change the correlation coefficient to a value of t (-3.70) and find p (0.0021). What is the null hypothesis, and what can you conclude about the null hypothesis?

d. Is the relationship between the two variables causal, or can you think of other variables that might explain the presence of the relationship?

12.19 When you rank the ten largest cities in the country in 1980 and in 1985, you find that $r_s = 0.95$ for the two sets of rankings.

a. If each city grew by the same number of people during a certain time period, would that growth alter the rankings of the cities?

b. If each city grew by the same percentage of people during a time period, would that growth alter the rankings of the cities?

c. What can you say about the growth of the various cities when you know that $r_s = 0.95$ for the two sets of rankings?

d. To see if the two sets of rankings could have occurred at random, you change r_s to a value of the *t*-variable and get $t = 8.75$ on 8 degrees of freedom, $p = 0.00001$. What can you conclude about the rankings?

12.20 The United Nations Development Program has developed under the guidance of the Pakistani economist Mahboub ul Haq what they call the human development index (HDI). This index goes beyond the old gross national product (GNP) and provides a measure based on a combination of purchasing power, life expectancy, and literacy. On a list of 130 countries with population more than 1 million people, Niger has the lowest HDI rank (1) and Japan the highest HDI rank (130). Figure 12.5 shows a scatterplot of the rankings of a random sample of 13 countries on the two variables DNP and HDI.

a. Describe the pattern you see in the scatterplot.

b. The correlation between the two sets of rankings is 0.89 with a p-value of 0.00002. Do you think that the observed relationship between the two sets of rankings could have occurred by chance alone?

c. If each country had the same ranking on the two variables, the points would lie on a 45-degree line in the scatterplot. What

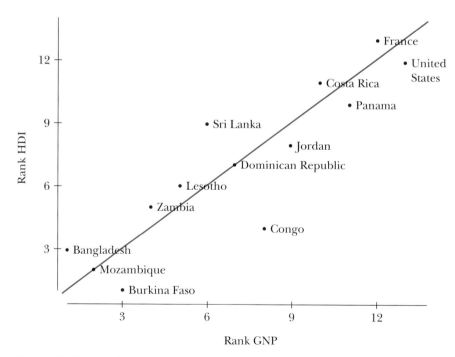

Figure 12.5 Random sample of 13 countries ranked on gross national product (GNP) and Human Development Index (HDI) (*Source:* The Economist, *May 26, 1990, p. 81.*)

might the reason be that the United States lies below such a line and France lies above such a line?

ANALYSIS (EXERCISES 12.21–12.34)

12.21 In a national survey the respondents were asked, among other things, what their stance was on abortion and also how important they thought the abortion issue was (Table 12.8).

a. Is there a relationship between stance and importance in these data?

b. How strong is the relationship?

c. Is there a relationship between the variables in the population of all adult Americans?

d. Is the relationship causal?

12.22 The data in Table 12.9 show the rankings of revenues spent on advertising for various types of products in general magazines and national farm magazines for 1950 and 1970. In 1950 the country was adjusting to peacetime life after World War II; by 1970 that adjustment was over. The smaller the ranking number, the more money was spent advertising that product.

a. From just looking at the data, has there been any change in the promotion of the products?

b. How strong would the relationship between the two sets of rankings be if there were no change from 1950 to 1970?

c. Make a scatterplot of the data and label teach type of product on the graph.

Table 12.8 Data for Exercise 12.21

		Abortion stance			
		Anti	Mixed	Pro	Total
	One of the most	85	167	89	341
Importance	Important	80	638	357	1075
of issue	Not very/not at all	56	583	366	1005
	Total	221	1388	812	2421

Source: John Scott and Howard Schuman, "Attitude strength and social action in the abortion dispute," American Sociological Review, vol. 53 (1988), p. 788.

Table 12.9 Data for Exercise 12.22

Type of product	Rank in 1950	Rank in 1970
Apparel, footwear, etc.	3	6
Automotive	2	2
Alcoholic beverages	5	1
Building materials	8	13
Consumer services	11	4
Food and food products	1	3
Household equipment	4	11
Household furnishings	6	9.5
Industrial materials	7	8
Insurance	13	12
Radio, TV, etc.	12	9.5
Smoking materials	9.5	5
Travel	9.5	7

Source: U.S. Bureau of the Census, Statistical Abstract of the United States: 1972, *93rd edition*, Washington, DC: U.S. Government Printing Office, 1972, p. 759.

   d. What kind of patterns do you see in the plot, and how do the patterns relate to what you know about the period from 1950 to 1970?

   e. How strong is the relationship between the two sets of rankings?

   f. Could this relationship have occurred by chance alone?

12.23 Find data from the most recent winter and summer Olympic Games.

   a. Rank the countries that received medals in both sets of games by the total number of medals they received.

   b. Find the rank order correlation coefficient for the two sets of rankings.

   c. Is the value of r_s statistically different from 0?

   d. How do the results of this analysis compare with the results in Exercise 12.12 on the summer games that were boycotted?

12.24 From the data in Exercise 10.57 on the Calabrian Mafia, it is possible to study whether Mafia groups (coscas) choose their leaders such that the age of each leader is related to the mean age of the cosca. The data may not be linearly related, so we change the original ages to ranks.

    a. Change the two age variables to ranks.

    b. Analyze the relationship between the two rank variables.

    c. Compare the results of the analysis of these ranks with the results of the analysis of the original data in Exercise 10.57. What could account for the differences in the two sets of results?

12.25 In Exercise 10.52 we study the relationship between number of divorces as the dependent variable and number of marriages as the independent variable over a span of 100 years. The scatterplot of the original data shows a strong nonlinear pattern, so we change the data to ranks by taking the logarithm of each observation. The data show the number of marriages and divorces in thousands for 1890 and every fifth year up to 1980. The year variable has been recorded to 1 for 1890, 2 for 1895, and up to 19 for 1980 to make it easier to enter the data in a computer.

Year	1	2	3	4	5	6	7
Marriages	570	620	709	842	948	1,008	1,274
Divorces	33	40	56	68	83	104	170

Year	8	9	10	11	12	13	14
Marriages	1,188	1,127	1,327	1,596	1,613	1,667	1,531
Divorces	175	196	218	264	485	385	377

Year	15	16	17	18	19
Marriages	1,523	1,800	2,159	2,153	2,413
Divorces	393	479	708	1036	1182

Source: National Center for Health Statistics, Public Health Service, in The World Almanac *1986, p. 779.*

An alternative approach would be to rank the observations on each of the two variables and then study the relationship between the two sets of ranks.

a. Rank the observations on the two variables from 1 to 19.

b. Make a scatterplot of the two sets of ranks.

c. Comment on the shape of the scatterplot.

d. Analyze the relationship between the two sets of ranks.

12.26 In Exercise 10.54 we study data on literacy and per capita income for a sample of 20 countries from among countries with per capita income less than $2,000. A scatterplot of the original data does not show a clear linear relationship between the two variables, so we rank each of the variables and then study the relationship between the two rank variables. The countries are Bangladesh, Botswana, Cambodia, Chile, Cuba, Egypt, Ghana, Guyana, Ivory Coast, North Korea, Madagascar, Mauritania, Mozambique, Pakistan, Philippines, Sao Tome, South Africa, Tanzania, Uganda, and Zaire.

Country	1	2	3	4	5	6	7
Percent literate	25	30	48	90	96	44	30
Per capita income	119	544	100	1,950	840	686	420

Country	8	9	10	11	12	13	14
Percent literate	86	24	99	53	17	14	24
Per capita income	457	1,100	570	279	466	220	280

Country	15	16	17	18	19	20
Percent literate	88	50	98	66	25	40
Per capita income	772	300	1,296	240	240	127

a. Rank the observations on the two variables from 1 to 20.

b. Make a scatterplot of the two sets of ranks.

c. Comment on the shape of the scatterplot. Would it make sense to do a correlation analysis of the relationship between the two sets of ranks?

d. Find the correlation between the two sets of ranks.

e. Is the rank correlation coefficient significantly different from zero?

12.27 Table 12.10 shows data on the age of women and the probability of getting breast cancer.

Table 12.10 Data for Exercise 12.27

Age	Probability of breast cancer
Young (to age 39)	0.0005
Young middle age (40–49)	0.015
Older middle age (50–59)	0.024
Old (60–69)	0.036
Oldest (70–80)	0.042

Source: National Cancer Institute, American Cancer Society, as reported in The Philadelphia Inquirer, *January 18, 1993, p. D1.*

a. If young is ranked 1 and oldest is ranked 5, and the probabilities are also ranked from 1 to 5, what is the value of the correlation coefficient measuring the strength of the relationship between these variables? Describe the finding in a sentence or two.

b. What does the table suggest about the media description of an "epidemic of breast cancer" among women in America?

c. Why might the number of breast cancer cases have gone up in this country in the last forty years without change in the probability of getting breast cancer at given age?

12.28 A national study of eighth-grade achievement indicated that children differed in the scores according to socioeconomic status. The variables are socioeconomic status (high and low) and ability (high, middle/mixed, and low). The percentages data are shown in Table 12.11.

Table 12.11 Data for Exercise 12.28

		Socioeconomic status (SES)	
		Low (%)	High (%)
	High	13	39
Ability	Middle/mixed	50	47
	Low	37	14
	Total	100	100

Source: U.S. Department of Education, National Center for Educational Statistics, National Education Longitudinal Study of 1988.

Table 12.12 Data for Exercise 12.29

Year	Hours per worker per year
1200	1,620
1300	1,440
1400	2,300
1500	3,200
1600	1,980
1700	—
1800	3,300
1900	1,900

Source: Juliet B. Schor, The Overworked American: The Unexpected Decline of Leisure, *New York: Basic Books,* 1991, p. 45.

a. Illustrate the data in a graph using circles (as in Figure 12.1c) or a bar graph (as in Figure 12.1a or b).

b. What can you conclude from the data about the relationship between the two variables?

12.29 How much have people worked at productive labor (which does not include housework, unfortunately) over the centuries? Estimates of the number of hours worked per year have been gathered by creative historians and demographers. Most of the data are gathered for British peasants and laborers in manufacturing. Some of the data is shown in Table 12.12.

a. Show the data in a scatterplot.

b. What does the scatterplot tell you?

c. Change the data to data on two rank variables

d. If you wanted to know if there was a change in working hours over the centuries, what would be the null hypothesis and how would you go about testing the null hypothesis, using the rank data?

e. Can you reject the null hypothesis?

f. What other statistical method could you use to study the data?

12.30 Collect data on two rank variables of your choice.

a. Is there a relationship between the two variables?

b. How strong is the relationship?

c. Is there a relationship between the two variables in the population from which your sample was drawn?

d. Is the relationship causal?

12.31 The International Association for the Evaluation of Educational Achievement published a study in 1991 on the performance of twelfth- and thirteenth-grade students in different countries in the sciences. The countries were ranked, and the data for biology and chemistry are shown in Table 12.13.

a. Make a scatterplot of the data using biology and chemistry as the two variables, and label each point with the name of the country.

b. Describe some of the patterns you see in the scatterplot, including, for example, countries that rank higher in chemistry than in biology.

Table 12.13 Data for Exercise 12.31

Country	Rank	
	Biology	Chemistry
Singapore	1	3
Britain	2	2
Hungary	3	5
Poland	4	7
Hong Kong	5	1
Norway	6	8
Finland	7	13
Sweden	8	9
Austria	9	6
Japan	10	4
Canada	11	12
Italy	12	10
United States	13	11

Source: International Association for the Evaluation of Educational Achievement.

    c. How strong is the relationship between the two sets of rankings?

    d. Could this relationship have occurred by chance alone?

    e. Is the relationship causal?

12.32 In a poll of statisticians working in government statistical organizations, international institutions, and other organizations that use international statistics, *The Economist* came up with a ranking of government statistical offices in 10 countries. In the same article, the magazine presents data on the number of statisticians per 10,000 population and the government statistics budget per head in dollars. These data are shown in Table 12.14.

    a. Change the last two columns to rank values where the highest value has rank 1, the next highest rank 2, and so on.

    b. Is the quality of the statistics office related to the number of statisticians?

    c. Is the quality of the statistics office related to the amount of money spent?

Table 12.14 Data for Exercise 12.32

Country	Rank of statistical office	Statisticians per 10,000	Government statistics budget (dollars per head)
Canada	1	1.6	8.20
Australia	2	2.0	9.00
Holland	3	2.0	7.60
France	4	1.7	6.00
Britain	5	0.9	4.20
Germany	6	1.9	8.00
United States	7	0.6	8.80
Italy	8	1.4	5.00
Spain	9	1.2	4.20
Belgium	10	1.3	3.60

Source: The Economist, *September 11, 1993, p. 65.*

12.33 Each week during the fall, college football teams are ranked both through the Associated Press by sports writers and through *USA Today* and CNN by coaches. The ratings change from week to week, and Table 12.15 shows how the rankings changed at the end of November 1994. These are the rankings for the top dozen teams after Texas A&M and Auburn were deleted from the AP list; the two teams were not included in the coaches' rankings because of penalties assessed by the NCAA.

 a. Make a scatterplot of the rankings.

 b. Comment on the pattern shown in the scatterplot.

 c. Compute the rank correlation to see how well the two sets of rankings agree.

 d. Could this arrangement of rankings have occurred by chance alone?

12.34 Much political tension was created as countries debated whether or not to join the European Common Market. In the 1960s Norway considered joining, and Table 12.16 shows the change of opinions from 1965 to 1969 in a sample of 286 Norwegians who expressed their opinions at the two points in time.

 a. Analyze the relationship between the two variables represented by the two dates.

Table 12.15 Data for Exercise 12.33

University	Rank Nov. 28	Rank Nov. 21
Nebraska	1	1
Penn State	2	2
Alabama	3	3
Miami	4	4
Colorado	5	5
Florida	6	6
Florida State	7	7
Colorado State	8	10
Kansas State	9	8
Oregon	10	9
Ohio State	11	11
Utah	12	12

Source: The New York Times, *November 28, 1994, p. C2.*

b. What do you learn when you compare the 45 + 15 + 23 = 83 people in the upper right triangle of the table with the 24 + 10 + 4 = 38 people in the lower left triangle of the table?

Table 12.16 Data for Exercise 12.34

		1965 Full membership	1965 Loose connection	1965 Stay out	Total
1969	Full membership	100	45	15	160
	Loose connection	24	35	23	82
	Stay out	4	10	30	44
	Total	128	90	68	286

Source: Henry Valen and Willy Martinussen, Velgere og politiske frontlinjer (Voters and Political Front Lines), *Oslo: Gyldendal Norsk Forlag, 1972, p. 214.*

CHAPTER 13

13.1 Partial phis: Three categorical variables

13.2 Multiple regression with metric variables

13.3 Multiple regression with a dummy variable

13.4 Two-way analysis of variance

13.5 Establishing causality

13.6 Summary

MULTIVARIATE ANALYSIS

Political scientists often find a relationship between people's gender and how they vote, but maybe it is not really gender that affects vote. Maybe there are other variables at work as well. In particular, in this chapter we examine income as a third categorical variable affecting vote.

The number of calories in a snack food is determined by many different factors. In Chapter 10 we looked at the effect of fat content on calories, and now we wonder if cholesterol and sodium also affect calories. A multiple regression analysis can help answer this question.

The commuting time to work differs for a commuter depending on which of two roads she takes. Her driving time also differs depending on whether it is rush hour or not. How does the choice of road and time of day affect her driving time? A two-way analysis of variance can help answer that question.

> A **multivariate statistical analysis** examines the relative impact of two or more independent variables on a dependent variable.

For most problems, the outcome, or dependent variable, is determined by the influences of more than a single independent variable. Therefore, in the statistical analysis of a dependent variable, we often use more than one independent variable. When we analyze the relative impact of several independent variables, we are doing a *multivariate statistical analysis*.

With several independent variables, we can always analyze the relationship between each independent variable and the dependent variable one at a time. In a study of voting preferences, we may look first at effect of age, then at gender, then race, and so on. But it is more efficient and more instructive to study the effects of all the independent variables together at the same time on the dependent variable. That way we can see the effect of a particular variable *with other variables present* in the analysis.

In multivariate analyses, the influence of the residual variable on the dependent variable is reduced. This occurs because we take the effects of all the independent variables out of the residual variable at the same time instead of one at a time.

In multivariate statistical analyses we want answers to the four questions about statistical relationships: Question 1, Does a particular variable have an effect in the data? Question 2, How strong is the relationship between the independent variables taken together and the dependent variable? Also, how large is the effect of each independent variable on the dependent variable? That way we can see which of the independent variables are more important and which are less important. Often we also consider Question 3, Is the relationship between each independent variable and the dependent variable statistically significant?

Up to this point, with only one independent variable and one dependent variable, we have not been able to do much with Question 4, Is the relationship between two variables causal? With multivariate analysis it is sometimes possible to determine whether a relationship between two variables is causal or not. Sometimes we find that a variable that at first seems to be related to the dependent variable actually has no effect when we bring in other variables and do a multivariate analysis. If a relationship between two variables disappears as a result of a multivariate analysis, then the original relationship was not a causal relationship.

As always, the choice of statistical method is governed by the nature of the variables involved. Categorical variables require one type of analysis, while metric variables require another type. In this chapter we consider three of the many multivariate analysis methods that statisticians have developed.

Dependent variable	*Independent variables*	*Method*
Categorical	Categorical	Partial phis
Metric	Metric	Multiple regression
Metric	Dummy	Multiple regression
Metric	Two categorical	Two-way analysis of variance

First we consider the case where all the variables are categorical variables in order to illustrate what a *partial phi coefficient* is. Most multivariate methods have been developed for metric dependent variables, however, and we discuss them here. The independent variables can be either metric or categorical. For metric independent variables we do *multiple regression analyses,* extensions of the simple regression methods described in Chapter 10. Here we consider a case with three independent variables, but it is possible to have more than three independent variables in one analysis. Multiple regression methods also work when so-called "dummy" variables are constructed to represent categorical variables; we can also do a multiple regression analysis with a combination of metric and dummy variables. Finally, when all the independent variables are categorical variables, we usually shift from multiple regression analysis to *analysis of variance.*

13.1 PARTIAL PHIS: THREE CATEGORICAL VARIABLES

Table 13.1 shows an example of the relationship between the two categorical variables gender and vote. In answer to Question 2, the strength of the relationship between the two variables is measured by the coefficient phi = 0.21. Since phi is not 0, we know that the two variables in these data are related.

Does this mean that there is a causal impact of gender on vote, or are other variables operating here so that what we see is not a causal relationship? For example, is income an underlying variable that determines how men and women vote? The 205 women Democrats in the upper left cell of the table may be there because they are women and they are Democrats, but they also may all have a certain income. The same goes for the other three cells in the table. Maybe what we really have are four different income groups. Maybe we have 205 low-

Table 13.1 Gender and vote

		Gender		
		Women	Men	Total
Vote	Democratic	205	118	323
	Republican	167	230	397
	Total	372	348	720

phi = 0.21

income people, and they turn out to be Democratic women. Maybe the 167 people in the next highest income are Republican women—and similarly for the other two groups.

Control for a third variable: The neutralizing game

To keep the example simple, let us define income as a variable with only the two values poor and rich. One way to examine the impact of income is to consider the relationship between gender and vote while we *control* for income. Is there then still a relationship between gender and vote?

To control for income means that we keep income constant while we examine the relationship between gender and vote. Even though income is a variable and therefore has different values, we can keep a variable constant by looking only at the people who have a particular value of the variable. Thus, we look first at only the observations that have the value poor on the income variable, and then we look separately at the observations that have the value rich.

> To **control for a third variable** in studying the relationship between two variables, the data are first divided into subgroups defined by the control variable. Then we study the relationship between the other two variables within each subgroup of the control variable. Creating subgroups of data according to a third variable means the same as keeping that third variable constant.

Phi is equal to 0.21 in the overall table (Table 13.1) that shows the relationship between gender and vote. Now we create two subtables for

13.1 Partial Phis: Three Categorical Variables

Table 13.2 Relationship between gender and vote, controlling for income

		Income = Poor					*Income = Rich*		
		Women	Men	Total			Women	Men	Total
Vote	Democratic	153	24	177	Vote	Democratic	52	94	146
	Republican	44	7	51		Republican	123	223	346
	Total	197	31	228		Total	175	317	492
		phi = 0.00					phi = 0.00		

Average phi for gender and vote, controlling for income, = 0.00

gender and vote in Table 13.2, one for poor people and one for rich people. People cannot be put in *different* cells in the "poor" table because of income because everybody in the table is poor. The effect of income is therefore the *same* for them all. The same is also true for the "rich" table, where everybody is rich.

Now that the effect of income has been controlled for, what can we conclude about the relationship between gender and vote? We find is that there is no relationship between gender and vote among the poor people, and there is similarly no relationship between the two variables among the rich people. In each of the two subtables, phi equals 0.00, which indicates no relationship between the variables in either table. Thus, when we divided up the data into income groups, the original relationship between the two variables within each group disappeared. Since the relationship between gender and vote disappeared, we conclude that the original relationship was not causal but spurious.

STOP AND PONDER 13.1

Can you think of an example with two variables where the effect of one on the other would disappear if you controlled for a third variable?

Partial phi

For each income group in Table 13.2 we computed a value of phi equal to 0.00. If the control variable has more than two values (e.g. poor, middle income, upper middle income, rich), then we make four subtables and compute a separate phi for each of them.

Because it is hard to interpret a large group of phi values, we summarize the phis from the subtables by finding an average value. Such an average coefficient from a set of subgroups is known as a *partial coefficient*. In our example, the average of the two phis, which each equal 0.00, is obviously 0.00. Thus, the partial phi for gender and vote when we control for income is 0.00.

> A **partial phi** gives the strength of the relationship between two categorical variables when one or more other variables are controlled for. A partial phi can be thought of as the average within-group phi for the two variables, where the groups are defined by the control variables.

The partial phi of 0.00 can also be called the *average within-group phi* for gender and vote when we control for income. This name applies because the people in the study are divided into income groups. We find phi for each group, and then we average the phis. A partial phi for the relationship between two variables, while controlling for a third variable, can be computed according to Formula 13.1.

To summarize these findings:

$$\text{phi(gender and vote)} = 0.21$$
$$\text{phi(gender and vote, controlling for income)} = 0.00$$

It is because the partial phi equals 0.00 that we say the relationship between gender and vote disappeared when we brought in income. The fact that the partial phi equals 0.00 is evidence that there is no causal relationship between gender and vote. Causal relationships do not disappear when we control for other variables. Therefore, the original relationship between the two variables was spurious.

We can also use income as an independent variable and study the relationship between income and vote, first alone and then controlling for gender. To study the relationship between income and vote, we first arrange the data in a table showing the frequencies for those two variables (Table 13.3); here phi is equal to 0.45 for income and vote. To control for gender, we create separate tables for the women and for the men (Table 13.4). We find phi for each of the two tables, and then we average the two phis. The partial phi (or within-group phi) for

13.1 Partial Phis: Three Categorical Variables

Table 13.3 Income and vote

		Income		Total
		Poor	Rich	
Vote	Democratic	177	146	323
	Republican	51	346	397
	Total	228	492	720

phi = 0.45

income and vote, controlling for gender, is 0.40. This average is the weighted average of phi = 0.48 for the women and phi = 0.29 for the men. (The main reason the average is a little closer to the phi for women is that there are more women than men in the example.)

The original phi for income and vote was 0.45. When we control for gender, the partial phi for income and vote stays at about the same level (0.40). Thus, income does not disappear; it remains an important variable, and there may be a causal relationship between income and vote. We say "may be" because there could be another control variable that would make the relationship disappear, indicating that the relationship is not causal after all.

The different analyses are summarized in Figure 13.1. Figure 13.1a shows the strengths of the separate relationships between gender and vote and between income and vote, and Figure 13.1b shows what happens when we do a multivariate analysis. For the multivariate analysis

Table 13.4 Relationship between income and vote, controlling for gender

		Gender = Women					Gender = Men		
		Poor	Rich	Total			Poor	Rich	Total
Vote	Democratic	153	52	205	Vote	Democratic	24	94	118
	Republican	44	123	167		Republican	7	223	230
	Total	197	175	372		Total	31	317	348

phi = 0.48   phi = 0.29

Average phi for gender and vote, controlling for income, = 0.40

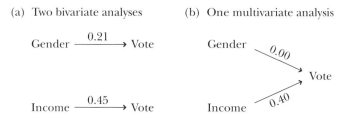

Figure 13.1 Phis for two bivariate analyses and partial phis for one multivariate analysis

each arrow shows phi for the corresponding variable *in the presence of* the other independent variable in the analysis. Thus, when gender is the only independent variable, then phi for the relationship with vote equals 0.21. But in the presence of income, the partial phi for gender and vote equals 0.00 and the relationship between the two variables disappears. Similarly, when income is the only independent variable, then the strength of the relationship with vote is 0.45, while in the presence of gender the strength becomes slightly reduced to 0.40.

The reason the partial phi for gender and vote, controlling for income, is different from the original phi has to do with the relationship between gender and income and the relationship between income and vote. In the two contingency tables in Table 13.2, we see in the first table that poor people tend to vote Democratic while rich people tend to vote Republican. In the second table we see that women tend to be poor and men tend to be rich. The possible causal mechanism at work here is therefore that gender influences income and then income influences vote; The observed relationship between gender and vote is a byproduct of the effect of gender on income and then the effect of income on vote. The relationships between the three variables are shown in Figure 13.2.

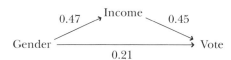

Figure 13.2 Strengths of relationships (phi) between gender, income, and vote

13.2 MULTIPLE REGRESSION WITH METRIC VARIABLES

With a metric dependent variable and several metric independent variables, we can use a statistical method called *multiple regression analysis* to analyze the relationship between the dependent variable and all the independent variables at the same time. Multiple regression analysis is today the most commonly used method in multivariate statistics. Because it requires large numbers of computations, it was not used very much in the days before statistical software programs for computers were readily available. Statistical software makes it possible to perform a multiple regression almost instantly, enabling us to study complicated relationships that were previously undetected. But multiple regression analysis is easily misused by people who do not have enough statistical knowledge. Many things can go wrong in a multiple regression analysis, and when the analysis is not done properly, the results can be very misleading.

Multiple regression is a statistical method used for the study of the relationship between several independent metric variables and one dependent metric variable.

Multiple regression has become very important for marketing, advertising, public relations, and many other research applications. For example, advertisers try to match the specific demographic characteristics of their clients' products with the audience characteristics of the media used for promotion. If Jaguar owners are affluent, over 50, and predominantly white males, *Playboy, Ebony,* or *Vogue* may not be the magazines in which to advertise. Multiple regression analysis helps clarify the roles age, income, and race play in buying a Jaguar, thus, influencing choice of advertising strategy.

As another example, research in the health professions benefits from regression analysis. Medical researchers use multiple regression analysis to pinpoint relevant characteristics in predicting disease rates, treatments, and recovery rates. Recent research has shown, for example, that for certain breast cancers, if other variables are held constant, lumpectomies are as successful as mastectomies in predicting long-term recovery rates.

Question 1. Relationship in the data?

In Chapter 10 we studied the relationship of fat content to calories in several snack foods. For the same foods we also have the cholesterol content and sodium content, so here we use fat, cholesterol, and sodium as independent variables in a multiple regression analysis with calories as the dependent variable. With three variables, we should get a better understanding of how calories are determined. Even though there are only a small number of observations, the example still illustrates issues that come up in multiple regression.

First we examine the data to see if they are appropriate for multiple regression and if there are relationships in the data. Each of the independent variables should be related to the dependent variable, and the independent variables should not be related to each other. One good way to do this is to look at the scatterplot for each pair of variables. Fat content has a fairly strong and positive relationship with calories, while cholesterol and calories do not have a very strong relationship. The third independent variable, sodium, has a positive relationship with calories, but it is weaker than the one for fat content. All three relationships show appropriately linear relationships, so we can go ahead with the regression analysis. Fat content is related to both cholesterol and sodium, and that may create some difficulties. (The name independent variables can be misleading, since the variables are related among themselves.)

Question 2b. Form of the relationship? Partial regression coefficients

When we do a multiple regression analysis to see how the three variables fat, cholesterol and sodium together determine calories, we find the equation

$$\text{calories} = 21.3 + 12.6 \text{ fat} - 0.11 \text{ cholesterol} + 0.18 \text{ sodium}$$

The equation tells us the way in which each of the three variables is related to the dependent variable when the other two variables are also included in the analysis. The first term, 21.3, is the intercept, and it indicates that we would expect a food with no fat, no cholesterol, and no sodium to have 21.3 calories. However, this imaginary food with all zero values lies outside the ranges of values for the other data, so in that sense 21.3 is a meaningless number. We need this number, however, to get the correct level of calories when we substitute actual values, for the three independent variables.

The remaining three numbers in the equation are *partial regression coefficients;* they tell something about the effect of the particular variable when the other two variables are also present in the analysis and controlled for. The partial regression coefficient 12.6 for the fat variable tells us that when we control for cholesterol and sodium, two foods that differ by one gram of fat will differ by 12.6 calories. Another way to say this is that when two foods have the same cholesterol and sodium content and differ by one gram of fat, then they differ on the average by 12.6 calories. We keep cholesterol and sodium constant by requiring the foods to have the same values on these two variables, and then we see what happens when the foods differ by one gram of fat. Since the coefficient 12.6 is positive, we know that the food with the higher fat content will also have a higher calorie count.

One way to convince ourselves that the two foods will differ by 12.6 calories is to work out the numbers. Suppose the two foods both have 100 milligrams of cholesterol and 300 milligrams of sodium, but one has 11 grams of fat and the other has 10 grams of fat. The predicted numbers of calories for the two foods are

$$\text{calories food 1} = 21.3 + 12.6(11) - 0.11(100) + 0.18(300) = 202.9$$
$$\text{calories food 2} = 21.3 + 12.6(10) - 0.11(100) + 0.18(300) = \underline{190.3}$$
$$\text{Difference} = 12.6$$

The predicted calorie count of the first food is 202.9 and of the second food is 190.3. The difference between the two foods of one gram of fat translates into a difference of 12.6 calories. The same result occurs for any other values of cholesterol and sodium.

Following is another way to think of the partial regression coefficient 12.6. To control for both cholesterol and sodium we divide the data into subgroups in such a way that within each group all the observations have the same values of cholesterol and of sodium. Then we make a scatterplot of calories against fat content in each group and do a regression analysis. That will give us a regression coefficient for fat for each group. When we average all the coefficients, the value of the average is 12.6.

Similarly, the partial regression coefficient -0.11 indicates the effect of cholesterol when we control for fat and sodium. Two foods with the same amount of fat and sodium that differ by one milligram of cholesterol will, on the average, differ by 0.11 calorie. This coefficient is negative, so the food with more cholesterol will have fewer calories. Finally, the partial regression coefficient 0.18 indicates the effect of sodium when we control for fat and cholesterol. When two foods have the same fat and cholesterol and they differ by one milligram of sodium, then the one with more sodium will, on the average, have 0.18 additional calorie.

A **partial regression coefficient** is the coefficient for a variable when we control for the other independent variables and keep them constant. It is the average, within-group regression coefficient when the groups are defined by the values of the other independent variables in the analysis.

Comparing regression coefficients The magnitudes of the three partial regression coefficients 12.6, −0.11, and 0.18 are very different, but it is not possible to compare them and conclude anything about which variables are more important than the others. The coefficients have different units; comparing the three coefficients is like comparing apples and oranges. There are 12.6 calories per fat unit, −0.11 calorie per cholesterol unit, and 0.18 calorie per sodium unit. Because of the different units, we do not know what the differences are in their magnitude.

Changing regression coefficients Figure 13.3 shows how the regression coefficients for the three variables change when we do a multiple regression analysis including all three variables compared to a simple regression analysis for each variable. The coefficient for fat goes down some, the coefficient for cholesterol almost vanishes, and the coefficient for sodium goes down. This shows that in the presence of the two other variables fat and sodium, cholesterol is almost not associated with calories. In addition, the figure shows the proportion of variation explained in each of the analyses (we return to this point in the subsection on the multiple correlation coefficient).

The reason the coefficients change is that the three independent variables fat, cholesterol, and sodium are correlated among themselves. We saw this correlation in the scatterplots for the independent variables

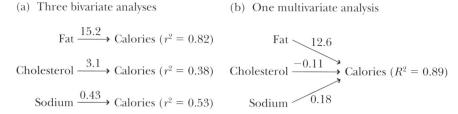

Figure 13.3 Regression coefficients for three bivariate analyses and partial regression coefficients for one multivariate regression analysis

in Figure 13.3. In calculating the strength of these relationships, we find these three correlation coefficients:

$$r \text{ for fat and cholesterol} = 0.69$$
$$r \text{ for fat and sodium} = 0.59$$
$$r \text{ for cholesterol and sodium} = 0.41$$

Whenever we introduce a variable that is correlated with the variable(s) already included in the analysis, the regression coefficients change. No variable has a single, unique value of the coefficient; the coefficient varies with whatever other variables we use. This phenomenon is called *collinearity* between the independent variables. We try to avoid collinearity, but it is often not possible to do so. One exception occurs in experimental situations where we can often choose the values of the independent variables in order to study the dependent variable. When we have a choice, we choose values so that the independent variables are not correlated.

> **Collinearity** exists when two or more of the independent variables are correlated among themselves.

Question 2a. Strength of the relationships? Partial correlation coefficients

Just as a partial regression coefficient tells about the effect of a particular variable when other variables are present, a *partial correlation coefficient* tells the strength of the relationship between two variables when we control for other variables. We can control for one or several other variables. The partial correlation coefficient for the relationship between two variables while controlling for a third variable can be computed according to Formula 13.1.

For the snack data, $r = 0.91$ for the strength of the relationship between calories and fat. When we control for cholesterol and sodium, we find that the partial correlation coefficient for calories and fat equals 0.82. Similarly, for calories and cholesterol $r = 0.62$, and when we control for the other two variables, the correlation slips to -0.05. Finally, for calories and sodium $r = 0.73$, and after controlling for the other two variables, the correlation equals 0.58. The relationship of calories to cholesterol almost disappears when we control for fat and sodium, and therefore it looks as if the relationship between calories and cholesterol was a spurious one. This is the same result the regression coefficients produced.

Question 2a. Strength of the overall relationship? Multiple correlation coefficient

Was it worth doing a multiple regression analysis for the snack food data? How well do the three variables fat, cholesterol and sodium together determine calories? These questions can be answered by looking at the so-called multiple correlation coefficient R.

A multiple regression analysis gives a partial regression coefficient for each of the independent variables. These coefficients can be put together with the variables to form an equation, as we saw on page 568. For all the snack foods

$$\text{calories} = 21.3 + 12.6 \text{ fat} - 0.11 \text{ cholesterol} + 0.18 \text{ sodium}$$

This equation looks impressive, but how well does it work for an actual snack food, such as one lowly, plain doughnut? When we plug in the values of fat, cholesterol, and sodium for the doughnut and multiply everything, is the result the right value for the calories in one doughnut? For a plain doughnut, fat = 8, cholesterol = 25, and sodium = 210. We put these numbers into the regression equation and get the predicted value

$$21.3 + (12.6)(8) - (0.11)(25) + (0.18)(210) = 156.7 \text{ calories}$$

But checking the calorie variable in Table 10.1, we see that a plain doughnut has 164 calories, not 156.7 calories, so we did not get exactly the right value. However, we came quite close. We can do the same computations for each of the other foods in the data file to get the predicted number of calories from the regression equation. The closer the predicted calorie values are to the actual values, the better our analysis is.

One way to examine how close the predicted values are to the actual values is to make a scatterplot of the two sets of numbers, shown in Figure 13.4. For each food we plot the actual and the predicted calories as a point. If the predicted value is equal to the actual, observed value, then the corresponding point will lie on the 45-degree line drawn on the graph. The greater the difference between the two numbers, the farther away from this line the point will fall.

The graph shows that most of the points cluster around the 45-degree line, with a couple of outliers. One of the outliers is the apple pie point at the right of the graph; it falls some distance below the line,

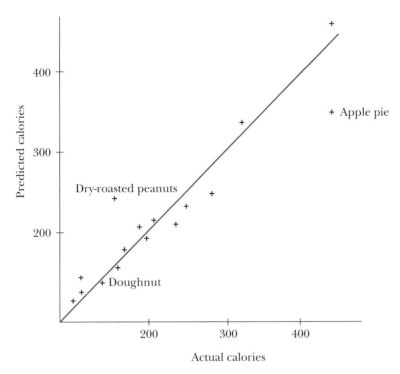

Figure 13.4 Scatterplot of predicted and actual calorie values of snack foods

indicating a predicted calorie value quite a bit lower than the actual calorie value. Thus, apple pie has more calories than can be predicted by its fat content, cholesterol, and sodium. The equation does not work particularly well in estimating what the calories should be for apple pie, implying that we need data on one or more additional independent variables (perhaps the culprit is sugar or cinnamon). In general, however, the results from the regression equation are pretty impressive: most of the predicted values are quite close to the observed values.

To measure how well the predicted values correspond to the observed values, we use the correlation coefficient for two metric variables introduced in Chapter 10. Recall that if the predicted values are equal to the observed values and the points fall on the 45-degree line, then the correlation coefficient would be equal to 1.00; the more scattered the points, the smaller the correlation coefficient. For the data in Figure 13.4, the correlation coefficient equals a high 0.94. This is the correlation between the observed and the predicted values of the dependent variable. Because the predicted values were computed from the three independent variables, we say that we have found the correlation between the observed dependent variable (calories) on the one hand and the combined effect of all the independent variables (fat, cholesterol, sodium) on the other hand. Because this is such a special correlation coefficient, it has its own name and its own symbol. It is known as the *multiple correlation coefficient,* and it is denoted by the capital letter R. (Recall that an ordinary correlation coefficient between any two variables is denoted by lower-case r.)

> The **multiple correlation coefficient** R measures the strength of the relationship between the observed values of the dependent variable and the predicted values of the dependent variable computed from the regression equation. Values of R range from 0 to 1.

The square of the multiple correlation coefficient R^2 is equal to 0.89, and it means that the three independent variables together explain 89% of the variation in the calorie values. The residual variable explains the remaining 11%. Another way to say this is that the amount of variation in the predicted calorie values is 89% of the variation in the observed calorie values. The reason the observed values differ more among themselves is that they also contain the effects of the residual variable.

The explained part of the variation in calories is shown in Figure 13.3. The figure shows that fat alone explains 82% of the variation in calories, cholesterol alone explains 38%, and sodium alone explains 53%. The three variables together explain more of the variation than does any one of the three variables alone, but the improvements in prediction are not large. For example, analyzed separately, fat explains 82% of the variation, and we would expect the other two variables to add another 7% to get the overall 89%. But the separately analyzed values for cholesterol and sodium are much larger than 7%, and the total of the three separately analyzed values 82%, 38%, and 53% far exceeds 100%. The reason we cannot add the separate percentages for the three variables to find their total impact is that the three variables are correlated among themselves, that is, they are collinear.

Question 3. Relationship in the population?

When we have data on only a sample of observations and want to draw conclusions about the population from which the sample was drawn, we have to do some form of statistical inference. As we did in earlier chapters, either we do hypothesis testing or we construct confidence intervals. Of the two, hypothesis testing is the more commonly used method with multiple regression analysis, and we do it for both the overall R and for the regression coefficient for each variable.

Hypothesis testing for the overall R The null hypothesis that the multiple correlation coefficient R equals 0 in the population says that the independent variables taken together have *no effect* on the dependent variable. It is a bit hard to imagine that we know so little about the variables that we chose independent variables that together have no effect on the dependent variable. This null hypothesis of no effect is therefore usually not very interesting, and it is usually rejected.

In the snack food example, the sample R is equal to 0.94. Using hypothesis testing, we now ask if a value of R that large or larger is possible in a sample from a population where the multiple correlation coefficient is 0. That is, is it true in the snack foods population that fat, cholesterol, and sodium have no effect on calories? We answer this question by finding the p-value for $R = 0.94$.

We need one of the four standard statistical variables to find the p-value. In the case of R, we use the F-variable. With 16 observations and 3 independent variables in the analysis, the R of 0.94 gives $F = 31.3$, with 3 and 12 degrees of freedom. (The formula for F together with the degrees of freedom is Formula 13.2 at the end of the chapter.) The probability that this F is 31.3 or larger is a very small 0.000006; if we took a million different samples from a population where the multiple regression coefficient equaled 0, only 6 samples would give an R of 0.94 or larger by chance alone. This is strong evidence against the null hypothesis of no effects of the independent variables, and we reject the null hypothesis.

Hypothesis testing for each variable The value for R is highly significant, but it does not tell us whether all three independent variables have a statistically significant effect or whether *only some or maybe only one* of them have an effect. To find out, we do a separate hypothesis test for each variable.

Does fat content seem to be related to calories in the population of snack foods? The partial regression coefficient for the fat content variable of snack foods equals 12.6. Could the same coefficient for the population data equal 0? With this question as the null hypothesis, we change 12.6 to a value of the statistical t-variable; the statistical software gives $t = 4.98$ with 12 degrees of freedom. The probability of getting this value or a larger value of t equals 0.00016. Thus, only in 16 of 100,000 samples from a population where the coefficient for fat content equals 0 would we get a regression coefficient of 12.6 or larger. A regression coefficient of 12.6 or more from a population where the corresponding coefficient equals 0 is therefore very unlikely, so we reject the null hypothesis and conclude that the fat content variable is related to the calories in the population. Since the population regression coefficient is different from 0, we can estimate the coefficient and find that 95% confidence interval for the partial population regression coefficient is 7.1 to 18.1.

Similarly, the cholesterol variable has a partial regression coefficient equal to -0.11. If we go through the same procedures as for the fat variable, we find that the p-value for the cholesterol coefficient equals 0.44. This p-value is so large that we do not reject the null hypothesis of no effect. The corresponding coefficient in the population could well be equal to 0, and we conclude that cholesterol may not be related to the calories. Finally, for the sodium variable, a partial regression coefficient of 0.18 translates to $t = 2.46$. This value of t has a p-value of 0.015, small enough to reject the null hypothesis and conclude that the sodium variable does seem to be related to the calories in snack foods.

Fortunately, all these computations can be done on a computer with the proper statistical software. We give no formulas for how to perform these computations by hand; it's just too much work!

We often leave nonsignificant variables in a multiple regression out of the analysis to keep the analysis as simple as possible. Leaving out the cholesterol variable, the remaining two variables, fat and sodium, produce almost the same value of R. Thus, it is no loss to study calories as dependent on fat and sodium only.

13.3 MULTIPLE REGRESSION WITH A DUMMY VARIABLE

According to published figures from the college administration, for the 1994–1995 school year the female full professors at Swarthmore College had a mean salary of $71,100 while the corresponding figure for the male full professors was $76,300. That is a difference of $5,200, and it raises the question of whether the college discriminates between men and women in paying its professors. The two means certainly are different. But before we jump to conclusions, let's control for other variables in our study of gender and salary.

The snack foods example involved only metric variables. Here we wish to combine metric with categorical independent variables, as we did in Chapter 10. Instead of studying the relationship between gender and salary by comparing the two means, we study the data using a dummy variable for gender. Just as we did in Chapter 10, we assign a value of 0 to each woman and a value of 1 to each man. A scatterplot of the data then looks like Figure 13.5. Salary is shown on the vertical axis in units of $1,000, and gender, with the two values 0 and 1, is shown on the horizontal axis. We do not know the actual pattern in the scatterplot because the college does not release individual salary figures.

The scatterplot shows the women's salaries as the points at the left and the men's salaries as the points at the right. The figure also shows the regression line through these points. The line cuts through the vertical salary scale at 71.1. This is the intercept of the regression line, and it equals the mean salary of the women. The slope of the line is the regression coefficient for gender. The slope equals 5.2, which is the difference in mean salaries between the men and the women.

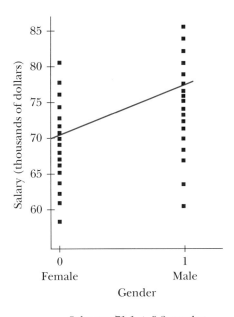

Salary = 71.1 + 5.2 gender

Figure 13.5 Scatterplot for salary and gender

When there are only two sets of points, as here, the regression line always goes through the means of the dependent variable for the two groups.

The difference in salaries shows that a relationship exists between the two variables gender and salary. Does this mean that the college discriminates in the salaries it pays? We won't know until we have found out whether other relevant variables have been controlled for.

Nothing in statistical theory guides us in the choice of control variables. The choice has to come from what we know about the subject we are studying and the availability of data. One possible control variable here is age. Maybe the men are older and therefore make more money because they have worked longer. Thus, to find the effect of gender when controlling for age, we should do a multiple regression analysis with salary as the dependent variable and both gender and age as independent variables. Of course, this analysis will also give us the effect of age when we control for gender.

Suppose we had data on individual faculty salaries as well as on age and gender and that a multiple regression analysis gave the following result:

$$\text{salary} = 40 + 0.0 \text{ gender} + 0.5 \text{ age}$$

The most striking feature of this result is that the partial regression coefficient for gender, when we control for age, equals 0.0. Before we controlled for age, the regression coefficient for gender was 5.2. Because the coefficient now is 0.0, the effect of gender on salary disappears, there is no difference in the salaries of a man and a woman of the *same* age.

To comprehend controlling for age, we could divide the data into age groups (all 40-year-olds, all 41-year-olds, all 42-year-olds, and so on) and make scatterplot gender and salary for each age group. Figure 13.6 shows a few of these scatterplots. Since we are looking at only a specific age group in each scatterplot, the number of observations in each scatterplot is smaller than the number of observations for the entire data set. However the data points are distributed in each scatterplot, age has nothing to do with it because all the people are of the same age.

For each scatterplot the regression line goes through the mean salaries for women and men. In each group the regression line is approximately horizontal, with a slope of 0.00, meaning that there is no relationship between gender and salary in the group. The average of the slopes in all the age groups equals 0. This is the partial regression

13.3 Multiple Regression with a Dummy Variable

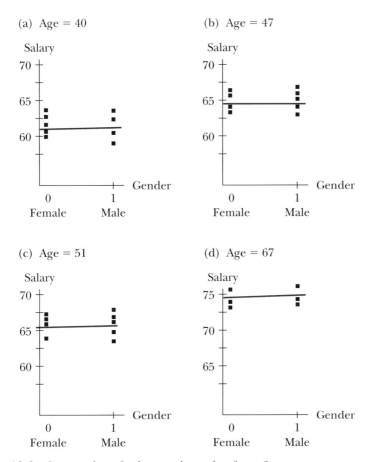

Figure 13.6 Scatterplot of salary and gender for a few age groups

coefficient for gender when we control for age. Since the average slope is 0, the average line is horizontal, and there is no difference in salary between men and women when we control for age. Sometimes it can be helpful to think of a partial regression coefficient as the average within-group regression coefficient, as we do for the partial phi.

Luckily, we do not have to actually divide the data into groups when we want to compute a partial regression coefficient. One reason we are lucky is that if the sample of observations is small to start with, then there would not be enough observations in each group to do an adequate analysis. Conversely, if the amount of data is large to start with, even small groups would be too cumbersome to handle. The procedure we describe here can be converted into mathematical equations and formulas derived for the computation of the partial regression coefficients. These formulas are cumbersome when there are many variables, but when they are programmed in a statistical software package, they are easy to use.

Another way to illustrate what is going on here is a scatterplot for age and salary with different symbols for the points for men and women (Figure 13.7). Within each group we can regress salary on age. To make the point, we have simplified the data so that all the women are younger than all the men. One way to control for age is to pick a man and a woman with the same age. Suppose we pick the overall mean age. The predicted salary for both the man and the woman would be the same if we extend both regression lines. We get the same predicted salary because the equations for the two lines in the figure are identical, with the same intercept and the same slope. The only difference is that the line for women is located below and to the left of the line for men.

The simple explanation for the original difference in salaries in this example is that on the average the men are older than the women and consequently make more money than the women. But sometimes it is not so easy to identify appropriate control variables or we realize too late which control variable we should have used. If we collect data from a sample survey and do not collect data on the variable we later

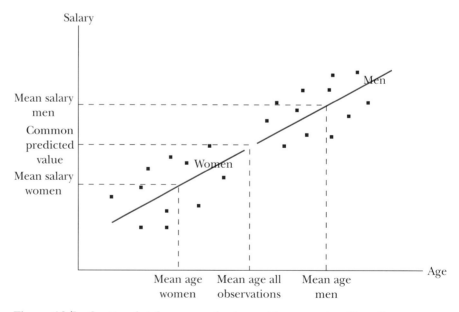

Figure 13.7 Scatterplot for age and salary with regression lines for women and men

would like to use as a control variable, it is difficult or even impossible to later go back to the respondents and collect the additional data. Thus, in planning to collect data, it is very important to think through the analysis as clearly as we can before we collect the data to make certain we collect data on all the control variables we plan to use in the analysis.

As you can see, dummy variables are useful in multiple regression analyses because they make it possible to analyze both metric and categorical variables as independent variables. The categorical variable we have looked at has only two categories, but of course many categorical variables, have more than two categories. Religious preference, for example, might have the four categories Catholic, Jew, Protestant, and other. For more than one category we use more than one dummy variable. Good statistical software programs construct dummy variables automatically.

13.4 TWO-WAY ANALYSIS OF VARIANCE

Imagine the following situation involving a woman driving to work. Sally Jones can take either Main Street or High Street to go to work, and she also has the choice of driving during rush hour or not. She wants to know how she can get to work most rapidly. Stated as a statistical problem, the two independent variables for this problem are route and time of day, while the dependent variable is length of time it takes to drive from home to work. Is one route better than another? Is one time of day better than the other? Or is one route perhaps better at one time of day and the other route better at another time of day? One way to study the effect of route and time of day on the length of the commute would be to drive each of the two routes at different times of day for a few days and measure how long each trip takes.

Route and time of day are both categorical variables, each with two values. Route has the values Main Street and High Street; time has the values rush hour and nonrush hour. Length of time it takes to drive to work is a metric variable, measured in minutes. When the independent variables are categorical and the dependent variable is metric, we can introduce dummy variables for the categorical variables and do a multiple regression analysis. However, we often instead do a two-way analysis of variance. In Chapter 12 we discussed analysis of variance with only one independent variable. Here the problem requires a two-way analysis of variance because there are two independent variables that

Two-way analysis of variance is analysis of the effects of two categorical independent variables on a metric dependent variable.

Table 13.5 Mean driving time for different routes and times of day

		Route		Mean
		Main Street	High Street	
Time of day	Rush hour	25 min.	21 min.	23 min.
	Nonrush hour	19 min.	15 min.	17 min.
	Mean	22 min.	18 min.	20 min.

both affect the dependent variable. With three independent categorical variables we would use a three-way analysis of variance, and so on.

To gather the data for her experiment, Ms. Jones randomly chose different routes and different times of day for her drive to work for the next four weeks. This schedule gave her 5 different trips for each of the possible combinations of drives. The raw data consist of traveling times for a total of 20 trips. (Having the same number of observations for each combination of the variables makes the two-way analysis much easier. We do not take up the case of unequal numbers of observations in this text).

The mean lengths of time the drives took are shown in Table 13.5. The table shows that the overall mean was 20 minutes for the drive to work. The mean for the trips along Main Street is 22 minutes, so that route took 2 minutes more than the overall mean. The trips along High Street had a mean of 18 minutes, so that route took 2 minutes less than the overall mean. Similarly, rush hour added 3 minutes to the drive for a mean of 23 minutes, and during nonrush hour it took 3 minutes less for a mean of 17 minutes for the drive.

One-way analysis with time of day only

Studying the relationship between time of day and how long it takes to drive to work is studying the relationship between a categorical independent variable and a metric dependent variable. From Chapter 11 we know that this requires a one-way analysis of variance. Table 13.6 shows the results of this analysis. The sum of squares for time equals 180.00 and the total sum of squares equals 315.98. That gives an $R^2 = 180.00/315.98 = 0.57$. Thus, time of day explains 57% of the variation in driving time. The square root of this number gives $R = 0.75$. Thus, we have a fairly strong relationship between the two variables.

Table 13.6 One-way analysis of variance table for time of day

Source	Degrees of freedom	Sum of squares	Mean square	F-ratio	p-value
Time of day	1	180.00	180.00	23.87	0.00012
Residual	18	135.98	7.554		
Total	19	315.98			

$R^2 = 180/315.98 = 0.57$ and $R = 0.75$

We know that it takes 3 minutes more to drive during rush hour and 3 minutes less to drive during nonrush hour. But we do not know whether these differences are statistically significant. Perhaps the differences are simply random variations. The F-ratio is equal to 23.87, and the p-value for this F equals a small 0.00012. Only 12 in 100,000 times would an F-value of this magnitude or larger occur if there were no difference in driving time. This means that the data are very unlikely if the null hypothesis of no difference is true, and we therefore reject the null hypothesis. The difference in driving time between the two times of day is statistically significant.

One-way analysis with route only

To study the relationship between route and how long it takes to drive to work, we also do a one-way analysis of variance. Table 13.7 shows the results of this analysis: $R^2 = 0.25$ and $R = 0.50$. Thus, route explains 25% of the variation in driving time, and there is a moderately strong relationship between the two variables.

Similarly, we know it takes 2 minutes more to drive Main Street and 2 minutes less to drive High Street. But we do not know whether these differences are statistically significant. Perhaps the differences are simply random variations. The F-ratio is equal to 6.10, and the p-value for this F equals a small 0.024. This is not quite as significant as for the time-of-day variable, but the data are still very unlikely if the null hypothesis of no difference between the two streets is true. We therefore reject this null hypothesis also. Thus, the difference in driving time between the two streets is statistically significant.

Table 13.7 One-way analysis of variance table for route

Source	Degrees of freedom	Sum of squares	Mean square	F-ratio	p-value
Route	1	80.00	80.00	6.10	0.024
Residual	18	235.98	13.110		
Total	19	315.98			

$R^2 = 80.0/315.98 = 0.25$ and $R = 0.50$

Two-way analysis with time of day and route

Instead of two one-way analyses, we can do one two-way analysis of the driving times. This is just like doing a multiple regression with two independent variables instead of two separate simple regressions each with only one independent variable. A two-way analysis that includes both time of day and route as independent variables reduces the effect of the residual variable. In the one-way analysis with time of day as the independent variable, route was one of the variables included in the residual variable. Similarly, in a one-way analysis with route as the independent variable, time of day was one of the variables included in the residual variable. With a two-way analysis of variance, both of the independent variables are brought out of the residual variable at the same time.

We can list the means of the driving times in a more organized fashion:

Overall: Mean length of time = 20 min.

Rows: Effect of rush hour = 23 min. − 20 min. = 3 min.
Effect of nonrush hour = 17 min. − 20 min. = −3 min.

Columns: Effect of Main Street = 22 min. − 20 min. = 2 min.
Effect of High Street = 18 min. − 20 min. = −2 min.

These differences are the effects on the driving time for each value of the two independent variables. They are also known as the row and the column effects for the two categorical variables. For example, rush hour is the first row in Table 13.5, and the effect of driving during rush hour is 3 minutes: driving to work at that time takes 3 minutes longer than the mean length of time for all the trips. Because the driving time is different at rush hour and nonrush hour, the time variable has an effect on the data. Similarly, because the driving time is different along the two streets, route has an effect as well.

If we know the overall mean, the two row effects, and the two column effects, we can find the means in the remaining four cells in Table 13.5. These computations are shown in Table 13.8. For example, driving during rush hour along Main Street takes $20 + 3 + 2 = 25$ minutes. This is because the mean driving time is 20 minutes, and then it takes 3 minutes more to drive at rush hour and 2 minutes more to drive Main Street for a total of 25 minutes. The same type of computations work for the other three cells as well.

Furthermore, if we take the mean of the two cells in the first row in the table, the $+2$ and -2 for Main and High Streets cancel out and give a mean of $20 + 3 = 23$ minutes for rush hour. If we take the mean of the two cells in the column for Main Street, we see that the $+3$ and -3 cancel out and the Main Street route takes $20 + 2 = 22$ minutes. The same type of computations can be made for the row for nonrush hour and the column for High Street.

The means can also be shown in a figure. In Figure 13.8, the two streets are on the horizontal axis and the four cell means are shown as small squares. To keep track of the means, the two rush-hour points are connected by one line, and the two nonrush-hour points by another line. Of course, the two times of day could have been marked off on the horizontal axis and the two means for each street connected by lines. The advantage of doing the graph the way it is done is that each mean point in the graph corresponds directly with the placement of the mean in Table 13.5.

Table 13.8 Mean driving times for different routes and different times of day found from the overall mean and the row and column effects

		Route		Mean
		Main Street	High Street	
Time of day	Rush hour	$20 + 3 + 2 = 25$ min.	$20 + 3 - 2 = 21$ min.	$20 + 3 = 23$ min.
	Nonrush hour	$20 - 3 + 2 = 19$ min.	$20 - 3 - 2 = 15$ min.	$20 - 3 = 17$ min.
	Mean	$20 + 2 = 22$ min.	$20 - 2 = 18$ min.	20 min.

586 Chapter 13 • Multivariate Analysis

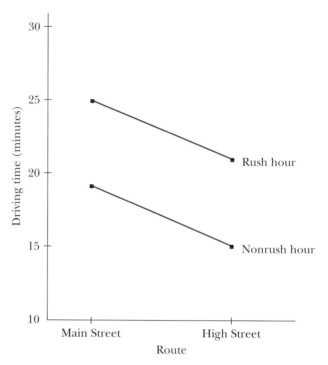

Figure 13.8 Mean driving times for the two routes and two times of day

The striking feature of the graph is that the two lines are parallel. There is no obvious reason for that; after all, the graph only shows the magnitudes of four means. If we connect the two pairs with lines, we have little reason to expect the lines to be parallel. In fact, sometimes the lines are parallel and sometimes they are not. We give a brief discussion of what nonparallel lines mean in the subsection in the second study with interaction effects.

The residual variable The various driving times and the means are shown in Table 13.9. We already know that the drives at rush hour along Main Street have a mean of 25 minutes, but we see from the data that each drive did not take exactly 25 minutes. The reason the observations are different from each other within each cell of Table 13.9 is that other variables besides street and time of day affect the driving time. Some days, for example, Ms. Jones got stuck behind a school bus, while other days the path was clear.

As we know, the combined effect of all these other variables is known as the residual variable. In the example, the first trip took 26

Table 13.9 Driving times for 20 different trips

		Route		Mean
		Main Street	High Street	
Time of day	Rush hour	26.0	18.8	23 min.
		24.2	20.0	
		26.5	22.7	
		24.1	22.3	
		24.2	21.2	
		(mean = 25)	(mean = 21)	
	Nonrush hour	19.9	17.4	17 min.
		16.7	17.3	
		20.7	12.7	
		20.5	15.6	
		17.2	12.0	
		(mean = 19)	(mean = 15)	
	Mean	22 min.	18 min.	20 min.

minutes instead of the mean of 25 minutes, so the residual variable added another minute to the trip. The effect of the residual variable on a particular trip is the difference between that observation time and the mean time in the cell. Thus, the effects of the residual variable on the other trips along Main Street at rush hour were -0.8, 1.5, -0.9 and -0.8 minutes. Similarly, we can find the effect of the residual variable for each of the other trips Sally Jones took.

Question 1. Are there any relationships in the data? We already know from the analyses of each of the two independent variables that it makes a difference whether Ms. Jones travels during rush hour or not and that it makes a difference which route she takes. We can therefore proceed with the next questions.

Question 2. How strong are the relationships? From Table 13.6, $R^2 = 0.57$ and $R = 0.75$ for the relationship between time of day and length of travel time. Similarly, from Table 13.7, $R^2 = 0.25$ and $R = 0.50$ for the relationship between route and length of travel time. In addition to the strengths of the relationships of each independent variable and the dependent variable, we can also find the strength of the relationship between the combined effect of the independent variables and the dependent variable. This is what in regression analysis gave the multiple correlation coefficient.

We have already found the sums of squares for the time-of-day variable (180.00) and the route variable (80.00). The combined effect of the two independent variables becomes 180.00 + 80.00 = 260.00. The two independent variables account for 260.00/313.98 = 82% of the variation in the driving times, while the residual variable accounts for the remaining 18% of the variation. Since $R^2 = 0.82$, the multiple correlation coefficient for the effect of both independent variables is $R = 0.91$. Thus, the two variables together have a strong relationship with the driving time.

The difference between the total sum of squares and the combined effect of the two independent variables equals 55.98, and that is the effect of the residual variable in the two-way analysis. If we had computed each of the residuals and squared them, that sum would also be equal to 55.98. It is not surprising that the residual sum of squares for the two-way analysis is smaller than either of the residual sums of squares for the one-way analyses. The residual variable now does not contain the effect of either of the two independent variables. These sums of squares are shown in Table 13.10.

Question 3. Are the relationships statistically significant? As in multiple regression, we can test to see whether the two independent variables together have a significant effect, and we can do separate tests for each of the variables.

Table 13.10 Two-way analysis of variance table for time of day and route

Source	Degrees of freedom	Sum of squares	Mean square	F-ratio	p-value
Time of day	1	180.00	180.00	54.66	0.000001
Route	1	80.00	80.00	24.29	0.00012
Residual	17	55.98	3.293		
Total	19	315.98			

For both variables, $R^2 = 0.82$, and that translates to a value of the *F*-variable of 39.48 with 2 and 17 degrees of freedom. The *p*-value for this *F* equals 0.0000004; that is, only 4 in 1 million values of *F* are that large or larger if there is no relationship between the two independent variables and the dependent variable. The *p*-value shows that R^2 would almost never equal 0.82 or more if no relationship exists. Thus, we reject the null hypothesis.

The separate test for each variable is shown in Table 13.10. The residual sum of squares has lost one degree of freedom compared with the two earlier ones, but the net effect is still that the residual mean square now is smaller than either separate residual mean square in Tables 13.6 and 13.7. Therefore, the values of *F* are larger now in the two-way analysis than they were in the two one-way analyses. Larger values of *F* produce smaller *p*-values. The *p*-value for time of day went from 0.00012 to 0.000001, and the *p*-value for route went down from 0.024 to 0.00012. Both variables have a statistically significant relationship to driving time, and the *p*-values from the two-way analysis are smaller than those in the one-way analyses.

A second study with interaction effects

The following year Sally Jones repeated her study. The new mean driving times are shown in Table 13.11. It still takes 3 minutes more than the mean to drive during rush hour and 3 minutes less during nonrush hour, and Main Street still takes 2 minutes more while High Street takes 2 minutes less than the overall mean. The row and column effects are unchanged.

Table 13.11 Mean driving time for different routes and times of day a year later

		Route		Mean
		Main Street	High Street	
Time of day	Rush hour	26 min.	20 min.	23 min.
	Nonrush hour	18 min.	16 min.	17 min.
	Mean	22 min.	18 min.	20 min.

But something has happened to the two routes and the two times of day compared to the earlier data. The difference lies in the four cell means. Driving Main Street during the rush hour now takes 26 minutes instead of the 25 minutes it took a year ago. Similarly, both High Street during rush hour and Main Street during nonrush hour take a minute less than they used to, while High Street during nonrush hour takes a minute more. Graphing the four cell means gives the picture shown in Figure 13.9.

The striking difference between Figure 13.9 and Figure 13.8 is that in Figure 13.8 the lines for the two times of day are parallel and in Figure 13.9 they are not parallel. In Figure 13.8 with parallel lines, the difference in driving time between rush hour and nonrush hour is the same 6 minutes for both Main Street and High Street. In Figure 13.9, the two lines indicate an 8-minute difference between the two times of day for Main Street and only a 4-minute difference for High Street.

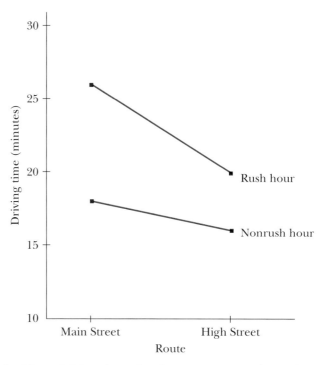

Figure 13.9 Mean driving times for the two routes and two times of day a year later

The combined effect of driving along Main Street during rush hour is now more than the sum of the separate effects of Main Street and rush hour. The particular combination of street and time of day adds another minute to the driving time. Such an additional effect is known as an *interaction effect*. The two variables act together to produce an additional effect on the dependent variable over and beyond their two separate effects.

In Figure 13.9, three variables affect driving time. As before, there is the route variable and the time of day variable, but now there is also the route/time of day interaction variable. The sum of squares for the interaction variable is found to equal 20.00, and it is displayed with the other sums of squares in Table 13.12. The interaction variable explains $20.00/335.98 = 6\%$ of the total variation in the driving times. Thus, the relationship between the interaction variable and driving time is not very strong, with $R = 0.24$. Formula 13.3 shows how to compute the various sums of squares needed for a two-way analysis of variance. Formula 13.4 shows how to find the appropriate degrees of freedom for each sum of squares.

Do each of the three variables have statistically significant relationships with driving time for the new data? Table 13.12 answers this question. All three variables have small p-values, and this means they all have a statistically significant relationship to driving time. The interaction variable has only a barely significant effect at the usual 5% level, while the other two variables have highly significant effects.

> An **interaction effect** occurs when two variables act together to produce an additional effect on the dependent variable beyond the sum of their two separate effects.

Table 13.12 Two-way analysis of variance of driving time for time of day, route, and time of day/route interaction

Source	Degrees of freedom	Sum of squares	Mean square	F-ratio	p-value
Time of day	1	180.00	180.00	54.66	0.000001
Route	1	80.00	80.00	24.29	0.00012
Time of day/route interaction	1	20.00	20.00	5.72	0.03
Residual	16	55.98	3.499		
Total	19	335.98			

13.5 ESTABLISHING CAUSALITY

Decisions about what constitutes a causal relationship depend on both a significant statistical result and the consensus of researchers and others involved with the problem. Most actual dependent variables we study using statistical methods are affected by several other variables and must be analyzed with multivariate statistical methods. A major feature of multivariate methods is that they enable us to see the effect of a particular independent variable in the presence of other independent variables. If a relationship found using an independent variable alone disappears in the presence of additional independent variables, the original relationship cannot have been causal. On the other hand, if a relationship does not disappear when we control for other variables, we still cannot consider the relationship causal because we can never control for every possible other variable; we do not have data on all other possible variables. Thus, proving causality remains an elusive task even using multivariate statistical methods.

STOP AND PONDER 13.2

A study by Yale psychiatry professor Kyle Pruett, as reported in a newspaper article *(Marc Schogol, "A father's hand,"* The Philadelphia Inquirer, *May 31, 1995. p. H-1)*, suggests that if fathers are involved in child care during their children's first six months, the youngsters will have better scores on intellectual and motor development tests in fourth grade. Professor Pruett studied infants and their families—including a follow-up with the families 10 years later—to test the children's achievements. Pruett implies that father involvement in the first six months of children's lives *causes* them to perform better in school and on the playground.

What criticisms of this result can you give, based on your knowledge of multiple regression analysis, two-way analysis of variance, and the problems of causality? Can you give alternative explanations for the results of this study? What other information would be critical in assessing the results? If Pruett had suggested that early father involvement were correlated with later developmental advantages, would it have taken care of the major objections you might have to this claim?

13.6 SUMMARY

Multivariate analysis is used to analyse the effects of several independent variables on a dependent variable in one procedure. As a result of the analysis, the independent variables can often be ranked in terms of their relative impact on the dependent variable.

In multivariate analysis it is sometimes possible to determine if a relationship between two variables is not causal. If a relationship between two variables disappears as a result of a multivariate analysis, then we presume it was not a causal relationship.

13.1 Partial phis: Three categorical variables

The strength of the relationship between two categorical variables is measured by the coefficient phi (Chapter 9). For each subgroup of a variable that is held constant it is also possible to compute a phi for an independent categorical variable and a dependent variable categorical variable. Each phi describes the strength of the relationship of an independent variable and the dependent variable for the observations that have a particular value of the control variable.

Controlling for a variable means taking away its influence in the study of the effect of other independent variables on the dependent variable. When we control for the effect of a variable we say that we are keeping the controlled variable constant. We keep a variable constant by dividing the data up into subgroups according to the values of that variable.

An average coefficient calculated from a set of phi coefficients from subgroups of the control variable is known as a partial coefficient. The partial coefficient expresses the overall strength of the relationship between two variables once the effect of a third (or more) variable has been controlled for.

13.2 Multiple regression with metric variables

For metric variables, multivariate analyses are done by computing partial regression and partial correlation coefficients. These coefficients are comparable to the regression and correlation coefficients discussed for two variables in Chapter 10.

The partial regression coefficients of two or more independent metric variables can be calculated while holding all but one of the independent variables constant at a time. The partial coefficients can be combined into one multiple regression equation to estimate the combined effect of the independent variables on the dependent variable.

The values of partial coefficients can change depending on which variables are included in the analysis. The change in the values of partial coefficients happens when the independent variables are correlated with each other as well as with the dependent variable. Collinearity is the name given to the correlation of independent variables among themselves.

The strength of the relationship between the predicted values and the actual values of the dependent variable in a multiple regression analysis is measured by the multiple correlation coefficient R. R^2 gives the amount of variation explained by all the independent variables together.

To find if the results of a multiple regression analysis are statistically significant, that is, if the sample results can be applied to the population from which the sample was drawn, hypothesis testing methods rather than confidence levels are usually used. The usual null hypothesis is that the multiple correlation coefficient in the population is 0. By converting the sample multiple correlation coefficient R to a value of an F-variable with the proper degrees of freedom, we can use statistical tables or software to find the p-value for the observed data. The p-value is the probability of finding the observed value of R or a larger value of R in a sample from a population where the multiple correlation coefficient equals 0.

In a multiple regression analysis, hypothesis testing can also be done separately on the regression coefficient for each of the independent variables.

13.3 Multiple regression with a dummy variable

Categorical independent variables can be used in regression analysis by converting the categorical variables to dummy variables. Two numerical values, often 0 and 1, are assigned to the values of a categorical variable. For example, for gender, man could equal 0 and woman 1.

13.4 Two-way analysis of variance

To study the effects of two independent categorical variables on a metric dependent variable, the statistical method called two-way analysis of variance can be used. One reason a two-way analysis of variance is superior to two one-way analyses of variance is that it can take into account the possible joint effect of the two categorical variables beyond their separate effects. The joint effect of the two independent variables on the dependent variable is called the interaction effect. By taking into account several independent variables and their interactions simultaneously, the effect of the residual variable is reduced, making it easier to establish statistical significance.

13.5 Establishing causality

Proving causality means testing all possible variables that might have an effect on a dependent variable. Since this is not possible, claiming a causal relationship is always a tentative decision, based on the knowledge of and the accessibility of relevant variables.

ADDITIONAL READINGS

Achen, Christopher. *Interpreting and Using Regression* (Sage University Paper Series on Quantitative Applications in the Social Sciences, series no. 07-029). Beverly Hills, CA: Sage, 1982. Uses of multiple regression.

Asher, Herbert. *Causal Modeling*, 2nd ed. (Sage University Paper Series on Quantitative Applications in the Social Sciences, series no. 07-003). Beverly Hills, CA: Sage, 1983. Using regression to examine possible causal models.

Berry, William D., and Stanley Feldman. *Multiple Regression in Practice* (Sage University Paper Series on Quantitative Applications in the Social Sciences, series no. 07-050). Newbury Park, CA: Sage, 1985. Uses of multiple regression.

Bray, James H., and Scott E. Maxwell. *Multivariate Analysis of Variance* (Sage University Paper Series on Quantitative Applications in the Social Sciences, series no. 07-054). Beverly Hills, CA: Sage, 1985. Introduction to multivariate analysis of variance.

Fox, John. *Regression Diagnostics* (Sage University Paper Series on Quantitative Applications in the Social Sciences, series no. 07-079). Newbury Park, CA: Sage, 1991. Using the data to see if they violate any of the underlying assumptions for the use of regression analysis.

Knoke, David, and Peter J. Burke. *Log-Linear Models* (Sage University Paper Series on Quantitative Applications in the Social Sciences, series no. 07-020). Beverly Hills, CA: Sage, 1980. Introduction to the multivariate analysis of categorical variables.

Wildt, Albert R, and Olli T. Ahtola. *Analysis of Covariance* (Sage University Paper Series on Quantitative Applications in the Social Sciences, series no. 07-012). Beverly Hills, CA: Sage, 1978. Regression analysis with both categorical and metric independent variables.

FORMULAS

PARTIAL r (OR PHI)

For three variables denoted $x1$, $x2$ and $x3$, the partial correlation coefficient of $x1$ and $x2$ while controlling for $x3$ (denoted $r_{12.3}$) is found from the three pairwise correlation coefficients r_{12}, r_{13}, and r_{23} according to the expression

$$r_{12.3} = \frac{r_{12} - r_{13}r_{23}}{\sqrt{(1 - r_{13}^2)(1 - r_{23}^2)}} \tag{13.1}$$

The same formula works substituting phis for the r's.

F-TEST FOR MULTIPLE CORRELATION COEFFICIENT

To test the null hypothesis that R equals 0 in the population, based on k independent variables and n observations

$$F = \frac{R^2}{1 - R^2} \frac{n - k - 1}{k} \qquad \text{d.f.} = k, n - k - 1 \tag{13.2}$$

From the computed value of F, statistical software or tables of the F-distributions can be used to find the corresponding p-value.

TWO-WAY ANALYSIS OF VARIANCE

Table 13.13 shows how the data can be displayed for a two-way analysis of variance when the numerical values of the dependent variable are replaced by symbols. A typical observation is denoted y with the three

Table 13.13 Numerical observations replaced by symbols

		Route		Mean
		Main Street	High Street	
Time of day	Rush hour	y_{111}	y_{121}	$\bar{y}_{1\cdot}$
		y_{112}	y_{122}	
		y_{113}	y_{123}	
		y_{114}	y_{124}	
		y_{115}	y_{125}	
		(mean = $\bar{y}_{11}$)	(mean = $\bar{y}_{12}$)	
	Nonrush hour	y_{211}	y_{221}	$\bar{y}_{2\cdot}$
		y_{212}	y_{222}	
		y_{213}	y_{223}	
		y_{214}	y_{224}	
		y_{215}	y_{225}	
		(mean = $\bar{y}_{21}$)	(mean = $\bar{y}_{22}$)	
	Mean	$\bar{y}_{\cdot 1}$	$\bar{y}_{\cdot 2}$	$\bar{y}$

subscripts i, j, and k. The i subscript indicates which row the observation is located in, the j subscript indicates the column, and the k subscript indicates which observation is under examination at in a particular cell. The overall mean has no subscript. The mean of the data in a row is denoted by y-bar with subscripts i and a dot, and the mean of the data in a column is denoted by y-bar with subscripts dot and j.

The various sums of squares are found the following way:

Row variable sum of squares = $\Sigma n_{i\cdot}(\bar{y}_{i\cdot} - \bar{y})^2$

Column variable sum of squares = $\Sigma n_{\cdot j}(\bar{y}_{\cdot j} - \bar{y})^2$

Interaction variable sum of squares = $\Sigma\Sigma n_{ij}(\bar{y}_{ij} - \bar{y}_{i\cdot} - \bar{y}_{\cdot j} + \bar{y})^2$ (13.3)

Residual variable sum of squares = $\Sigma\Sigma\Sigma(y_{ijk} - \bar{y}_{ij})^2$

Total sum of squares = $\Sigma\Sigma\Sigma(y_{ijk} - \bar{y})^2$

The n's refer to the number of observations in the various rows, columns, and cells. These formulas can be used directly for very small data

Table 13.14 Data for a two-way analysis of variance set up in a computer file

y	Row	Column
y_{111}	1	1
y_{112}	1	1
y_{113}	1	1
y_{114}	1	1
y_{115}	1	1
y_{121}	1	2
y_{122}	1	2
y_{123}	1	2
y_{124}	1	2
y_{125}	1	2
y_{211}	2	1
y_{212}	2	1
y_{213}	2	1
y_{214}	2	1
y_{215}	2	1
y_{221}	2	2
y_{222}	2	2
y_{223}	2	2
y_{224}	2	2
y_{225}	2	2

sets; otherwise, a two-way analysis of variance is best done on a computer.

With r rows, c columns, and m observations in each cell, the various degrees of freedom are found from the expressions

$$\text{Row variable degrees of freedom} = r - 1$$
$$\text{Column variable degrees of freedom} = c - 1$$
$$\text{Interaction variable degrees of freedom} = (r-1)(c-1) \quad (13.4)$$
$$\text{Residual variable degrees of freedom} = rcm - rc$$
$$\text{Total degrees of freedom} = rcm - 1$$

The mean squares are found by dividing the sums of squares by their corresponding degrees of freedom. The F-ratios are found by dividing each mean square by the residual mean square. All these numbers are typically displayed in an analysis of variance table like the one in Table 13.12.

When the analysis is done on a computer, the data are typically set up in the computer file as shown in Table 13.14. For each observed value of the dependent variable y, we enter the row and the column the observation is located in. The software program takes care of the construction of the interaction variable.

EXERCISES

REVIEW (EXERCISES 13.1–13.17)

13.1 a. How does a multivariate statistical analysis differ from a single-variable statistical analysis?

b. Why is one multivariate statistical analysis often more useful than several single-variable analyses?

13.2 If a relationship between two variables disappears as the result of a multivariate analysis, what can you presume about the relationship?

13.3 What is meant by *controlling* for a variable in an analysis?

13.4 If a phi for a relationship between two variables such as gender and vote was 0.32, and after controlling for a third variable it became 0.00, what would you conclude about the two variables?

13.5 a. What is a partial phi coefficient?

b. If the control variable has four values, e.g., high, medium, low, none, how do you find the partial phi coefficient for the relationship between the two independent variables?

c. If there are many more highs than the other values in the sample, does this affect the partial phi?

13.6 When can you use a multiple regression analysis?

13.7 If independent variables in a data set are correlated, _____ is said to exist between them.

13.8 If a regression line passing through the data points of two groups in a scatterplot is horizontal, what can you say about the effect of being in one of the groups as opposed to the other on the dependent variable (for example, the effect of gender on visual accuracy)?

13.9 a. What is a dummy variable?

b. Give an example from the chapter or one of your own construction.

c. Why is it useful to create a dummy variable?

d. What is the important restriction on its use?

13.10 a. Define and describe what a correlation called R must be.

b. What does R^2 tell us?

c. What does $1 - R^2$ tell us?

13.11 In order to discover whether a finding from a sample is applicable to an entire population, what must you do?

13.12 a. How would you explain a two-way analysis of variance to a nonstatistical friend?

b. Give an example of an imaginary study using a two-way analysis of variance.

c. In what respect is a two-way analysis of variance better than two one-way analyses of variances?

13.13 What is the residual variable as defined for a two-way analysis of variance?

13.14 Suggest a problem where more than a two-way analysis of variance would be appropriate—for example, one with three independent variables.

13.15 a. What does interaction effect mean in a two-way analysis of variance?

b. Give an example of an interaction effect.

13.16 a. Why has multiple regression analysis become popular in recent years?

b. Why is it said that this procedure can be "dangerous"?

13.17 Why is it so difficult to determine whether or not an independent variable is a causal variable, even if a multivariate analysis with several important independent variables is done?

INTERPRETATION (EXERCISES 13.18–13.37)

13.18 When we regress larceny rates on robbery rates for the 48 contiguous states we find

$$\text{larceny} = 2682 + 1.49 \text{ robbery} \qquad (t = 2.05 \text{ with } 46 \text{ d.f.}, p = 0.023)$$

To study whether this relationship could be causal, we want to control for the per capita state income. When we regress larceny rates on robbery rates and per capita income, we find

$$\text{larceny} = 3880 + 2.23 \text{ robbery} - 0.06 \text{ income}$$
$$(t = 2.75, \qquad (t = -1.86,$$
$$p = 0.004) \qquad 45 \text{ d.f.},$$
$$p = 0.035)$$

(*Source: Bureau of the Census,* Statistical Abstracts of the United States: 1995, *115th ed., Washington, D.C., 1995.*)

a. In principle, how do we find the coefficient 2.03 for larceny when we control for income?

b. What does this second analysis tell us about the relationship between larceny and robbery rates?

13.19 In Exercise 9.9, we studied the relationship between two votes in the House of Representatives, and phi was equal to 0.62. When we analyze the same data controlling for a third variable, political party, we find that the partial phi for the two votes equals 0.26.

a. How do we go about controlling for political party?

b. How do we find the partial phi?

c. What does the value 0.26 of the partial phi tell us about the original relationship between the two votes?

13.20 Age-adjusted melanoma rates from the Connecticut Tumor Registry indicate an increase in melanoma incidences from 1936 to 1972, and the rates also seem to vary with the relative number of sunspots each year. Melanoma incidences are measured as the number of cases per 10,000 population, and the range of the values is 8 to 46. The time variable is rescored for each year with 1936 as 1, 1937 as 2, and so on up to 1972 as 37. The sunspot variable ranges in values from 5 to 190. Various regression analyses give the following results:

melanoma incidences = 26.0 + 0.03 sunspots	$r = 0.13$
melanoma incidences = 7.1 + 1.10 time	$r = 0.96$
melanoma incidences = 6.2 + 0.02 sunspots + 1.10 time	$R = 0.97$

(Source: A. Houghton, E. W. Munster, and M. V. Viola, "Increased incidence of malignant melanoma after peaks of sunspot activity," The Lancet, April 8, 1978, pp. 759–760, as reported in D. F. Andrews and A. M. Herzberg, Data: A Collection of Problems from Many Fields for the Student and Research Worker, New York: Springer-Verlag, 1985, p. 201.)

a. Describe the relationship between melanoma incidences and sunspots.

b. Describe the relationship between melanoma incidences and time.

c. Describe the relationship between melanoma incidences and both sunspots and time.

d. What is the advantage of doing a multivariate analysis of these data?

13.21 Members of Congress are some of the few people who can determine their own salaries. However, voting for increases may be disapproved of by the electorate, and evidence indicates that the closer the next election is, the less likely a member is to vote for a pay increase. In a roll call on a pay increase in 1991, senators were classified by whether they voted for or against the pay increase, whether they were running for reelection at the next general election in 1992 (not all senators are up for election at the same time), and their party. These are all categorical variables with two categories, so we can find phis and partial phis for the study of the relationship between them:

phi(running again, vote) = 0.37

phi(running again, vote | controlling for party) = 0.37

The vote was such that senators running for reelection tended to vote against the pay increase. *(Source: Roll call as reported in* The New York Times, *July 19, 1991, p. A13.)* What do the two values of phi tell you?

13.22 The study of the quality of wines has a long history and is based on much subjective judgment. Professor Orley Ashenfelter, a Princeton economist, studied the relationship between auction prices of wines as a measure of wine quality, the winter rainfall from October through March in millimeters, the mean temperature in centigrade degrees during the growing season April through September, and the rainfall during the harvest period August and September for each of several years. A regression analysis for Bordeaux wines produced the following result:

quality = −12.1 + 0.0012 winter rain
+ 0.62 temperature − 0.004 harvest rain

(Source: Article in The New York Times, *March 4, 1990, pp. A1, A22. This work was done for wines up through 1989, and according to the equation the Bordeaux wines of 1989 should be of an excellent quality. This wine was still too young to be judged at that time, and it is thought that one test of this analysis will be how well the wines of 1989 actually turn out to be when they reach maturity.)*

a. What are some of the things this equation tells you about how the quality of Bordeaux wines relates to the temperature and the rainfall variables?

b. What are some of the things this equation does not tell you about the relationships between these variables?

13.23 A two-way analysis of variance to find whether the raters of the flavor found any differences between chocolate and vanilla desserts as well as between the three types—ice milk, frozen yogurt, and frozen dessert—gives the results in Table 13.15. What do you conclude about the flavors of the desserts from these results?

13.24 In Exercises 10.36 and 10.37 we looked separately at how the percentage of people below the poverty level is related to the percentage of people with a ninth-grade or less education and to the percent-

Table 13.15 Data for Exercise 13.23

Source	Sum of squares	Degrees of freedom	Mean square	F-ratio	p-value
Type	9,248	2	4,624	31.81	0.0000
Chocolate/vanilla	14	1	14	0.10	0.75
Interaction	477	2	238	1.64	0.21
Total	5,524	38	145		
Total	16,613	43			

Source: "Low-fat frozen desserts: Better for you than ice cream?" Consumer Reports, vol. 57, no. 8 (August 1992), pp. 483–487.

age of people with college or more education, using data on the 50 states:

$$\text{percent poor} = 4.6 + 0.8 \text{ percent low education}$$
$$(r = 0.70, \ t = 6.72 \text{ on } 49 \text{ d.f.}, \ p\text{-value} < 0.0001)$$

and

$$\text{percent poor} = 26.9 - 0.7 \text{ percent college or more}$$
$$(r = -0.62, \ t = -5.54 \text{ on } 48 \text{ d.f.}, \ p < 0.0001)$$

When both of the education variables are analyzed at the same time in a multiple regression, the following results occur:

percent poor = 14.6 + 0.6 percent ninth grade − 0.4 percent college
($t = 4.41$,   ($t = -3.02$, $R = 0.75$
$p = 0.0001$) $p = 0.002$)

(Source: 1990 U.S. Census data. Reported in The Chronicle of Higher Education, vol. 34, no. 1 (August 26, 1992), p. 4.)

a. Why are the regression coefficients in this analysis different from the corresponding coefficients in Exercises 10.36 and 10.37?

b. Does the improvement in the correlation from each of the simple regressions to this multiple regression seem to be large enough to make a multivariate analysis worthwhile? (We are looking for your intuitive sense, not a statistical answer.)

c. What other variables might you want to use to understand better what determines the percentages of poor people in the various states?

13.25 The scatterplot in Figure 13.10 shows data on an independent and a dependent variable for a sample of students from all four class years. The data points are labeled "fr" for freshman, "so" for sophomore, "jr" for junior, and "sr" for senior. For each question, explain your answer in some detail.

a. When you do a regression analysis of the dependent on the independent variable, will the regression coefficient be positive or negative?

b. Will the corresponding correlation coefficient be positive or negative?

c. Will the same partial correlation coefficient be small, medium, or large?

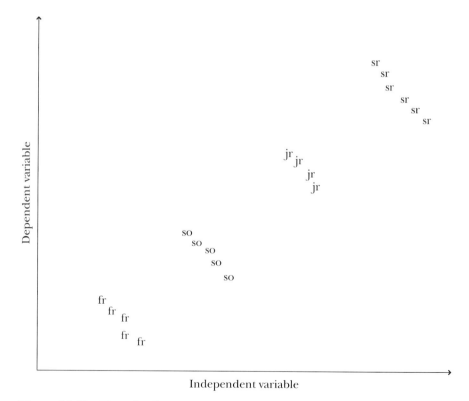

Figure 13.10 Data for Exercise 13.25

d. When you study the relationship between the independent and the dependent variable and control for class year, will the partial regression coefficient for the independent variable be positive or negative?

e. Will the corresponding partial correlation coefficient be positive or negative?

f. Will the same correlation coefficient be small, medium, or large?

13.26 According to the *Bulletin of the American Association of University Professors* (vol. 79, no. 2 (March/April 1993), p. 71) the mean salary for the 7 female full professors at Swarthmore College was $63,700 while the same mean for the 59 male full professors was $70,900 for the school year 1991–1992. If we change gender to a dummy variable with 0 for women and 1 for men, the regression line from regressing income on gender would have the equation

$$\text{income} = 63.7 + 7.2 \text{ gender}$$

From this equation and the means themselves, it looks as if the college has different pay scales for its female and male professors. But before we accept that explanation for the difference between the two means, we need to control for other variables that may be relevant. Suppose we control for age of the professors in a multiple regression analysis, and suppose the coefficient for gender equals 0.0 in the analysis of income on both gender and age.

a. Explain in some detail to your intelligent but nonstatistical friend what it means to control for a third variable and how we found the value 0.0 for the partial regression coefficient for gender while controlling for age.

b. What do we learn from the fact that the regression coefficient for gender changed from 7.2 in the simple regression analysis to 0.0 in the multiple regression analysis?

13.27 A statistical relationship has been found between the days of the week and the number of childhood accidents in Nashville, Tennessee.

a. Name a variable you think would cancel out this relationship if it were used as a control variable. Explain your reasoning.

b. What might this analysis indicate about the causal effect of day of the week on the dependent variable?

13.28 Look at a regression analysis of how the cost of textbooks is related to the number of pages in the book and the number of copies printed, and answer the following questions.

$$\text{cost of textbook} = 30 + 0.05 \text{ total pages} - 0.0001 \text{ copies printed}$$

    a. What does the 30 stand for?

    b. What are the two numbers 0.05 and 0.0001 called?

    c. Explain what 0.05 and 0.0001 indicate about the cost of a book in terms of the size of the book and how many copies are printed.

    d. If you did a separate simple regression analysis of the cost of textbooks on the number of pages, would you find the same coefficient 0.05 as in the multiple analysis?

13.29 Why is it said that trying to compare the magnitudes of partial regression coefficients in an analysis is like comparing apples and oranges? (You may wish to use the example about gas mileage to explain your answer.)

13.30 Full-time faculty members on the five regional campuses of a large university charge the administration with discrimination because their salaries are $10,000 less, on the average, than those of the full-time faculty on the main campus.

    a. Working for the board of managers as a statistical consultant, what are three variables you might wish to study in order to decide whether region alone accounts for the difference in salary?

    b. What hunches do you have about the outcome of such a study? Do you think the difference might be explained by other variables?

    c. Do you think it would make a difference if you did bivariate instead of multivariate analyses?

    d. Which analysis would you recommend? Why?

13.31 How do we know what variables to control for when we wish to study the nature of the statistical relationship between two variables, for example, between region and salary, as in Exercise 13.30?

13.32 For an R of 0.84, you obtain an F equal to 22.20 with 4 and 15 degrees of freedom. The probability of finding an F this large or larger is very small ($p < 0.0001$).

   a. Should you reject the null hypothesis of no effect?
   b. What does this finding strongly suggest about the population from which the sample data were drawn?

13.33 If a multiple regression analysis with three independent variables (e.g., fat, protein, sodium) has an overall R that is significantly different from 0, does that mean that each individual variable also has a significant effect? Explain your answer.

13.34 For a two-way analysis of variance, the differences between the observations and the mean in the cell were found to be the following:

$$-1.5 \quad 2.0 \quad -0.5 \quad 0.0 \quad -2.0 \quad 1.0 \quad 1.0$$

Why are not all the observations in a cell equal to the cell mean, such that these differences would all equal 0.0?

13.35 In a comparison of six European cars, a car magazine had several people evaluate several aspects, such as engine, gearbox, quietness, seats, and so on. All the evaluations were done on a scale from 0 to 10, and the scores were averaged to find the mean. For example, the Alfa Romeo 164L received a mean score of 9.4 for its engine. Each car had 21 mean scores, and they are shown in Figure 13.11. How do these cars differ among themselves? Is any car better than the others? Is any car worse than the others?

   a. Describe some of the patterns you see in the figure.
   b. The Mercedes has the highest overall mean at 8.32, and the VW has the lowest at 8.05. An analysis of variance gives the results in Table 13.16. What can you conclude about the differences between the cars?
   c. These data are like paired data in that there is one observation for each car on each feature and the features are not scored on the same scale. For example, the mean score for body

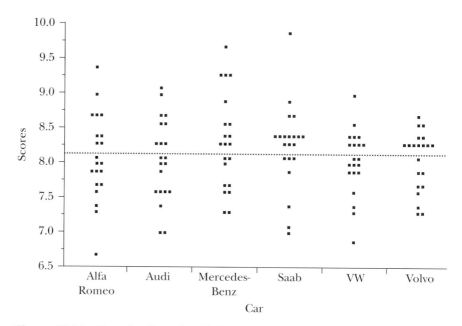

Figure 13.11 Data for Exercise 13.35 *(Source: "European influence," Road and Track, August 1991, pp. 64–84.)*

Table 13.16 Data for Exercise 13.35b

Source	Degrees of freedom	Sum of squares	Proportion	Mean square	F-ratio	p-value
Car	5	140	0.03	28.05	0.85	0.52
Residual	120	3977	0.97	33.14		
Total	125	4117	1.00			

Table 13.17 Data for Exercise 13.35c

Source	Degrees of freedom	Sum of squares	Proportion	Mean square	F-ratio	p-value
Car	5	140	0.03	28.05	0.86	0.51
Feature	20	712	0.17	35.60	1.09	
Residual	100	3265	0.80	32.65		
Total	125	4117	1.00			

structure is 8.67, while the mean score for engine is 7.71. These differences are included in the residual degrees of freedom and sums of squares in Table 13.16 and should be removed. Because there are six observations, not just two, for each feature, you cannot look at differences the way you do for paired data. But you can do a two-way analysis of variance without interaction for the results in Table 13.17. Why does removing the variation due to the difference in features from the residual variable not seem to have any effect for the cars?

13.36 In an analysis of sulfur dioxide as the dependent variable and annual mean temperature and population size in thousands as independent variables for a sample of 41 cities, using government data,

sulfur dioxide = 91.7 − 1.3 temperature + 0.02 population size $R = 0.64$

For the temperature coefficient −1.3, $t = -3.23$ on 38 d.f. and $p = 0.0013$. For the population size variable 0.02, $t = 3.74$ on 38 d.f. and $p = 0.0003$. What do these numbers tell you?

13.37 Using the manufacturers' data on a sample of 1996 model cars, you find that when you regress city mileage on the weight of the car (measured in thousands of pounds) and on the horsepower separately and together,

city mpg = 40.5 − 6.27 weight   ($r^2 = 0.77$)

city mpg = 29.2 − 0.05 horsepower   ($r^2 = 0.56$)

city mpg = 40.4 − 6.23 weight − 0.0004 horsepower   ($R^2 = 0.77$)

What do these results tell you about the relationship between city mileage and the weight and horsepower of cars?

ANALYSIS (EXERCISES 13.38–13.47)

13.38 In Exercise 9.42 we found that phi = 0.10 for the relationship between gender and college graduation for athletes in Division I schools. In the same study, phi = 0.22 for the relationship between race and graduation, and phi = 0.14 for the relationship between race and gender.

Table 13.18 Data for Exercise 13.39

		Flavor		
		Chocolate	Vanilla	Overall
Type	Frozen yogurt	71.3	54.1	67.3
	Ice milk	55.2	64.0	61.1
	Frozen dessert	31.0	25.5	27.3
	Overall	60.4	55.5	57.4

Source: "Low-fat frozen desserts: Better for you than ice cream?" *Consumer Reports,* vol. 57, no. 8 (August 1992), pp. 483–487.

a. Find the partial phi for the relationship between gender and graduation, controlling for race.

b. Find the partial phi for the relationship between race and graduation, controlling for gender.

c. What do these phis and partial phis tell you about the relationship among the three variables?

13.39 The results in Exercise 13.23 do not give the full story on how the various desserts tasted. One additional step in the analysis consists of looking at the mean flavor score for each group (Table 13.18).

a. Show the six cell means in the table in a graph similar to Figure 13.10.

b. How does the graph add to what we learn from Table 13.15 for Exercise 13.23?

Table 13.19 Data for Exercise 13.40

Y	X	D
1	1	0
1	2	0
1	3	0
3	5	1
3	6	1
3	7	1

13.40 The data matrix in Table 13.19 shows data on a dependent metric variable Y, an independent metric variable X, and an independent dummy variable D for a categorical variable with two categories.

a. Make a scatterplot of Y versus X. (A regression analysis of Y versus X results in a line with equation $Y = 0.3 + 0.4X$.)

b. What does it mean to study the relationship between Y and X and control for the variable D?

c. For these data, find the numerical value for the regression coefficient b_1 when you do a multiple regression analysis of Y

on both *X* and *D* and the analysis results in the regression equation

$$Y = 1.0 + b_1 X + 2.0D$$

d. What does the value of the regression coefficient for *X* tell you about the relationship between *Y* and *X* when you control for *D* versus when you do not control for *D*?

13.41 In a multiple regression analysis of salary on four independent variables,

salary = 30,000 + 2,500 years of college + 400 years of service
− 5,000 hourly/salaried (1,0) + 1,500 man/woman (1,0)

a. For two salaried men with 10 years of service, one of whom had 3 years of college and the other 4 years of college, what would you expect the difference in salaries to be?

b. What would you expect the difference to be if the two people were women instead of men?

c. Which seems to be more important for the salary, years of college or years of service?

13.42 Men and women raters were asked to rate scholarly papers on a scale from 1 (best) to 5 (worst). John T. McKay was cited as the author of a third of the papers, Joan T. McKay as the author of another third, and J. T. McKay as the author of the remaining papers. The papers with John cited as the author received mean ratings of 1.9 by the male reviewers and 2.3 by the female reviewers, the papers with Joan cited as the author received mean ratings of 3.0 by both sets of reviewers, and the papers with J. T. cited as the author received mean ratings of 2.7 by the male and 2.6 by the female reviewers. *(Source: Quoted in L. Billard, "A different path into print," Academe: Bulletin of the American Association of University Professors, vol. 79, no. 3 (May/June 1993), pp. 28–29, from M. A. Paludi and W. D. Bauer, "Goldberg revisited: What's in an author's name," Sex Roles, 9 (1983), 287–390.)*

a. Why should you use a two-way analysis of variance for this problem?

b. Display the means in a table like Table 13.5.

c. Do the means provide evidence of an interaction effect for gender of rater and gender of author?

d. Is there any observed difference in the mean ratings given by female and male raters?

e. Is there any observed difference in the mean ratings given to papers with John, Joan, and J. T. cited as author?

(There is not enough information in the *Academe* article to perform tests of statistical significance in this exercise.)

13.43 In Exercise 10.61, we did a separate analysis for each gender on uptake and incorporated amounts of oleic acid for female and male rats. Here we do one multivariate analysis of the data. Livers from 4 female and 4 male rats were given oleic acid. Table 13.20 shows the uptake and the amount incorporated into keotone bodies.

a. Change the gender variable to a dummy variable with 0 for the female rats and 1 for the male rats.

b. Do a multivariate analysis for amount incorporated with uptake and gender as the two independent variables.

c. How does the coefficient for uptake compare with the two separate coefficients in Exercise 10.61?

d. Substitute 0 for gender and find the separate regression line for the female rats. Substitute 1 for gender and find the separate regression line for the male rats.

Table 13.20 Data for Exercise 13.43

Uptake	Incorporated	Gender
29.3	1.82	Female
25.5	0.84	Female
26.3	1.09	Female
31.0	1.45	Female
20.6	1.56	Male
17.9	0.93	Male
23.6	1.54	Male
25.4	1.76	Male

Table 13.21 Data for Exercise 13.44

		Protein source	
		Beef	Cereal
Protein amount	Low	90, 76, 90, 64, 86, 51, 72, 90, 95, 78	107, 95, 97, 80, 98, 74, 74, 67, 89, 58
	High	73, 102, 118, 104, 81, 107, 100, 87, 117, 111	98, 74, 56, 111, 95, 88, 82, 77, 86, 92

Source: George W. Snedecor and William C. Cochran, Statistical Methods, 6th edition, Ames: Iowa University Press, 1967, p. 347.

    e. How do these two lines compare with the two original regression lines in Exercise 10.61?

    f. How large is the vertical distance between the two lines in part d?

    g. Why can the distance between the two lines be interpreted as the effect of gender when we control for uptake?

13.44 You are interested in whether protein source (beef or cereal) and protein amount (low or high) have effects on weight gain in rats. Analyze the data in Table 13.21 showing weight gains of four groups of rats with ten rats in each group.

13.45 In a statistical study, law professor David Baldus and statistics professor George Woodworth analyzed data from 2,475 cases before the courts in the state of Georgia. Among the variables they considered were the race of victim and defendant together with whether or not the death penalty was given (Table 13.22).

    a. Construct a table that shows the relationship between the race of the defendant and the race of the victim.

    b. How strong is the relationship between the race of the defendant and the race of the victim?

    c. Controlling for whether the death sentence was given or not, find the strength of the relationship between the race of the defendant and the race of the victim.

    d. What do the answers to parts b and c together tell you?

    e. What other analyses of these data might be of interest?

Table 13.22 Data for Exercise 13.45

(a) Black defendant

		Race of victim		
		Black	White	Total
Death sentence	Yes	18	50	68
	No	1420	178	1598
	Total	1,438	228	1,666

(b) White defendant

		Race of victim		
		Black	White	Total
Death sentence	Yes	2	58	60
	No	62	687	749
	Total	64	745	809

Source: Chance, vol. 1, no. 1, p. 7, 1988.

13.46 In Exercise 11.48 we compared the number of hours teachers spend teaching in lower and upper secondary schools. Now we also include the primary level (Table 13.23). Analyze the data using a two-way analysis of variance (without interaction).

13.47 Table 13.24 shows the data on the Chinese foods from Table 3.5. Analyze the Chinese food data and compare the results with the multivariate analyses of the snack food data in Section 13.2.

Table 13.23 Data for Exercise 13.46

Nation	Primary	Lower secondary	Upper secondary	Mean
Germany	790	761	673	741
Ireland	951	792	792	845
Italy	748	612	612	657
Norway	749	666	627	681
Spain	900	900	630	810
Sweden	624	576	528	576
United States	1093	1042	1019	1051
Mean	836	764	697	766

Source: OECD, from The New York Times May 28, 1995, p. E7.

Table 13.24 Data for Exercise 13.47

Dish (number of cups)	Calories	Fat (grams)	Percent calories from fat	Sodium (milligrams)
Egg roll (1 roll)	190	11	52	463
Moo shu pork (4)	1,228	64	47	2,593
Kung Pao chicken (5)	1,620	76	42	2,608
Sweet and sour pork (4)	1,613	71	39	818
Beef with broccoli (4)	1,175	46	35	3,146
General Tso's chicken (5)	1,597	59	33	3,148
Orange (crispy) beef (4)	1,766	66	33	3,135
Hot and sour soup (1)	112	4	32	1,088
House lo mein (5)	1,059	36	31	3,460
House fried rice (4)	1,484	50	30	2,682
Chicken chow mein (5)	1,005	32	28	2,446
Hunan tofu (4)	907	28	27	2,316
Shrimp in garlic sauce (3)	945	27	25	2,951
Stir-fried vegetables (4)	746	19	22	2,153
Szechuan shrimp (4)	927	19	18	2,457

Source: Center for Science in the Public Interest, as given in The Philadelphia Inquirer, *September 2, 1993, p. F7.*

CHAPTER 14

14.1 Stepping stones to statistical sophistication

14.2 Approaching numbers with care

14.3 Data and statistical methods

14.4 How things can go wrong

14.5 Statistics and Big Brother

14.6 Ending on the upbeat

STATISTICS IN EVERYDAY LIFE

Often, when we are blazing a trail through the woods in unfamiliar territory, we get so caught up in cutting through the surrounding brambles that we lose track of the bigger forest in which we are (possibly) lost. Now that we have arrived at a clearing and have survived the challenge of finding our way, we can take stock of what we have accomplished and look forward to future prospects.

In this chapter we have two major purposes. First we retrace our steps so that you can see how far you have come on the path to statistical literacy. Each chapter is a stepping stone to the next ones in terms of the skills and understandings acquired; you cannot read a statistics book backward, as you might a book of poetry. In Section 14.1, we review the statistical knowledge you have accumulated. These touchstones are here to mark your progress from what may have been a form of social illiteracy, in this case statistical, to a higher level of sophistication. Unlike many introductory textbook writers with grandiose dreams, we hope you've gained an appreciation for the work statisticians do, but we do not expect you to become an ersatz expert in the field. The second portion of the chapter is designed to remind you of some of the problems inherent in developing research designs and how statistics can be used and abused in public forums.

> ### TOOTH FAIRIES AND THE PRICE OF A MOLAR
>
> Most of our statistical information is piecemeal, unreliable, and open to grandiose claims on the part of journalists and commentators. A survey reported in *Family Circle* magazine found that the tooth fairy on average pays about $1 to $2 per tooth but sometimes as much as $20 per incisor and bicuspid. The survey, conducted by the American Academy of Pediatric Dentistry, found the greatest generosity in Houston, where at least one child had gotten $50 for a tooth. *(Source: The Philadelphia Inquirer, May 24, 1995, Family section, p. 1.)*
>
> Do these statistics help an anxious parent wanting to do the right thing for a child with a loose tooth? From this informational swatch it is difficult to know what to conclude. How good was the sample of tooth fairies? What type of measurement (mean, median, mode, rough guess) yielded the $1 to $2 average? Should the data have been classified by type of tooth? How much of an outlier compared to the tooth fairies in other cities is Houston in general and the $50 tooth fairy in particular?
>
> For every statistical situation, we need to assess how good our "patchwork quilt" report of reality is. To find out, we must know how statistical methods are used and misused. If we are knowledgeable consumers of statistics, then we are equipped to understand and evaluate the many applications of statistics and the conclusions drawn from statistical studies.

14.1 STEPPING STONES TO STATISTICAL SOPHISTICATION

Much of the work statisticians do focuses on whether one variable affects another. We frame the discussion of this focus in terms of four questions:

- Question 1. Is there a relationship between the variables in the data?

- Question 2. How strong is the relationship between the variables?

- Question 3. Is there a relationship in the population?

- Question 4. Is the observed relationship a causal relationship?

Throughout the book, we develop skills necessary to answer these questions in a variety of situations. To lay the groundwork for answering the questions, the first half of the book points out many critical facets of the statistical numbers game. The second half of the book combines these concepts in various ways to illustrate several critical forms of data analysis.

In Chapter 1, statistics is defined as the search for regularity in the face of randomness. Later chapters are anticipated by descriptions of the three parts of statistical work: data collection, data analysis, and making inferences from data. The important concept of a variable and its values measured on some set of elements is introduced, and the chapter ends with a discussion of the kinds of people who use statistics.

The focus of Chapter 2 is data collection. The critical nature of getting a "good" sample is emphasized, and the notion of sampling error was introduced. Learning more than you wanted to about scurvy, you are presented with distinctions between observational and experimental research. The data matrix and data file are also described.

Displaying data in visual form is the central theme of Chapter 3. Throughout the text, we suggest that data be visualized before being analyzed. Tufte's requirements for graphical excellence—less is more—were introduced.

In Chapter 4 data analysis is introduced via measures of central tendency—mean, median, and mode—and measures of variability, primarily the standard deviation and variance. Familiarity with the standard error of the mean and standard scores paves the way for various forms of more complex data analysis. The tension between losing information and gaining simplicity in data analyses is explicitly confronted.

Chapter 5 covers probabilities. The crucial standard, normal, or bell-shaped curve is introduced, along with its unique properties concerning the proportion of the total area found under the curve within each standard deviation segment. These and similar curves form the basis of evaluating the significance of the collected statistics. Brief mention is made of the four major theoretical variables used later in the book: z, t, chi-square, and F. What a p-value means and how people make decisions about data based on how large the probability of an event occurring by chance might be set the stage for hypothesis testing.

Chapter 6, on drawing conclusions, distinguishes sample statistics from population parameters and the way in which we estimate parameters from sample statistics. Both the point and interval methods of estimating parameters are discussed. The notion of the confidence interval sets up ways of judging how good an estimate of a parameter is.

Chapter 7 explores in greater detail how hypothesis testing methods are used to draw conclusions about population parameters from sample data. Topics discussed include the reasoning behind testing null hypotheses; the types of errors that can be made in deciding whether or not to reject the null hypothesis; how to find and use the *p*-value; how to find the proper degrees of freedom. These skills are applied to problems using the *t*-test or *z*-scores. Doing hypotheses testing is compared with developing confidence intervals. In hypothesis testing we ask if the parameter could possibly be equal to a particular value; with confidence intervals we estimate the actual value of the parameter by getting a range of values that we hope contain the true value of the parameter.

Chapter 8 emphasizes how we proceed to answer the four critical questions about statistical relationships. For question 1, we look at the patterns in the sample data. If we find a relationship, then we ask question 2. To answer question 2, we calculate the strength of the relationship between the variables. To answer question 3, we set up a null hypothesis that there is no relationship between the variables and test the hypothesis to see if we can reject it or not. It is usually difficult to answer question 4 about causal relationships. A relationship (even a strong one) can exist between two variables without any causal connection between the two variables. However, even if two variables are not causally related, it is possible to predict values of one variable if we know the values of the other. The strength of the relationship tells us the degree to which we are able to predict from one variable to the other, even though it tells us little or nothing about causality.

In Chapter 9 the analysis of categorical variables using contingency tables and chi-square analysis is discussed. Ways of answering the four questions are demonstrated with various examples. Whether there is a relationship between variables in the "real world" requires hypothesis testing. To find the *p*-value for a sample, we transform the phi or *V* coefficient (a statistic used to assess the strength of the relationship between the variables) to a value of the chi-square variable. To find the *p*-value associated with a chi-square, we need to know the degrees of freedom associated with the contingency table.

Correlation and regression analysis for metric variables are explored in Chapter 10. A scatterplot indicates whether there is a positive or negative relationship between the variables, and a correlation coefficient measures the strength of the relationship. Correlation coefficients range between −1 and +1. Correlation coefficients tell us how well we can predict, but they are not used for assessing causation between two variables. A regression analysis involves drawing a line through the middle of the data points on the scatterplot. The slope of the line tells us how much one variable changes with the other variable. Regression equations produced by the slope and the intercept of the line can be used to predict the value of the dependent variable from a value of the independent variable. Regression analysis can be used to study the relationship between a categorical and a metric variable by constructing a dummy variable (e.g., 0, 1) for the categorical variable.

In Chapter 11, analysis of variance, or anova, is introduced as a method of studying the effect of a categorical independent variable on a metric dependent variable. If the effect of the independent variable on the dependent variable is large relative to residual effects, then we can reject the null hypothesis of no relationship. A value of an F-variable must be calculated, with its degrees of freedom, to find the p-value. Once we have found a statistically significant relationship, then we examine which means of the dependent variables are significantly different from one another, if there are more than two. The simple sign test is useful for the study of differences in paired data.

Special methods for analyzing rank variables are highlighted in Chapter 12. Gamma, a statistic measuring the strength of relationship between two rank variables with words as rank labels, is introduced. As with other analyses, the relationship between the variables is tested using a null hypothesis of no relationship. The p-value is found by converting gamma to z and using the tables for the normal bell curve. When the variables have numerical ranks, the strength of the relationship is measured by the Spearman rank order correlation coefficient. This coefficient is similar to the Pearson correlation coefficient for metric variables. The statistical significance of the Spearman r_s is evaluated by converting it to a t-score and then finding the p-value. (As usual, correlations does not imply causation.)

Chapter 13 gives a brief introduction to multivariate analysis, which is used to analyze the effects of several independent variables on a dependent variable in a single procedure. The independent variables can often be ranked in terms of impact on the dependent variable. In multivariate analysis, if a relationship between two variables disappears as a result of holding a third variable constant, then we presume that the original relationship was not a causal relationship. An average coefficient from a set of subgroups defined by the control variable(s) is known as a partial coefficient. The partial coefficient expresses the

> **CREATIVE USES OF STATISTICS:
> MUSIC AND MYSTERIES**
>
> People in many diverse professions use statistics to entertain and inform their audiences. The late composer John Cage used computers to generate random patterns that he incorporated into his compositions. Author Michael Crichton has used statistical applications to enhance the pleasures of movie-goers who enjoy being terrified by such films as *The Andromeda Strain, The Terminal Man, Jurassic Park,* and *Coma.*

overall relationship between two variables once the effect of a third or more variables has been taken out. Regression equations can be created by combining the partial regression coefficients from each variable in the analysis. If the independent variables are well-chosen, multiple regression is very powerful in predicting real-world outcomes of the dependent variable.

Chapter 13 also reviews two-way analysis of variance, where two categorical variables are simultaneously evaluated for their independent and interactive effects on a metric dependent variable. As with multivariate regression analysis, the two-way anova improves the accuracy of the results over the one-way analysis by simultaneously comparing the variables with each other and thereby decreasing the effects of the residual variable.

Issues of causation are discussed throughout the last six chapters. The safest statement to make about causation is a negative one: it is easier to show that something is not or may not be causal than that something is. Any presumed causal variable is always vulnerable to being challenged by a new control variable.

14.2 APPROACHING NUMBERS WITH CARE

Familiarity with statistical methods helps us evaluate and understand the results of a statistical analysis. Being statistically literate also helps us to know when to be skeptical about statistical claims. When we read that the unemployment percentage of the month is 6.7%, that women make $7,000 dollars in income less than men, or that Saggitarians have more fun, we should be cautious about accepting these numbers as facts. We have learned enough to realize that many limitations, oversights, and errors can accumulate along the way to a final statistically based claim and that the results of a statistical study most likely are not equal to the exact, true values that would have been obtained in a statistically perfect world.

We can characterize why an observed value of a sample statistic is not equal to the true value of the population parameter in a formal way:

observed value = true value + nonstatistical mistakes + randomness

The observed value on the left side of the equation may be a percentage, a difference between two means, or any other value computed from the data. On the right side of the equation are three items that control the observed value outcome. The true value is the parameter of our statistical dreams, an imaginary score that is unaffected by randomness and mistakes. The factors of nonstatistical mistakes and randomness are very different in character. Because statistics originated as a mathematical science that deals primarily with formulas and equations, nonstatistical mistakes have traditionally been viewed as outside the realm of problems with which a statistician must deal. It had been assumed that these are the concerns of people who work with methodological issues within a particular discipline. Psychologists, for example, have always been more interested than statisticians in how the race, gender, or age of an experimenter might affect the responses of a subject. Today, concerns about data collection are no longer outside the realm of the statistician's world. In today's quest for number-crunching, statisticians cannot avoid some of these "nonstatistical" issues when they get involved with practical applications of their formulas and equations. Statisticians cannot bury their heads in the sand and let others wrestle with the prickly cacti in the methodology field.

Randomness, on the other hand, is an expected impediment to the true value that statisticians endure with the knowledge that there is no way to avoid it. Randomness is a part of the statistical world, with a degree of mathematical respectability that other mistakes do not have. And statisticians, as you realize, have built elaborate defenses and cautions against the inevitability of randomness.

14.3 DATA AND STATISTICAL METHODS

By now it is clear that the results of any statistical analysis are based on (1) the data collected and (2) the statistical method used (Figure 14.1). This may be the most important message of this book.

Suppose we see a headline in the newspaper which says that women make $7,000 less in income than men. This is the result of a statistical analysis. Where did the result come from? The $7,000 difference is not an objective fact of life that just exists out there; it is not a description of the world the way it really is just because we saw the fact in print. The result is based on the particular data that were collected and the particular methods that were used to analyze the data. With other data and other methods, a different result might have occurred.

To understand a statistical result, we first need to know how the data were collected. Did the data come from a random sample of all adults, or did they come, say, from all tax returns filed with the Internal Revenue Service a certain year? Each separate mode of collection would influence the results. Besides knowing where the data came from, we need to know what was actually measured for each individual in the study. Do the income figures consist of earned income from jobs, or do they include interest and dividends from bonds and stocks as well? Some very wealthy people do not *earn* any of their income; would they be excluded from the analysis?

When we know who was studied and how the income figures were determined, we need to know how the data were analyzed and therefore exactly what the result means. Does every man make $7,000 more than every woman? This cannot be the case because we know that incomes vary a great deal for both men and women. Maybe the $7,000

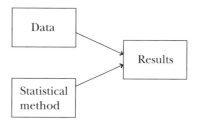

Figure 14.1 Factors influencing the results of a statistical study

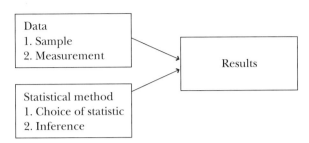

Figure 14.2 Detailed factors influencing the results of a statistical study

difference is a difference between an average income for men and an average income for women. If that is so, did the average refer to the mean or the median or perhaps some other average? Whatever average was used, the researchers would have obtained another value for the difference if they had used another type of average; the difference between the mean income for men and the mean income for women is not the same as the difference between the corresponding medians.

We also need to know if the difference is statistically significant. The amount of $7,000 sounds large and meaningful, but we cannot really tell how important it is until we know the results of a test of statistical inference of one kind or another. Perhaps it is a chance difference caused by randomness.

With this background we can flesh out the boxes in Figure 14.1 as shown in Figure 14.2. This figure is a framework for examining any statistical results we are exposed to; we can use the figure to frame questions about a result before we accept it. The figure encourages us to question the source of the data and how the variables were measured. It also makes us question the statistical method that was used. What statistics were computed from the data, were there any controls for other variables, and would the result be statistically significant using different statistical tests?

The figure can also be used to examine what might have gone wrong in a study that would have led to a questionable result. Because statistical results depend on both the data and the method, the results will be wrong if something is wrong with either the data or the method or both. If serious mistakes are made in the data collection or the method of analysis, then the results should not be accepted.

14.4 HOW THINGS CAN GO WRONG

Statistics is a field in which many abuses are possible. Things can go wrong in every step, from the formulation of the original ideas for the problem to the printing of the final report. Most mistakes and misuses of statistics are not intentional, but results can be purposely skewed, using questionable methods.

Dangers in the collection of data

Collecting data is a two-step process. The first step consists of selecting the elements on which to measure our variables. This may mean a sample of elements from a larger population, or it may mean the entire population itself. The task consists of identifying the people who will be asked questions in a survey or the elements that will be used in an experiment. Of course, the elements do not have to be people. We may be studying any type of unit, for example, animals, plots of land, light bulbs, baseball games, or countries.

The second step is to actually collect the data. Sometimes we collect data on a specific group, and we want our results to apply only to that group. Having data on everyone in that group eliminates the selection process of the study and the problems that selection entails. If, for example, the owners of the National Football League want to find out how much each of the players is paid, they can make a list of all the players and their salaries in a given time period. The problem of inference from sample to population is irrelevant; the data *are* the population, and instead of statistics the owners have parameters.

Things change if we want the results to apply to elements beyond the elements on which we have collected data. If we want to generalize to a larger population using confidence intervals and/or hypothesis testing, then it is imperative that we select elements using proper statistical procedures. For experiments, this means random allocation of subjects to treatment and control groups. For surveys, this means drawing a sample randomly.

Even with a proper random sample, the results apply to the sample and only through the use of statistical inference to the population from which the sample was drawn. This fact can impose serious limits on studies. Suppose a pharmaceutical company sends sample products only to doctors who are stockholders in the company and later surveys a sample of these doctors to see how many prefer the company products to competitors' products. The company should not claim that "76% of doctors in a survey prefer our products" without including information about the very specific population from which they drew their sample. The results do not apply to the population of all doctors, and the reporting should not imply that they do.

We believe that researchers are aware of the need for randomness, but achieving it is another matter. In many studies, the condition of randomness is almost never truly satisfied. A majority of psychological studies, for example, rely on college students for their research results. (Critics have suggested that modern psychology should be called the psychology of the college sophomore.) Are college students a random sample of the adult population or even the adolescent population? Not likely. Yet they are convenient, literate, cooperative, and interested subjects, and it is a great temptation to overlook the possible biases that can occur by using them. Due to pressures on scholars to publish as speedily and efficiently as possible, researchers may not be willing to pay the costs for a truly random sample of experimental subjects.

Psychology is not the only field where subjects are recruited on the basis of their convenience or cooperation. Much of the research in medicine depends on the availability of patients in various hospitals connected with research facilities. Healthy people, people treated with unconventional health practices, or sick people who do not get medical attention are not included in the research plan. Thus, significant control groups are often not included in the research design. As a result, doctors do not know how often people with various medical conditions may be self-curing without medical treatment or can live with various conditions without medical intervention. The question of whether medical intervention may impede as well as extend life satisfaction and longevity is just beginning to be asked. In the case of prostate cancer, for example, the answer seems to be that often it is better to leave it alone than to do surgery. Yet until medical research plans are created that pay more attention to the drawing of random samples, these important questions cannot be answered.

Despite the desirability of random sampling, from time to time we authors have applied hypothesis testing to data that we did not think were obtained from a random sample from some specific, underlying population. In those cases the null hypothesis states that the pattern in the data were created by chance alone, not that some parameter has a given value. When such a null hypothesis is rejected, the conclusion is that some factor(s) other than chance are at work. This can be a helpful finding, even when a sample is imperfect.

Special problems of survey research

Surveys are also subject to serious problems of data collection. It is often very difficult and expensive to draw proper random samples. One difficulty is to properly define the population. Suppose you want to interview teenage mothers in Philadelphia. This sounds like a well-defined group, but how do you create a list of all the young women belonging to this group as a population from which the sample could be drawn? Drawing a sample can also pose problems even if one population is known. If you watch the market research surveyors the next time they appear in your local mall, you will be able to see that they cannot persuade everyone who passes to participate. They may also avoid approaching shoppers who look harried, disheveled, or hostile. Statisticians can often help overcome this problem by compensating for underrepresented and overrepresented groups in a survey. Recent suggestions for improving the national census involved adjustments of the enumerations to compensate for the undercounting that occurs in certain areas and for certain types of people. The details of this precise operation are beyond the scope of the chapter, but it is comforting to know that surveys with problematic samples can be improved with careful manipulations of the data.

While many studies rely on data from a sample that is not collected in the proper random fashion, researchers usually do not intentionally set out to mislead the public. There are often just too many obstacles standing in the way of obtaining a random sample of respondents. Telephone surveys, relied on as the primary source of data collection by most survey groups today, must contend with the fact that millions of people do not have telephones. No amount of random dialing will find this last group, which composes about 6% of the total population.

Context of the survey The data collection situation itself can have problems. In interview studies, for example, differences in responses have been found depending on whether the interviewer seems to be similar or different from the respondent in such aspects as gender, ethnicity, and personal preferences. We know that who asks the question, and in what tone of voice also affects the answer people give. A smoker who knows that the surveyor also smokes will be freer to admit how much she enjoys smoking than if her questioner does not smoke, especially if he is a militant ex-smoker. The place of the interview is also important. Respondents may be more willing to talk at length when they are in a comfortable and private setting and less so in shopping malls or on the telephone or on the street. Contextual effects cannot be overcome totally and must be accepted as a facet of the data collection process.

What is the question and when was it asked? Survey and experimental questions always have some effect on the respondent. Question formulation and placement influence the results. A smoker who is asked, "How much would you say you enjoy smoking?" might answer more positively than one who is asked, "Would you say you like or dislike smoking?"

Questionnaires that assume the sample is informed about the subject get more definite answers than those that assume a lack of knowledge. Yet the data gathered might lack validity. For example, in answering the question "How do you think the U.S. Congress should react to the latest UN peace-keeping missions in the Middle East?" respondents may feel forced to "fake" knowledge in a manner that would not occur if they were first told about the UN missions or were allowed to save face by saying they knew nothing of the situation. In terms of question placement, questions that ask about religious affiliation early in the interview format may create more religiously influenced responses than the same questions asked later. A study of college students asking about social life and religious beliefs indicated that when the subjects were first asked to specify their religious affiliation, they were much more conservative about their views than when they were asked later. *(Source: W. W. Charters and T. M. Newcomb, "Some attitudinal effects of experimentally increased salience of a membership group," in E. Maccoby, T. Newcomb, and E. Hartley (Eds.),* Readings in Social Psychology, *3rd ed., New York: Holt, Rinehart & Winston, 1958.)* When a question intentionally leads respondents in a certain direction, then it is incorrect to analyze the answers statistically as if the question were expressed in a neutral fashion. (Of course, we might ask what a truly neutral question is.) These types of issues are more psychological and sociological than statistical, but they are relevant when we try to understand the results of the statistical analysis of the data.

What are the variables selected for analysis? To conduct a survey, the researchers must decide what questions to ask. The decisions depend to a great extent on the theoretical orientation of the study as well as other factors, such as the historical background of the study—the questions that have been asked before and in general the work of other researchers—what do other research groups do, the equipment that is available to test the hypotheses, and the traditions of the particular field. In creating statistical analyses for company annual reports, for example, comparison of quarterly sales results, operating profits, and market share, as well as comparisons with other companies, are typically made. However, these analyses do not typically assess such variables as what percentage of the products are new, how quickly a product has gone from research to development, or how long it takes to complete a business transaction. But today these variables are the lifeblood of a successful organization and perhaps should be measured and included in the report. Innovative companies such as Minnesota Mining and Manufacturing have made such variables a part of their stockholders' reports, part of their company's mission, and an important gauge of their company's success.

How are results coded and stored? Problems do not stop with the asking. Once a question is answered, the words are coded in a category system designed by the researcher. The researcher might rank a response such as "I like to smoke about as much as the next person" as a 4 on a scale from 1 to 7 yet not know precisely what the respondent meant by the remark. Even if the respondent herself scores her smoking pleasure as a 4, the researcher still doesn't really know how the respondent feels about her smoking enjoyment or how to compare her 4 with another person's 4. To make matters worse, coded answers can be recorded erroneously by the data entry person creating the computer data file. What if this type of error happens 10% of the time? Fortunately, survey researchers recheck data entries on a fairly regular basis to try to catch these errors.

In addition, the answers people give tell us only what they answered, not necessarily what they do. A study that asked about toothbrushing habits found that on the basis of what people said they did, the toothpaste consumption in this country should have been three times larger

"PC and Pixel." 1996, Washington Post Writers Group. Reprinted with permission.

than the amount of toothpaste that is actually sold. The bottom line is that people exaggerate in order to make the most of the research situation, whatever that may mean to them.

Misuses of analysis methods

In Chapter 3 several possible problems in the creation of graphic displays were introduced. We saw, for example, that tables can be misleading when the rows and columns are not chosen properly or when the numbers in the tables have too many decimal places. Statistical graphs can have too much useless information—chart junk—that obscures the major messages of the data. Bar graphs with moving baselines can be misleading and hard to read. Bars in odd shapes, such as human figures or oil drums, may have the proper relative heights but incorrect areas and be misleading indeed.

Computations must be done with correct quantities. As you recall, skewed distributions—such as income distributions—are best represented by medians, and it is a misuse of statistical methods to use means. Yet means are very often used because the public's conception of an average does not discriminate among mean, median, and mode, and statisticians can produce more sophisticated statistical results when the mean is used.

In general, the results of a statistical analysis depend on the data that goes into the computations (which we have already discussed at length) and the statistical methods used to analyze them. For example, in regression analysis the strength of the relationship between two variables is found using least squares methods. If absolute values instead of squares are used, a different number results. Thus, the statistical method as well as data itself contributes to conclusions.

Here a word about computer programs is in order. Because of the extremely helpful simplifying capacities of statistical computer packages, the computer will analyze any data it is given, right or wrong. There is no checks-and-balances mechanism to intervene between the automatic processing of the program and the user and suggest caution or warn of possible misuse of the data, so results can be computed that bear little resemblance to the actual information that exists in the data. One such "disaster" occurred when the research assistant of author Gergen incorrectly entered into the first nine columns of a data file the social security numbers of the subjects—totally false information for the first nine variables that gave wrong values to each of the subsequent variables because they were out of place. The computer had no trouble crunching the numbers, although a careful reader was disturbed to discover a reported mean age of 50, with a standard deviation of 15, for a college-age sample. To adapt a familiar computer saying, Wrong data in, wrong results out!

Most standard statistical computer packages use formulas that assume the data have been collected as simple random samples. Yet data for large, national studies are often collected using more complicated sampling methods and should not be analyzed using standard software packages.

Misuses of statistical inference

Both hypothesis testing and confidence intervals take into account that conclusions are sometimes wrong. As you recall, statistical conventions accept wrong conclusions 5 times or less out of 100 ($p < 0.05$) when we deal with correct null hypotheses. Strictly speaking, these errors are not misuses of statistics, but it is important to realize that we are wrong from time to time. The difference between statistics and some other forms of mathematics is that in statistics we expect to be wrong sometimes, and the method itself makes it possible to state how often we are wrong if we repeat the study many times over. Unlike other disciplines, statistics does not try to perfect itself: Statistics means never having to say you're certain.

For example, a p-value of 0.05 admits that conclusions can be wrong 5 times in 100. In an analysis of 100 hypotheses in which all the null hypotheses are correct, if 5 results are found to be significant (and the null hypotheses are rejected) then these results are due to chance alone. Of course, we cannot be definite about this either, because we do not know whether a particular null hypothesis is correct or not. There is no position from which we can snatch off the veil of uncertainty surrounding the hidden body called Truth. Even though we know how often we are wrong, we do not know *when* we are wrong.

Misinterpretation of numbers

The Mercedes-Benz company advertises that 97% of all their cars registered over the last 15 years are still in operation, and that this is more than any other comparable car make sold during that time period. How do we interpret 97%?

A number tells a story just like a sentence of words. As with any story, we interpret the number in a context that makes sense to us. What is the story the car maker wants us to hear from this percentage? They want us to think that this is a high percentage and that most of the cars they have sold are still being driven, reflecting well on the quality and desirability of owning their cars. In the end, they want us to buy their cars. The advertisement does not say all this, but this is the story they want us to develop.

From a statistical point of view things may not be quite as simple. First of all, how does the company get information of this kind? Each state has a motor vehicle registration office, and the company would have to contact all 50 offices for information on how many of their cars made in this 15-year period are registered in each state. Since the company knows how many cars it sold in the same time period, it can then find the percentage of cars sold that are still registered. Since the records are computerized, they would presumably be as accurate as the data entered into the computer. Yet the collection and retrieval process within each state is unique. It may not be so simple to get the registration information from each state or to combine the available data to create the summary statistic.

Second, how can 97% of the cars sold in a 15-year period still be in operation? It makes a large difference when these cars were sold. If only a few cars were sold in the first 10 years and many more cars were sold within the last 5 years, then it would not be very surprising that most of the cars were still on the road. If there had been a steady growth in the number of cars sold each year, then many of the cars would still be relatively new and most of them would still be driveable. If an average car is driven 12,000 miles a year and can be expected to last until it reaches 100,000 miles, then the average life of a car is about 8 years.

Thus, without knowing more about the sales pattern over the 15-year period, it is not at all clear how we should interpret this advertising claim. And claims like this one appear over and over again: we are told a few numbers with the intention of leading us to certain conclusions. But when we think critically about how good the numbers are and what alternative explanations might be, then what the numbers are telling us is no longer clear.

14.5 STATISTICS AND BIG BROTHER

Going beyond skepticism to a perhaps graver social concern, statistical knowledge has the power to regulate lives. A significant drawback of having a strong and unified system of statistical collection is that people can be easily put under surveillance by the government and private business interests. Big Brother can become a reality with the help of integrated and elaborate statistical networks. Historically, statistical analysis has been used by the elite to monitor citizens for the benefit of the state, especially for purposes of taxation and conscription. For Christians, the New Testament version of the birth of Jesus begins with a story of statistics: Joseph is required to go to Bethlehem to be enrolled for the census in the house of David.

Traditionally, citizens in liberal democracies have been reluctant to allow governments centralized power to ascertain the status of individuals. Organizations such as the American Civil Liberties Union are dedicated to the preservation of individual rights against intrusive interests of the majority. In recent years this sensitivity seems to have weakened as people become accustomed to constant monitoring—from bank teller machines that take one's picture without permission, telephone companies that keep extremely detailed information on personal calls, stores that track customers' movements with hidden cameras and one-way mirrors, and school records that give test performance profiles from kindergarten to graduate school to talk show hosts who elicit intimate and traumatic details of the personal lives of their guests before millions of viewers.

> **STOP AND PONDER 14.1**
>
> Statistical information can have strong implications for personal lives. How might the form in which the following information is presented influence your thoughts about sexuality, birth control measures, and abortion?
>
> In the United States, approximately 55 million women are between the ages of thirteen and forty-five. Eleven million of them prevent pregnancy by sexual abstinence, and most of these are teenagers and young adults. Each year, of the 44 million fertile women who are sexually active, fewer than six million want to become pregnant, according to the study by the Alan Guttmacher Institute concerning pregnancy intentions of U.S. women.
>
> The remaining 38 million women of childbearing age who are sexually active do not want to become pregnant. Each year, over 90 percent of them will succeed in preventing pregnancy during each menstrual cycle by using various methods of birth control. Overall, that is a phenomenal achievement. By abstinence and contraception, the fertile women of America successfully avoid fertilization of an egg and resultant unwanted pregnancy nearly 600 million times every year.
>
> Only 8 percent of all fertile women who do not want children fail in their family planning goals. But even that small percentage means that over four million women experience unwanted pregnancy each year. Statistically speaking, the chances are that almost every woman in the United States either has or will experience an unwanted pregnancy. *(Source: Biology and Gender Study Group, "The Importance of Feminist Critique for Contemporary Biology," unpublished manuscript, 1993.)*

While statistics themselves are modest servants of intelligent users, they can become instruments of oppression when used by powerful leaders of important groups. Among these are government agencies such as the police and tax agencies, corporations, and insurers. Medical records, aptitude tests, and other inventories of personal skills, personality traits, character assessments, and interests are forms of statistical measurement that can be stored and used in a variety of ways. Recognition codes for security systems, such as fingerprints, retinal images, and voice patterns, are based on statistical evaluations and can be stored and retrieved to track one's mobility. The type and frequency of long-distance calls one makes create a profile that alerts a statistically based system that will automatically cancel a credit card if the profile is violated. Criminal checks, including the analysis of blood, hair, semen, and skin, depend on statistical inputs and become part of permanent, accessible records. A governmental suggestion to do background checks on all airplane passengers in order to create terrorist profiles from statistical compilations of data is being raised as a way to combat sabotage. There is no end to the possibilities for intervention and surveillance of private lives, once statistically sophisticated measures are initiated.

Part of the task of being an educated consumer of statistical information is to ask what the boundaries should be on the collection, storage, and promulgation of statistical information. Who should be able to find out what, and under what circumstances? And when and how should people be safeguarded against the excesses of a computerized age, in which statistics can be used to regulate and control our lives.

14.6 ENDING ON THE UPBEAT

Bemoaning the dangers to society if statistical information is used to threaten personal freedom does not change anything. Knowing as much about the power and limits of statistics as we citizens can helps. Thus, we understand that we live in a seemingly random and chaotic world, but there are regularities in all that randomness, and statistical methods can be used to explore the regularities. We also know that we should be critical of the quality of the data used in statistical analyses, and that the results of any statistical analysis are influenced by the quality of the data as well as the particular methods used to analyze the data.

We appreciate how statistically informed inferences can be much more comprehensive, logically consistent, and rigorously undogmatic in allowing conclusions than are the results of personal experience, focus group commentary, informal surveys, or personal logic (including drawing conclusions on the basis of analogies, common sense, or principles of authority-based knowledge). While each of these forms of rhetoric has an important place in the advancement of persuasive arguments, each is flawed in ways that statistical methods and inferences are not.

Statistical methods have made it possible to tap public opinion and help set public policy for the direction the country takes. Statistics has played an enormous role in the development of the many goods and services we purchase. New model cars, for example, do not break down nearly as often as earlier ones did because they are better made, thanks to the sophisticated statistical quality control that now exists. Statistics has had a great influence on the practice of medicine and the availability of pharmaceutical drugs to fight diseases. From glancing through the interpretation and analysis exercises at the end of each chapter we can see that statistics is used in a very wide field of applications. In short, any situation in which empirical data are collected has a need for statistical methods. Statistics is so integral to the well-being of society that it is impossible to imagine how we would function in a world without it.

We leave you with a hope and conviction that you now are better able to make sound decisions about the value of the ever-increasing amount of statistical information you encounter. We think this is an important part of the education of a person who is entering a new millennium.

ADDITIONAL READINGS

Crossen, Cynthia. *Tainted Truths: The Manipulation of Facts in America.* Simon & Schuster, 1994. Good book on the interpretation of numerical results.

Eberstadt, Nicholas. *The Tyranny of Numbers: Mismeasurement and Misrule.* Washington, DC: AEI Press, 1995. On the role of government statistics.

Hooke, Robert. *How to Tell the Liars from the Statisticians.* New York: Marcel Dekker, 1983. Entertaining book on misuses of statistics.

Jaffe, A. J., and Herbert F. Spirer. *Misused Statistics: Straight Talk for Twisted Numbers.* New York: Marcel Dekker, 1987. On the many things that can go wrong with a statistical investigation.

Paulos, John A. *A Mathematician Reads the Newspaper.* New York: Basic Books, 1995. Entertaining and instructive book on the interpretation of numbers in the newspapers.

EXERCISES

14.1 Find a statistical study reported in a newspaper, news magazine, journal, or book and comment on it in light of some of the points in this chapter.

14.2 What are some of the major problems with the collection of data that cannot be easily overcome, even with careful and conscientious planning?

14.3 a. Statistics has been described as an essential element in a well-functioning state. Give a historical example.

b. Are there ways in which statistics may contribute to a *reduction* in the level of social personal "goods"? Give a historical example.

14.4 a. Why is it so crucial that people be literate in understanding statistical reports such as those found in newspapers and magazines?

b. Give an example from your daily life when a poor understanding of statistical findings led to negative outcomes for people.

14.5 Describe an instance where you believe advertisers or other media producers intentionally created untrustworthy statistical results.

14.6 Examine the statistical reports and graphics that have been used to demonstrate a national, regional, or local trend of significance to the citizens of the area. Experiment with changing the character of these findings so as to significantly alter the meaning of the data for readers. If possible, do it in more than one way. Write a report showing how certain perspectives aiming for certain outcomes were privileged by the published data and how alterations in the presentations or in the statistical methods, data collection, and so on could have affected these outcomes.

14.7 Create a research proposal involving an issue of some importance to yourself. (If it is possible within the limits of your course, actually collect some data, or generate a data set based on imaginary subjects. This exercise will also enhance the "hands-on" feeling that comes from working with "real" numbers.) Develop a rationale for why this is an important issue to study, hypotheses, and ways to test your hypotheses. Describe how you would select your sample, develop your research instruments, create the proper setting for your research, and consider the ethical and social issues surrounding your endeavor. Describe how you would collect your data, how you would organize it, how you might present graphic representations of it, and how you would analyze the data statistically.

Stat's all, folks!

GLOSSARY

Alternative hypothesis: Possible values of a parameter other than the one specified in the null hypothesis.

Analysis of variance: Statistical method for the analysis of the relationship between one or more categorical independent variables and a metric dependent variable.

Analysis of variance table: Table showing sums of squares, degrees of freedom, mean squares, F-ratios, and p-values.

Average absolute deviation: Mean of the absolute values of the differences between the observation and the mean.

Bar graph: Graph in which bars show the number of observations for each value of the variable.

Binomial distribution: Theoretical distribution for a dichotomy giving the probability of x outcomes of one kind and $n - x$ outcomes of the other kind.

Boxplot: Plot showing the largest observation, the 75th and 25th percentile values defining a box, the 50th percentile value as a line through the box, and the smallest observation.

Categorical variable: Variable where two values are different from each other. The values cannot be ordered, and we cannot say that one value is more of something than another value.

Census: Process of collecting data on an entire population.

Central value: Single value of the variable used to represent all the observations.

Chartjunk: Extra elements of a graph that carry no information.

Chi-square distribution: Theoretical distribution used to make inference from sample data.

Confidence interval: Interval that hopefully contains the value of the population parameter.

Constant: Quantity that takes on only one value, usually a parameter.

Contingency table: Table of frequencies for two categorical variables.

Control group: Randomly selected subset of elements in an experiment that are not manipulated in the way the experimental group is.

Control variable: Additional variable brought in to see if two variables could be related causally.

Correlation coefficient r: Measure of the strength of the relationship between two metric variables.

Critical value: One or more predetermined values of a sample statistic used to reject the null hypothesis if the observed statistic is more extreme than the predetermined value.

Cramer's V: See V.

Data analysis: Simplifying the data through graphs, tables, and computations.

Data file (data matrix): Table of data in which a column contains observations on a variable and a row contains observations on an element.

Data density: Number of observations per square inch in a graph.

Degrees of freedom: Smallest number of observations needed to find all the observations.

Dependent variable: Variable that is influenced by one of more independent variables.

Dummy variable: Variable with only two values, usually 0 and 1; used for categorical variables.

Element: Unit on which we measure a variable.

Error of type I: Error that comes from rejecting a true null hypothesis.

Error of type II: Error that comes from not rejecting a false null hypothesis.

Estimation: Trying to find the value of a parameter.

Expected frequency: Frequency in each cell of a contingency table computed so that there is no relationship between the two categorical variables.

Experimental data: Data collected when we control the values of some of the variables.

Experimental design: Branch of statistical theory dealing with how to plan experiments and analyze the data from such experiments.

F-distribution: Theoretical distribution used to make inference from sample data.

Fiftieth percentile: Value of the variable that divides the data in two equal groups; all the observations in one group are less than this value and all the observations in the other group are larger than this value.

Frequency distribution: Set of pairs in which the first entry is a value of the variable and the other is the number of observations with that value, often shown as a histogram.

Gamma: Measure of association between two rank variables used when there are many observations of the same values of the variables.

Graphical excellence: Giving the viewer of a graph the greatest number of ideas in the shortest time with the least ink in the smallest space; communicating complex ideas in a graph with clarity, precision, and efficiency.

Group sum of squares: Sum of squares measuring the effect of an independent variable in analysis of variance.

Histogram: Graph showing the distribution of a metric variable in rectangles whose areas represent the frequencies of the values.

Hypothesis testing: Trying to find out if a parameter has a specific value.

Independent variable: Variable that precedes and is thought to influence the dependent variable.

Inference: Generalizations about populations made from sample data.

Intercept: Predicted value of a dependent variable when the independent variable(s) equals zero.

Interaction effect: Joint effect of two independent variables over and beyond their separate effects.

Interquartile range: 75th percentile value of a variable minus the 25th percentile value.

Interval estimate: See Confidence interval.

Interval variable: See Metric variable.

Lineplot: Line marked with values of a variable and each observation marked as a point above the line.

Logistic regression: Regression analysis with a categorical variable as a dependent variable.

Mean: Value of a variable when we add up all observations and divide the sum by the number of observations.
Measure of association: Number that measures on a scale from -1 to $+1$ the strength of the relationship between two variables.
Median: Value of a variable that divides all the observations into halves: one-half of the observations is smaller than this value, and the other half is larger.
Metric variable: Variable with a unit of measurement; we can say how much more or less one value is than another value.
Mode: Value of a variable that occurs most often.
Multiple correlation coefficient R: Correlation coefficient that measures the strength of the relationship between the observed and predicted values of a dependent variable.
Multivariate analysis: Study of a dependent variable affected by two or more independent variables.

Nominal variable: See Categorical variable.
Nonresponse error: Error that results when not everyone in the sample responded to parts or all of the survey.
Normal distribution: Particular unimodal, symmetrical theoretical distribution used extensively in the theory of statistics.
Null hypothesis: Statement about the value of a parameter.

Observational data: Data collected from observing the world as it is.
Odds: Ratio of numbers in which the numerator is the number of times an event fails to take place and the denominator is the number of times the event does take place.
One-sided (one-tailed) test: Test where the null hypothesis is rejected when the sample statistic differs from the population parameter in one specific direction.

p-value: Probability of observing the sample statistic or a more extreme sample statistic.
Parameter: Constant such as a mean, variance, or regression coefficient characterizing one or more variables in a population; usually designated by a Greek letter.
Partial coefficient: Coefficient measuring the relationship between two variables, controlling for one or more other variables.

Percentile: Value of a variable that divides the observations into two groups so that a certain percentage of observations are smaller than this value.

Phi: Measure of the strength of the relationship between two nominal variables, each with only two values.

Pie graph: Circle graph showing the distribution of a variable in "slices" sized according to the number of observations of each value.

Point estimate: Single numerical value as the estimate of a population parameter.

Poisson distribution: Theoretical distribution showing the probabilities for the number of occurrences of unlikely events.

Population: Collection of all elements under study.

Predicted value: Value of the dependent variable predicted by the independent variable; in regression analysis, a point on the regression line.

Probability: Long-run proportion of times that an event occurs.

Random sample: Sample selected so that every element has a known (sometimes equal) probability of being included in the sample.

Range: Difference between the largest and the smallest observation.

Rank order correlation: Coefficient used to measure the strength of the relationship between two rank variables using ranking numbers as their values.

Rank variable: Variable where the values are ordered, but we cannot measure how much more one value is than another.

Ratio variable: See Metric variable.

Regression analysis: Statistical method for the analysis of relationships among metric variables.

Regression coefficient: Slope of the regression line; shows how different on the average are two values of the dependent variable when they differ by one unit on the independent variable.

Regression line: Line summarizing the relationship between two metric variables.

Regression sum of squares: Sum of squares measuring the effect of the independent variable.

Residual: Vertical distance from an observed point to a predicted point; measures the effect on the dependent variable of all variables other than the independent variable(s).

Residual mean square: Variance of the residuals.

Residual sum of squares: Sum of squares measuring the effect of the residual variable.

Residual variable: Combined effect on the dependent variable of all variables other than the independent variable(s); the "everything else" variable.

Response error: Error in response resulting from something such as the formulation of the question, the placement of the question, on the effect of the interviewer on the respondent.

Sample: Set of elements selected from a population.

Sampling error: How far from the true population value 19 of 20 different sample results will fall if many different samples were selected.

Scatterplot: Graph showing observations on two metric variables as points, one for each pair of observations.

Sign test: Test to see if there has been any change in a variable measured twice.

Significance level: Predetermined probability of rejecting a null hypothesis that is really true.

Simple random sample: Sample collected in such a way that every element in the population has the same chance of being selected.

Spurious relationship: Observed relationship between two variables that are not causally related.

Standard deviation: Average distance of the observations from the mean; found as the square root of the variance, measured with the same unit as the variable itself.

Standard error: Standard deviation of a statistic computed from many different samples.

Standard score: Score obtained for an observation by subtracting the mean and dividing this difference by the standard deviation.

Standard normal variable: Variable that has the normal distribution and where the mean equals 0 and the standard deviation equals 1.

Statistic: Number such as a mean, variance, or correlation coefficient computed from the observations in a sample.

Statistical significance: When the sample result is such that the null hypothesis is rejected.

Statistics: Set of concepts, rules, and methods for collecting, analyzing, and drawing conclusions from data; the search for regularities in the face of randomness.

Stemplot: Plot showing the larger part of an observation on the left side of a line and the smallest integer on the right side of the line.

Table: Array for one or more variables showing frequencies, percentages, or probabilities of the various observations of the variable(s).

t-distribution: Unimodal, symmetrical theoretical distribution for the t-variable, related to the normal distribution.

Total sum of squares: Sum of squares measuring the effect of all variables.

Two-way analysis of variance: Study of the relationship between a metric dependent variable and two categorical independent variables.

Two-sided (two-tailed) test: Test in which the null hypothesis is rejected when the sample statistic is either much smaller or much larger than the value of the population parameter specified in the null hypothesis.

Unbiased estimate: Estimate for which the mean of the sample estimates from many samples equals the population parameter.

V: Statistic measuring the strength of the relationship between two categorical variables when one or both variables have three or more values.

Values: Categories we assign to a given variable.

Variable: Characteristic, trait, or attribute that can take on two or more values.

Variance: Average squared deviation of the observations from the mean, measured with the square of the unit of the variable itself; the square of the standard deviation.

Variation: Amount by which a set of observed values of a metric variable differ from each other.

STATISTICAL TABLES

Statistical Table 1a Values of z for tail probabilities from 0.50 to 0.01

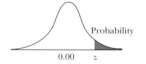

Probability	0.50	0.40	0.25	0.15	0.10	0.05	0.025	0.010	0.005	0.001	0.0001
z	0.00	0.25	0.67	1.04	1.28	1.64	1.96	2.33	2.58	3.09	3.72

Table 1b Tail probabilities for values of z

z	.00	.01	.02	.03	.04	.05	.06	.07	.08	.09
0.0	0.5000	0.4960	0.4920	0.4880	0.4840	0.4801	0.4761	0.4721	0.4681	0.4641
0.1	0.4602	0.4562	0.4522	0.4483	0.4443	0.4404	0.4364	0.4325	0.4286	0.4247
0.2	0.4207	0.4168	0.4129	0.4090	0.4052	0.4013	0.3974	0.3936	0.3897	0.3859
0.3	0.3821	0.3783	0.3745	0.3707	0.3669	0.3632	0.3594	0.3557	0.3520	0.3483
0.4	0.3446	0.3409	0.3372	0.3336	0.3300	0.3264	0.3228	0.3192	0.3156	0.3121
0.5	0.3085	0.3050	0.3015	0.2981	0.2946	0.2912	0.2877	0.2843	0.2810	0.2776
0.6	0.2743	0.2709	0.2676	0.2643	0.2611	0.2578	0.2546	0.2514	0.2483	0.2451
0.7	0.2420	0.2389	0.2358	0.2327	0.2296	0.2266	0.2236	0.2206	0.2177	0.2148
0.8	0.2119	0.2090	0.2061	0.2033	0.2005	0.1977	0.1949	0.1922	0.1894	0.1867
0.9	0.1841	0.1814	0.1788	0.1762	0.1736	0.1711	0.1685	0.1660	0.1635	0.1611
1.0	0.1587	0.1562	0.1539	0.1515	0.1492	0.1469	0.1446	0.1423	0.1401	0.1379
1.1	0.1357	0.1335	0.1314	0.1292	0.1271	0.1251	0.1230	0.1210	0.1190	0.1170
1.2	0.1151	0.1131	0.1112	0.1093	0.1075	0.1056	0.1038	0.1020	0.1003	0.0985
1.3	0.0968	0.0951	0.0934	0.0918	0.0901	0.0885	0.0869	0.0853	0.0838	0.0823
1.4	0.0808	0.0793	0.0778	0.0764	0.0749	0.0735	0.0721	0.0708	0.0694	0.0681
1.5	0.0668	0.0655	0.0643	0.0630	0.0618	0.0606	0.0594	0.0582	0.0571	0.0559
1.6	0.0548	0.0537	0.0526	0.0516	0.0505	0.0495	0.0485	0.0475	0.0465	0.0455
1.7	0.0446	0.0436	0.0427	0.0418	0.0409	0.0401	0.0392	0.0384	0.0375	0.0367
1.8	0.0359	0.0351	0.0344	0.0336	0.0329	0.0322	0.0314	0.0307	0.0301	0.0294
1.9	0.0287	0.0281	0.0274	0.0268	0.0262	0.0256	0.0250	0.0244	0.0239	0.0233
2.0	0.0228	0.0222	0.0217	0.0212	0.0207	0.0202	0.0197	0.0192	0.0188	0.0183
2.1	0.0179	0.0174	0.0170	0.0166	0.0162	0.0158	0.0154	0.0150	0.0146	0.0143
2.2	0.0139	0.0136	0.0132	0.0129	0.0125	0.0122	0.0119	0.0116	0.0113	0.0110
2.3	0.0107	0.0104	0.0102	0.0099	0.0096	0.0094	0.0091	0.0089	0.0087	0.0084
2.4	0.0082	0.0080	0.0078	0.0075	0.0073	0.0071	0.0069	0.0068	0.0066	0.0064
2.5	0.0062	0.0060	0.0059	0.0057	0.0055	0.0054	0.0052	0.0051	0.0049	0.0048
2.6	0.0047	0.0045	0.0044	0.0043	0.0041	0.0040	0.0039	0.0038	0.0037	0.0036
2.7	0.0035	0.0034	0.0033	0.0032	0.0031	0.0030	0.0029	0.0028	0.0027	0.0026
2.8	0.0026	0.0025	0.0024	0.0023	0.0023	0.0022	0.0021	0.0021	0.0020	0.0019
2.9	0.0019	0.0018	0.0018	0.0017	0.0016	0.0016	0.0015	0.0015	0.0014	0.0014
3.0	0.0013	0.0013	0.0013	0.0012	0.0012	0.0011	0.0011	0.0011	0.0010	0.0010
3.1	0.0010	0.0009	0.0009	0.0009	0.0008	0.0008	0.0008	0.0008	0.0007	0.0007
3.2	0.0007	0.0007	0.0006	0.0006	0.0006	0.0006	0.0006	0.0005	0.0005	0.0005
3.3	0.0005	0.0005	0.0005	0.0004	0.0004	0.0004	0.0004	0.0004	0.0004	0.0003
3.4	0.0003	0.0003	0.0003	0.0003	0.0003	0.0003	0.0003	0.0003	0.0003	0.0002
3.5	0.0002	0.0002	0.0002	0.0002	0.0002	0.0002	0.0002	0.0002	0.0002	0.0002
3.6	0.0002	0.0002	0.0001	0.0001	0.0001	0.0001	0.0001	0.0001	0.0001	0.0001
3.7	0.0001	0.0001	0.0001	0.0001	0.0001	0.0001	0.0001	0.0001	0.0001	0.0001
3.8	0.0001	0.0001	0.0001	0.0001	0.0001	0.0001	0.0001	0.0001	0.0001	0.0001

Statistical Table 2 Values of t for tail probabilities from 0.50 to 0.001

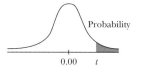

Degrees of freedom	Probability									
	0.50	0.40	0.25	0.15	0.10	0.05	0.025	0.010	0.005	0.001
1	0.00	0.32	1.00	1.96	3.08	6.31	12.71	31.82		
2	0.00	0.29	0.82	1.39	1.88	2.92	4.30	6.96	9.92	
3	0.00	0.28	0.76	1.25	1.64	2.35	3.18	4.54	5.84	10.21
4	0.00	0.27	0.74	1.19	1.53	2.13	2.78	3.75	4.60	7.17
5	0.00	0.27	0.73	1.16	1.48	2.02	2.57	3.36	4.03	5.89
6	0.00	0.26	0.72	1.13	1.44	1.94	2.45	3.14	3.71	5.21
7	0.00	0.26	0.71	1.12	1.41	1.89	2.36	3.00	3.50	4.78
8	0.00	0.26	0.71	1.11	1.40	1.86	2.31	2.90	3.36	4.50
9	0.00	0.26	0.70	1.10	1.38	1.83	2.26	2.82	3.25	4.30
10	0.00	0.26	0.70	1.09	1.37	1.81	2.23	2.76	3.17	4.14
11	0.00	0.26	0.70	1.09	1.36	1.80	2.20	2.72	3.11	4.02
12	0.00	0.26	0.70	1.08	1.36	1.78	2.18	2.68	3.05	3.93
13	0.00	0.26	0.69	1.08	1.35	1.77	2.16	2.65	3.01	3.85
14	0.00	0.26	0.69	1.08	1.35	1.76	2.14	2.62	2.98	3.79
15	0.00	0.26	0.69	1.07	1.34	1.75	2.13	2.60	2.95	3.73
16	0.00	0.26	0.69	1.07	1.34	1.75	2.12	2.58	2.92	3.69
17	0.00	0.26	0.69	1.07	1.33	1.74	2.11	2.57	2.90	3.65
18	0.00	0.26	0.69	1.07	1.33	1.73	2.10	2.55	2.88	3.61
19	0.00	0.26	0.69	1.07	1.33	1.73	2.09	2.54	2.86	3.58
20	0.00	0.26	0.69	1.06	1.33	1.72	2.09	2.53	2.85	3.55
21	0.00	0.26	0.69	1.06	1.32	1.72	2.08	2.52	2.83	3.53
22	0.00	0.26	0.69	1.06	1.32	1.72	2.07	2.51	2.82	3.50
23	0.00	0.26	0.69	1.06	1.32	1.71	2.07	2.50	2.81	3.48
24	0.00	0.26	0.68	1.06	1.32	1.71	2.06	2.49	2.80	3.47
25	0.00	0.26	0.68	1.06	1.32	1.71	2.06	2.49	2.79	3.45
26	0.00	0.26	0.68	1.06	1.31	1.71	2.06	2.48	2.78	3.43
27	0.00	0.26	0.68	1.06	1.31	1.70	2.05	2.47	2.77	3.42
28	0.00	0.26	0.68	1.06	1.31	1.70	2.05	2.47	2.76	3.41
29	0.00	0.26	0.68	1.06	1.31	1.70	2.05	2.46	2.76	3.40
30	0.00	0.26	0.68	1.05	1.31	1.70	2.04	2.46	2.75	3.39
32	0.00	0.26	0.68	1.05	1.30	1.69	2.04	2.45	2.74	3.37
34	0.00	0.26	0.68	1.05	1.30	1.69	2.03	2.44	2.73	3.35
36	0.00	0.26	0.68	1.05	1.30	1.69	2.03	2.43	2.72	3.33
38	0.00	0.26	0.68	1.05	1.30	1.69	2.02	2.43	2.71	3.32
40	0.00	0.26	0.68	1.05	1.30	1.68	2.02	2.42	2.70	3.31

Statistical Table 2 Values of *t* for tail probabilities from 0.50 to 0.001 *(continued)*

Degrees of freedom	Probability									
	0.50	0.40	0.25	0.15	0.10	0.05	0.025	0.010	0.005	0.001
42	0.00	0.25	0.68	1.05	1.30	1.68	2.02	2.42	2.70	3.30
44	0.00	0.25	0.68	1.05	1.30	1.68	2.02	2.41	2.69	3.29
46	0.00	0.25	0.68	1.05	1.30	1.68	2.02	2.41	2.69	3.28
48	0.00	0.25	0.68	1.05	1.30	1.68	2.01	2.41	2.68	3.27
50	0.00	0.25	0.68	1.05	1.30	1.68	2.01	2.40	2.68	3.26
55	0.00	0.25	0.68	1.05	1.30	1.67	2.00	2.40	2.67	3.25
60	0.00	0.25	0.68	1.05	1.30	1.67	2.00	2.39	2.66	3.23
65	0.00	0.25	0.68	1.04	1.29	1.67	2.00	2.39	2.65	3.22
70	0.00	0.25	0.68	1.04	1.29	1.67	1.99	2.38	2.65	3.21
75	0.00	0.25	0.68	1.04	1.29	1.67	1.99	2.38	2.64	3.20
80	0.00	0.25	0.68	1.04	1.29	1.66	1.99	2.37	2.64	3.20
90	0.00	0.25	0.68	1.04	1.29	1.66	1.99	2.37	2.63	3.18
100	0.00	0.25	0.68	1.04	1.29	1.66	1.98	2.36	2.63	3.17
200	0.00	0.25	0.68	1.04	1.29	1.65	1.97	2.35	2.60	3.13
500	0.00	0.25	0.67	1.04	1.28	1.65	1.96	2.33	2.59	3.11
Infinity	0.00	0.25	0.67	1.04	1.28	1.64	1.96	2.33	2.58	3.09

Statistical Table 3 Values of chi-square for tail probabilities from 0.50 to 0.001

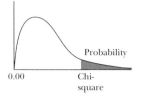

Degrees of freedom	Probability									
	0.50	0.40	0.25	0.15	0.10	0.05	0.025	0.010	0.005	0.001
1	0.45	0.71	1.32	2.07	2.71	3.84	5.02	6.63	7.88	10.83
2	1.39	1.83	2.77	3.79	4.61	5.99	7.38	9.21	10.60	13.82
3	2.37	2.95	4.11	5.32	6.25	7.81	9.35	11.34	12.84	16.27
4	3.36	4.04	5.39	6.74	7.78	9.49	11.14	13.28	14.86	18.47
5	4.35	5.13	6.63	8.12	9.24	11.07	12.83	15.09	16.75	20.52
6	5.35	6.21	7.84	9.45	10.64	12.59	14.45	16.81	18.55	22.46
7	6.35	7.28	9.04	10.75	12.02	14.07	16.01	18.48	20.28	24.32
8	7.34	8.35	10.22	12.03	13.36	15.51	17.53	20.09	21.95	26.12
9	8.34	9.41	11.39	13.29	14.68	16.92	19.02	21.67	23.59	27.88
10	9.34	10.47	12.55	14.53	15.99	18.31	20.48	23.21	25.19	29.59
11	10.34	11.53	13.70	15.77	17.28	19.68	21.92	24.72	26.76	31.26
12	11.34	12.58	14.85	16.99	18.55	21.03	23.34	26.22	28.30	32.91
13	12.34	13.64	15.98	18.20	19.81	22.36	24.74	27.69	29.82	34.53
14	13.34	14.69	17.12	19.41	21.06	23.68	26.12	29.14	31.32	36.12
15	14.34	15.73	18.25	20.60	22.31	25.00	27.49	30.58	32.80	37.70
16	15.34	16.78	19.37	21.79	23.54	26.30	28.85	32.00	34.27	39.25
17	16.34	17.82	20.49	22.98	24.77	27.59	30.19	33.41	35.72	40.79
18	17.34	18.87	21.60	24.16	25.99	28.87	31.53	34.81	37.16	42.31
19	18.34	19.91	22.72	25.33	27.20	30.14	32.85	36.19	38.58	43.82
20	19.34	20.95	23.83	26.50	28.41	31.41	34.17	37.57	40.00	45.31
21	20.34	21.99	24.93	27.66	29.62	32.67	35.48	38.93	41.40	46.80
22	21.34	23.03	26.04	28.82	30.81	33.92	36.78	40.29	42.80	48.27
23	22.34	24.07	27.14	29.98	32.01	35.17	38.08	41.64	44.18	49.73
24	23.34	25.11	28.24	31.13	33.20	36.42	39.36	42.98	45.56	51.18
25	24.34	26.14	29.34	32.28	34.38	37.65	40.65	44.31	46.93	52.62
26	25.34	27.18	30.43	33.43	35.56	38.89	41.92	45.64	48.29	54.05
27	26.34	28.21	31.53	34.57	36.74	40.11	43.19	46.96	49.64	55.48
28	27.34	29.25	32.62	35.71	37.92	41.34	44.46	48.28	50.99	56.89
29	28.34	30.28	33.71	36.85	39.09	42.56	45.72	49.59	52.34	58.30
30	29.34	31.32	34.80	37.99	40.26	43.77	46.98	50.89	53.67	59.70
32	31.34	33.38	36.97	40.26	42.58	46.19	49.48	53.49	56.33	62.49
34	33.34	35.44	39.14	42.51	44.90	48.60	51.97	56.06	58.96	65.25
36	35.34	37.50	41.30	44.76	47.21	51.00	54.44	58.62	61.58	67.99
38	37.34	39.56	43.46	47.01	49.51	53.38	56.90	61.16	64.18	70.70
40	39.34	41.62	45.62	49.24	51.81	55.76	59.34	63.69	66.77	73.40

Statistical Table 3   Values of chi-square for tail probabilities from 0.50 to 0.001 *(continued)*

Degrees of freedom	Probability									
	0.50	0.40	0.25	0.15	0.10	0.05	0.025	0.010	0.005	0.001
42	41.34	43.68	47.77	51.47	54.09	58.12	61.78	66.21	69.34	76.08
44	43.34	45.73	49.91	53.70	56.37	60.48	64.20	68.71	71.89	78.75
46	45.34	47.79	52.06	55.92	58.64	62.83	66.62	71.20	74.44	81.40
48	47.34	49.84	54.20	58.14	60.91	65.17	69.02	73.68	76.97	84.04
50	49.33	51.89	56.33	60.35	63.17	67.50	71.42	76.15	79.49	86.66

For values of chi-square with more degrees of freedom than this table shows, use the fact that the quantity

$$\frac{\text{chi-square} - \sqrt{\text{d.f.}}}{\sqrt{2 \text{ d.f.}}} = z$$

approximately follows the standard normal distribution.

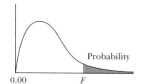

Statistical Table 4 Values of *F* for tail probabilities from 0.10 to 0.001

Denominator d.f.	Probability	\multicolumn{9}{c}{Degrees of freedom for the numerator}

Denominator d.f.	Probability	1	2	3	4	5	6	7	8	9
1	0.100	39.86	49.50	53.59	55.83	57.24	58.20	58.91	59.44	59.86
	0.050	161.45	199.50	215.71	224.58	230.16	233.99	236.77	238.88	240.54
	0.025	647.79	799.50	864.16	899.58	921.85	937.11	948.22	956.66	963.28
	0.010	4,052.2	4,999.5	5,403.4	5,624.6	5,763.6	5,859.0	5,928.4	5,981.1	6,022.5
	0.001	405,284	500,000	540,379	562,500	576,405	585,937	592,873	598,144	602,284
2	0.100	8.53	9.00	9.16	9.24	9.29	9.33	9.35	9.37	9.38
	0.050	18.51	19.00	19.16	19.25	19.30	19.33	19.35	19.37	19.38
	0.025	38.51	39.00	39.17	39.25	39.30	39.33	39.36	39.37	39.39
	0.010	98.50	99.00	99.17	99.25	99.30	99.33	99.36	99.37	99.39
	0.001	998.50	999.00	999.17	999.25	999.30	999.33	999.36	999.37	999.39
3	0.100	5.54	5.46	5.39	5.34	5.31	5.28	5.27	5.25	5.24
	0.050	10.13	9.55	9.28	9.12	9.01	8.94	8.89	8.85	8.81
	0.025	17.44	16.04	15.44	15.10	14.88	14.73	14.62	14.54	14.47
	0.010	34.12	30.82	29.46	28.71	28.24	27.91	27.67	27.49	27.35
	0.001	167.03	148.50	141.11	137.10	134.58	132.85	131.58	130.65	129.86
4	0.100	4.54	4.32	4.19	4.11	4.05	4.01	3.98	3.95	3.94
	0.050	7.71	6.94	6.59	6.39	6.26	6.16	6.09	6.04	6.00
	0.025	12.22	10.65	9.98	9.60	9.36	9.20	9.07	8.98	8.90
	0.010	21.20	18.00	16.69	15.98	15.52	15.21	14.98	14.80	14.66
	0.001	74.14	61.25	56.18	53.44	51.71	50.53	49.66	49.00	48.47
5	0.100	4.06	3.78	3.62	3.52	3.45	3.40	3.37	3.34	3.32
	0.050	6.61	5.79	5.41	5.19	5.05	4.95	4.88	4.82	4.77
	0.025	10.01	8.43	7.76	7.39	7.15	6.98	6.85	6.76	6.68
	0.010	16.26	13.27	12.06	11.39	10.97	10.67	10.46	10.29	10.16
	0.001	47.18	37.12	33.20	31.09	29.75	28.83	28.16	27.65	27.24
6	0.100	3.78	3.46	3.29	3.18	3.11	3.05	3.01	2.98	2.96
	0.050	5.99	5.14	4.76	4.53	4.39	4.28	4.21	4.15	4.10
	0.025	8.81	7.26	6.60	6.23	5.99	5.82	5.70	5.60	5.52
	0.010	13.75	10.92	9.78	9.15	8.75	8.47	8.26	8.10	7.98
	0.001	35.51	27.00	23.70	21.92	20.80	20.03	19.46	19.03	18.69
7	0.100	3.59	3.26	3.07	2.96	2.88	2.83	2.78	2.75	2.72
	0.050	5.59	4.74	4.35	4.12	3.97	3.87	3.79	3.73	3.68
	0.025	8.07	6.54	5.89	5.52	5.29	5.12	4.99	4.90	4.82
	0.010	12.25	9.55	8.45	7.85	7.46	7.19	6.99	6.84	6.72
	0.001	29.25	21.69	18.77	17.20	16.21	15.52	15.02	14.63	14.33

Degrees of freedom for the numerator

10	12	15	20	30	40	50	75	120	1000
60.19	60.71	61.22	61.74	62.26	62.53	62.69	62.90	63.06	63.30
241.88	243.91	245.95	248.01	250.10	251.14	251.77	252.62	253.25	254.19
968.63	976.71	984.87	993.10	1,001.4	1,005.6	1,008.1	1,011.5	1,014.0	1,017.8
6,055.8	6,106.3	6,157.3	6,208.7	6,260.6	6,286.8	6,302.5	6,323.6	6,339.4	6,362.7
605,621	610,668	615,764	620,908	626,099	628,712	630,285	632,390	633,972	636,301
9.39	9.41	9.42	9.44	9.46	9.47	9.47	9.48	9.48	9.49
19.40	19.41	19.43	19.45	19.46	19.47	19.48	19.48	19.49	19.49
39.40	39.41	39.43	39.45	39.46	39.47	39.48	39.48	39.49	39.50
99.40	99.42	99.43	99.45	99.47	99.47	99.48	99.49	99.49	99.50
999.40	999.42	999.43	999.45	999.47	999.47	999.48	999.49	999.49	999.50
5.23	5.22	5.20	5.18	5.17	5.16	5.15	5.15	5.14	5.13
8.79	8.74	8.70	8.66	8.62	8.59	8.58	8.56	8.55	8.53
14.42	14.34	14.25	14.17	14.08	14.04	14.01	13.97	13.95	13.91
27.23	27.05	26.87	26.69	26.50	26.41	26.35	26.28	26.22	26.14
129.25	128.32	127.37	126.42	125.45	124.96	124.66	124.27	123.97	123.53
3.92	3.90	3.87	3.84	3.82	3.80	3.80	3.78	3.78	3.76
5.96	5.91	5.86	5.80	5.75	5.72	5.70	5.68	5.66	5.63
8.84	8.75	8.66	8.56	8.46	8.41	8.38	8.34	8.31	8.26
14.55	14.37	14.20	14.02	13.84	13.75	13.69	13.61	13.56	13.47
48.05	47.41	46.76	46.10	45.43	45.09	44.88	44.61	44.40	44.09
3.30	3.27	3.24	3.21	3.17	3.16	3.15	3.13	3.12	3.11
4.74	4.68	4.62	4.56	4.50	4.46	4.44	4.42	4.40	4.37
6.62	6.52	6.43	6.33	6.23	6.18	6.14	6.10	6.07	6.02
10.05	9.89	9.72	9.55	9.38	9.29	9.24	9.17	9.11	9.03
26.92	26.42	25.91	25.39	24.87	24.60	24.44	24.22	24.06	23.82
2.94	2.90	2.87	2.84	2.80	2.78	2.77	2.75	2.74	2.72
4.06	4.00	3.94	3.87	3.81	3.77	3.75	3.73	3.70	3.67
5.46	5.37	5.27	5.17	5.07	5.01	4.98	4.94	4.90	4.86
7.87	7.72	7.56	7.40	7.23	7.14	7.09	7.02	6.97	6.89
18.41	17.99	17.56	17.12	16.67	16.44	16.31	16.12	15.98	15.77
2.70	2.67	2.63	2.59	2.56	2.54	2.52	2.51	2.49	2.47
3.64	3.57	3.51	3.44	3.38	3.34	3.32	3.29	3.27	3.23
4.76	4.67	4.57	4.47	4.36	4.31	4.28	4.23	4.20	4.15
6.62	6.47	6.31	6.16	5.99	5.91	5.86	5.79	5.74	5.66
14.08	13.71	13.32	12.93	12.53	12.33	12.20	12.04	11.91	11.72

Statistical Table 4 Values of F for tail probabilities from 0.10 to 0.001 (continued)

Denominator d.f.	Probability	\multicolumn{9}{c}{Degrees of freedom for the numerator}
		1	2	3	4	5	6	7	8	9
8	0.100	3.46	3.11	2.92	2.81	2.73	2.67	2.62	2.59	2.56
	0.050	5.32	4.46	4.07	3.84	3.69	3.58	3.50	3.44	3.39
	0.025	7.57	6.06	5.42	5.05	4.82	4.65	4.53	4.43	4.36
	0.010	11.26	8.65	7.59	7.01	6.63	6.37	6.18	6.03	5.91
	0.001	25.41	18.49	15.83	14.39	13.48	12.86	12.40	12.05	11.77
9	0.100	3.36	3.01	2.81	2.69	2.61	2.55	2.51	2.47	2.44
	0.050	5.12	4.26	3.86	3.63	3.48	3.37	3.29	3.23	3.18
	0.025	7.21	5.71	5.08	4.72	4.48	4.32	4.20	4.10	4.03
	0.010	10.56	8.02	6.99	6.42	6.06	5.80	5.61	5.47	5.35
	0.001	22.86	16.39	13.90	12.56	11.71	11.13	10.70	10.37	10.11
10	0.100	3.29	2.92	2.73	2.61	2.52	2.46	2.41	2.38	2.35
	0.050	4.96	4.10	3.71	9.48	3.33	3.22	3.14	3.07	3.02
	0.025	6.94	5.46	4.83	4.47	4.24	4.07	3.95	3.85	3.78
	0.010	10.04	7.56	6.55	5.99	5.64	5.39	5.20	5.06	4.94
	0.001	21.04	14.91	12.55	11.28	10.48	9.93	9.52	9.20	8.96
11	0.100	3.23	2.86	2.66	2.54	2.45	2.39	2.34	2.30	2.27
	0.050	4.84	3.98	3.59	3.36	3.20	3.09	3.01	2.95	2.90
	0.025	6.72	5.26	4.63	4.28	4.04	3.88	3.76	3.66	3.59
	0.010	9.65	7.21	6.22	5.67	5.32	5.07	4.89	4.74	4.63
	0.001	19.69	13.81	11.56	10.35	9.58	9.05	8.66	8.35	8.12
12	0.100	3.18	2.81	2.61	2.48	2.39	2.33	2.28	2.24	2.21
	0.050	4.75	3.89	3.49	3.26	3.11	3.00	2.91	2.85	2.80
	0.025	6.55	5.10	4.47	4.12	3.89	3.73	3.61	3.51	3.44
	0.010	9.33	6.93	5.95	5.41	5.06	4.82	4.64	4.50	4.39
	0.001	18.64	12.97	10.80	9.63	8.89	8.38	8.00	7.71	7.48
13	0.100	3.14	2.76	2.56	2.43	2.35	2.28	2.23	2.20	2.16
	0.050	4.67	3.81	3.41	3.18	3.03	2.92	2.83	2.77	2.71
	0.025	6.41	4.97	4.35	4.00	3.77	3.60	3.48	3.39	3.31
	0.010	9.07	6.70	5.74	5.21	4.86	4.62	4.44	4.30	4.19
	0.001	17.82	12.31	10.21	9.07	8.35	7.86	7.49	7.21	6.98
14	0.100	3.10	2.73	2.52	2.39	2.31	2.24	2.19	2.15	2.12
	0.050	4.60	3.74	3.34	3.11	2.96	2.85	2.76	2.70	2.65
	0.025	6.30	4.86	4.24	3.89	3.66	3.50	3.38	3.29	3.21
	0.010	8.86	6.51	5.56	5.04	4.69	4.46	4.28	4.14	4.03
	0.001	17.14	11.78	9.73	8.62	7.92	7.44	7.08	6.80	6.58

Degrees of freedom for the numerator

10	12	15	20	30	40	50	75	120	1000
2.54	2.50	2.46	2.42	2.38	2.36	2.35	2.33	2.32	2.30
3.35	3.28	3.22	3.15	3.08	3.04	3.02	2.99	2.97	2.93
4.30	4.20	4.10	4.00	3.89	3.84	3.81	3.76	3.73	3.68
5.81	5.67	5.52	5.36	5.20	5.12	5.07	5.00	4.95	4.87
11.54	11.19	10.84	10.48	10.11	9.92	9.80	9.65	9.53	9.36
2.42	2.38	2.34	2.30	2.25	2.23	2.22	2.20	2.18	2.16
3.14	3.07	3.01	2.94	2.86	2.83	2.80	2.77	2.75	2.71
3.96	3.87	3.77	3.67	3.56	3.51	3.47	3.43	3.39	3.34
5.26	5.11	4.96	4.81	4.65	4.57	4.52	4.45	4.40	4.32
9.89	9.57	9.24	8.90	8.55	8.37	8.26	8.11	8.00	7.84
2.32	2.28	2.24	2.20	2.16	2.13	2.12	2.10	2.08	2.06
2.98	2.91	2.85	2.77	2.70	2.66	2.64	2.60	2.58	2.54
3.72	3.62	3.52	3.42	3.31	3.26	3.22	3.18	3.14	3.09
4.85	4.71	4.56	4.41	4.25	4.17	4.12	4.05	4.00	3.92
8.75	8.45	8.13	7.80	7.47	7.30	7.19	7.05	6.94	6.78
2.25	2.21	2.17	2.12	2.08	2.05	2.04	2.02	2.00	1.98
2.85	2.79	2.72	2.65	2.57	2.53	2.51	2.47	2.45	2.41
3.53	3.43	3.33	3.23	3.12	3.06	3.03	2.98	2.94	2.89
4.54	4.40	4.25	4.10	3.94	3.86	3.81	3.74	3.69	3.61
7.92	7.63	7.32	7.01	6.68	6.52	6.42	6.28	6.18	6.02
2.19	2.15	2.10	2.06	2.01	1.99	1.97	1.95	1.93	1.91
2.75	2.69	2.62	2.54	2.47	2.43	2.40	2.37	2.34	2.30
3.37	3.28	3.18	3.07	2.96	2.91	2.87	2.82	2.79	2.73
4.30	4.16	4.01	3.86	3.70	3.62	3.57	3.50	3.45	3.37
7.29	7.00	6.71	6.40	6.09	5.93	5.83	5.70	5.59	5.44
2.14	2.10	2.05	2.01	1.96	1.93	1.92	1.89	1.88	1.85
2.67	2.60	2.53	2.46	2.38	2.34	2.31	2.28	2.25	2.21
3.25	3.15	3.05	2.95	2.84	2.78	2.74	2.70	2.66	2.60
4.10	3.96	3.82	3.66	3.51	3.43	3.38	3.31	3.25	3.18
6.80	6.52	6.23	5.93	5.63	5.47	5.37	5.24	5.14	4.99
2.10	2.05	2.01	1.96	1.91	1.89	1.87	1.85	1.83	1.80
2.60	2.53	2.46	2.39	2.31	2.27	2.24	2.21	2.18	2.14
3.15	3.05	2.95	2.84	2.73	2.67	2.64	2.59	2.55	2.50
3.94	3.80	3.66	3.51	3.35	3.27	3.22	3.15	3.09	3.02
6.40	6.13	5.85	5.56	5.25	5.10	5.00	4.87	4.77	4.62

Statistical Table 4 Values of F for tail probabilities from 0.10 to 0.001 *(continued)*

Denominator d.f.	Probability	\multicolumn{9}{c}{Degrees of freedom for the numerator}

Denominator d.f.	Probability	1	2	3	4	5	6	7	8	9
15	0.100	3.07	2.70	2.49	2.36	2.27	2.21	2.16	2.12	2.09
	0.050	4.54	3.68	3.29	3.06	2.90	2.79	2.71	2.64	2.59
	0.025	6.20	4.77	4.15	3.80	3.58	3.41	3.29	3.20	3.12
	0.010	8.68	6.36	5.42	4.89	4.56	4.32	4.14	4.00	3.89
	0.001	16.59	11.34	9.34	8.25	7.57	7.09	6.74	6.47	6.26
16	0.100	3.05	2.67	2.46	2.33	2.24	2.18	2.13	2.09	2.06
	0.050	4.49	3.63	3.24	3.01	2.85	2.74	2.66	2.59	2.54
	0.025	6.12	4.69	4.08	3.73	3.50	3.34	3.22	3.12	3.05
	0.010	8.53	6.23	5.29	4.77	4.44	4.20	4.03	3.89	3.78
	0.001	16.12	10.97	9.01	7.94	7.27	6.80	6.46	6.19	5.98
17	0.100	3.03	2.64	2.44	2.31	2.22	2.15	2.10	2.06	2.03
	0.050	4.45	3.59	3.20	2.96	2.81	2.70	2.61	2.55	2.49
	0.025	6.04	4.62	4.01	3.66	3.44	3.28	3.16	3.06	2.98
	0.010	8.40	6.11	5.18	4.67	4.34	4.10	3.93	3.79	3.68
	0.001	15.72	10.66	8.73	7.68	7.02	6.56	6.22	5.96	5.75
18	0.100	3.01	2.62	2.42	2.29	2.20	2.13	2.08	2.04	2.00
	0.050	4.41	3.55	3.16	2.93	2.77	2.66	2.58	2.51	2.46
	0.025	5.98	4.56	3.95	3.61	3.38	3.22	3.10	3.01	2.93
	0.010	8.29	6.01	5.09	4.58	4.25	4.01	3.84	3.71	3.60
	0.001	15.38	10.39	8.49	7.46	6.81	6.35	6.02	5.76	5.56
19	0.100	2.99	2.61	2.40	2.27	2.18	2.11	2.06	2.02	1.98
	0.050	4.38	3.52	3.13	2.90	2.74	2.63	2.54	2.48	2.42
	0.025	5.92	4.51	3.90	3.56	3.33	3.17	3.05	2.96	2.88
	0.010	8.18	5.93	5.01	4.50	4.17	3.94	3.77	3.63	3.52
	0.001	15.08	10.16	8.28	7.27	6.62	6.18	5.85	5.59	5.39
20	0.100	2.97	2.59	2.38	2.25	2.16	2.09	2.04	2.00	1.96
	0.050	4.35	3.49	3.10	2.87	2.71	2.60	2.51	2.45	2.39
	0.025	5.87	4.46	3.86	3.51	3.29	3.13	3.01	2.91	2.84
	0.010	8.10	5.85	4.94	4.43	4.10	3.87	3.70	3.56	3.46
	0.001	14.82	9.95	8.10	7.10	6.46	6.02	5.69	5.44	5.24
21	0.100	2.96	2.57	2.36	2.23	2.14	2.08	2.02	1.98	1.95
	0.050	4.32	3.47	3.07	2.84	2.68	2.57	2.49	2.42	2.37
	0.025	5.83	4.42	3.82	3.48	3.25	3.09	2.97	2.87	2.80
	0.010	8.02	5.78	4.87	4.37	4.04	3.81	3.64	3.51	3.40
	0.001	14.59	9.77	7.94	6.95	6.32	5.88	5.56	5.31	5.11

| \multicolumn{10}{c}{Degrees of freedom for the numerator} |
10	12	15	20	30	40	50	75	120	1000
2.06	2.02	1.97	1.92	1.87	1.85	1.83	1.80	1.79	1.76
2.54	2.48	2.40	2.33	2.25	2.20	2.18	2.14	2.11	2.07
3.06	2.96	2.86	2.76	2.64	2.59	2.55	2.50	2.46	2.40
3.80	3.67	3.52	3.37	3.21	3.13	3.08	3.01	2.96	2.88
6.08	5.81	5.54	5.25	4.95	4.80	4.70	4.57	4.47	4.33
2.03	1.99	1.94	1.89	1.84	1.81	1.79	1.77	1.75	1.72
2.49	2.42	2.35	2.28	2.19	2.15	2.12	2.09	2.06	2.02
2.99	2.89	2.79	2.68	2.57	2.51	2.47	2.42	2.38	2.32
3.69	3.55	3.41	3.26	3.10	3.02	2.97	2.90	2.84	2.76
5.81	5.55	5.27	4.99	4.70	4.54	4.45	4.32	4.23	4.08
2.00	1.96	1.91	1.86	1.81	1.78	1.76	1.74	1.72	1.69
2.45	2.38	2.31	2.23	2.15	2.10	2.08	2.04	2.01	1.97
2.92	2.82	2.72	2.62	2.50	2.44	2.41	2.35	2.32	2.26
3.59	3.46	3.31	3.16	3.00	2.92	2.87	2.80	2.75	2.66
5.58	5.32	5.05	4.78	4.48	4.33	4.24	4.11	4.02	3.87
1.98	1.93	1.89	1.84	1.78	1.75	1.74	1.71	1.69	1.66
2.41	2.34	2.27	2.19	2.11	2.06	2.04	2.00	1.97	1.92
2.87	2.77	2.67	2.56	2.44	2.38	2.35	2.30	2.26	2.20
3.51	3.37	3.23	3.08	2.92	2.84	2.78	2.71	2.66	2.58
5.39	5.13	4.87	4.59	4.30	4.15	4.06	3.93	3.84	3.69
1.96	1.91	1.86	1.81	1.76	1.73	1.71	1.69	1.67	1.64
2.38	2.31	2.23	2.16	2.07	2.03	2.00	1.96	1.93	1.88
2.82	2.72	2.62	2.51	2.39	2.33	2.30	2.24	2.20	2.14
3.43	3.30	3.15	3.00	2.84	2.76	2.71	2.64	2.58	2.50
5.22	4.97	4.70	4.43	4.14	3.99	3.90	3.78	3.68	3.53
1.94	1.89	1.84	1.79	1.74	1.71	1.69	1.66	1.64	1.61
2.35	2.28	2.20	2.12	2.04	1.99	1.97	1.93	1.90	1.85
2.77	2.68	2.57	2.46	2.35	2.29	2.25	2.20	2.16	1.09
3.37	3.23	3.09	2.94	2.78	2.69	2.64	2.57	2.52	2.43
5.08	4.82	4.56	4.29	4.00	3.86	3.77	3.64	3.54	3.40
1.92	1.87	1.83	1.78	1.72	1.69	1.67	1.64	1.62	1.59
2.32	2.25	2.18	2.10	2.01	1.96	1.94	1.90	1.87	1.82
2.73	2.64	2.53	2.42	2.31	2.25	2.21	2.16	2.11	2.05
3.31	3.17	3.03	2.88	2.72	2.64	2.58	2.51	2.46	2.37
4.95	4.70	4.44	4.17	3.88	3.74	3.64	3.52	3.42	3.28

Statistical Table 4 Values of F for tail probabilities from 0.10 to 0.001 (continued)

Denominator d.f.	Probability	\multicolumn{9}{c}{Degrees of freedom for the numerator}

Denominator d.f.	Probability	1	2	3	4	5	6	7	8	9
22	0.100	2.95	2.56	2.35	2.22	2.13	2.06	2.01	1.97	1.93
	0.050	4.30	3.44	3.05	2.82	2.66	2.55	2.46	2.40	2.34
	0.025	5.79	4.38	3.78	3.44	3.22	3.05	2.93	2.84	2.76
	0.010	7.95	5.72	4.82	4.31	3.99	3.76	3.59	3.45	3.35
	0.001	14.38	9.61	7.80	6.81	6.19	5.76	5.44	5.19	4.99
23	0.100	2.94	2.55	2.34	2.21	2.11	2.05	1.99	1.95	1.92
	0.050	4.28	3.42	3.03	2.80	2.64	2.53	2.44	2.37	2.32
	0.025	5.75	4.35	3.75	3.41	3.18	3.02	2.90	2.81	2.73
	0.010	7.88	5.66	4.76	4.26	3.94	3.71	3.54	3.41	3.30
	0.001	14.20	9.47	7.67	6.70	6.08	5.65	5.33	5.09	4.89
24	0.100.	2.93	2.54	2.33	2.19	2.10	2.04	1.98	1.94	1.91
	0.050	4.26	3.40	3.01	2.78	2.62	2.51	2.42	2.36	2.30
	0.025	5.72	4.32	3.72	3.38	3.15	2.99	2.87	2.78	2.70
	0.010	7.82	5.61	4.72	4.22	3.90	3.67	3.50	3.36	3.26
	0.001	14.03	9.34	7.55	6.59	5.98	5.55	5.23	4.99	4.80
25	0.100	2.92	2.53	2.32	2.18	2.09	2.02	1.97	1.93	1.89
	0.050	4.24	3.39	2.99	2.76	2.60	2.49	2.40	2.34	2.28
	0.025	5.69	4.29	3.69	3.35	3.13	2.97	2.85	2.75	2.68
	0.010	7.77	5.57	4.68	4.18	3.85	3.63	3.46	3.32	3.22
	0.001	13.88	9.22	7.45	6.49	5.89	5.46	5.15	4.91	4.71
26	0.100	2.91	2.52	2.31	2.17	2.08	2.01	1.96	1.92	1.88
	0.050	4.23	3.37	2.98	2.74	2.59	2.47	2.39	2.32	2.27
	0.025	5.66	4.27	3.67	3.33	3.10	2.94	2.82	2.73	2.65
	0.010	7.72	5.53	4.64	4.14	3.82	3.59	3.42	3.29	3.18
	0.001	13.74	9.12	7.36	6.41	5.80	5.38	5.07	4.83	4.64
27	0.100	2.90	2.51	2.30	2.17	2.07	2.00	1.95	1.91	1.87
	0.050	4.21	3.35	2.96	2.73	2.57	2.46	2.37	2.31	2.25
	0.025	5.63	4.24	3.65	3.31	3.08	2.92	2.80	2.71	2.63
	0.010	7.68	5.49	4.60	4.11	3.78	3.56	3.39	3.26	3.15
	0.001	13.61	9.02	7.27	6.33	5.73	5.31	5.00	4.76	4.57
28	0.100	2.89	2.50	2.29	2.16	2.06	2.00	1.94	1.90	1.87
	0.050	4.20	3.34	2.95	2.71	2.56	2.45	2.36	2.29	2.24
	0.025	5.61	4.22	3.63	3.29	3.06	2.90	2.78	2.69	2.61
	0.010	7.64	5.45	4.57	4.07	3.75	3.53	3.36	3.23	3.12
	0.001	13.50	8.93	7.19	6.25	5.66	5.24	4.93	4.69	4.50

Degrees of freedom for the numerator

10	12	15	20	30	40	50	75	120	1000
1.90	1.86	1.81	1.76	1.70	1.67	1.65	1.63	1.60	1.57
2.30	2.23	2.15	2.07	1.98	1.94	1.91	1.87	1.84	1.79
2.70	2.60	2.50	2.39	2.27	2.21	2.17	2.12	2.08	2.01
3.26	3.12	2.98	2.83	2.67	2.58	2.53	2.46	2.40	2.32
4.83	4.58	4.33	4.06	3.78	3.63	3.54	3.41	3.32	3.17
1.89	1.84	1.80	1.74	1.69	1.66	1.64	1.61	1.59	1.55
2.27	2.20	2.13	2.05	1.96	1.91	1.88	1.84	1.81	1.76
2.67	2.57	2.47	2.36	2.24	2.18	2.14	2.08	2.04	1.98
3.21	3.07	2.93	2.78	2.62	2.54	2.48	2.41	2.35	2.27
4.73	4.48	4.23	3.96	3.68	3.53	3.44	3.32	3.22	3.08
1.88	1.83	1.78	1.73	1.67	1.64	1.62	1.59	1.57	1.54
2.25	2.18	2.11	2.03	1.94	1.89	1.86	1.82	1.79	1.74
2.64	2.54	2.44	2.33	2.21	2.15	2.11	2.05	2.01	1.94
3.17	3.03	2.89	2.74	2.58	2.49	2.44	2.37	2.31	2.22
4.64	4.39	4.14	3.87	3.59	3.45	3.36	3.23	3.14	2.99
1.87	1.82	1.77	1.72	1.66	1.63	1.61	1.58	1.56	1.52
2.24	2.16	2.09	2.01	1.92	1.87	1.84	1.80	1.77	1.72
2.61	2.51	2.41	2.30	2.18	2.12	2.08	2.02	1.98	1.91
3.13	2.99	2.85	2.70	2.54	2.45	2.40	2.33	2.27	2.18
4.56	4.31	4.06	3.79	3.52	3.37	3.28	3.15	3.06	2.91
1.86	1.81	1.76	1.71	1.65	1.61	1.59	1.57	1.54	1.51
2.22	2.15	2.07	1.99	1.90	1.85	1.82	1.78	1.75	1.70
2.59	2.49	2.39	2.28	2.16	2.09	2.05	2.00	1.95	1.89
3.09	2.96	2.81	2.66	2.50	2.42	2.36	2.29	2.23	2.14
4.48	4.24	3.99	3.72	3.44	3.30	3.21	3.08	2.99	2.84
1.85	1.80	1.75	1.70	1.64	1.60	1.58	1.55	1.53	1.50
2.20	2.13	2.06	1.97	1.88	1.84	1.81	1.76	1.73	1.68
2.57	2.47	2.36	2.25	2.13	2.07	2.03	1.97	1.93	1.86
3.06	2.93	2.78	2.63	2.47	2.38	2.33	2.26	2.20	2.11
4.41	4.17	3.92	3.66	3.38	3.23	3.14	3.02	2.92	2.78
1.84	1.79	1.74	1.69	1.63	1.59	1.57	1.54	1.52	1.48
2.19	2.12	2.04	1.96	1.87	1.82	1.79	1.75	1.71	1.66
2.55	2.45	2.34	2.23	2.11	2.05	2.01	1.95	1.91	1.84
3.03	2.90	2.75	2.60	2.44	2.35	2.30	2.23	2.17	2.08
4.35	4.11	3.86	3.60	3.32	3.18	3.09	2.96	2.86	2.72

Statistical Table 4 Values of F for tail probabilities from 0.10 to 0.001 (continued)

Denominator d.f.	Probability	\multicolumn{9}{c}{Degrees of freedom for the numerator}
		1	2	3	4	5	6	7	8	9
29	0.100	2.89	2.50	2.28	2.15	2.06	1.99	1.93	1.89	1.86
	0.050	4.18	3.33	2.93	2.70	2.55	2.43	2.35	2.28	2.22
	0.025	5.59	4.20	3.61	3.27	3.04	2.88	2.76	2.67	2.59
	0.010	7.60	5.42	4.54	4.04	3.73	3.50	3.33	3.20	3.09
	0.001	13.39	8.85	7.12	6.19	5.59	5.18	4.87	4.64	4.45
30	0.100	2.88	2.49	2.28	2.14	2.05	1.98	1.93	1.88	1.85
	0.050	4.17	3.32	2.92	2.69	2.53	2.42	2.33	2.27	2.21
	0.025	5.57	4.18	3.59	3.25	3.03	2.87	2.75	2.65	2.57
	0.010	7.56	5.39	4.51	4.02	3.70	3.47	3.30	3.17	3.07
	0.001	13.29	8.77	7.05	6.12	5.53	5.12	4.82	4.58	4.39
35	0.100	2.85	2.46	2.25	2.11	2.02	1.95	1.90	1.85	1.82
	0.050	4.12	3.27	2.87	2.64	2.49	2.37	2.29	2.22	2.16
	0.025	5.48	4.11	3.52	3.18	2.96	2.80	2.68	2.58	2.50
	0.010	7.42	5.27	4.40	3.91	3.59	3.37	3.20	3.07	2.96
	0.001	12.90	8.47	6.79	5.88	5.30	4.89	4.59	4.36	4.18
40	0.100	2.84	2.44	2.23	2.09	2.00	1.93	1.87	1.83	1.79
	0.050	4.08	3.23	2.84	2.61	2.45	2.34	2.25	2.18	2.12
	0.025	5.42	4.05	3.46	3.13	2.90	2.74	2.62	2.53	2.45
	0.010	7.31	5.18	4.31	3.83	3.51	3.29	3.12	2.99	2.89
	0.001	12.61	8.25	6.59	5.70	5.13	4.73	4.44	4.21	4.02
50	0.100	2.81	2.41	2.20	2.06	1.97	1.90	1.84	1.80	1.76
	0.050	4.03	3.18	2.79	2.56	2.40	2.29	2.20	2.13	2.07
	0.025	5.34	3.97	3.39	3.05	2.83	2.67	2.55	2.46	2.38
	0.010	7.17	5.06	4.20	3.72	3.41	3.19	3.02	2.89	2.78
	0.001	12.22	7.96	6.34	5.46	4.90	4.51	4.22	4.00	3.82
60	0.100	2.79	2.39	2.18	2.04	1.95	1.87	1.82	1.77	1.74
	0.050	4.00	3.15	2.76	2.53	2.37	2.25	2.17	2.10	2.04
	0.025	5.29	3.93	3.34	3.01	2.79	2.63	2.51	2.41	2.33
	0.010	7.08	4.98	4.13	3.65	3.34	3.12	2.95	2.82	2.72
	0.001	11.97	7.77	6.17	5.31	4.76	4.37	4.09	3.86	3.69
80	0.100	2.77	2.37	2.15	2.02	1.92	1.85	1.79	1.75	1.71
	0.050	3.96	3.11	2.72	2.49	2.33	2.21	2.13	2.06	2.00
	0.025	5.22	3.86	3.28	2.95	2.73	2.57	2.45	2.35	2.28
	0.010	6.96	4.88	4.04	3.56	3.26	3.04	2.87	2.74	2.64
	0.001	11.67	7.54	5.97	5.12	4.58	4.20	3.92	3.70	3.53

			Degrees of freedom for the numerator						
10	12	15	20	30	40	50	75	120	1000
1.83	1.78	1.73	1.68	1.62	1.58	1.56	1.53	1.51	1.47
2.18	2.10	2.03	1.94	1.85	1.81	1.77	1.73	1.70	1.65
2.53	2.43	2.32	2.21	2.09	2.03	1.99	1.93	1.89	1.82
3.00	2.87	2.73	2.57	2.41	2.33	2.27	2.20	2.14	2.05
4.29	4.05	3.80	3.54	3.27	3.12	3.03	2.91	2.81	2.66
1.82	1.77	1.72	1.67	1.61	1.57	1.55	1.52	1.50	1.46
2.16	2.09	2.01	1.93	1.84	1.79	1.76	1.72	1.68	1.63
2.51	2.41	2.31	2.20	2.07	2.01	1.97	1.91	1.87	1.80
2.98	2.84	2.70	2.55	2.39	2.30	2.25	2.17	2.11	2.02
4.24	4.00	3.75	3.49	3.22	3.07	2.98	2.86	2.76	2.61
1.79	1.74	1.69	1.63	1.57	1.53	1.51	1.48	1.46	1.42
2.11	2.04	1.96	1.88	1.79	1.74	1.70	1.66	1.62	1.57
2.44	2.34	2.23	2.12	2.00	1.93	1.89	1.83	1.79	1.71
2.88	2.74	2.60	2.44	2.28	2.19	2.14	2.06	2.00	1.90
4.03	3.79	3.55	3.29	3.02	2.87	2.78	2.66	2.56	2.40
1.76	1.71	1.66	1.61	1.54	1.51	1.48	1.45	1.42	1.38
2.08	2.00	1.92	1.84	1.74	1.69	1.66	1.61	1.58	1.52
2.39	2.29	2.18	2.07	1.94	1.88	1.83	1.77	1.72	1.65
2.80	2.66	2.52	2.37	2.20	2.11	2.06	1.98	1.92	1.82
3.87	3.64	3.40	3.14	2.87	2.73	2.64	2.51	2.41	2.25
1.73	1.68	1.63	1.57	1.50	1.46	1.44	1.41	1.38	1.33
2.03	1.95	1.87	1.78	1.69	1.63	1.60	1.55	1.51	1.45
2.32	2.22	2.11	1.99	1.87	1.80	1.75	1.69	1.64	1.56
2.70	2.56	2.42	2.27	2.10	2.01	1.95	1.87	1.80	1.70
3.67	3.44	3.20	2.95	2.68	2.53	2.44	2.31	2.21	2.05
1.71	1.66	1.60	1.54	1.48	1.44	1.41	1.38	1.35	1.30
1.99	1.92	1.84	1.75	1.65	1.59	1.56	1.51	1.47	1.40
2.27	2.17	2.06	1.94	1.82	1.74	1.70	1.63	1.58	1.49
2.63	2.50	2.35	2.20	2.03	1.94	1.88	1.79	1.73	1.62
3.54	3.32	3.08	2.83	2.55	2.41	2.32	2.19	2.08	1.92
1.68	1.63	1.57	1.51	1.44	1.40	1.38	1.34	1.31	1.25
1.95	1.88	1.79	1.70	1.60	1.54	1.51	1.45	1.41	1.34
2.21	2.11	2.00	1.88	1.75	1.68	1.63	1.56	1.51	1.41
2.55	2.42	2.27	2.12	1.94	1.85	1.79	1.70	1.63	1.51
3.39	3.16	2.93	2.68	2.41	2.26	2.16	2.03	1.92	1.75

Statistical Table 4 Values of F for tail probabilities from 0.10 to 0.001 *(continued)*

Denominator d.f.	Probability	\multicolumn{9}{c}{Degrees of freedom for the numerator}

Denominator d.f.	Probability	1	2	3	4	5	6	7	8	9
100	0.100	2.76	2.36	2.14	2.00	1.91	1.83	1.78	1.73	1.69
	0.050	3.94	3.09	2.70	2.46	2.31	2.19	2.10	2.03	1.97
	0.025	5.18	3.83	3.25	2.92	2.70	2.54	2.42	2.32	2.24
	0.010	6.90	4.82	3.98	3.51	3.21	2.99	2.82	2.69	2.59
	0.001	11.50	7.41	5.86	5.02	4.48	4.11	3.83	3.61	3.44
200	0.100	2.73	2.33	2.11	1.97	1.88	1.80	1.75	1.70	1.66
	0.050	3.89	3.04	2.65	2.42	2.26	2.14	2.06	1.98	1.93
	0.025	5.10	3.76	3.18	2.85	2.63	2.47	2.35	2.26	2.18
	0.010	6.76	4.71	3.88	3.41	3.11	2.89	2.73	2.60	2.50
	0.001	11.15	7.15	5.63	4.81	4.29	3.92	3.65	3.43	3.26
1,000	0.100	2.71	2.31	2.09	1.95	1.85	1.78	1.72	1.68	1.64
	0.050	3.85	3.00	2.61	2.38	2.22	2.11	2.02	1.95	1.89
	0.025	5.04	3.70	3.13	2.80	2.58	2.42	2.30	2.20	2.13
	0.010	6.66	4.63	3.80	3.34	3.04	2.82	2.66	2.53	2.43
	0.001	10.89	6.96	5.46	4.65	4.14	3.78	3.51	3.30	3.13

			Degrees of freedom for the numerator						
10	12	15	20	30	40	50	75	120	1000
1.66	1.61	1.56	1.49	1.42	1.38	1.35	1.32	1.28	1.22
1.93	1.85	1.77	1.68	1.57	1.52	1.48	1.42	1.38	1.30
2.18	2.08	1.97	1.85	1.71	1.64	1.59	1.52	1.46	1.36
2.50	2.37	2.22	2.07	1.89	1.80	1.74	1.65	1.57	1.45
3.30	3.07	2.84	2.59	2.32	2.17	2.08	1.94	1.83	1.64
1.63	1.58	1.52	1.46	1.38	1.34	1.31	1.27	1.23	1.16
1.88	1.80	1.72	1.62	1.52	1.46	1.41	1.35	1.30	1.21
2.11	2.01	1.90	1.78	1.64	1.56	1.51	1.44	1.37	1.25
2.41	2.27	2.13	1.97	1.79	1.69	1.63	1.53	1.45	1.30
3.12	2.90	2.67	2.42	2.15	2.00	1.90	1.76	1.64	1.43
1.61	1.55	1.49	1.43	1.35	1.30	1.27	1.23	1.18	1.08
1.84	1.76	1.68	1.58	1.47	1.41	1.36	1.30	1.24	1.11
2.06	1.96	1.85	1.72	1.58	1.50	1.45	1.36	1.29	1.13
2.34	2.20	2.06	1.90	1.72	1.61	1.54	1.44	1.35	1.16
2.99	2.77	2.54	2.30	2.02	1.87	1.77	1.62	1.49	1.22

Statistical Table 5 Binomial distribution

The probability of x successes in n trials for different values of π. For π larger than 0.5, the table shows the probability of $n - x$ failures for different values of $1 - \pi$.

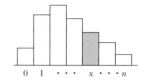

		π										
n	x	0.025	0.05	0.10	0.15	0.20	0.25	0.30	0.35	0.40	0.45	0.50
2	0	.9506	.9025	.8100	.7225	.6400	.5625	.4900	.4225	.3600	.3025	.2500
	1	.0488	.0950	.1800	.2550	.3200	.3750	.4200	.4550	.4800	.4950	.5000
	2	.0006	.0025	.0100	.0225	.0400	.0625	.0900	.1225	.1600	.2025	.2500
3	0	.9269	.8574	.7290	.6141	.5120	.4219	.3430	.2746	.2160	.1664	.1250
	1	.0713	.1354	.2430	.3251	.3840	.4219	.4410	.4436	.4320	.4084	.3750
	2	.0018	.0071	.0270	.0574	.0960	.1406	.1890	.2389	.2880	.3341	.3750
	3		.0001	.0010	.0034	.0080	.0156	.0270	.0429	.0640	.0911	.1250
4	0	.9037	.8145	.6561	.5220	.4096	.3164	.2401	.1785	.1296	.0915	.0625
	1	.0927	.1715	.2916	.3685	.4096	.4219	.4116	.3845	.3456	.2995	.2500
	2	.0036	.0135	.0486	.0975	.1536	.2109	.2646	.3105	.3456	.3675	.3750
	3	.0001	.0005	.0036	.0115	.0256	.0469	.0756	.1115	.1536	.2005	.2500
	4			.0001	.0005	.0016	.0039	.0081	.0150	.0256	.0410	.0625
5	0	.8811	.7738	.5905	.4437	.3277	.2373	.1681	.1160	.0778	.0503	.0313
	1	.1130	.2036	.3280	.3915	.4096	.3955	.3602	.3124	.2592	.2059	.1563
	2	.0058	.0214	.0729	.1382	.2048	.2637	.3087	.3364	.3456	.3369	.3125
	3	.0001	.0011	.0081	.0244	.0512	.0879	.1323	.1811	.2304	.2757	.3125
	4			.0004	.0022	.0064	.0146	.0284	.0488	.0768	.1128	.1562
	5				.0001	.0003	.0010	.0024	.0053	.0102	.0185	.0312
6	0	.8591	.7351	.5314	.3771	.2621	.1780	.1176	.0754	.0467	.0277	.0156
	1	.1322	.2321	.3543	.3993	.3932	.3560	.3025	.2437	.1866	.1359	.0938
	2	.0085	.0305	.0984	.1762	.2458	.2966	.3241	.3280	.3110	.2780	.2344
	3	.0003	.0021	.0146	.0415	.0819	.1318	.1852	.2355	.2765	.3032	.3125
	4		.0001	.0012	.0055	.0154	.0330	.0595	.0951	.1382	.1861	.2344
	5			.0001	.0004	.0015	.0044	.0102	.0205	.0369	.0609	.0938
	6					.0001	.0002	.0007	.0018	.0041	.0083	.0156
7	0	.8376	.6983	.4783	.3206	.2097	.1335	.0824	.0490	.0280	.0152	.0078
	1	.1503	.2573	.3720	.3960	.3670	.3115	.2471	.1848	.1306	.0872	.0547
	2	.0116	.0406	.1240	.2097	.2753	.3115	.3177	.2985	.2613	.2140	.1641
	3	.0005	.0036	.0230	.0617	.1147	.1730	.2269	.2679	.2903	.2918	.2734
	4		.0002	.0026	.0109	.0287	.0577	.0972	.1442	.1935	.2388	.2734
	5			.0002	.0012	.0043	.0115	.0250	.0466	.0774	.1172	.1641
	6				.0001	.0004	.0013	.0036	.0084	.0172	.0320	.0547
	7						.0001	.0002	.0006	.0016	.0037	.0078

Statistical Table 5 Binomial distribution *(continued)*

n	x	0.025	0.05	0.10	0.15	0.20	0.25	0.30	0.35	0.40	0.45	0.50
8	0	.8167	.6634	.4305	.2725	.1678	.1001	.0576	.0319	.0168	.0084	.0039
	1	.1675	.2793	.3826	.3847	.3355	.2670	.1977	.1373	.0896	.0548	.0313
	2	.0150	.0515	.1488	.2376	.2936	.3115	.2965	.2587	.2090	.1569	.1094
	3	.0008	.0054	.0331	.0839	.1468	.2076	.2541	.2786	.2787	.2568	.2188
	4		.0004	.0046	.0185	.0459	.0865	.1361	.1875	.2322	.2627	.2734
	5			.0004	.0026	.0092	.0231	.0467	.0808	.1239	.1719	.2188
	6				.0002	.0011	.0038	.0100	.0217	.0413	.0703	.1094
	7					.0001	.0004	.0012	.0033	.0079	.0164	.0312
	8							.0001	.0002	.0007	.0017	.0039
9	0	.7962	.6302	.3874	.2316	.1342	.0751	.0404	.0207	.0101	.0046	.0020
	1	.1837	.2985	.3874	.3679	.3020	.2253	.1556	.1004	.0605	.0339	.0176
	2	.0188	.0629	.1722	.2597	.3020	.3003	.2668	.2162	.1612	.1110	.0703
	3	.0011	.0077	.0446	.1069	.1762	.2336	.2668	.2716	.2508	.2119	.1641
	4		.0006	.0074	.0283	.0661	.1168	.1715	.2194	.2508	.2600	.2461
	5			.0008	.0050	.0165	.0389	.0735	.1181	.1672	.2128	.2461
	6			.0001	.0006	.0028	.0087	.0210	.0424	.0743	.1160	.1641
	7					.0003	.0012	.0039	.0098	.0212	.0407	.0703
	8						.0001	.0004	.0013	.0035	.0083	.0176
	9								.0001	.0003	.0008	.0020
10	0	.7763	.5987	.3487	.1969	.1074	.0563	.0282	.0135	.0060	.0025	.0010
	1	.1991	.3151	.3874	.3474	.2684	.1877	.1211	.0725	.0403	.0207	.0098
	2	.0230	.0746	.1937	.2759	.3020	.2816	.2335	.1757	.1209	.0763	.0439
	3	.0016	.0105	.0574	.1298	.2013	.2503	.2668	.2522	.2150	.1665	.1172
	4	.0001	.0010	.0112	.0401	.0881	.1460	.2001	.2377	.2508	.2384	.2051
	5		.0001	.0015	.0085	.0264	.0584	.1029	.1536	.2007	.2340	.2461
	6			.0001	.0012	.0055	.0162	.0368	.0689	.1115	.1596	.2051
	7				.0001	.0008	.0031	.0090	.0212	.0425	.0746	.1172
	8					.0001	.0004	.0014	.0043	.0106	.0229	.0439
	9							.0001	.0005	.0016	.0042	.0098
	10									.0001	.0003	.0010
11	0	.7569	.5688	.3138	.1673	.0859	.0422	.0198	.0088	.0036	.0014	.0005
	1	.2135	.3293	.3835	.3248	.2362	.1549	.0932	.0518	.0266	.0125	.0054
	2	.0274	.0867	.2131	.2866	.2953	.2581	.1998	.1395	.0887	.0513	.0269
	3	.0021	.0137	.0710	.1517	.2215	.2581	.2568	.2254	.1774	.1259	.0806
	4	.0001	.0014	.0158	.0536	.1107	.1721	.2201	.2428	.2365	.2060	.1611
	5		.0001	.0025	.0132	.0388	.0803	.1321	.1830	.2207	.2360	.2256
	6			.0003	.0023	.0097	.0268	.0566	.0985	.1471	.1931	.2256
	7				.0003	.0017	.0064	.0173	.0379	.0701	.1128	.1611
	8					.0002	.0011	.0037	.0102	.0234	.0462	.0806
	9						.0001	.0005	.0018	.0052	.0126	.0269
	10								.0002	.0007	.0021	.0054
	11										.0002	.0005

Statistical Table 5 Binomial distribution *(continued)*

n	x	0.025	0.05	0.10	0.15	0.20	0.25	0.30	0.35	0.40	0.45	0.50
12	0	.7380	.5404	.2824	.1422	.0687	.0317	.0138	.0057	.0022	.0008	.0002
	1	.2271	.3413	.3766	.3012	.2062	.1267	.0712	.0368	.0174	.0075	.0029
	2	.0320	.0988	.2301	.2924	.2835	.2323	.1678	.1088	.0639	.0339	.0161
	3	.0027	.0173	.0852	.1720	.2362	.2581	.2397	.1954	.1419	.0923	.0537
	4	.0002	.0021	.0213	.0683	.1329	.1936	.2311	.2367	.2128	.1700	.1208
	5		.0002	.0038	.0193	.0532	.1032	.1585	.2039	.2270	.2225	.1934
	6			.0005	.0040	.0155	.0401	.0792	.1281	.1766	.2124	.2256
	7				.0006	.0033	.0115	.0291	.0591	.1009	.1489	.1934
	8				.0001	.0005	.0024	.0078	.0199	.0420	.0762	.1208
	9					.0001	.0004	.0015	.0048	.0125	.0277	.0537
	10							.0002	.0008	.0025	.0068	.0161
	11								.0001	.0003	.0010	.0029
	12										.0001	.0002
13	0	.7195	.5133	.2542	.1209	.0550	.0238	.0097	.0037	.0013	.0004	.0001
	1	.2398	.3512	.3672	.2774	.1787	.1029	.0540	.0259	.0113	.0045	.0016
	2	.0369	.1109	.2448	.2937	.2680	.2059	.1388	.0836	.0453	.0220	.0095
	3	.0035	.0214	.0997	.1900	.2457	.2517	.2181	.1651	.1107	.0660	.0349
	4	.0002	.0028	.0277	.0838	.1535	.2097	.2337	.2222	.1845	.1350	.0873
	5		.0003	.0055	.0266	.0691	.1258	.1803	.2154	.2214	.1989	.1571
	6			.0008	.0063	.0230	.0559	.1030	.1546	.1968	.2169	.2095
	7			.0001	.0011	.0058	.0186	.0442	.0833	.1312	.1775	.2095
	8				.0001	.0011	.0047	.0142	.0336	.0656	.1089	.1571
	9					.0001	.0009	.0034	.0101	.0243	.0495	.0873
	10						.0001	.0006	.0022	.0065	.0162	.0349
	11							.0001	.0003	.0012	.0036	.0095
	12									.0001	.0005	.0016
	13											.0001
14	0	.7016	.4877	.2288	.1028	.0440	.0178	.0068	.0024	.0008	.0002	.0001
	1	.2518	.3593	.3559	.2539	.1539	.0832	.0407	.0181	.0073	.0027	.0009
	2	.0420	.1229	.2570	.2912	.2501	.1802	.1134	.0634	.0317	.0141	.0056
	3	.0043	.0259	.1142	.2056	.2501	.2402	.1943	.1366	.0845	.0462	.0222
	4	.0003	.0037	.0349	.0998	.1720	.2202	.2290	.2022	.1549	.1040	.0611
	5		.0004	.0078	.0352	.0860	.1468	.1963	.2178	.2066	.1701	.1222
	6			.0013	.0093	.0322	.0734	.1262	.1759	.2066	.2088	.1833
	7			.0002	.0019	.0092	.0280	.0618	.1082	.1574	.1952	.2095
	8				.0003	.0020	.0082	.0232	.0510	.0918	.1398	.1833
	9					.0003	.0018	.0066	.0183	.0408	.0762	.1222
	10						.0003	.0014	.0049	.0136	.0312	.0611
	11							.0002	.0010	.0033	.0093	.0222
	12								.0001	.0005	.0019	.0056
	13									.0001	.0002	.0009
	14											.0001

Statistical Table 5 Binomial distribution *(continued)*

		π										
n	x	0.025	0.05	0.10	0.15	0.20	0.25	0.30	0.35	0.40	0.45	0.50
15	0	.6840	.4633	.2059	.0874	.0352	.0134	.0047	.0016	.0005	.0001	
	1	.2631	.3658	.3432	.2312	.1319	.0668	.0305	.0126	.0047	.0016	.0005
	2	.0472	.1348	.2669	.2856	.2309	.1559	.0916	.0476	.0219	.0090	.0032
	3	.0052	.0307	.1285	.2184	.2501	.2252	.1700	.1110	.0634	.0318	.0139
	4	.0004	.0049	.0428	.1156	.1876	.2252	.2186	.1792	.1268	.0780	.0417
	5		.0006	.0105	.0449	.1032	.1651	.2061	.2123	.1859	.1404	.0916
	6			.0019	.0132	.0430	.0917	.1472	.1906	.2066	.1914	.1527
	7			.0003	.0030	.0138	.0393	.0811	.1319	.1771	.2013	.1964
	8				.0005	.0035	.0131	.0348	.0710	.1181	.1647	.1964
	9				.0001	.0007	.0034	.0116	.0298	.0612	.1048	.1527
	10					.0001	.0007	.0030	.0096	.0245	.0515	.0916
	11						.0001	.0006	.0024	.0074	.0191	.0417
	12							.0001	.0004	.0016	.0052	.0139
	13								.0001	.0003	.0010	.0032
	14										.0001	.0005
16	0	.6669	.4401	.1853	.0743	.0281	.0100	.0033	.0010	.0003	.0001	
	1	.2736	.3706	.3294	.2097	.1126	.0535	.0228	.0087	.0030	.0009	.0002
	2	.0526	.1463	.2745	.2775	.2111	.1336	.0732	.0353	.0150	.0056	.0018
	3	.0063	.0359	.1423	.2285	.2463	.2079	.1465	.0888	.0468	.0215	.0085
	4	.0005	.0061	.0514	.1311	.2001	.2252	.2040	.1553	.1014	.0572	.0278
	5		.0008	.0137	.0555	.1201	.1802	.2099	.2008	.1623	.1123	.0667
	6		.0001	.0028	.0180	.0550	.1101	.1649	.1982	.1983	.1684	.1222
	7			.0004	.0045	.0197	.0524	.1010	.1524	.1889	.1969	.1746
	8			.0001	.0009	.0055	.0197	.0487	.0923	.1417	.1812	.1964
	9				.0001	.0012	.0058	.0185	.0442	.0840	.1318	.1746
	10					.0002	.0014	.0056	.0167	.0392	.0755	.1222
	11						.0002	.0013	.0049	.0142	.0337	.0667
	12							.0002	.0011	.0040	.0115	.0278
	13								.0002	.0008	.0029	.0085
	14									.0001	.0005	.0018
	15										.0001	.0002

Statistical Table 5 Binomial distribution *(continued)*

		π									
n	x	0.025	0.05	0.10	0.15	0.20	0.25	0.30	0.35	0.40	0.45	0.50
17	0	.6502	.4181	.1668	.0631	.0225	.0075	.0023	.0007	.0002		
	1	.2834	.3741	.3150	.1893	.0957	.0426	.0169	.0060	.0019	.0005	.0001
	2	.0581	.1575	.2800	.2673	.1914	.1136	.0581	.0260	.0102	.0035	.0010
	3	.0075	.0415	.1556	.2359	.2393	.1893	.1245	.0701	.0341	.0144	.0052
	4	.0007	.0076	.0605	.1457	.2093	.2209	.1868	.1320	.0796	.0411	.0182
	5		.0010	.0175	.0668	.1361	.1914	.2081	.1849	.1379	.0875	.0472
	6		.0001	.0039	.0236	.0680	.1276	.1784	.1991	.1839	.1432	.0944
	7			.0007	.0065	.0267	.0668	.1201	.1685	.1927	.1841	.1484
	8			.0001	.0014	.0084	.0279	.0644	.1134	.1606	.1883	.1855
	9				.0003	.0021	.0093	.0276	.0611	.1070	.1540	.1855
	10					.0004	.0025	.0095	.0263	.0571	.1008	.1484
	11					.0001	.0005	.0026	.0090	.0242	.0525	.0944
	12						.0001	.0006	.0024	.0081	.0215	.0472
	13							.0001	.0005	.0021	.0068	.0182
	14								.0001	.0004	.0016	.0052
	15									.0001	.0003	.0010
	16											.0001
18	0	.6340	.3972	.1501	.0536	.0180	.0056	.0016	.0004	.0001		
	1	.2926	.3763	.3002	.1704	.0811	.0338	.0126	.0042	.0012	.0003	.0001
	2	.0638	.1683	.2835	.2556	.1723	.0958	.0458	.0190	.0069	.0022	.0006
	3	.0087	.0473	.1680	.2406	.2297	.1704	.1046	.0547	.0246	.0095	.0031
	4	.0008	.0093	.0700	.1592	.2153	.2130	.1681	.1104	.0614	.0291	.0117
	5	.0001	.0014	.0218	.0787	.1507	.1988	.2017	.1664	.1146	.0666	.0327
	6		.0002	.0052	.0301	.0816	.1436	.1873	.1941	.1655	.1181	.0708
	7			.0010	.0091	.0350	.0820	.1376	.1792	.1892	.1657	.1214
	8			.0002	.0022	.0120	.0376	.0811	.1327	.1734	.1864	.1669
	9				.0004	.0033	.0139	.0386	.0794	.1284	.1694	.1855
	10				.0001	.0008	.0042	.0149	.0385	.0771	.1248	.1669
	11					.0001	.0010	.0046	.0151	.0374	.0742	.1214
	12						.0002	.0012	.0047	.0145	.0354	.0708
	13							.0002	.0012	.0045	.0134	.0327
	14								.0002	.0011	.0039	.0117
	15									.0002	.0009	.0031
	16										.0001	.0006
	17											.0001

Statistical Table 5 Binomial distribution *(continued)*

n	x	0.025	0.05	0.10	0.15	0.20	0.25	0.30	0.35	0.40	0.45	0.50
19	0	.6181	.3774	.1351	.0456	.0144	.0042	.0011	.0003	.0001		
	1	.3011	.3774	.2852	.1529	.0685	.0268	.0093	.0029	.0008	.0002	
	2	.0695	.1787	.2852	.2428	.1540	.0803	.0358	.0138	.0046	.0013	.0003
	3	.0101	.0533	.1796	.2428	.2182	.1517	.0869	.0422	.0175	.0062	.0018
	4	.0010	.0112	.0798	.1714	.2182	.2023	.1491	.0909	.0467	.0203	.0074
	5	.0001	.0018	.0266	.0907	.1636	.2023	.1916	.1468	.0933	.0497	.0222
	6		.0002	.0069	.0374	.0955	.1574	.1916	.1844	.1451	.0949	.0518
	7			.0014	.0122	.0443	.0974	.1525	.1844	.1797	.1443	.0961
	8			.0002	.0032	.0166	.0487	.0981	.1489	.1797	.1771	.1442
	9				.0007	.0051	.0198	.0514	.0980	.1464	.1771	.1762
	10				.0001	.0013	.0066	.0220	.0528	.0976	.1449	.1762
	11					.0003	.0018	.0077	.0233	.0532	.0970	.1442
	12						.0004	.0022	.0083	.0237	.0529	.0961
	13						.0001	.0005	.0024	.0085	.0233	.0518
	14							.0001	.0006	.0024	.0082	.0222
	15								.0001	.0005	.0022	.0074
	16									.0001	.0005	.0018
	17										.0001	.0003
20	0	.6027	.3585	.1216	.0388	.0115	.0032	.0008	.0002			
	1	.3091	.3774	.2702	.1368	.0576	.0211	.0068	.0020	.0005	.0001	
	2	.0753	.1887	.2852	.2293	.1369	.0669	.0278	.0100	.0031	.0008	.0002
	3	.0116	.0596	.1901	.2428	.2054	.1339	.0716	.0323	.0123	.0040	.0011
	4	.0013	.0133	.0898	.1821	.2182	.1897	.1304	.0738	.0350	.0139	.0046
	5	.0001	.0022	.0319	.1028	.1746	.2023	.1789	.1272	.0746	.0365	.0148
	6		.0003	.0089	.0454	.1091	.1686	.1916	.1712	.1244	.0746	.0370
	7			.0020	.0160	.0545	.1124	.1643	.1844	.1659	.1221	.0739
	8			.0004	.0046	.0222	.0609	.1144	.1614	.1797	.1623	.1201
	9			.0001	.0011	.0074	.0271	.0654	.1158	.1597	.1771	.1602
	10				.0002	.0020	.0099	.0308	.0686	.1171	.1593	.1762
	11					.0005	.0030	.0120	.0336	.0710	.1185	.1602
	12					.0001	.0008	.0039	.0136	.0355	.0727	.1201
	13						.0002	.0010	.0045	.0146	.0366	.0739
	14							.0002	.0012	.0049	.0150	.0370
	15								.0003	.0013	.0049	.0148
	16									.0003	.0013	.0046
	17										.0002	.0011
	18											.0002

ANSWERS TO ODD-NUMBERED

EXERCISES

CHAPTER 1 STATISTICS: RANDOMNESS AND REGULARITY

1.1 Early statistical activities centered around state governments collecting demographic data on their citizens, and statistics was thought of as "matters of state."

1.3 a. Randomness occurs when the value of the next observation cannot be predicted accurately.

b. The patterns found with many observations.

c. Statistics looks through the variability and obtains regularities.

1.5 Probabilities show how often various results occur when a study is repeated. Assuming randomness only, an observed difference may have a small probability of occurring.

1.7 a. A characteristic that can take on different values.

b. Empirical variables have observed data, while theoretical variables are constructed so that they follow certain distributions.

1.9 January 1, January 2, . . . ; Florida, Texas, . . . ; 100 mph, 101 mph, . . . ; $10 million, $11 million, . . . ; 0, 1, 2, ...

1.11 A constant does not change its value when observed several times.

1.13 The Census Bureau and the Bureau of Labor Statistics.

1.15 We never know whether a particular property will be destroyed because of the randomness of the storm. Yet, for a storm of a certain strength, the financial losses are about the same, showing the regularity in the phenomenon.

1.17 a. Computers have made it possible to keep track of a great variety of statistical results.

b. Easier for viewers to make comparisons.

c. Surveys have made it possible to measure the popularity of music, films, politicians, and so on. In turn, the surveys have a great impact on what will occur in the future.

1.19 Student project. d. Shakespeare may have used longer sentences.

1.21 a. Infant mortality has gone down over time. Whites and non-whites have different levels of infant mortality.

b. Same as for infant mortality.

c. The maternal mortality rates since the numerical values are smaller.

d. A death must be reported before it can become part of a table. The further back in time, the fewer deaths were reported.

e. Older data may be more inaccurate. Birth data may be more inaccurate, as not all births are reported.

1.23 Student project.

CHAPTER 2 COLLECTION OF DATA

2.1 Should the population of students include students in two-year and four-year colleges? Should part-time students be included? Should both female and male students be included?

2.3 The elements that are not exposed to the experimental treatment and that are measured on the response variable.

2.5 It can be controlled, and most of the time we can find how large it is.

2.7 a. Assign to each student a number and use a table of random numbers to chose the required number of students. From a list of all students use a random start and select every kth student.

b. The population list may not be complete. There could be refusals to respond or wrong answers.

c. Biased results.

d. Spend time and energy seeking out and getting cooperation from those selected and asking the questions as well as possible.

2.9 No. Sampling error cannot be avoided. It refers to how much variability there would be in the results if the study is repeated many times.

2.11 a. The value we would get from data on the entire population.

b. Applies to the entire population.

2.13 Student project.

2.15 a. They did not say they favored the decision. They have no opinion or are opposed or do not know about the decision.

b. Is it a proper random sample? How many people answered the question? How was the question formulated? Where in the questionnaire was the question located?

c. In 19 out of 20 repetitions of this study, the sample percentage lies within 2 percentage points of the true population value. If this sample is one of the 19, then the true population percentage lies between 54% and 58%.

2.17 The response could have been created by the stimulus or some other factor.

2.19 a. Impossible.

b. It could lead the respondents to answer in ways that may not represent how they feel.

2.21 A table that displays the data collected for a study.

2.23 Given the many problems with data collection, it may be better to think of bad and worse data.

2.25 a. No. We cannot generalize from these data. Among those who chose to participate in this survey, the Philadelphia/Trenton participants had a lower percentage than observed nationally.

b. The sample is not a random sample. It is self-selected.

2.27 a. The survey results apply only to firefighters in one part of the county, and the results cannot be generalized to the entire county.

b. Zero.

2.29 Other services may have higher percentages. On what basis was the percentage computed and what types of people does it apply to?

2.31 Experimenting on people without their consent is always troublesome.

2.33 a. Not a random sample.

b. Asking for volunteers makes the sample nonrandom.

c. No. The women did not volunteer as much.

d. May be well received by the men.

e. Lack of statistical knowledge.

2.35 a. The actual numbers.

b. Same.

2.37 Salary for what time period? Quality of the movie may not fit a three-point scale. Two questions on how good the movie was. Compared to what movie? May not have seen *Some Like It Hot*. Repeat question about what we like best. Imprecise car question: sedan versus

Table A.1 Gender and life expectancy table for Exercise 2.41a

Year	Gender	Life expectancy
1789	m	34.5
1789	f	36.5
1850	m	38.3
1850	f	40.5
1890	m	42.5
1890	f	46.0
1910	m	54.0
1910	f	56.6
1930	m	59.3
1930	f	62.6

station wagon or other type, year? Does not say which direction the scales run.

2.39 Student project.

2.41 a. Table A.1.

b. Life expectancy has increased. Women have higher life expectancies than men through the entire period.

CHAPTER 3 DESCRIPTION OF DATA: GRAPHS AND TABLES

3.1 Gain in simplicity versus loss of information.

3.3 a. A line representing the variable with its scale and each observation marked with a symbol, such as x or $*$.

b. Shows where there are clusters of observations and where there are only a few observations.

c. Not good for many observations.

3.5 a. A. Names of the geographical areas. B. Frequency figures. C. Geographical area. D. About 17,000.

b. As areas of the bars.

c. The areas, and not the heights, of the sailors have to represent the frequencies. Hard to read from such a graph how many sailors are in each area.

d. Most of the troops are in Europe and East Asia/Pacific; only a few are in the Western Hemisphere, Africa, and South Asia.

e. The names of the regions are confusing and could be specified in greater detail.

3.7 By using the proper vertical scale, the bars do not have to be very large even if they represent many observations.

3.9 Unimodal.

3.11 One half of the histogram is not a mirror image of the other.

3.13 a. The population went up and the price index went down from the beginning to the end of the century.

b. It shows how the two variables behave over a century.

c. We could show a scatterplot for the two variables.

3.15 The same as for a graph. Decimals can be dropped. The rows and columns should be arranged so that the larger numbers are toward the upper left corner of the table. Numbers to be compared should be arranged in columns.

3.17 A good graph conveys a message about the data from the person who made the graph.

3.19 a. No loss of information. The original observations can be recovered.

b. When the sample is large.

c. Not many large scores and not many successes.

d. Information on other variables, such as age and gender.

3.21 Student project.

3.23 a. The medians show how the regions differ. The lengths of the boxes show the variation in rates across the regions.

b. Yes. The graph gives a clear and precise view of the data.

3.25 a. Figure 3.8 is more true to the data with the scale starting at 0 inches. Easier to read the exact inches for each year in Figure 3.7.

b. In what way is the graph true to the data? How easy is it to read the graph?

3.27 Stemplots cannot be used for categorical variables since there are no numeric observations.

3.29 Stemplots are good for small samples, and boxplots are good for large samples. Boxplots show five specific numbers that can be compared across several groups of data. A stemplot gives only an overall sense of the center and spread of the data. Stemplots show the original data. The original data cannot be recovered from a boxplot.

3.31 a. "Best" may refer to taste and have little to do with fat.

b. The foods could be ranked by any of the other variables.

3.33 a. Saturday is best, Sunday and Wednesday are worst.

b. They stretch the vertical scale and do not include the origin.

c. The days would look more alike.

d. Hard to know since we know very little about the scale.

e. Things may not seem so bad when looking back.

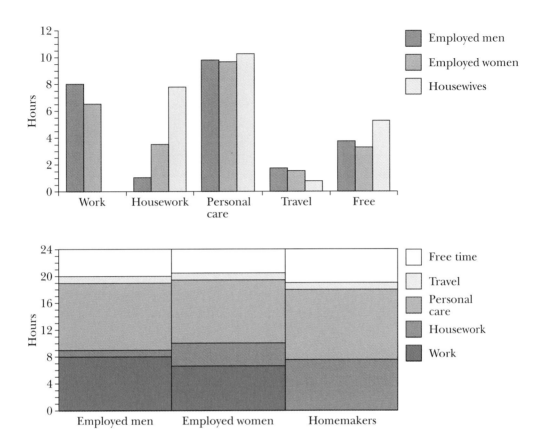

Figure A.1 Graphs for Exercise 3.35a

3.35 a. Figure A.1.

b. Among employed men and women the personal and work time take up most of their time, with some time left over for free time and housework. Homemakers spend most of their time on personal time, with housework and free time as the next two categories.

c. Personal time is approximately the same for all three groups. Employed women work not quite as long as employed men and spend more time on housework.

3.37 a. Figure A.2.

b. Unimodal and almost symmetric.

c. There may be other considerations about when to have children.

d. The total number of mothers in each age group.

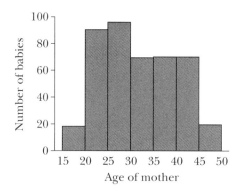

Figure A.2 Histogram for Exercise 3.37a

3.39 a. A bar graph for small, medium, and large will show how many there are of each size.

b. How the 10 sizes relate to the three values small, medium, and large.

3.41 Student project.

3.43 Student project.

3.45 a. Figure A.3.

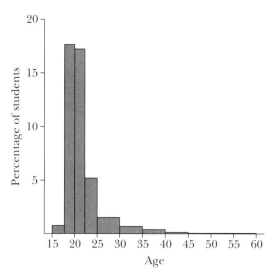

Figure A.3 Histogram for Exercise 3.45a

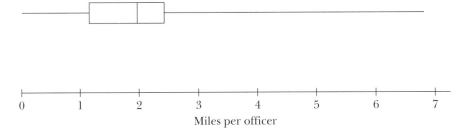

Figure A.4 Boxplot for Exercise 3.47b

b. The skewed shape shows most of the observations between 18 and 25 years.

c. The number of older students is surprising.

3.47 a. Very skewed, with many more small observations than large ones.

b. Figure A.4.

c. The boxplot is simpler and easier to read but does not carry as much information. The boxplot shows the 25th, 50th, and 75th percentiles and the smallest and largest observations.

3.49 a. Figure A.5.

b. The numbers of black and white victims are not known.

3.51 Student project.

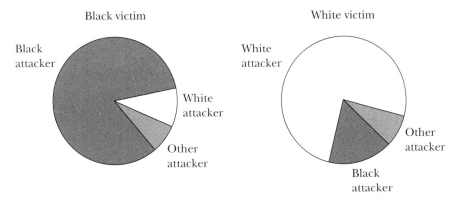

Figure A.5 Pie charts for Exercise 3.49a

Table A.2 New version of Table 3.6 for Exercise 3.53

Type of event	France	Austria	United States	Norway	Italy	Netherlands
Transportation	24	35	23	17	23	18
Natural	31	30	16	25	19	18
Homicide	1	2	10	1	1	1
Other	22	9	11	5	4	4
Total	78	75	61	48	47	40

3.53 Table A.2. Table 3.6 is transposed to compare causes. Countries are arranged by death rate instead of alphabetically. No decimals for ease of comparisons.

CHAPTER 4 DESCRIPTION OF DATA:
COMPUTATION OF SUMMARIES

4.1 Gain in simplicity and little loss of information.

4.3 The mode is the value of the variable that occurs most often. The median is the value of the variable that divides the observations in two groups so that half the observations are smaller than this value and the other half is larger than this value. The mean is the center of gravity of the histogram of the data.

4.5 Height of students in a class, with one mode mainly determined by the women and the other by the men.

4.7 a. Most income distributions are skewed, and the median is a better summary for such distributions.

b. If that were the income for the majority.

4.9 Student project.

4.11 a. Any observation other than the largest and the smallest.

b. The largest and smallest observations.

4.13 s.

4.15 About two thirds.

4.17 The variance.

4.19 A high mean score tells us that the pizza was evaluated to be good. A small standard deviation tells us that most of the scores are close to each other and therefore high.

4.21 A sample has some small and some large observations and their effects on the mean will cancel out. Therefore the mean will not vary as much across different samples as the observations themselves vary.

4.23 a. To see whether an observation is unusual or not. Raw scores on different variables can be compared by changing to standard scores.

b. Student example.

4.25 From -2 to $+2$.

4.27 t(ea)-values.

4.29 The observations in a sample are different because of the randomness of the data, and the standard deviation measures how different the observations are.

4.31 The observations are equal without any variation.

4.33 Income distributions are usually skewed, and the median is better for skewed distributions.

4.35 The mean does not have to be equal to any observed value.

4.37 Divides the data into two equal halves.

4.39 a. Accurate according to the production figures, misleading because the headline does not mention the exclusion of certain types of workers.

b. The rating scale does not say what is high and what is low.

c. It may be harder to measure how much those people produce. Difficult to say how the countries would compare for those types of workers.

4.41 a. The mean for the Braves is less than the mean for the Phillies.

b. The standard deviation for the Phillies is smaller, and they make about the same number of errors in each game. The range of errors for the Braves is larger.

c. The Braves have a large standard deviation.

d. With a mean of 2.0 and a standard deviation of 0.3, there cannot be many observations equal to 0.

4.43 a. Good. The scores in both academic subjects are high.

b. Yes. The score indicates average music understanding.

4.45 Duluth had warmer weather than Hibbing.

4.47 No. You would need the median, the interquartile range, and possibly a histogram showing the distribution of incomes.

4.49 a. Media has a high mean and a small standard deviation such that all the increases are large. Rose Valley has a large standard deviation such that there is greater variability in the increases.

b. Two standard deviations above the mean equals $15,000 in Rose Valley, and that is about the largest increase anywhere.

4.51 Very skewed, with only a few players making a very large salary.

4.53 a. Table A.3.

b. Figure A.6.

c. Mean for batting percentage; median for the others.

d. Some of the distributions are skewed, some have outlier observations.

4.55 a. Use the standard deviation.

b. The group in Exercise 3.34 has $s = 10.9$, and the new group has $s = 14.6$.

Table A.3 Statistics for Exercise 4.53a

Variable	Mean	Median	Standard deviation	Range
Batting percentage	0.262	0.264	0.039	0.228
Number of times at bat	319	290	180	665
Number of runs	47.3	36.5	33.1	140
Number of hits	87.6	76.0	55.6	224
Number of home runs	10.4	6.0	11.0	52
Number of runs batted in	44.8	35.0	34.5	150

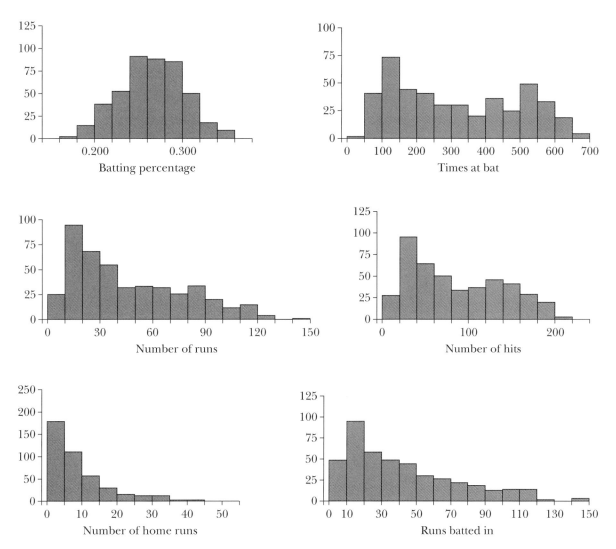

Figure A.6 Frequency histograms for Exercise 4.53b

c. 14.6/10.9 = 1.3; one standard deviation is 30% larger than the other.

4.57 a. 2.18.

b. The standard score is unusually large, and chocolate ice cream is different from the other desserts.

4.59 a. Figure A.7.

686 Answers to Odd-Numbered Exercises

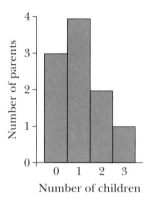

Figure A.7 Histogram for Exercise 4.59a

b. 1.

c. 0 for the men, 1 for the women.

4.61 a. Figure A.8.

b. Mode 77, median 77, mean 80.4.

c. The mode gives the most common value, the median gives the value such that half of the observations are smaller than this value and half are larger. The mean gives the center of gravity in the distribution.

d. The number of medical schools went down in this period.

e. Schools grew in number of students.

4.63 Figure A.9. All five scores have the same value—5.

4.65 a. 48% smoked cigarettes, 32% smoked marijuana, and 38% drank alcohol at least once in the past month.

b. The numbers do not seem to agree.

```
9 | 0 5 6 6

8 | 0 0 0 0 1 3 5 5

7 | 6 6 6 6 6 7 7 7 7 7 7 7 7 7 7 9 9
```

Figure A.8 Stemplot for Exercise 4.61a

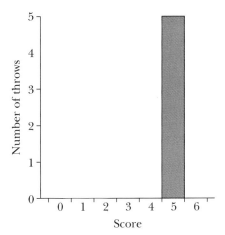

Figure A.9 Histogram for Exercise 4.63

4.67 a. $10.00.

b. $10.36.

c. $10.50.

d. $12.13.

e. The median is not affected by changes in magnitudes as long as observations remain above or below (or equal) the median. The mean is affected by numerical changes in any of the observations.

4.69 a. Mean 21.9, standard deviation 5.2.

b. Nine.

c. Mean equals 71.3, standard deviation 11.2.

d. Seven.

4.71 Brides 27 years, grooms 30 years. The grooms are on the average three years older than the brides.

CHAPTER 5 PROBABILITY

5.1 Likelihood, odds, chances.

5.3 A number between 0 and 1 that describes how often an event occurs. With probability 0 the event never occurs and with 1 the event always occurs.

5.5 If an experiment has n different and equally likely outcomes and we want the probability of k of them occurring, then k/n is the probability.

5.7 $1.00 - 0.12 = 0.88$.

5.9 a. A subjective, personal probability.

b. The event is unique and occurs only once.

5.11 The *two* possibilities.

5.13 a. The average (expected) number of girls in families with 4 children.

b. μ.

5.15 σ.

5.17 a. The normal, t-, chi-square, and F-variables.

b. The t-variable.

5.19 a. The t-distribution.

b. Mr. Gosset used the pseudonym Student because his employer did not permit employees to write scientific papers.

5.21 The degrees of freedom.

5.23 a. No chi-square variable has a negative value.

b. Need to know the number of degrees of freedom first.

5.25 Since getting the sample data or more extreme data is so unlikely if the electorate is evenly split, we conclude that the electorate is not evenly split.

5.27 The statement takes off on a similar statement about love, and a statistician almost never finds a probability of anything being equal to 1.

5.29 Student project.

5.31 Trudi is the surest thing, and Rod is the big killing.

5.33 Student project.

5.35 There are two outcomes with constant probabilities and n observed independent events.

5.37 It is better to use the binomial distribution. Many samples of 100 families with 4 children would yield the correct answer. The data from a single sample have a sampling error attached.

5.39 Find the probabilities of 0, 1, 2, and 3 girls to find the mean. Same answer as $\mu = np = 3 * 0.49 = 1.47$.

5.41 9 of 1,000 children would break a bone; half the undergraduates would be involved in an accident; 1 in 100,000 BMWs would be carjacked.

5.43 These data belong to a set of data that have a high probability, so there is nothing unusual about the data.

5.45 a. The probability of getting all new members in the clean-up crew is very, very small if the drawing were truly random. Since the crew consists of all new members, it is hard to imagine that the drawing was random.

b. Table A.4. It seems strange that nobody was chosen from the old members, while almost all the new members were chosen.

5.47 a. 2 among 100,000 people would be hit by lightning.

b. It is easier to understand a frequency such as 2 in 100,000 than a very small number such as 0.00002.

5.49 a. Assess which charity has the greatest probability of having an impact.

b. They are not always entirely objective.

5.51 The data could have come from populations with equal means.

5.53 a. $\frac{1}{6}$.

b. $\frac{1}{6}$.

c. $\frac{1}{6} + \frac{1}{6} = \frac{2}{6}$.

Table A.4 Distribution for Exercise 5.45b

		Member		
		Old	New	Total
Member	Chosen	0	5	5
	Not chosen	52	2	54
	Total	52	7	59

5.55 a. $\frac{2}{36}$.

b. $\left(\frac{2}{36}\right)\left(\frac{1}{35}\right) = \frac{2}{1260} = \frac{1}{630}$.

5.57 Student project.

5.59 Student project.

5.61 a. $(7 + 1)/256 = 0.031$.

b. It is so small that 0.5 may not be the correct probability.

5.63 a. 0.034.

b. 0.114.

5.65 a. 0.5.

b. $0.5^{10} = \frac{1}{1024} = 0.001$.

c. The income distribution is skewed, and the mean would be unrepresentably large.

5.67 $0.33^4 = 0.012$.

5.69 a. $(0.25)(0.08) = 0.02$.

b. Intuitively it seems that two things happening at the same time ought to have a smaller probability than only one or the other happening.

c. Thunder and lightning are not independent events, and the probabilities cannot simply be multiplied.

5.71 a. $0.07 + 0.29 = 0.36$.

b. $(0.07)(0.29) = 0.02$.

c. $(1 - 0.07)(1 - 0.29) = 0.66$.

5.73 $(0.05)(0.10) = 0.005$.

5.75 a. $2.5\% + 2.5\% = 5.0\%$.

b. 2.5%.

c. 47.5%.

5.77 The probability that a new member is chosen first is $\frac{7}{52}$. The probability that the next person is new is $\frac{6}{51}$. The remaining probabilities become $\frac{5}{50}$, $\frac{4}{49}$, and $\frac{3}{48}$. The product of these probabilities gives the overall probability that all five are new members.

Figure A.10 Histogram for Exercise 5.79a

5.79 a. Figure A.10.
b. $\mu = 0(0.05) + 1(0.19) + 2(0.31) + 3(0.28) + 4(0.14) + 5(0.04) + 6(0.001) = 2.42$.
c. $6(0.40) = 2.40$.
d. Yes.

CHAPTER 6 ESTIMATION

6.1 To learn about values of parameters in a population.

6.3 Parameter.

6.5 The sample mean, percentage, and standard deviation are statistics; the population mean, percentage, and standard deviation are parameters.

6.7 From many different samples from the same population, compute a statistic from each sample, and the mean of these statistics equals the population parameter.

6.9 a. $10 - 2 = 8$ to $10 + 2 = 12$.

b. This is one of many possible confidence intervals from different samples, and hopefully it is one of the 95% of all confidence intervals that contain the true population mean.

6.11 A large sample, a low degree of confidence.

6.13 We can generalize from the data in the sample to the entire population without having to measure every element in the population.

6.15 a. The standard deviation of the sample means (the standard error of the means) will be smaller than the standard deviation of the sample medians (the standard error of the medians).

b. A small standard deviation tells us that the observations are not very different from each other and that they are all close to the mean. When an estimate has a small standard deviation (standard error), then each statistic is close to the value of the population parameter.

6.17 Because we cannot find the parameter value, we use sample data to estimate the parameter.

6.19 a. The tanks must be distributed randomly to the various battlefields and captured randomly.

b. If the latest batch of tanks were sent to a particular front and many tanks were captured there, the estimate would be too high. Similarly, if older tanks were sent to a particular front and not many tanks were captured there, the estimate would be too low.

6.21 A high degree of confidence for the interval.

6.23 a. They could be opposed or not have any opinion.

b. Whether the sample was drawn in a proper way, how many people responded to the survey, whether the questioning was done over the telephone or face to face, how the question was formulated, where in the interview the question was placed, etc.

c. The confidence interval goes from $56\% - 2\% = 54\%$ to $56\% + 2\% = 58\%$. This interval is constructed according to a method such that with many samples and therefore many intervals, 95% of these intervals would contain the true population percentage and 5% of the intervals would not. We do not know if this particular interval is one of the many or one of the flukey few.

6.25 Not true. Hopefully, the interval from 70 to 75 is one of the many intervals that contain the population percentage, but it may be one of the few that do not contain the true value.

6.27 a. A sample of students.

b. A population of clarinet players.

c. A population of voters.

d. A sample of war deaths.

6.29 a. The confidence interval from 18 to 38 is constructed according to a method such that if we had many different intervals from different samples, then 95% of all these intervals would contain the true difference and 5% of the intervals would not contain the difference. We do not know whether this particular interval is one of the few or one of the many.

b. Since the confidence interval for the difference does not contain 0, it does not seem as if the two percentages are equal in the population.

6.31 a. This interval from 6 to 18 is constructed according to a method such that if we had many intervals from many different samples, then 95% of all the intervals would contain the true difference and 5% of the intervals would not. Whether this particular interval is one of the many or one of the few, we do not know.

b. Since the interval does not contain the value 0, it does not look as if the difference could be 0 and that the two percentages could be equal.

6.33 a. This interval from 5.22 to 6.70 is constructed according to a method such that if we had many intervals from many different samples, then 95% of all the intervals would contain the true difference and 5% of the intervals would not. Whether this particular interval is one of the many or one of the few, we do not know.

b. Since the confidence interval goes from 5.22 to 6.70, the claim of a mean flow of at least 5.00 seems justified.

6.35 a. This interval from 0.90 to 1.28 is constructed according to a method such that if we had many intervals from many different samples, then 95% of all the intervals would contain the true difference and 5% of the intervals would not. Whether this particular interval is one of the many or one of the few, we do not know.

b. It could be that these data come from a population where the mean equals 1.00 ppm, since that value is included within the confidence interval.

6.37 Most confidence intervals are found by computing the sampling error and finding the sample statistic plus and minus the sampling error.

6.39 a. Each poll represents a different sample. It is not surprising that different samples give different results.

b. The samples did not have the same number of observations.

c. 38 to 46, 35 to 43, 38 to 44, 36 to 44; Figure A.11.

d. 38 to 43.

e. Since the sampling error is approximately equal to $100/\sqrt{n}$, there are about 600 people in the polls with a 4% sampling error and 1,100 people in the poll with a 3% sampling error.

f. The total Bush vote becomes $(0.42)(600) + (0.39)(600) + (0.41)(1,100) + (0.40)(600) \approx 1,170$. The total number in the four polls becomes approximately 2,900. His percentage becomes $1,170/2,900 = 40$ and with sampling error of $100/\sqrt{2,900} = 2$.

g. 38 to 42.

h. It compares well, and the differences may well result from rounding errors.

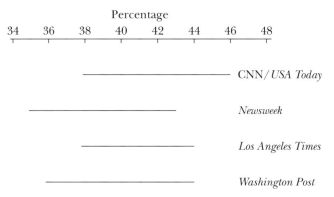

Figure A.11 Confidence intervals for Exercise 6.39c

6.41 a. $7.0 \pm (2.01)(2.5/\sqrt{49}) = 6.3$ to 7.7.

b. This interval from 6.3 days to 7.7 days is constructed according to a method such that if we had many intervals from many different samples, then 95% of all the intervals would contain the population mean number of days and 5% of the intervals would not. Whether this particular interval is one of the many or one of the few, we do not know.

6.43 a. $12.3 \pm (2.086)(0.3) = 11.7$ years to 12.9 years.

b. We are 95% confident that the population mean lies between 11.7 years and 12.9 years.

6.45 a. $56 \pm 1.96 \sqrt{\dfrac{61 * 39}{41} + \dfrac{5 * 95}{807}} = 56 \pm 15 = 41$ to 71.

b. This interval is constructed according to a method such that if we had many intervals from many different samples, then 95% of all the intervals would contain the population value of the difference between the two percentages and 5% of the intervals would not contain the population difference.

c. The interval does not contain 0, and it seems as if the two percentages are different in the populations of young smokers.

6.47 a. $61 \pm 100/\sqrt{531} = 57$ to 65. $33 \pm 100/\sqrt{531} = 29$ to 37.

b. The number of companies that expected improved customer service is larger than the number that found this to have been the case.

6.49 a. $0.56 * 502 = 281$ "morning people" and $0.44 * 502 = 221$ "night owls."

b. $53 \pm 100/\sqrt{281} = 53 \pm 6.0 = 57.0$ to 59.0;
$39 \pm 100/\sqrt{221} = 39 \pm 6.7 = 32.3$ to 45.7.

c. $45 \pm 6.0 = 39.0$ to 51.0; $37 \pm 6.7 = 30.3$ to 43.7.

d. $74 \pm 6.0 = 68.0$ to 80.0; $64 \pm 6.7 = 57.3$ to 70.3.

e. In b the confidence intervals do not overlap, indicating that the two groups are different; in c and d the intervals overlap, indicating that the two groups do overlap.

f. When we use a sample size of 502 the sampling error becomes 4.4.

g. Using 4.4 makes a difference in d.

6.51 a. $3.96 \pm 2.064(0.35) = 3.96 \pm 0.72 = 3.24$ to 4.68.

b. The interval does not contain the value 5, which implies that 5 is not a possible value of the mean number of heads when spinning a coin 10 times.

CHAPTER 7 HYPOTHESIS TESTING

7.1 The sample data lead to rejection of a null hypothesis.

7.3 a. Whether a parameter is equal to a particular value.

b. The alternative hypothesis asks whether the parameter is equal to all possible values not specified in the null hypothesis.

c. H_0 and H_a.

7.5 We reject a null hypothesis when the sample statistic is very different from the value of the parameter specified in the null hypothesis.

7.7 a. The probability of getting the observed data or more extreme data from a population when the null hypothesis is true.

b. The data belong to a very unlikely set of data coming from the population where the null hypothesis is true. Since we do not believe the data are particularly unlikely, the only explanation is that the data come from another population. Thus, we reject the null hypothesis.

c. A significance level is a small probability chosen by us before we look at the data. A p-value is computed from the data.

7.9 a. 0.05.

b. Among all the true hypotheses tested, 5% of them will erroneously be rejected.

7.11 a. α.

b. Alpha.

7.13 a. $H_0: \pi = 0.5$.

b. $H_0: \Pi = 50\%$.

7.15 The significance level and the sizes of the two samples.

7.17 a. There is no difference in the population mean scores for the two groups.

b. The mean score for the outdoor group is larger than the mean score for the indoor group.

7.19 a. The population mean salary in 1995 equals $43,000—a hypothesis stating that there has been no change.

b. Because of inflation and possible salary raises, the mean salary in 1995 may be higher than in 1985.

7.21 The probability of getting our data or more extreme data is 0.50, given the null hypothesis.

7.23 In a one-tailed test the alternative states that the parameter is larger (smaller) than a given value. In a two-tailed test the alternative hypothesis states that the parameter is different from a given value.

7.25 a. Not every jar could weigh exactly 18 oz.

b. The weight of the next jar would be different.

c. Buy a random sample of jars and find the mean weight to use in a test of the null hypothesis that the population mean equals 18 oz.

7.27 a. Because of the small *p*-value.

b. The difference between the national value of 53% and the sample value of 45% is not very large.

7.29 The null hypothesis often states that there is no difference or there has been no change.

7.31 a. The probability is 0.025 of getting our data or more extreme data from a population where the parameter has the value specified in the null hypothesis.

b. Because the *p*-value is so small, we reject the null hypothesis.

7.33 a. We reject the null hypothesis of equal mean goodness because we do not believe our data belong to such an unusual set of possible data.

b. It could be that the null hypothesis is correct and that we got some very unusual data in our particular sample. The probability of the observed data or more extreme data equals 0.001.

c. A difference of 0.4 point on a 7-point scale may not be very interesting.

7.35 When the null hypothesis is true, then 5% of the samples will be so extreme that we reject the null hypothesis. 5% equals 1 in 20, a small number according to Fisher.

7.37 A result can be statistically significant and at the same time substantively very trivial.

7.39 a. Is the population mean speed equal to 5.0? A sample has mean speed of 5.96. This translates to a *t*-value of 2.49 with 47 degrees of freedom. We get this or a larger value of *t* only 49 out of 10,000 times in samples from a population where the mean equals 5.0.

b. We do not think our sample came from that population, and we reject the null hypothesis. The population mean is larger than 5.0.

7.41 a. The difference between the two perceived discrimination means 17.3 and 11.4 is statistically significant. The probability of getting this observed difference or more from a population where the means are equal is less than 0.001. The difference between the mean concerns about starting a new family is statistically significant, with *p*-value less than 0.05.

b. The women who had experienced discrimination had higher depression scores.

7.43 We could test the null hypothesis that the percentage of support for the president is 50, and we might not be able to reject it.

7.45 a. They vary around 1.4 because of the random variation from one sample to the next.

b. $t = (2.0 - 1.4)/0.5 = 1.2$.

c. It is not unusually large, and our sample could have come from a population where the mean equals 1.4.

7.47 a. $H_0: \mu = 5.1$ days.

b. $t = (7.0 - 5.1)/(2.5/\sqrt{49}) = 5.32$ with 48 d.f., $p < 0.001$.

c. It is very unlikely to get a sample mean of 7.0 or more if the population mean equals 5.1; reject the null hypothesis. The workers are sick more often.

7.49 a. $H_0: \Pi = 73.2\%$.

b. $z = (67 - 73.2)/\sqrt{67 * 33/300} = 2.18$ and $p = 0.015$. This

p-value is so small that we reject the null hypothesis. Significantly fewer city employees drive alone.

c. The reduction does not seem very large.

7.51 H_0: $\pi = 0.5$, where π is the probability that a randomly chosen person is a Baptist. $z = (103 - 0.5 * 190)/\sqrt{190 * 0.5 * (1 - 0.5)} = 1.16$. Since $z < 1.96$, we cannot reject the null hypothesis. The number of Baptists and Methodists in the population may be the same.

7.53 $z = (98 - 80)/\sqrt{98 * 2/50 + 80 * 20/200} = 6.74$, $p < 0.0001$. Reject the null hypothesis. Since the null hypothesis states that the two percentages are equal in the population, an estimate of this common value is the total number of Democrats divided by the total number of car owners, $209/250 = 83.6\%$. When we use this common percentage in the denominator, then $z = 3.07$ with $p = 0.001$.

7.55 H_0: $\Pi = 50\%$. $z = (134 - 200 * 0.5)/\sqrt{200 * 0.5 * (1 - 0.5)} = 4.81$, $p < 0.001$; reject the null hypothesis. More men ride the roller coaster.

7.57 The binomial distribution with $\pi = 0.5$ and $n = 10$ gives $p = (120 + 45 + 10 + 1)/1{,}024 = 176/1{,}024 = 0.17$. We cannot reject the null hypothesis that $\pi = 0.5$, and it may be that the probability is the same for girls and boys.

7.59 $z = \dfrac{77.8 - 44.4}{\sqrt{61.1(100 - 61.1)} \sqrt{\frac{1}{36} + \frac{1}{36}}} = -2.91$

with $p = 0.0018$. Reject the null hypothesis; there is a difference between female and male college students.

CHAPTER 8 RELATIONSHIPS BETWEEN VARIABLES

8.1 Values of one variable correspond to certain values of the other variable.

8.3 There may be no relationship between the two variables in the population from which the data came.

8.5 Is the relationship statistically significant?

8.7 When it is not causal and can be explained away by one or more other variables.

8.9 a. Could the relationship have been produced by another variable?

b. We know enough about the two variables to conclude that temperature could be another variable producing the observed relationship.

c. It might be hard to identify other variables that might have produced the observed relationship.

d. Historically there have been many relationships that were thought to be causal. For example, witches produced illnesses.

8.11 0.17 indicates a weak relationship, 0.87 indicates a strong relationship.

8.13 The relationship exists not only in the sample but also in the population from which the sample came.

8.15 False. We can always predict from one variable to another without the variables necessarily being causally related.

8.17 The best would be knowing that one variable actually causes another. Often the independent variable is the one that comes first in time.

8.19 There are different statistical methods for different types of variables, and we need to determine the variable types before choosing the proper method.

8.21 a. X is continent and Y is country GDP.

b. Mean number of runs per game and number of games won in a season for baseball teams.

c. Ranking of the Big Ten football teams one week and the next week.

8.23 Multivariate statistical methods.

8.25 A relationship can be spurious or causal. The data on the two variables alone cannot be used to determine whether the relationship is spurious or causal.

8.27 With observational data it is very difficult to prove causality, and it can be done only if all other possible variables are brought in to see what happens to the observed relationship. *All* other possible variables can never be considered.

8.29 Other variables must be brought in to see if they can account for the observed relationship.

8.31 Smoking and size of baby may be a spurious or a causal relationship. Without information on other variables, or running an experiment, you cannot conclude what type of relationship it is. There may be other variables that make women smoke and also make them have smaller babies.

8.33 a. They must have asked women questions about how the women felt about their bodies. The conflicting percentages may have something to do with the formulations of the questions.

b. It may be spurious. It would be hard to prove that the relationship was causal.

8.35 a. Table A.5.

b. 214 of 665 is 32%.

c. 175 of 772 is 23%.

d. 32 − 23 = 9 tells what percentage more were bald.

e. Baldness is the independent variable, and we compare percentages for categories of the independent variable.

f. The people are self-selected, since only those with heart attacks came to the hospital. It would have been preferable to follow one group of bald men and another group that were not bald and see how many in each group had heart attacks.

g. Nothing.

8.37 a. The choices of brands is different.

Table A.5 Distribution for Exercise 8.35a

		Bald		
		Yes	No	Total
Ailment	Heart	214	451	665
	Other	175	597	772
	Total	389	1048	1437

Table A.6 Percentage distributions for Exercise 8.37b

		Race		
		Black	White	Total
Brand	Marlboro	9	71	68
	Newport	56	6	8
	Kool	9	1	1
	Other	27	22	24
	Total	101	100	101

b. Table A.6. Most blacks prefer Newports, most whites prefer Marlboros.

c. The percentages are very different, indicating a fairly strong relationship.

d. The choices may have something to do with how the brands are advertised.

8.39 a. There are more higher percentages in the at home row.

b. We would make the wrong prediction for the last three years, for 52%, 58%, and 58%.

c. Most women stayed at home that year.

d. 32%.

e. She is in the labor force.

f. 42% of the time.

g. For each year we could make a more accurate prediction.

h. There has been an increase of mothers in the labor force over this time.

i. We cannot tell about causality from these data alone.

CHAPTER 9 CHI-SQUARE ANALYSIS FOR TWO CATEGORICAL VARIABLES

9.1 Student project.

9.3 A table with rows and columns used to display frequencies.

9.5 Phi.

9.7 The size of chi-square has nothing to do with causality. It is only used to find the p-value.

9.9 a. Yes. The two columns are different.

b. 0.62, moderate.

c. No, the null hypothesis is rejected because of the small p-value.

d. Maybe.

9.11 The relationship is very weak. It could have been produced by chance alone.

9.13 a. Yes. The column percentages are different.

b. Weak.

c. Yes, small p-value.

d. We cannot tell about causality from these data alone.

9.15 a. Baldness, since it comes first in time.

b. A relationship exists in these data, but it is weak. The null hypothesis of no relationship in the population is rejected. We cannot tell about causality.

9.17 a. Yes. The column percentages are different.

b. The relationship seems weak.

c. Statistically significant.

9.19 a. There are more and more no answers with increased schooling. More children have an opinion with increased schooling.

b. There is a weak but statistically very significant relationship.

9.21 a. Table A.7.

Table A.7 Distribution for Exercise 9.21a

		Sentence		
		Course	Jail	Total
New crimes	Yes	6	18	24
	No	26	22	48
	Total	32	40	72

Table A.8 Distribution for Exercise 9.25a

		Gender		
		Men	Women	Total
Elected	Yes	36	13	49
	No	41	7	48
	Total	77	20	97

   b. There is a weak but statistically significant relationship.

   c. It is not clear that the sample is a proper random sample.

9.23 Most delinquent boys do not wear glasses and most nondelinquent boys wear glasses, so there is relationship between the two variables in these data. It is moderately strong, and we reject the null hypothesis of no relationship in the population of all boys.

9.25 a. Table A.8.

   b. There is a relationship in these data. It is weak and not statistically significant. You cannot conclude anything about causality.

9.27 There is a relationship in these data. It is moderately strong and statistically significant. A p-value would determine how significant it is. From these data alone you cannot conclude anything about causality.

9.29 a. Table A.9.

   b. The two columns are different, and there is a relationship in these data.

   c. Chi-square $= \dfrac{366(73 * 74 - 109 * 110)^2}{183 * 183 * 182 * 184}$

Table A.9 Distribution for Exercise 9.29a

		Birthday		
		January–June	July–December	Total
Draft number	1–183	91	92	183
	186–366	91	92	183
	Total	182	184	366

Table A.10 Distribution for Exercise 9.31a

		IQ		
		Low	High	Total
Crimes	0, 1	806	1158	1964
	2+	382	161	543
	Total	1188	1319	2507

d. Reject the null hypothesis that chance alone produced the observed table. The lottery does not seem random.

9.31 a. Table A.10.

b. Yes. The columns of percentages will not be the same.

c. Phi = 0.24.

d. Chi-square = 146.59 on 1 d.f. and $p < 0.0001$. Reject the null hypothesis of no relationship in the population.

e. Among the low IQ men there are more with 2 or more crimes than among the high IQ men. We do not know from these data alone whether the relationship is causal.

9.33 a, b. Table A.11.

c. Yes. The two percentage distributions are different.

d. $V = 0.33$.

e. Student project.

Table A.11 Frequencies and percentages for Exercise 9.33a, b

		Location				N.A.	S.A.
		N.A.	S.A.	Total			
Religion	Catholic	190	310	500		38%	69%
	Jew	10	10	20		2	2
	Protestant	120	30	150		24	7
	Other	180	100	280		36	22
	Total	500	450	950		100%	100%

f. Chi-square = 103.31 on 3 d.f. and $p < 0.0001$. Reject the null hypothesis of no relationship in the population.

g. Chi-square = 103.27 on 2 d.f. and $p < 0.0001$. Still reject the null hypothesis.

h. The difference between the two chi-squares equals 0.04 and 1 d.f.

9.35 a. Yes. It looks as if the percentage columns would be different.

b. Weak.

c. Chi-square = 82.07 on 3 d.f. and $p < 0.0001$. Reject the null hypothesis of no relationship between the variables in the population of all registered voters.

d. $V = 0.14$.

9.37 a. Yes.

b. $V = 0.34$.

c. Chi-square = 17.64 on 2 d.f. and $p = 0.0001$. Reject the null hypothesis that these data could have occurred by chance alone.

d. There are more women in the lower ranks because there are now more women working as college faculty and they have not reached the top rank yet.

9.39 a. Yes.

b. $V = 0.30$.

c. Chi-square = 11.69 on 2 d.f. and $p = 0.003$. The relationship could not have occurred by chance alone.

d. Proportionally, East Germany's women got the most medals and the Soviet Union's women got the fewest medals.

9.41 There is a relationship in these data. Phi = 0.21. Chi-square = 4.63 on 1 d.f. and $p = 0.031$. The relationship is statistically significant.

9.43 $(3 - 1)(4 - 1) = 6$ d.f.

9.45 Table A.12. Phi = 0.19. Chi-square = 9.45 on 1 d.f. and $p = 0.002$.

9.47 There is a relationship. $V = 0.46$. Chi-square = 181.97 on 1 d.f. and $p < 0.0001$. Reject the null hypothesis of no relationship. You cannot tell about causality from these data alone.

9.49 a. Table A.13.

Table A.12 Distribution for Exercise 9.45

		Car		
		Saab	Volvo	Total
Vote	Democrat	49	160	209
	Republican	1	40	41
	Total	50	200	250

Table A.13 Distribution for Exercise 9.49a

		Proteinuria		
		Yes	No	Total
Hypertension	Yes	28	21	49
	No	82	286	368
	Total	110	307	417

b. There is a relationship in these data. Phi = 0.25. Chi-square = 27.06 on 1 d.f. and $p < 0.0001$. Reject the null hypothesis of no relationship in the population.

9.51 There is a relationship in these data. $V = 0.26$. Chi-square = 75.89 on 6 d.f. and $p < 0.0001$. Reject the null hypothesis of no relationship in the larger population.

9.53 a. They are different.

b. Chi-square = 5.61 on 5 d.f. and $p = 0.35$. $V = 0.14$. The relationship is weak and not significant.

9.55 a. There is a relationship in these data. Phi = 0.12. Chi-square = 1.31 on 1 d.f. with $p = 0.29$.

b. Here there should be no difference because of the alphabetic division.

9.57 The percentages are reported for the rows. This implies that whether or not the nuns developed Alzheimer's disease is the independent variable. Since their linguistic ability was measured first, it makes sense to use linguistic ability as the independent variable.

708 Answers to Odd-Numbered Exercises

b. There is a relationship in these data. Phi = 0.76. Chi-square = 14.31 on 1 d.f. with $p = 0.0002$. The null hypothesis is rejected. You cannot tell about causality.

9.59 Student project.

CHAPTER 10   REGRESSION AND CORRELATION FOR TWO METRIC VARIABLES

10.1 A measure of strength of the relationship between two metric variables, indicating how close the points in the scatterplot are to the regression line.

10.3 One independent variable.

10.5 a. Student project.

b. Slope and intercept.

10.7 a. The independent variable along the x-axis and the dependent variable along the y-axis.

b. Number of commercials.

c. The independent variable.

d. The points fall in a scatter from the lower left corner to the upper right corner of the graph.

10.9 Isolated points have a large impact. The correlation coefficient would increase.

10.11 All the points fall on a (45-degree) line.

10.13 a. +1, if one believes that wealthier people should pay more tax.

b. −1. As education goes up, illiteracy should go down.

c. 0. There should be no correlation. (There may be a positive correlation in our society.)

10.15 a. 36.1 calories.

b. This is an extrapolation from the observed data. If the relationship follows the same line, then a food with no fat is predicted to have 36.1 calories.

10.17 $\hat{y}$.

10.19 a. The regression sum of squares and the residual sum of squares.

b. The regression sum of squares is found by squaring the distance from the regression line to the mean line for each observation and adding the squares. The residual sum of squares is found by squaring the distance from the observed point to the regression line for each observation and adding the squares.

10.21 a. All the effect on the dependent variable comes from the independent variable.

b. No effect.

c. Plus or minus one.

10.23 a, b. Regression gives the form of the relationship, while correlation gives the strength of the relationship. Both are needed.

10.25 Positive.

10.27 Use the regression or the correlation coefficient.

10.29 a. The correlation coefficient becomes larger, given that everything else stays the same.

b. A larger correlation coefficient gives a smaller p-value.

10.31 a. When the dependent variable is a dummy variable and the independent variable is a metric variable.

b. Purchase of a Chevrolet or a Cadillac as a function of income.

10.33 a. A horizontal with an intercept of 183 and slope of 0. The correlation coefficient is expected to equal 0.

b. The line has a negative slope, meaning that early months have a larger mean draft number and late months have a smaller mean draft number. The correlation is strong between the two variables.

c. A t-value of that magnitude has a small p-value. Reject the null hypothesis that the relationship between the variables is due to chance only.

10.35 a. The intercept equals the mean of the chocolate desserts. The slope equals the difference between the two means.

b. No, t is too small to be statistically significant.

10.37 a. The more people who have been to college and therefore have higher incomes, the fewer poor people there would be.

b. No. The p-value is so small that we reject the null hypothesis that the relationship occurred by chance alone.

c. We cannot tell from these data.

d. Washington, D.C., is unusual because as the nation's capital it has many people with college degrees working for the government, while at the same time it has many poor people.

10.39 a. Sam's line will be steeper.

b. Only for Sam's data.

c. Yes, it would seem so. It may depend on the correlation coefficients between $Y1$ and X and between $Y2$ and X.

10.41 a. The higher the dose, the more animals responded.

b. With a one-unit increase in dosage, the proportion responding increases by 0.13. The relationship is statistically significant. Reject the null hypothesis that the two variables are not related in the larger population.

c. You do not know the strength of the relationship, and you do not know about causality.

10.43 a. A decrease of 3.47 seconds.

b. A decrease of 13.525 seconds.

c. His time was $3:44.39 - 3:43.82 = 0.57$ seconds slower than predicted.

d. $76.07 - 0.3468 * 93 = 43.82$ or 3:43.82 minutes.

10.45 a. There is a strong positive relationship. Two ships that differ by one ton will on the average differ by 0.00062 crew members. The relationship is statistically significant, and there is every reason to think that the relationship is causal.

b. 6.2 crew members.

c. When tonnage = 192, predicted crew size becomes 10.7 men. When tonnage = 3246, predicted crew size equals 29.7 men.

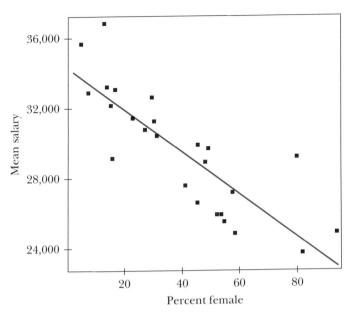

Figure A.12 Scatterplot and regression line for Exercise 10.47a

10.47   a. Figure A.12.

b. It is negative, strong, and statistically significant.

c. Faculty ranks, not all schools may have faculty in all the professions, etc.

d. The points above the line represent more service-oriented professions, while the professions below the line are more academically oriented professions.

10.49   a. $r = 0.40$.

b. $t = 2.23$ on 26 d.f. and $p = 0.017$. Significant.

c. Number of wins = $45.8 + 7.0$ mean number of runs.

d. One more run per game can be expected to produce 7 more wins, on the average.

10.51   a. Road distance = $72.6 + 1.11$ direct distance.

b. If the distances are the same, the points will lie on a line with slope 1 through the origin.

c. $t = 1.40$ on 13 d.f., so $p = 0.09$. You cannot reject the null hypothesis that the intercept equals 0.

d. The United States has many east-west and north-south roads, while roads in England tend to connect cities more directly.

10.53 a, b.

Regression	Correlation
number of wins = 118 − 8.15 ERA	−0.54
number of wins = 45.8 + 6.98 number runs	0.40
number of wins = 71.2 + 0.056 number of home runs	0.22
number of wins = −16 + 360 batting average	0.46

Improving (lowering) the ERA by one run would add 8.5 more wins. Scoring one more run per game would lead to 7 more wins. Scoring 100 more home runs in the season would lead to 5.6 more wins. Improving the batting average by 0.1 would lead to 36 more wins.

10.55 a. Mean for batting percentage; median for the others.

b.
number of hits = −9.0 + 0.30 number of times at bat	$r = 0.98$
number of runs = −2.2 + 0.57 number of hits	$r = 0.95$
number of runs = −7.1 + 0.17 number of times at bat	$r = 0.93$
number of home runs = −2.7 + 0.29 number of runs batted in	$r = 0.91$
number of home runs = −21.1 + 120.1 batting percentage	$r = 0.42$

These are the regression equations for the four highest and the lowest correlation coefficient.

10.57 The scatterplot looks linear, with a positive relationship. A regression analysis gives

$$\text{age of chief} = -51.2 + 2.7 \text{ mean age of members}$$

For two groups where the mean ages of the members differ by one year, on the average the ages of the chiefs differ by 2.7 years. The strength of the relationship is $r = 0.70$. A test for the null hypothesis that the relationship occurred by chance alone gives $t = 2.62$ on 7 d.f. and $p = 0.03$. This p-value is borderline significant, mainly because the sample is small.

10.59 a. Student project.

b. Student project.

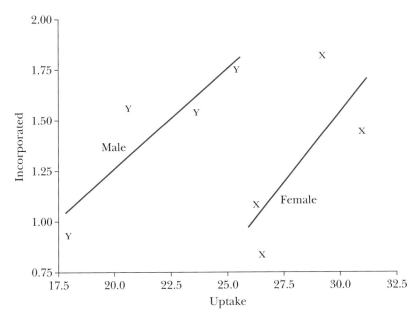

Figure A.13 Scatterplots and regression lines for Exercise 10.61a

c. Student project.

d. Gives an approximate fit for every person.

e. The predicted graduating GPA is $0.6 + 0.74 * 1.9 = 2.0$; it looks as if Elmer will make it.

f. The equation would stay the same, and the correlation would be weaker.

10.61 a. Figure A.13.

b. Both relationships are positive. The male relationship has a smaller slope than the female relationship. The correlations seems to be about the same.

c. Female: incorporated $= -2.70 + 0.14$ uptake
   Male: incorporated $= -0.67 + 0.10$ uptake

d. The line for the female rats is steeper than the line for the male rats, and the line for the females therefore has a smaller intercept than the line for the male rats.

e. The predicted values are $-2.70 + 0.14 * 25 = 0.80$ and $-0.67 + 0.10 * 25 = 1.83$, so the difference becomes $1.83 - 0.80 = 1.03$.

f. Even though the means are not very different, for a fixed uptake the male rats have a higher incorporated value.

10.63 a. Because the states have different numbers of hospital beds and for a variety of other reasons, Medicaid percentages differ from one state to the next.

b. Figure A.14.

c. The three southern states plus Indiana and North Dakota have more hospital beds per population than can be explained by the percentage of the population receiving Medicaid assistance. The states with fewer beds are all in the northern part of the country. North Dakota has the largest residual and is therefore the most unusual state.

d. Beds per 100,000 population = 286 + 9.4 percent receiving Medicaid, $r = 0.44$, $t = 1.38$ on 8 d.f., and p-value = 0.10. Two states that differ by one percentage point of people receiving Medicaid differ on the average by 9.4 hospital beds per 100,000 population. The relationship is moderately strong. The p-value is so large that the relationship may have occurred by chance alone.

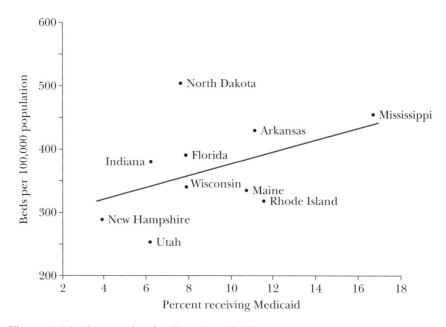

Figure A.14 Scatterplot for Exercise 10.63b

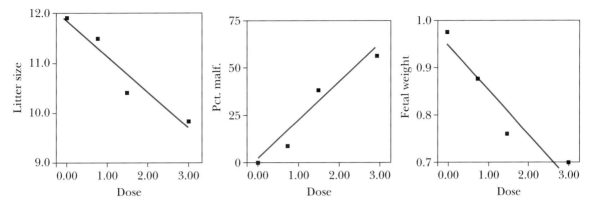

Figure A.15 Scatterplot of three variables versus dosage.

10.65 a. Figure A.15.

b. As the dosage goes up, the litter size goes down, the percentage malformed goes up, and the fetal weight goes down.

c. litter size = 11.85 − 0.75 dose $r = -0.96$ $p = 0.02$
percent malformed = 0.3 + 19.9 dose $r = 0.97$ $p = 0.02$
fetal weight = 0.95 − 0.09 dose $r = -0.96$ $p = 0.02$

The relationships are all strong and statistically significant.

d. No. They are measured in different units.

e. Each of the three analyses are based on four observations. With the individual data there would have been about 100 observations in each analysis. There would have been variation around each of the mean points, and the effect of the residual variable would have been larger. But with larger samples, the *p*-values might be smaller.

10.67 The scatterplot shows a linear, positive relationship.

$$\text{percent fat} = 3.2 + 0.55 \text{ age} \qquad r = 0.79$$

This gives $t = 5.19$ on 16 d.f. and $p = 0.0001$, so the relationship is statistically significant. You do not know if the relationship is causal.

10.69 a. The scatterplot shows a relationship.

$$\text{mortality} = -21.8 + 2.4 \text{ temperature} \qquad r = 0.87$$

This gives $t = 6.76$ on 14 d.f. and $p < 0.0001$. The relationship is significant.

b. From these data alone we can only guess about causality.

10.71 Student project.

10.73 a. Slope 1 and intercept 0.

b. Slope 1 and intercept 5.

c. Slope 1.1 and intercept 0.

d. Groom = $-1.6 + 1.13$ bride.

e. The grooms tend to be older than the brides, particularly in older couples.

f. In the scatterplot each couple can be identified, while two stemplots show the two distributions separately.

CHAPTER 11 ANALYSIS OF VARIANCE FOR A CATEGORICAL AND A METRIC VARIABLE

11.1 a. Divide the number of reported crimes by the population size.

b. Larger states would have more crimes just because the states are larger.

c. We could look at individual states and ask if they differ from other states, but then there would be no independent variable to explain why the assaults differ. We think region has an effect on crimes, so we make region the independent variable.

11.3 a. Gender is the independent variable and crime rate is the dependent variable.

b. Categorical and metric.

11.5 Student project.

11.7 The rates would all be the same in each state.

11.9 Question 3.

11.11 a. Five.

b. $50 - 6 = 44$.

11.13 Student project.

11.15 a. The one variable we are interested in is pulled out as the independent variable. All other variables are combined into the residual variable.

b. When an element is measured, the difference between the observed value and the true value becomes the error in the measurement.

11.17 If there is no difference within each pair, then the two values are alike. Plotting such a point, it will lie on the 45-degree line through the origin.

11.19 a. $56.6 - 48.6 = 8.0$.

b. Yes.

11.21 a. Region.

b. All independent variables other than region.

c. If R^2 is the proportion explained by the independent variable, then $1 - R^2$ is the proportion explained by the residual variable.

d. The total sum of squares.

e. $R^2 = 1.00$.

f. They would all be equal.

11.23 There is a weak relationship between the variables.

11.25 a. The strength of the relationship between university and score is $R = \sqrt{0.40} = 0.63$. There are significant differences between the universities.

b. Which universities are different from others and which are not?

11.27 a. There is a relationship between the country variable and number of trips in these data. The relationship is statistically highly significant.

b. The strength of the relationship.

11.29 a. 109.6 mm for the females and 113.4 mm for the males.

b. $109.6 - 113.4 = -4.8$ gives $t = -3.484$ on 18 d.f. with $p = 0.0013$ (Table A.14).

Table A.14 Analysis of variance for Exercise 11.49b

Source	Sum of squares	Degrees of freedom	Mean square	F-ratio	p-value
Gender	115.2	1	115.2	12.14	0.0026
Residual	170.8	18	9.489		
Total	286.0	19			

Note that $(-3.484)^2 = 12.14$ since we have F on 1 and something degrees of freedom. This means that the p-value for F equals the probability that $t < -3.484$ or $t > 3.484$.

11.31 a. Figure A.16. From the scatterplot we see that there is little or no difference between the frozen yogurts and the ice milks, while the frozen desserts do not have as much flavor.

b. 64.2, 64.0, 27.2.

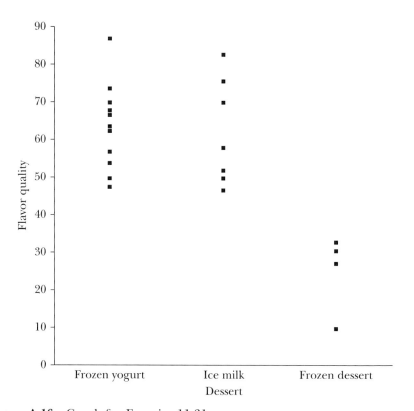

Figure A.16 Graph for Exercise 11.31a

c. $R = \sqrt{6364/9395} = 0.82$, for a strong relationship. $F = 3182.1/126.3 = 25.20$ on 2 and 24 d.f., which gives $p = 0.000001$. We reject the null hypothesis of equal means.

11.33 a. When a southern city has a low/high score, the corresponding northern city is also low/high.

b. On the average, the northern city is 4.8 points higher than the corresponding southern city.

c. For a paired test, $t = 1.47$ on 9 d.f. and $p = 0.09$. The mean difference is not statistically different from zero.

11.35 a. Type of sport.

b. Height prediction.

c. $t = (5.48 - 8.00)/\sqrt{(0.32^2 + 0.50^2)} = -4.24$ on 41 d.f. and $p = 0.00006$.

d. The exact value of p tells us that the results are very significant. If we know only that $p < 0.05$, then p can be anything less than 0.05. If p is close to that value, then the result is only borderline significant. If p is much smaller than that value, then the result is very significant.

11.37 a. Table A.15.

b. The strength of the relationship is $R = \sqrt{0.41} = 0.64$. Because of the small p-value the relationship is statistically significant, and there is a difference between zones.

c. We can make pairwise comparisons of the zones to see which ones are different from each other.

11.39 a. There are 9 positive differences and 1 negative difference. Using the binomial distribution with sample size $n = 10$ and

Table A.15 Analysis of variance for Exercise 11.37a

Source	Degrees of freedom	Effect	Proportion	Mean square	F-ratio	p-value
Climate zone	7	88,866	0.41	12,695	3.93	0.002
Residual	40	129,165	0.59	3,229		
Total	47	218,031	1.00			

probability $\pi = 0.5$, $p = (10 + 1)/1024 = 0.001$. This p-value is so small that we reject the null hypothesis that $\pi = 0.5$. Thus, positive and negative differences are not equally likely.

b. They both give statistically significant results, and the p-value for the sign test is here smaller than the p-value for the t-test. There may be a question of whether the differences are distributed normally enough for the t-test.

11.41 The four means are 6.69 for center, 5.34 for home, 7.36 for nanny, and 5.06 for relative. Because the means are different, there is a relationship between cost and type of care for these data. $R = 0.99$, and the relationship is very strong. We reject the null hypothesis that all four population means are equal because $F = 173.96$ on 3 and 12 d.f. and $p < 0.0001$.

11.43 A t-test for the difference between the two means gives $t = 4.02$ on 37 d.f. and $p = 0.0001$. There is a statistically significant difference between the groups.

11.45 a. The mean of the differences is 67. Testing the null hypothesis of 0 mean gives $t = 1.87$ on 6 d.f. and $p = 0.055$, borderline significant.

b. As unpaired data $t = 0.78$ on 12 d.f. and $p = 0.34$.

CHAPTER 12 RANK METHODS FOR TWO RANK VARIABLES

12.1 a. A variable where the values can be ranked from more to less, but we cannot tell how much more or less one value is from another.

b. Student project.

12.3 a. A coefficient that measures the strength of the relationship between two rank variables with words for the values.

b. -1 to $+1$.

c. r_s.

d. Most observations lie on the main diagonal of the table from the lower left to the upper right, and gamma will be positive.

12.5 Gamma needs to be changed to z. Use Statistical Table 1 or statistical software to find the p-value.

12.7 Student project.

12.9 a. Rank variable.

b. With class rank we know only that one student is better than another. With GPA we know how much better one student is than another when performance is measured by grades. (Grades themselves may be a rank variable, even though it is treated as a metric variable for the computation of a grade point average.)

12.11 Ordinal (rank) variable.

12.13 Families at the upper end have more children than families at the lower end. The relationship is weak and not statistically significant.

12.15 a. Change gamma to a value of the z-variable and find the p-value.

b. If the sample is not too small, the p-value will be small and the null hypothesis rejected.

c. The variables have no relationship in the population of all grapes.

12.17 a. Teams that were good in one season are also good in the other season. Teams that were bad in one season are also bad in the other season.

b. Smaller.

12.19 a. No city could overtake another and the ordering would not change.

b. No city could overtake another and the ordering would not change.

c. Most cities have the same ranking, but a few cities grew more than other cities.

d. Reject the null hypothesis. The orderings did not occur by chance alone.

12.21 a. Table A.16. Since the percentage distributions are different, there is a relationship between the two variables in these data.

b. $G = \dfrac{407336 - 595409}{407336 + 595409} = -0.19.$

Table A.16 Percentage distributions for Exercise 12.21a

		Abortion stance			Total
		Anti	Mixed	Pro	
Importance	One of the most	38%	12%	11%	14%
	Important	36	46	44	44
	Not very/not at all	25	42	45	42
	Total	99%	100%	100%	100%

c. $z = \dfrac{-0.19}{\sqrt{4(3+1)(3+1)/9(2421)(3-1)(3-1)}} = -7.01$.

The *p*-value is very small and we reject the null hypothesis of no relationship in the population.

d. We cannot tell from these data alone whether the relationship is causal.

12.23 Student project.

12.25 a. Table A.17.

b. Figure A.17.

c. The data display a very linear relationship. There is nothing left of the bend in the original data.

d. $r_s = 0.97$, with $t = 17.03$ on 17 d.f. and $p < 0.0001$. The relationship is very strong and statistically significant. We do not know whether it is causal.

Table A.17 Ranked years for Exercise 12.25a

Year	1	2	3	4	5	6	7	8	9	10
Marriages	1	2	3	4	5	6	9	8	7	10
Divorces	1	2	3	4	5	6	7	8	9	10

Year	11	12	13	14	15	16	17	18	19
Marriages	13	14	15	12	11	16	18	17	19
Divorces	11	16	13	12	14	15	17	18	19

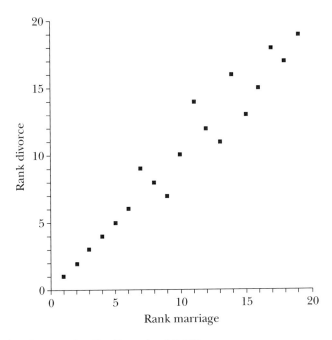

Figure A.17 Scatterplot for Exercise 12.25b

12.27 $r_S = 1.00$. The rank on one variable is identical to the rank on the other variable.

b. The probabilities are small. Most women do not get breast cancer.

c. The population has grown and that more women are exposed.

12.29 a. Figure A.18.

b. The hours increased for a while and then decreased. Since the data may not be linear, we cannot use ordinary regression and correlation analyses.

c. Table A.18.

d. This relationship occurred by chance alone. Find r_S, change that value to t, and find the corresponding p-value.

e. $r_S = 0.50$, $t = 1.29$ on 5 d.f., and $p = 0.13$. We cannot reject the null hypothesis.

f. The variables are both metric variables, and it may be possible to use correlation and regression analyses in some form.

Table A.18 Rank of years and hours worked for Exercise 12.29c

Rank year	Rank hours per year
1	2
2	1
3	5
4	6
5	4
6	7
7	3

724 Answers to Odd-Numbered Exercises

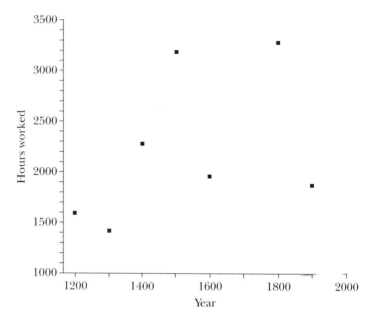

Figure A.18 Scatterplot for Exercise 12.29a

12.31 a. Figure A.19.

b. Countries good in one subject are also good in the other subject, and countries not as good in one subject are not as good in the other subject. Hong Kong, Japan, Austria, Italy, and the United States are better in chemistry than they are in biology.

c. $r_S = 0.65$.

d. $t = 2.82$ on 11 d.f. and $p = 0.008$. The relationship could not have occurred by chance alone.

e. We cannot tell about causality.

12.33 a. Figure A.20 (page 726).

b. The scatterplot shows little change from one week to the next.

c. $r_S = 0.98$.

d. $t = 15.19$ on 10 d.f., so $p < 0.0001$. This does not seem to be a pattern that could have occurred by chance alone.

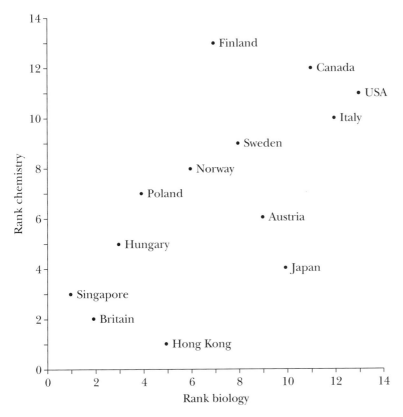

Figure A.19 Scatterplot for Exercise 12.31a

CHAPTER 13 MULTIVARIATE ANALYSIS

13.1 a. Several independent variables are used to study the dependent variable.

b. With one independent variable, the residual variable contains the effects of all other variables. By pulling all the independent variables out at the same time, we lessen the effect of the residual variable.

13.3 Holding the control variable constant by looking at the relationship between the dependent and other independent variables within subgroups defined by the control variable.

13.5 a. The coefficient we get when we divide the data into groups according to the control variable, compute a phi for each group, and average all the phis.

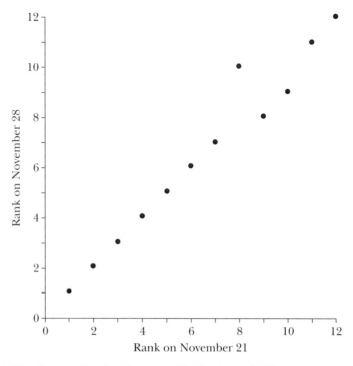

Figure A.20 Scatterplot for Exercise 12.33a (page 724)

b. Divide the data into four groups. Find phi in each group and average the phis.

c. The number of observations in the groups is taken care of in the averaging process.

13.7 Collinearity.

13.9 a. A variable created from a categorical variable with two categories such that one numerical value is assigned to all observations of one category and another numerical value is assigned to the observations of the other category.

b. Vote, with 0 for Democrats and 1 for Republicans.

c. A categorical variable can be included in a regression analysis.

d. The categorical variable can have only two categories.

13.11 Hypothesis testing or confidence interval estimation.

13.13 The combined effect of all variables other than the two categorical variables used in the analysis.

13.15 a. The combined effect of the two variables over and beyond their separate effects.

b. Smoking (yes/no) and drinking (yes/no) as the two independent variables and life expectancy as the dependent variable.

13.17 We never know if we have included the proper control variables.

13.19 a. Divide the data into Democrats and Republicans and do a separate analysis within each group.

b. Average the two phis from the separate analyses.

c. The original relationship may still be causal, since the partial phi is not equal to zero.

13.21 Phi does not change with control for party, so the original relationship may be causal.

13.23 Type has a strong and significant relationship to flavor. The chocolate/vanilla variable and the interaction variable have weak and nonsignificant relationships to flavor.

13.25 a. Positive.
b. Positive.
c. Large.
d. Negative.
e. Negative.
f. Large.

13.27 a. Whether a day is a school day or a weekend day, since children play more on weekends.

b. We may find that there is no difference in accidents between school days and no differences between weekend days. Thus, the original relationship is not causal.

13.29 The unit for a regression coefficient is in Y per X. If Y is miles per gallon and X is horsepower, then the unit for the coefficient is mpg per hp. When X is weight, then the unit for the coefficient is mpg per thousand pounds. Because the units are different for the two coefficients, they cannot be compared.

13.31 The choice of variables to control for is a nonstatistical question, and the choice must come from people who know about the substantive issue being studied.

13.33 The null hypothesis rejects the notion that all three regression coefficients in the population equal zero. The opposite of three numbers being equal to zero is that at least one is different from zero.

13.35 a. For each car there is a wide variation in scores for different features of the car, so the residual variable has a large effect. Mercedes has the largest mean score, and maybe VW has the lowest mean. Volvo shows the smallest effect of the residual variable.

b. The strength of the relationship between type of car and score is a weak $R = \sqrt{140/4117} = 0.18$. There are no statistically significant differences between the cars because of the large p-value.

c. The differences in the feature scores are not large enough to matter.

13.37 Weight may still have a causal effect on miles per gallon, and horsepower has a spurious relationship to miles per gallon.

13.39 a. Figure A.21.

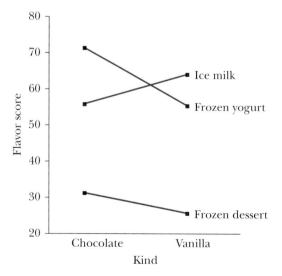

Figure A.21 Graph for Exercise 13.39a

b. Table 13.15 in Exercise 13.23 shows a significant difference between the types, but it does not tell which type is high and which is low. The interaction effect does not tell us that ice milk is different from the other two.

13.41 a. $2,500, since that is the coefficient for years of college.

b. The difference would be the same for two women.

c. Years of college gives a difference of $2,500 for two people who are the same on the other variables and differ by one year, while years of service gives a difference of $400 for two people who are the same on the other variables and differ by one year. Years of college goes from 0 to 4, and the most this variable can contribute to the salary is $10,000 for a person with 4 years of college. For a person who has worked 40 years, the variable years of service contributes $400 * 40 = $16,000 to the overall salary.

13.43 a. Student project.

b. incorporated = −1.79 + 0.11 uptake + 0.83 gender

c. For the separate analyses, the coefficients are 0.10 and 0.14. The weighted average of these coefficients in the multiple analysis is 0.11.

d. gender = 0 gives incorporated
= −1.79 + 0.11 uptake + 0.83 * 0
= −1.79 + 0.11 uptake gender = 1 gives incorporated
= −1.79 + 0.11 uptake + 0.83 * 1
= −0.96 + 0.11 uptake

e. These two lines are parallel, while the lines in Exercise 10.61 are not parallel.

f. 0.83.

g. Controlling for uptake means keeping uptake constant. For a fixed value of uptake, the predicted score for a male rat is 0.83 higher than the predicted score for a female rat. Thus, the effect of gender is 0.83.

13.45 a. Table A.19.

b. Phi = 0.72.

c. Partial phi for the races, controlling for sentence, = 0.70.

d. The relationship between the races of victims and defendants may be causal.

Table A.19 Race of defendant and victim for Exercise 13.45a

		Race of defendant		
		Black	White	Total
Race of victim	Black	1438	64	1502
	White	228	745	973
	Total	1666	809	2475

e. We could look at the relationship between sentence and race of the victim, controlling for race of the defendant, and the relationship between sentence and race of the defendant, controlling for race of the victim.

13.47 A simple regression analysis gives

$$\text{calories} = 287 + 19.8 \text{ fat} \qquad r^2 = 0.85$$

A multiple regression analysis gives

$$\text{calories} = 688 + 22.2 \text{ fat} - 17.2 \text{ calories} + 0.03 \text{ sodium} \qquad R^2 = 0.97$$

The simple analysis gives a slightly larger value for the regression coefficient for these foods. The multiple analysis shows that sodium can be dropped.

CHAPTER 14 ROLE OF STATISTICS

14.1 Student project.

14.3 Student project.

14.5 Student project.

14.7 Student project.

Index

A
Alternative hypothesis, 268
American Statistical Association, 3
Analysis of variance, 475
  one-way, 492
  two-way, 581
Analysis of variance table
  one-way, 492, 505
  regression, 428
  two-way, 591
Average
  mean, 138–140, 160
  median, 134–138, 140, 159
  mode, 131–134, 140

B
Bar graph, 77–80
  two variables, 348–351, 366
Beta, 426
Bias, 236
Bimodal, 86, 132
Binomial distribution, 187–190, 213–214
  hypothesis testing, 290

Boxplot, 82, 480
Bureau of Labor Statistics, 3, 17
Bureau of the Census, 17, 101

C
Categorical sum of squares, 483, 505
Categorical variable, 75, 324
Causality, 326–330
Census, 33
Chartjunk, 98
Chi-square variable, 15, 199–200, 358, 369–370
Class size, 53
Collinearity, 571
Confidence interval, 238
  for mean, 249
  for percentage, 249
  for difference of two means, 246, 250
  for difference of two proportions, 245, 250
  for regression coefficient, 426, 441

Confidence interval (*cont.*)
    versus hypothesis testing, 286–287
Confidence level, 239
Constant, 17
Consumer Price Index, 16
Contingency table, 346
Control, 562
Control group, 46
Convenience sample, 36
Correlation analysis, 398
Correlation coefficient, 406, 439
Critical values, 280

D
Data
    experimental, 45
    observational, 32
Data analysis, 3
Data density, 98
Data file, 55
Decision analysis, 207–210
Degrees of freedom, 197
    analysis of variance, 506
    chi-square, 359, 361
    t, 274
    two-way analysis of variance, 581
Dependent variable, 323
Distributions
    binomial, 187–190, 213–214
    chi-square, 199–200
    F, 201–202
    standard normal z, 194–196
    Poisson, 190–191, 214–215
    t, 196–199
Draft lottery, 12
Dummy variable, 431
    in multiple regression, 577

E
Error type I, 270
Error type II, 270
Errors in sampling, 39
Error variable, 483
Estimation, 232
    interval, 238–246
    point, 235–238
Expected frequency, 369
Experiments, 45
Explained amount of variation, 486–490

F
F-variable, 15, 201–202
F-ratio, 506
Fisher, Sir Ronald, 52
Four questions, 316

G
Galton, Francis, 398
Gamma, 530, 539
Graphical excellence, 96–101
Graphs, 73–101

H
Hawthorne effect, 48
Histogram, 84–89
Hypergeometric distribution, 192, 215
Hypothesis testing, 232, 267
    analysis of variance, 491
    contingency table, 355–360
    correlation coefficient, 427, 441
    difference of two means, 295
    difference of two proportions, 284, 297
    gamma, 532, 540
    mean, 294
    multiple R, 575

partial regression coefficient, 576
proportion, 281–284, 296
rank correlation, 537, 542
simple regression coefficient, 427–428, 441
versus confidence intervals, 286–287

I
Independent variable, 323
Inference, 232
Interaction sum of squares, 591
Interaction variable, 591
Intercept, 414, 440
Interquartile range, 143
Interval estimate, 238
Interval variable, 80

L
Least squares, 415
Lineplot, 80
Linear relationship, 404–406
Literary Digest, 8
Logistic regression, 435

M
Maps, 94–95
Matched pair analysis, 498, 506
Mean, 138–140, 160
for probability distribution, 193, 216
Mean squares, 506
Mean absolute deviation, 146
Median, 134–138, 140, 159
Metric variable, 80, 324
Misuses, 631–634
Mode, 131–134, 140
Multiple correlation coefficient, 574
Multiple regression, 567

Multivariate analyses, 560

N
Nonresponse error, 41
Normal distribution, 194–196
Null hypothesis, 268, 491

O
Odds, 185–186
Odds and probabilities, 213
One-sided test, 280

P
p-value, 206, 271
one-sided and two-sided, 271
Paired data, 498, 506
Parameter, 17, 233
Partial correlation coefficient, 571, 596
Partial phi, 564, 596
Partial regression coefficient, 568–570
Pearson correlation coefficient, 406
Pearson, Karl, 406
Percentiles, 135
phi, 353–355, 367–368
Pie chart, 76
Point estimate, 235
Poisson distribution, 190–191, 214–215
Polio, 11
Population, 32
Prediction, 322, 416
Probability, 13, 177
finding, 180–184
computing with, 184–185
Pythagoras's triangle, 487–488

R
r, 406, 439

R

R
  in analysis of variance, 486
  in multiple regression, 572
r^2, 411
  interpretation of, 417–422
R^2
  in analysis of variance, 485
  in multiple regression, 574
Randomness, 9
Range, 142
Rank correlation r_s, 535, 542
Rank variable, 324, 526
Ratio variable, 80
Regression analysis, 398
Regression coefficient, 414, 425, 439
Regression equation, 414
Regression line, 412
Regression sum of squares, 420, 441
Relationship, 314
Residual sum of squares
  in analysis of variance, 483, 505
  in regression, 420, 441
Residual value, 483
Residual variable, 417, 483
Response errors, 42

S

Sample, 32
  convenience, 36
  random, 35
  simple random, 36
Sampling error, 39
Scatterplot, 89–94, 402, 407, 478
Sign test, 501
Significance level, 279
Simple regression, 399
Skewed distribution, 86
Slope, 414
Spearman rank correlation, 535, 542
Spurious relationship, 317
Standard deviation, 143–148, 160
Standard error, 149
  for mean, 149, 161
  for regression coefficient, 441
Standard score, 150–152, 161
Statistic, 233
Statistical inference, 232
Statistics, 10
Stemplot, 84
Strength of relationship
  two categorical variables, 353, 360, 367–368
  two metric variables, 406
Subjective probability, 183
Summary number, 130
Surveys, 628–631
Symmetric distribution, 86

T

t-variable, 15, 196–199
Tables, 101–104
Test of significance, 232
Time series plot, 91–94
Total sum of squares
  in analysis of variance, 484, 505
  in regression, 420, 441
Two-sided test, 280
Two-way analysis of variance, 581, 596–598
Type I error, 270
Type II error, 270

U

Unbiased estimate, 236

V

V, 360, 368
Value, 14
Variable, 14, 31
   categorical, 75, 324
   empirical, 14
   metric, 80, 324
   rank, 324, 526
   theoretical, 15
Variance, 148, 160
   from probability distribution, 216

Variation, 10, 141–150, 419
   interquartile range, 143
   range, 142
   standard deviation, 143–148, 160
   standard error, 149
   variance, 148, 160

Z

z-variable, 15, 194–196